BIOSTATISTICS: A GUIDE TO DESIGN, ANALYSIS, AND DISCOVERY

Biostatistics: A Guide to Design, Analysis, and Discovery

Ronald N. Forthofer
Boulder County, Colorado

Eun Sul Lee
Division of Biostatistics
Department of Public Health and Preventive Medicine
Oregon Health & Science University, Portland, Oregon

Mike Hernandez
Department of Biostatistics and Applied Mathematics
M.D. Anderson Cancer Center, Houston, Texas

ELSEVIER
ACADEMIC
PRESS

AMSTERDAM • BOSTON • HEIDELBERG • LONDON
NEW YORK • OXFORD • PARIS • SAN DIEGO
SAN FRANCISCO • SINGAPORE • SYDNEY • TOKYO

Academic Press is an imprint of Elsevier
30 Corporate Drive, Suite 400, Burlington, MA 01803, USA
525 B Street, Suite 1900, San Diego, California 92101-4495, USA
84 Theobald's Road, London WC1X 8RR, UK

This book is printed on acid-free paper.

Library of Congress Cataloging-in-Publication Data
Application Submitted

British Library Cataloguing-in-Publication Data
A catalogue record for this book is available from the British Library.

ISBN 13: 978-0-12-369492-8
ISBN 10: 0-12-369492-2

For information on all Academic Press publications
visit our Web site at www.books.elsevier.com

Printed in the United States of America
06 07 08 09 10   9 8 7 6 5 4 3 2 1

To Mary, Chong Mahn, and the late Lucinda S.H.
Without their love and support,
this book would not have been possible.

Supporting Website

This text is supported by a website, www.biostat-edu.com. This site contains data sets, program notes, solutions to example exercises, and links to other resources. The material on this site will be updated as new versions of software become available.

Table of Contents

Preface

Our first edition, *Introduction to Biostatistics: A Guide to Design, Analysis, and Discovery*, was published in 1995 and was well received both by reviewers and readers. That book broke new ground by expanding the range of methods covered beyond what typically was included in competing texts. It also emphasized the importance of understanding the context of a problem — the why and what — instead of considering only the how of analysis.

Although the past several years have seen much interest in a second edition, our involvement with numerous other projects prevented us from tackling a new edition. Now that the stars are in alignment (or whatever), we finally decided to create a second edition. We are excited that Mike Hernandez has agreed to collaborate with us on this edition.

This new edition builds on the strengths of the first effort while including several new topics reflecting changes in the practice of biostatistics. Although parts of the second edition still serve as an introduction to the world of biostatistics, other parts break new ground compared to competing texts. For some of these relatively more advanced topics, we strongly advise the reader to consult with experts in the field before setting out on the analyses.

This revised and expanded edition continues to encourage readers to consider the full context of the problem being examined. This context includes understanding what the goal of the study is, what the data actually represent, why and how the data were collected, how to choose appropriate analytic methods, whether or not one can generalize from the sample to the target population, and what problems occur when the data are incomplete due to people refusing to participate in the study or due to the researcher failing to obtain all the relevant data from some sample subjects. Although many biostatistics textbooks do a very good job in presenting statistical tests and estimators, they are limited in their presentations of the context. In addition, most textbooks do not emphasize the relevance of biostatistics to people's lives and well-being. We have written and revised this textbook to address these deficiencies and to provide a good guide to statistical methods.

This textbook also differs from some of the other texts in that it uses real data for most of the exercises and examples. For example, instead of using data resulting from tossing dice or dealing cards, real data on the relation between prenatal care and birth weight are used in the definition of probability and in the demonstration of the rules of probability. We then show how these rules are applied to epidemiologic measures and the life table, major tools used by health analysts. Other major differences between this and other texts are found in Chapters 11, 14, and 15. In Chapter 11 we deal with the analysis of the follow-up life table; its use in survival analysis is considered in Chapter 14. In Chapter 15 we present strategies for analyzing survey data from complex sample designs. Survey data are used widely in public health and health services research, but most biostatistics texts do not deal with sample weights or methods for estimating the sample variance from complex surveys.

We also include material on tolerance and prediction intervals, topics generally ignored in other texts. We demonstrate in which situations these intervals should be used and how they provide different information than that provided by confidence intervals. In addition, we discuss the randomized response technique and the general linear model for analysis of data sets with an unequal number of observations in each cell, topics generally not covered in other texts. The randomized response technique is one way of dealing with response bias associated with sensitive questions, and it also illustrates the importance of statistical design in the data collection process.

Although we did not write this book with the assumption that readers have prior knowledge of statistical methods, we did assume that readers are not the type to be rendered unconscious by the sight of a formula. When presenting a formula, we first explain the concept that underlies the formula. We then show how the formula is a translation of the concept into something that can be measured. The emphasis is on when and how to apply the formula, not on its derivation. We also provide a review of some mathematical concepts that are used in our explanations in Appendix A. A website is provided that demonstrates the use of statistical software in carrying out the analyses shown in the text. As new versions of statistical packages become available, the website material will be updated.

The textbook is designed for a two-semester course for the first-year graduate student in health sciences. It is also intended to serve as a guide for the reader to discover and learn statistical concepts and methods more or less by oneself. If used for a one-semester course, possible deletions include the sections on the geometric mean, the Poisson distribution, the distribution-free approach to intervals, the confidence interval and test of hypothesis for the variance and coefficient of correlation, the Kruskal-Wallis and Friedman tests, the trend test for r by 2 contingency tables, the two-way ANOVA, ANOVA for unbalanced designs, the linear model representation of the ANOVA, the ordered and conditional logistic regression, the proportional hazards regression, and the analysis of survey data.

Several appendices are at the end of the book. Appendix A presents some basic mathematical concepts that are essential to understanding the statistical methods presented in this book. Appendix B contains several statistical tables referenced in the text. Appendix C is a listing of major governmental sources of health data, and Appendix D presents solutions to selected exercises.

Acknowledgments

Our special thanks go to Professor Anthony Schork of University of Michigan for his review of the first edition and his suggestions for the contents of the second edition.

There are several other anonymous reviewers of the contents of the second edition who made helpful comments and suggestions. We also wish to acknowledge those who provided with useful suggestions and comments for the first edition including Joel A. Harrison, Mary Forthofer, Herbert Gautschi, Irene Easling, Anna Baron, Mary Grace Kovar, and the students at the University of Texas School of Public Health Regional Program in El Paso, who reviewed parts or all of the first edition. Any problems in the text are the responsibility of the authors and not of the reviewers.

Introduction

Chapter Outline

1.1 What Is Biostatistics?

Biostatistics is the application of statistical methods to the biological and life sciences. Statistical methods include procedures for (1) designing studies, (2) collecting data, (3) presenting and summarizing data, and (4) drawing inferences from sample data to a population. These methods are particularly useful in studies involving humans because the processes under investigation are often very complex. Because of this complexity, a large number of measurements on the study subjects are usually made to aid the discovery process; however, this complexity and abundance of data often mask the underlying processes. It is in these situations that the systematic methods found in statistics help create order out of the seeming chaos. These are some of the areas of application:

1. Collection of vital statistics — for example, mortality rates — used to *inform* about and to *monitor* the health status of the population
2. Analysis of accident records to *find* out the times during the year when the greatest number of accidents occurred in a plant and *decide* when the need for safety instruction is the highest
3. Clinical trials to *determine* whether or not a new hypertension medication performs better than the standard treatment for mild to moderate essential hypertension
4. Surveys to *estimate* the proportion of low-income women of child-bearing age with iron deficiency anemia
5. Studies to *investigate* whether or not exposure to electromagnetic fields is a risk factor for leukemia

Biostatistics aids administrators, legislators, plant managers, and researchers in answering questions. The questions of interest are explicit in examples 2, 3 and 5 above — do seasonal patterns of accidents give any clues for reducing their occurrence?; is the new drug more effective than the standard?; and is exposure to the electromagnetic field a risk factor? In examples 1 and 4, the values or estimates obtained are measurements at a point in time that could be used with measures at other time points to determine whether or not a policy change, for example, a 10 percent increase in Medicaid funding in each state, had an effect.

The words in italics in the preceding list suggest that statistics is not a body of substantive knowledge but a body of methods for obtaining, organizing, summarizing, and presenting information and drawing inferences. However, whenever we draw an inference, there is a chance of being wrong. Fortunately, statistical methods incorporate probability ideas that allow us to determine the chance of making a wrong inference. As Professor C.R. Rao (1989) suggested, "Statistics is putting chance to work."

1.2   Data — The Key Component of a Study

Much of the material in this book relates to methods that are used in the analysis of data. It is necessary to become familiar with these methods and their use because it will help you to understand reports of studies, design studies, and conduct studies. However, readers should not feel overwhelmed by the large number of methods of analysis and the associated calculations presented in this book. More important than the methods used in the analysis are the use of the appropriate study design and the proper definition and measurement of the study variables. *You cannot have a good study without good data!* The following examples demonstrate the importance of the data.

Example 1.1

Sometimes, due to an incomplete understanding of the data or of possible problems with the data, the conclusion from a study may be problematic. For example, consider a study to determine whether circumcision is associated with cervical cancer. One issue the researcher must establish is *how to determine the circumcision status.* The easiest way is to just ask the male if he has been circumcised; however, Lilienfeld and Graham (1958) found that 34 percent of 192 consecutive male patients they studied gave incorrect answers about their circumcision status. Most of the incorrect responses were due to the men not knowing that they had been circumcised. Hence, the use of a direct question instead of an examination may lead to an incorrect conclusion about the relation between circumcision and cancer of the cervix.

Example 1.2

In Example 1.1, reliance on the study subject's memory or knowledge could be a mistake. Yaffe and Shapiro (1979) provide another example of potential problems when the study subjects' responses are used. They examined the accuracy of subjects' reports of health care utilization and expenditures for seven months compared to that shown in their medical and insurance records for two geographical areas. In a Baltimore area that provided data from approximately 375 households, subjects reported only 73 percent of the identified physician office visits and only 54 percent of the clinic visits. The results for Washington County, Maryland which were based on about 315 households, showed 84 percent accuracy for physician office visits but only 39 percent accuracy for clinic visits. Hence, the reported utilization of health services by subjects can greatly underestimate the actual utilization, and, perhaps more importantly, the accuracy can vary by type of utilization and by population subgroups. Note that this conclusion is based on the assumption that the medical/insurance records are more accurate than the subjects' memories.

Example 1.3

One example of how a wrong conclusion could be reached because of a failure to understand how data are collected comes from Norris and Shipley (1971). Figure 1.1 shows the infant mortality rates calculated conventionally as the ratio of the number of infant deaths to the number of live births during the same period multiplied by 1000 for different racial groups in California and the United States in 1967.

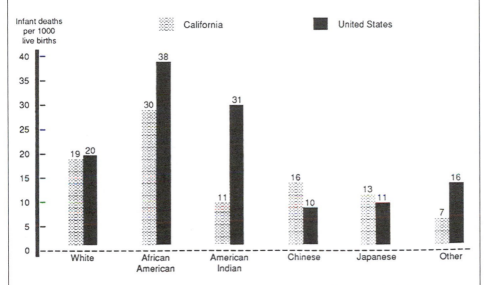

Figure 1.1   Infant mortality rates per 1000 live births by race for California and the United States in 1967.

Norris and Shipley questioned the accuracy of the rate for American Indians because it was much lower than the corresponding U.S. American Indian rate and even lower than the rates for Chinese and Japanese Americans in California. Therefore, they used a cohort method to recalculate the infant mortality rates. The cohort rate is based on following all the children who were born in California during a year and observing how many of those infants died before they reached one year of age. Some deaths were missed — for example, infants who died outside California — but it was estimated that almost 97 percent of the infant deaths of the cohort were captured in the California death records.

Norris and Shipley used three years of data in their reexamination of the infant mortality to provide better stability for the rates. Figure 1.2 shows the conventional and the cohort rates for the 1965–1967 period by race. The use of data from three years has not changed the conventional rates much. The conventional rate for American Indians in California is still much lower than the U.S. rate for American Indians, although now it is slightly above the Chinese and Japanese rates. However, the cohort rate for American Indians is now much closer to the corresponding rate found in the United States. The rates for the Chinese and Japanese Americans and other races have also increased substantially when the cohort method of calculation is used. What is the explanation for this discrepancy in results between these methods of calculating infant mortality rates?

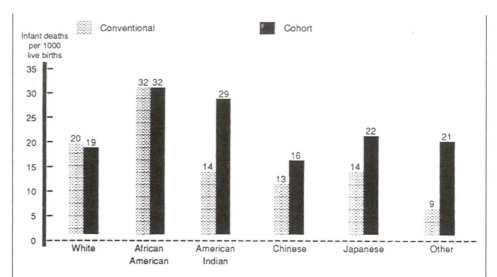

Figure 1.2 Infant mortality rates per 1000 live births by conventional and cohort methods by race for California, 1965–1967.

Norris and Shipley attributed much of the difference to how birth and death certificates, used in the conventional method, were completed. They found that a birth certificate is typically filled out by hospital staff who deal primarily with the mother, so the birth certificate usually reflects the race of the mother. The funeral director is responsible for completing the death record and usually deals with the father, who may be of a different racial group than the mother. Hence, the racial identification of an infant can vary between the birth and death records — a mismatch of the numerator (death) and the denominator (birth) in the calculation of the infant death rate. The cohort method is not affected by this possible difference because it uses only the child's race from the birth certificate.

Since the 1989 data year, the National Center for Health Statistics (NCHS) uses primarily the race of the mother taken from the birth certificate in tabulating data on births. This change should remove the problem caused by having parents from two racial groups in the use of the conventional method of calculating infant mortality rates.

So we can see that data rarely speak clearly and usually require an interpreter. The interpreter — someone like Norris and Shipley — is familiar with the subject matter, understands what the data are supposed to represent, and knows how the data were collected.

1.3 Design — The Road to Relevant Data

As we said, you cannot have a good study without good data. Obtaining relevant data requires a carefully drawn plan that identifies the population of interest, the procedure used to randomly select units for study, and the process used in the observation/measurement of the attributes of interest. Two standard methods of data collection are sample surveys and experiments (that may involve sampling).

Sample survey design deals with ways to select a random sample that is representative of the population of interest and from which a valid inference can be made. Unfortunately, it is very easy to select nonrepresentative samples that lead to misleading conclusions about the population, emphasizing the need for the careful design of sample surveys.

Experimental design involves the creation of a plan for determining whether or not there are differences between groups. The design attempts to control extraneous factors so that the only reason for any observed differences between groups is the factor under study. Since it is very difficult to take all extraneous factors into account in a design, we also use the random allocation of subjects to the groups. We hope that through the use of the random assignment, we can control for factors that have not been included in the design itself. Experimental design is also concerned with determining the appropriate sample size for the study.

Sometimes we also analyze data that were already collected. In this case, we need to understand how the data were collected in order to determine the appropriate methods of analysis. The following examples demonstrate the importance of design.

Example 1.4

The *Literary Digest* (1936) failed to correctly predict the 1936 presidential election after correctly predicting every presidential election between 1912 and 1932. In 1936 the *Digest* mailed out 10 million questionnaires and received 2.3 million replies. Based on the returned questionnaires, the *Digest* predicted that Alfred M. Landon would be elected. Actually, Franklin D. Roosevelt won in a huge landslide. What went wrong? The survey design was the problem. The questionnaires were initially mailed to magazine and telephone subscribers and automobile owners. The list clearly overrepresented people with high incomes. Given that there was a strong relation between income and party preference in the 1936 election, probably stronger than in previous elections, the embarrassingly wrong outcome should not have been a surprise. Another problem with the survey was the low response rate; bias due to a high nonresponse rate is a potential problem that must be considered when analyzing surveys. Another point to be made from this example is that the sheer size of a sample is no guarantee of an accurate inference. The design of the survey is far more important than the absolute size of the sample.

Example 1.5

Careful statistical plans were lacking in early American census-taking procedures from the inception of the decennial census in 1790 until 1840. The discovery of numerous errors in the 1840 census led to statistical reforms in the 1850 census, which accelerated the government's use of modern statistical procedures (Regan 1973). Two important players in the statistical reforms were Edward Jarvis and Lemuel Shattuck. As a physician, Jarvis was interested in mental illness and began to look at census data on "insane and idiots." To his surprise, he discovered numerous errors in census reports. For example, Waterford, Connecticut, listed no "Negro" population but showed "seven Negro insane and idiots." He published his findings in the journal and urged correction (Jarvis 1844). Together with Shattuck he brought the issue to the American Statistical Association, and the Association submitted a petition to Congress. Among other reform measures, the collection of vital statistics had begun with the 1850 decennial census. However, Shattuck protested against the plan. He believed that vital statistics could be better collected through a registration system rather than through a census (Shattuck 1948). Shattuck successfully persuaded Massachusetts to adopt the vital statistics registration system and provided a pattern for other states to follow. The U.S. vital registration system was complete when Texas came on board in 1933.

1.4   Replication — Part of the Scientific Method

Even though most of the examples and problems in this book refer to the analysis of data from a single study, the reader must remember that one study rarely tells the complete story.

Statistical analysis of data may demonstrate that there is a high probability of an *association* between two variables. However, a single study rarely provides proof that such an association exists. Results must be *replicated* by additional studies that eliminate other factors that could have accounted for the relationship observed between the study variables. The following examples demonstrate the importance of replication.

Example 1.6

There have been many, many studies examining the role of cigarette smoking in lung cancer and other diseases. Many of the early studies were retrospective case-control studies — that is, studies in which a group of people with, for example, lung cancer (the cases) is compared to another group without the disease (the controls). Factors such as smoking status that precede the outcome (disease or no disease) are then compared to determine whether or not there is a relationship between smoking status and lung cancer. This type of study can show association but cannot be used to prove causation. The retrospective study is often used to generate hypotheses of interest that may be tested in a follow-up study. Some leading statisticians (e.g., Joseph Berkson and Sir Ronald Fisher) raised methodological concerns about conclusions drawn from these retrospective studies, and tobacco companies chimed in. Due to these concerns, a different design, the prospective study that was long-term in nature, was used. Large numbers of smokers and nonsmokers, usually matched on a number of other factors, were followed over a lengthy period, and the proportion of lung cancers were compared between the groups at the end of the study. The consistent results of these prospective studies helped establish the causative linkage between smoking and lung cancer and smoking and other diseases (Brown 1972; Gail 1996).

Example 1.7

The Food and Drug Administration (FDA) used to require several studies that demonstrated the efficacy and safety of a drug before it was approved for general use. The FDA did not believe that a single trial provided sufficient evidence of the drug's efficacy and safety. The story of thalidomide illustrates the case in point (Insight Team 1979). A German pharmaceutical company, Gruenenthal, developed thalidomide and marketed it as a tranquilizer. Its U.S. counterpart, Richardson-Merrell, submitted an application to the FDA to obtain a U.S. license. Dr. Frances Kelsey questioned the validity of the submitted clinical data and demanded more complete and detailed evidence of the safety of the drug. As a result of her professionalism, the birth of thousands of deformed babies in the United States was prevented. Unfortunately, things have changed at the FDA in recent years (FDA 2006).

1.5   Applying Statistical Methods

The application of statistical methods requires more than the ability to use statistical software. In this text, we give priority to understanding the context for the use of statistical procedures. This context includes the study's goal, the data, and how the data are collected and measured. We do not focus on the derivation of formulas. Instead, we present the rationale for the different statistical procedures and the when and why of their use. We would like the reader to *think* instead of simply memorizing formulas.

EXERCISES

1.1 Provide an example from your area of interest in which data collection is problematic or data are misused, and discuss the nature of the problem.

1.2 Since 1972 the National Institute on Drug Abuse has periodically conducted surveys in the homes of adolescents on their use of cigarettes, alcohol, and marijuana. In the early surveys, respondents answered the questions aloud. Since 1979 private answer sheets were provided for the alcohol questions. Why do you think the agency made this change? What effect, if any, do you think this change might have had on the proportion of adolescents who reported consuming alcohol during the last month? Would you believe the reported values for the early surveys?

1.3 The infant mortality rate for Pennsylvania for the 1983–1985 period was 10.9 per 1000 live births compared to a rate of 12.5 for Louisiana. Is it appropriate to conclude that Pennsylvania had a better record than Louisiana related to infant mortality? What other variable(s) might be important to consider here? The infant mortality rate for whites in Pennsylvania was 9.4, and it was 20.9 for blacks there. This is contrasted with rates of 9.1 and 18.1 for whites and blacks, respectively, in Louisiana (National Center for Health Statistics 1987). Hence, the race-specific rates were lower in Louisiana than in Pennsylvania, but the overall rate was higher in Louisiana. Explain how this situation could arise.

1.4 Read the article by Bathlomew (1995) and prepare a short report commenting on the following points: What is *your* definition of statistics? Is the field of statistics broader than what you believed? How would you effectively study statistics?

REFERENCES

Bartholomew, D. "What Is Statistics?" *Journal of Royal Statistical Society*, A, 158, Part 1, 1–20, 1995.

Brown, B. W. "Statistics, Scientific Method, and Smoking." In *Statistics: A Guide to the Unknown*, edited by J. M. Taner, F. Mosteller, W. H. Kruskal, R. F. Link, R. S. Pieters, and G. R. Rising, 40–51. San Francisco: Holden-Day, Inc., 1972.

Food and Drug Administration (FDA), http://www.fda.gov/cder/guidance/105-115.htm#SEC. %20115 and http://www.os.dhhs.gov/asl/testify/t970423a.html, 2006.

Gail, M. H. "Statistics in Action." *Journal of American Statistical Association* 91:1–13, 1996.

The Insight Team of *The Sunday Times* of London, *Suffer the Children: The Story of Thalidomide*. New York: The Viking Press, 1979.

Jarvis, E. "Insanity among the Colored Population of the Free States." *American Journal of the Medical Sciences* 7:71–83, 1844.

Lilienfeld, A. M., and S. Graham. "Validity of Determining Circumcision Status by Questionnaire as Related to Epidemiological Studies of Cancer of the Cervix." *Journal of the National Cancer Institute* 21:713–720, 1958.

Literary Digest. "Landon, 1,293,669: Roosevelt, 972,897." 122(October 31): 5–6; and "What Went Wrong with Polls?" 122(November 14):7–8, 1936.

National Center for Health Statistics. *Health, United States, 1987.* Hyattsville, MD: Public Health Service. DHHS Pub. No. 88-1232, 1988, Table 15.

Norris, F. D., and P. W. Shipley. "A Closer Look at Race Differentials in California's Infant Mortality, 1965–67." *HSMHA Health Reports* 86:810–814, 1971.

Rao, C. R. *Statistics and Truth: Putting Chance to Work*. Burtonsville, MD: International Co-operative Publishing House, 1989.

Regan, O. G. "Statistical Reforms Accelerated by Sixth Census Errors." *Journal American Statistical Association* 68:540–546, 1973.

Shattuck, L. *Report of the Sanitary Commission of Massachusetts*. Cambridge: Harvard University Press, 1948.

Yaffe, R., and S. Shapiro. "Reporting Accuracy of Health Care Utilization and Expenditures in a Household Survey as Compared with Provider Records and Insurance Claims Records." Presented at Spring Meeting of the Biometric Society, ENAR, April 1979.

Data and Numbers

2

Appropriate use of statistical procedures requires that we understand the data and the process that generated them. This chapter focuses on data, specifically: (1) the link between numbers and phenomena, (2) types of variables, (3) data reliability and validity, and (4) ways data quality can be compromised.

2.1   Data: Numerical Representation

Any record, descriptive accounts, or symbolic representation of an attribute, event, or process may constitute a data point. Data are usually measured on a numerical scale or classified into categories that are numerically coded. Here are three examples:

1. Blood pressure (diastolic) is measured for all middle and high school students in a school district to learn what percent of students have a diastolic blood pressure reading over 90 mm Hg. [data = blood pressure reading]
2. All employees of a large company are asked to report their weight every month to evaluate the effects of a weight control program. [data = self-reported weight measurement]
3. The question "Have you ever driven a car while intoxicated?" was asked of all licensed drivers in a large university to build the case for an educational program. [data = yes (coded as 1) or no (coded as 0)]

We try to understand the real world — for example, blood pressure, weight, and the prevalence of drunken driving — through data recorded as or converted to numbers. This numerical representation and the understanding of the reality, however, do not occur automatically. It is easy for problems to occur in the conceptualization and measurement processes that make the data irrelevant or imprecise. Referring to the preceding examples, inexperienced school teachers may measure blood pressure inaccurately; those employees who do not measure their weight regularly each month may report inaccurate values; and some drivers may be hesitant to report drunken driving. Therefore, we must not draw any conclusions from the data before we determine whether or not any problems exist in the data and, if so, their possible effects. Guarding against misuse of data is as important as learning how to make effective use of data. Repeated exposure to misuses of data may lead people to distrust data altogether. A century ago, George Bernard Shaw (1909) described people's attitudes toward statistical data this way:

> The man in the street. . . . All he knows is that "you can prove anything by figures," though he forgets this the moment figures are used to prove anything he wants to believe.

The situation is certainly far worse today as we are constantly exposed to numbers purported to be important in advertisements, news reporting, and election campaigns. We need to learn to use numbers carefully and to examine critically the meaning of the numbers in order to distinguish fact from fiction.

2.2   Observations and Variables

In statistics, we observe or measure characteristics, called *variables*, of study subjects, called *observational units*. For each study subject, the numerical values assigned to the variables are called *observations*. For example, in a study of hypertension among school-children, the investigator measures systolic and diastolic blood pressures for each pupil. *Systolic and diastolic blood pressure are the variables, the blood pressure readings are the observations, and the pupils are the observational units.* We usually observe more than one variable on each unit. For example, in a study of hypertension among 500 school children, we may record each pupil's age, height, and weight in addition to the two kinds of blood pressure readings. In this case we have a data set of 500 students with observations recorded on each of five variables for each student or observational unit.

2.3   Scales Used with Variables

There are four scales used with variables: *nominal*, *ordinal*, *interval*, and *ratio*. The scales are defined in terms of the information conveyed by the numeric values assigned to the variable. The distinction between the scales is not terribly important. These scale types have frequently been used in the literature, so we are presenting them to be sure the reader understands them.

In some cases the numbers are simply indicators of a category. For example, when considering gender, 1 may be used to indicate that the person is female and 2 to indicate that the person is male. When the numbers merely indicate to which category a person belongs, a *nominal* scale is being used. Hence, gender is measured on a nominal scale, and it makes no difference what numeric values are used to represent females and males.

In other cases the numbers represent an ordering or ranking of the observational units on some variable. For example, from a worker's job description or work location, it may be possible to estimate the exposure to asbestos in the workplace, with 1 representing low, 2 representing medium, and 3 representing high exposure. In this case, the exposure to asbestos variable is measured on the *ordinal* scale. Values of 10, 50, and 100 could have been used instead of 1, 2, and 3 for representing the categories of low, medium, and high. The only requirement is that the order is maintained.

Other variables are measured on a scale of equal units — for example, temperature in degrees Celsius (*interval* scale) or height in centimeters (*ratio* scale). There is a subtle distinction between interval and ratio scales: A ratio scale has a zero value, which means there is none of the quantity being measured. For example, zero height means there is no height, whereas zero degrees Celsius does not mean there is no temperature. When a variable is measured on a ratio scale, the ratio of two numbers is meaningful. For example, a boy 140 centimeters tall is 70 centimeters taller and also twice as tall as a boy 70 centimeters tall. However, temperature in degrees Celsius is an interval variable but not a ratio variable because an oven at 300° is not twice as hot as one at 150°. This distinction between interval and ratio scales is of little importance in statistics, and both are measured on a scale continuously marked off in units.

These different scales measure three types of data: *nominal* (categorical), *ordinal* (ordered), and *continuous* (interval or ratio). The scale used often depends more on the method of measurement or the use made of it than on the property measured. The same property can be measured on different scales; for example, age can be measured in years (ratio scale), placed into young, middle-aged, and elderly age groups (ordinal scale), or classified as economically productive (ages 16 to 64) and dependent (under 16 and over 64) age groups (nominal scale). It is possible to convert a higher-level scale (ratio or interval) into a lower-level scale (ordinal and nominal scales) but not to convert from a lower level to a higher level. One final point is that all recorded measurements themselves are discrete. Age, for example, can be measured in years, months, or even in hours, but it is still measured in discrete steps. It is possible to talk about a continuous variable, yet actual measurements are limited by the measuring instruments.

2.4   Reliability and Validity

Data are collected by direct observation or measurement and from responses to questions. For example, height, weight, and blood pressure of school children are directly measured in a health examination. The investigator is concerned about accurate measurement. The measurement of height and weight sounds easy, but the measurement process must be well defined and used consistently. For example, we measure an individual's height without shoes on. Therefore, to understand any measurement, we need to know the operational definition — that is, the actual procedures used in the measurement. In measuring blood pressure, the investigator must specify what instrument is to be used, how much training will be given to the measurers, at what time of the day the blood pressure should be measured, what position the person should be in (sitting or standing), and how many times it should be measured.

There are two issues in specifying operational definitions: reliability and validity. *Reliability* requires that the operational definition should be sufficiently precise so that all persons using the procedure or repeated use of the procedure by the same person will have the same or approximately the same results. If the procedures for measuring height and weight of students are reliable, then the values measured by two observers — say, the teacher and the nurse — will be the same. If the person reading the blood pressure is hard of hearing, the diastolic blood pressure values, recorded at the point of complete cessation of the Korotkoff's sounds, or, if no cessation, at the point of muffling, may not be reliable.

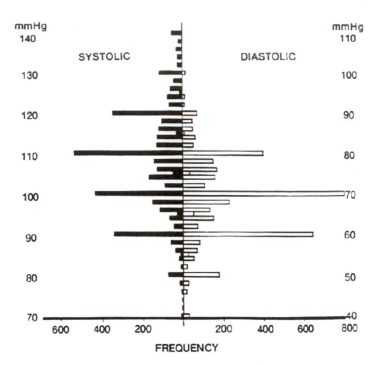

FREQUENCY

Figure 2.1 Blood pressure values (first reading) for 4053 children and adolescents in NHANES II. From Forthofer (1991).

Validity is concerned with the appropriateness of the operational definition — that is, whether or not the procedure measures what it is supposed to measure. For example, if a biased scale is used, the measured weight is not valid, even though the repeated measurements give the same results. Another example of a measurement that may not be valid is the blood pressure reading obtained when the wrong size cuff is used. In addition, the person reading the blood pressures may have a digit preference that also threatens validity. The data shown in Figure 2.1, from Forthofer (1991), suggests that there may have been a digit preference in the blood pressure data for children and adolescents in the second National Health and Nutrition Examination Survey (NHANES II). This survey, conducted by the NCHS from 1976 to 1980, provides representative health and nutrition data for the noninstitutionalized U.S. population. In this survey, the blood pressure values ending in zero have a much greater frequency of occurrence than the other values.

The reliability and validity issues are not only of concern for data obtained from measurements but also for data obtained from questionnaires. In fact, the concern may be greater because of the larger number of ways that problems threatening data accuracy can be introduced with questionnaires (Juster 1986; Marquis, Marquis, and Polich 1986; Suchman and Jordan 1990). One problem is that the question may be misinterpreted, and thus a wrong or irrelevant response may be elicited. For example, in a mail survey, a question used the phrase "place of death" instead of instructing the respondent to provide the county and state where a relative had died. One person responded that the deceased had died in bed. Problems like this one can be avoided or greatly reduced if careful thought goes into the design of questionnaires and into the preparation of instructions for the interviewers and the respondents. However, even when there are no obvious faults in the question, a different phrasing may have obtained a different response. For example, age can be ascertained by asking age at the last birthday or date of birth. It is known that the question about the date of birth tends to obtain the more accurate age.

Another problem often encountered is that many people are uncomfortable in appearing to be out of step with society. As a result, these people may provide a socially acceptable but false answer about their feelings on an issue. A similar problem is that many people are reluctant to provide accurate information regarding personal matters, and often the respondent refuses to answer or intentionally distorts the response. Some issues are particularly sensitive — for example, questions about whether a woman has had an abortion or if a person has ever attempted suicide. The responses, if any are obtained, to these sensitive questions are of questionable accuracy. Now we'll look at some ways to obtain accurate data on sensitive issues.

2.5   Randomized Response Technique

There is a statistical technique that allows investigators to ask sensitive questions, for example, about drug use or driving under the influence of alcohol, in a way that should elicit an honest response. It is designed to protect the privacy of individuals and yet provide valid information. This technique is called randomized response (Campbell and Joiner 1973; Warner 1965) and has been used in surveys about abortions, drinking and driving, drug use, and cheating on examinations.

In this technique, a sensitive question is paired with a nonthreatening question, and the respondent is told to answer only one of the questions. The respondent uses a chance mechanism — for example, the toss of a coin — to determine which question is to be answered, and only the respondent knows which question was answered. The interviewer records the response without knowing which question was answered. It may appear that these answers are of little value, but the following example demonstrates how they can be useful.

In the drinking and driving situation, the sensitive question is "Have you driven a car while intoxicated during the last six months?" This question is paired with an unrelated, nonthreatening question, such as "Were you born in either September or October?" Each respondent is asked to toss a coin and not to reveal the outcome; those with heads are asked to answer the sensitive question and those with tails answer the nonthreatening question. The interviewer records the yes or no response without knowing which question is being answered. Since only the respondent knows which question has been answered, there is less reason to answer dishonestly.

Suppose 36 people were questioned and 12 gave "yes" answers. At first glance, this information does not seem very useful, since we do not know which question was answered. However, Figure 2.2 shows how we can use this information to estimate the proportion of the respondents who had been driving while intoxicated during the past six months.

Since each respondent tossed a fair coin, we expect that half the respondents answered the question about drunk driving and half answered the birthday question. We also expect that 1/6 (2 months out of 12) of those who answered the birthday question will give a yes response. Hence, the number of yes responses from the birthday question

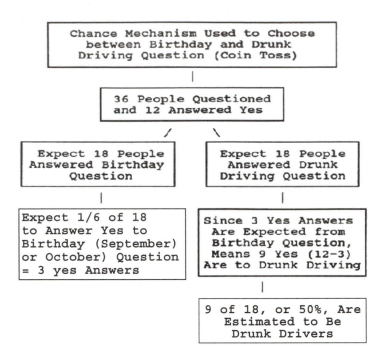

Figure 2.2 Use of randomized response information.

should be 3 [(36/2) * (1/6)]; the expected number of yes responses to the drinking and driving question then is 9 (the 12 yes answers minus the 3 yes answers from the birthday question). Then the estimated proportion of drunk drivers is 50 percent (= 9/18).

There is no way to prove that the respondents answered honestly, but they are more likely to tell the truth when the randomized response method was used rather than the conventional direct question. Note that the data gathered by the randomized response technique cannot be used without understanding the process by which the data were obtained. Individual responses are not informative, but the aggregated responses can provide useful information at the group level. Of course, we need to include a sufficiently large number of respondents in the survey to make the estimate reliable.

2.6 Common Data Problems

Examination of data can sometimes provide evidence of poor quality. Some clues to poor quality include many missing values, impossible or unlikely values, inconsistencies, irregular patterns, and suspicious regularity. Data with too many missing values will be less useful in the analysis and may indicate that something went wrong with the data collection process. Sometimes data contain extreme values that are seemingly unreasonable. For example, a person's age of 120 would be suspicious, and 200 would be impossible. Missing values are often coded as 99 or 999 in the data file, and these may be mistakenly interpreted as valid ages. The detection of numerous extreme ages in a data set would cast doubt on the process by which the data were collected and recorded, and hence on all other observations, even if they appear reasonable. Also, inconsistencies are often present in the data set. For example, a college graduate's age of 15 may appear inconsistent with the usual progress in school, but it is difficult to attribute this to an error. Some inconsistencies are obvious errors. The following examples illustrate various problems with data.

Example 2.1

As described in Example 1.5, Edward Jarvis (1803–1884) discovered that there were numerous inconsistencies in the 1840 Population Census reports; for example, in many towns in the North, the numbers of black "insane and idiots" were larger than the total numbers of blacks in those towns. He published the results in medical journals and demanded that the federal government take remedial action. This demand led to a series of statistical reforms in the 1850 Population Census (Regan 1973).

Example 2.2

A careful inspection of data sometimes reveals irregular patterns. For example, ages reported in the 1945 census of Turkey have a much greater frequency of multiples of 5 than numbers ending in 4 or 6 and more even-numbered ages than odd-numbered ages (United Nations 1955), as shown in Figure 2.3. This tendency of digit preference in age reporting is quite common. Even in the U.S. Census we can find a slight clumping or heaping at age 65, when most of the social benefit programs for the elderly begin. The same phenomenon of digit preference is often found in laboratory measurements as we just saw with the blood pressure measurements in NHANES II.

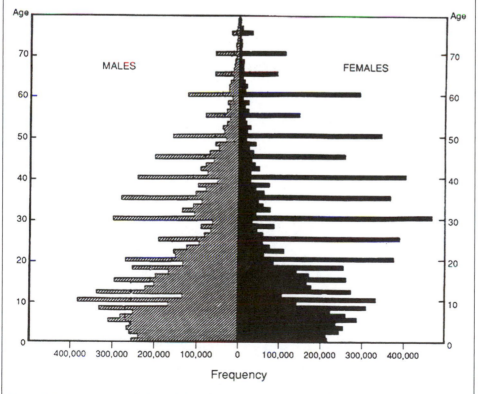

Figure 2.3 Population of Turkey, 1945, by sex and by single years of age.

Example 2.3

Large and consistent differences in the values of a variable may indicate that there was a change in the measurement process that should be investigated. An example of large differences is found in data used in the *Report of the Second Task Force on Blood Pressure Control in Children* (NHLBI Task Force 1987). Systolic blood pressure values for 5-year-old boys averaged 103.5 mmHg in a Pittsburgh study compared to 85.6 mmHg in a Houston study. These averages were based on 61 and 181 boys aged 5 in the Pittsburgh and Houston studies, respectively. Hence, these differences were not due to small sample sizes. Similar differences were seen for 5-year-old girls and for 3- and 4-year-old boys and girls as well. There are large differences between other studies also used by this Task Force, but the differences are smaller for older children. These incredibly large differences between the Pittsburgh and Houston studies were likely due to a difference in the measurement process. In the Houston study, the children were at the clinic at least 30 minutes before the blood pressure was measured compared to a much shorter wait in the Pittsburgh study. Since the measurement processes differed, the values obtained do not reflect the same variable across these two studies. The use of data from these two studies without any adjustment for the difference in the measurement process is questionable.

Example 2.4

The use of data from laboratories is another area in which it is crucial to monitor constantly the measurement process — in other words, the equipment and the personnel who use the equipment. In large multicenter trials that use different laboratories, or even with a single laboratory, referent samples are routinely sent to the laboratories to determine if the measurement processes are under control. This enables any problems to be detected quickly and prevents subjects from being either unduly alarmed or falsely comforted. It also prevents erroneous values from being entered into the data set. The Centers for Disease Control (CDC) has an interlaboratory program, and data from it demonstrate the need for monitoring. The CDC distributes samples to about 100 laboratories throughout the United States. The April 1980 results of measuring lead concentration in blood are shown in Figure 2.4

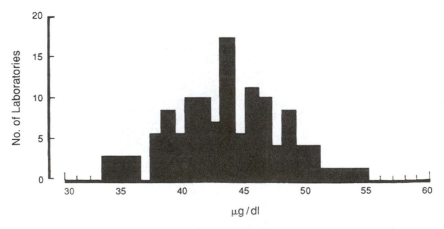

Figure 2.4 Distribution of measurements of blood lead concentration by separate laboratories, Centers for Disease Control.

(Hunter 1980). The best estimate of the blood lead concentration in the distributed sample was 41 micrograms per deciliter ($\mu g/dL$), but the average reported by all participating laboratories was 44 $\mu g/dL$. The large variability from the value of 41 shown in Figure 2.4 is a reason for concern, particularly since the usual value in human blood lies between 15 and 20 $\mu g/dL$.

Example 2.5

Of course, the lack of inconsistencies and irregularities does not mean that there are no problems with the data. Too much consistency and regularity sometimes is grounds for a special inquiry into its causes. Scientific frauds have been uncovered in some investigations in which the investigator discarded data that did not conform to theory. Abbe Gregor Mendel, the 19th-century monk who pioneered modern gene theory by breeding and crossbreeding pea plants, came up with such perfect results that later investigators concluded he had tailored his data to fit predetermined theories. Another well-known fabrication of data in science is the case of Sir Cyril Burt, a British pioneer of applied psychology. In his frequently cited studies of intelligence and its relation to heredity, he reported the same correlation in three studies of twins with different sample sizes (0.771 for twins reared apart and 0.944 for twins reared together). The consistency of his results eventually raised concern as it is highly unlikely that the exact same correlations would be found in studies of humans with different sample sizes. Science historians generally agree that his analyses were creations of his imagination with little or no data to support them (Gould 1981).

Example 2.6

Fabrication of data presents a real threat to the integrity of scientific investigation. The *San Francisco Chronicle* reported a case of data fabrication under the headline "Berkeley lab found research fabricated: Scientist accused of misconduct fired" (SFC 2002). A physicist at Lawrence Berkeley National Laboratory claimed the discovery of two new elements in the structure of the atomic nucleus in May 1998. Energy Secretary Bill Richardson called it "a stunning discovery which opens the door to further insights into the structure of the atomic nucleus." However, in follow-up experiments, outside labs were unable to replicate the results. The Lawrence Berkeley Laboratory retracted the finding after independent scientists were unable to duplicate the results. In announcing the laboratory's decision, director Charles V. Shank acknowledged that the false claim was "a result of fabricated research data and misconduct by one individual." He further reported, "The most elementary checks and data archiving were not done."

Conclusion

Data are a numerical representation of a phenomenon. By assigning numerical values to occurrences of the phenomenon, we are thus able to describe and analyze it. The assignment of the numerical values requires an understanding of the phenomenon and careful measurement. In the measurement process, some unexpected problems may be introduced, and the data then contain the intended numerical facts as well as the unintended fictions. Therefore, we cannot use data blindly. The meanings of data and their implications have been explored in a number of examples in this chapter.

EXERCISES

2.1 Identify the scale used for each of the following variables:
   a. Calories consumed during the day
   b. Marital status
   c. Perceived health status reported as poor, fair, good, or excellent
   d. Blood type
   e. IQ score

2.2 A person's level of education can be measured in several ways. It could be recorded as years of education, or it could be treated as an ordinal variable — for example, less than high school, high school graduate, and so on. Is it always better to use years of education than the ordinal variable measurement of education? Explain your answer.

2.3 In a health interview survey, a large number of questions are asked. For the following items, discuss (1) how the variable should be defined operationally, (2) whether nonresponse is likely to be high or low, and (3) whether reliability is likely to be high or low. Explain your answers.
   a. Weight
   b. Height
   c. Family income
   d. Unemployment
   e. Number of stays in mental hospitals

2.4 The pulse is usually reported as the number of heartbeats per minute, but the actual measurement can be done in several different ways — for example:
   a. Count the beats for 60 seconds
   b. Count for 30 seconds and multiply the count by 2
   c. Count for 20 seconds and multiply the count by 3
   d. Count for 15 seconds and multiply the count by 4
   Which procedure would you recommend to be used in clinics, considering accuracy and practicality?

2.5 Two researchers coded five response categories to a question differently as follows:

Response Category	Researcher A	Researcher B
Strongly agree	1	2
Agree	2	1
Undecided	3	0
Disagree	4	−1
Strongly disagree	5	−2

   a. What type of scale is illustrated by Researcher A?
   b. What type of scale is illustrated by Researcher B?
   c. Which coding scheme would you use and why?

2.6 The first U.S. Census was taken in 1790 under the direction of Thomas Jefferson. The task of counting the people was given to 16 federal marshals, who in turn hired enumerators to complete the task in nine months. In October 1791, all of the census reports had been turned in except the one from South Carolina, which was not received until March 3, 1792. As can be expected, the marshalls encountered many obstacles and the counting was incomplete. The first census revealed a population of 3,929,326. This result was viewed as an undercount, as is indicated in the following excerpt from a letter written by Jefferson:

> I enclose you also a copy of our census, written in black ink so far as we have actual returns, and supplied by conjecture in red ink, where we have no returns; but the conjectures are known to be very near the truth. Making very small allowance for omissions, we are certainly above four millions. . . . (Washington 1853)

Discuss what types of obstacles they might have encountered and what might have led Jefferson to believe there was an undercounting of the people.

2.7 The National Center for Health Statistics matched a sample of death certificates in 1960 with the 1960 population census records to assess the quality of data and reported the following results (NCHS 1968):

Agreement and Disagreement in Age Reporting, 1960					
		White		Nonwhite	
	Total	Male	Female	Male	Female
Agreement	68.8%	74.5%	67.9%	44.7%	36.9%
Disagreement					
1 year difference	17.8	16.6	18.8	20.8	20.2
2+ year difference	13.4	8.9	13.3	34.5	42.9

Do you think that the age reported in the death certificate is more accurate than that reported in the census? How do you explain the differential agreement by gender and race? How do you think these disagreements affect the age-specific death rates calculated by single years and those computed by five-year age groups?

2.7 Discuss possible reasons for the digit preference in the 1945 population census of Turkey that is shown in Figure 2.3. Why was the digit preference problem more prominent among females than among males? How would you improve the quality of age reporting in census or surveys? How do you think the digit preference affects the age-specific rates calculated by single years of age and those computed by five-year age groups?

2.8 Get the latest vital statistics report for your state from a library and find out the following:

a. Are residents of your state who died in a foreign country included in the report?

b. Are the data from your state report consistent with the data from the *National Vital Statistics Report* from the National Center for Health Statistics?

c. Is an infant born to a foreign student couple in your state included in the report?

REFERENCES

Campbell, C., and B. L. Joiner. "How to Get the Answer without Being Sure You've Asked the Question." *The American Statistician* 27:229–231, 1973.

Forthofer, R. N. "Blood Pressure Standards in Children." Paper presented at the American Statistical Association Meeting, August 1991. See Appendix C for details of the National Health and Nutrition Examination Survey.

Gould, S. J. *The Mismeasure of Man.* New York W. W. Norton, 1981.

Hunter, J. S. "The National System of Scientific Measurement." *Science* 210:869–874, 1980.

Juster, F. T. "Response Errors in the Measurement of Time Use." *Journal of the American Statistical Association* 81:390–402, 1986.

Marquis, K. H., M. S. Marquis, and J. M. Polich. "Response Bias and Reliability in Sensitive Topic Surveys." *Journal of the American Statistical Association* 81:381–389, 1986.

National Center for Health Statistics, *Vital and Health Statistics, Series 2*, November 29, 1968.

The NHLBI Task Force on Blood Pressure Control in Children. "The Report of the Second Task Force on Blood Pressure Control in Children, 1987." *Pediatrics* 79:1–25, 1987.

Regan, O. G. "Statistical Reforms Accelerated by Sixth Census Errors." *Journal of the American Statistical Association* 68:540–546, 1973. See Appendix D for details of the population census.

San Francisco Chronicle. July 13, 2002, p. 1.

Shaw, S. "Preface on Doctors." In *The Doctor's Dilemma.* New York: Brentano's, 1909.

Suchman, L., and B. Jordan. "Interactional Troubles in Face-to-Face Survey Interviews." *Journal of the American Statistical Association* 85:232–241, 1990.

United Nations. *Methods of Appraisal of Quality of Basic Data for Population Estimates, Population Studies, No. 23.* New York: United Nations Dept. of Economic and Social Affairs, 1955, p. 34.

Warner, S. L. "Randomized Response: A Survey Technique for Eliminating Evasive Answer Bias." *Journal of the American Statistical Association* 60:63–69, 1965.

Washington, H. A., ed. *The Writings of Thomas Jefferson*, III, 287. Washington, DC: Taylor & Maury, 1853.

Descriptive Methods

3

The Scotsman William Playfair is credited with being the first to publish graphics such as the bar chart, line graph, and pie charts that are commonly used in statistics today (Kennedy 1984). This chapter focuses on the summarization and display of data using the techniques Playfair first published along with several other useful procedures. We will rely on both numerical and pictorial procedures to describe data. We use charts and other procedures because they may capture features in the data that are often overlooked when using summary numerical measures alone. Although the utility of graphical methods has been well established and can be seen in all walks of life, the visual representation of data was not always common practice. According to Galvin Kennedy, the first 50 volumes of the *Journal of the Royal Statistical Society* contain only 14 charts, with the first one appearing in 1841.

3.1   Introduction to Descriptive Methods

The data we use in this section come from the Digitalis Investigation Group (DIG) trial (DIG 1997). The DIG trial was a multicenter trial with 302 clinical centers in the United States and Canada participating. (Its study design features will be discussed in a later chapter.) The purpose of the trial was to examine the safety and efficacy of Digoxin in treating patients with congestive heart failure in sinus rhythm. Subjects were recruited from those who had heart failure with a left ventricular ejection fraction of 0.45 or less and with normal sinus rhythm. The primary endpoint in the trial was to evaluate the effects of Digoxin on mortality from any cause over a three- to five-year period. Basic demographic and physiological data were recoded at the entry to the trial, and outcome related data were recorded during the course of the trial. The data presented in this chapter consists of baseline and outcome variables from 200 patients (100 on Digoxin treatment and 100 on placebo) randomly selected from the multicenter trial dataset.*

*This trial was conducted and supported by the National Heart, Lung, and Blood Institute in cooperation with the study investigators. The NHLBI has employed statistical methods to make components of the full datasets anonymous in order to provide selected data as a teaching resource. Therefore, the data are inappropriate for any publication purposes. The authors would like to thank the NHLBI, study investigators, and study participants for providing the data.

Table 3.1   Digoxin clinical trial data for 40 participants.

ID	Treatment[a]	Age[b]	Race[c]	Sex[d]	Body Mass Index[e]	Serum Creatinine[f]	Systolic Blood Pressure[g]
4995	0	55	1	1	19.435	1.600	150
2312	0	78	2	1	22.503	2.682	104
896	0	50	1	1	27.406	1.300	140
3103	0	60	1	1	29.867	1.091	140
538	1	31	1	1	27.025	1.159	120
1426	0	70	1	1	19.040	1.250	150
4787	1	46	1	1	28.662	1.307	140
5663	0	59	2	1	27.406	1.705	152
1109	0	68	1	2	27.532	1.534	144
666	0	65	1	1	28.058	2.000	120
2705	1	66	1	2	28.762	0.900	150
5668	0	74	1	1	29.024	1.227	116
999	1	47	1	2	30.506	1.386	120
1653	1	63	1	1	28.399	1.100	105
764	1	63	2	2	28.731	0.900	122
3640	0	79	1	1	18.957	2.239	150
1254	1	73	1	1	26.545	1.300	144
2217	1	65	1	1	23.739	1.614	170
4326	0	65	1	1	29.340	1.200	170
5750	1	76	1	1	39.837	1.455	140
6396	0	83	1	1	26.156	1.489	116
2289	0	76	1	1	30.586	1.700	130
1322	1	45	1	2	43.269	0.900	115
4554	1	58	1	2	28.192	1.352	130
6719	1	34	1	1	20.426	1.886	116
1954	1	77	1	1	26.545	1.307	140
5001	1	70	1	1	19.044	1.200	110
1882	0	50	1	1	25.712	1.034	140
5368	1	38	1	1	30.853	0.900	134
787	0	58	2	2	27.369	0.909	100
4375	0	61	1	1	32.079	1.273	128
5753	1	75	1	1	37.590	1.300	138
6745	0	45	1	1	22.850	1.398	130
6646	0	61	1	1	27.718	1.659	128
5407	1	50	1	2	24.176	1.000	130
4181	0	44	2	2	26.370	1.148	124
3403	0	55	1	2	21.790	1.170	130
2439	1	49	1	1	15.204	1.307	140
4055	0	71	1	1	22.229	1.261	100
3641	0	64	1	1	21.228	0.900	130

[a]Treatment group (0 = on placebo; 1 = on Digoxin)
[b]Age in years
[c]Race (1 = White; 2 = Nonwhite)
[d]Sex (1 = Male; 2 = Female)
[e]Body mass index (weight in kilograms/height in meters squared)
[f]Serum creatinine (mg/dL)
[g]Systolic blood pressure (mmHg)

We refer to this working dataset as DIG200 in this book. The DIG200 dataset is reduced to create a smaller dataset including 7 baseline variables from 40 patients referred to as DIG40. Table 3.1 displays the DIG40 dataset. Both data files are available on the supplementary website.

3.2   Tabular and Graphical Presentation of Data

The one- and two-way frequency tables and several types of figures (line graphs, bar charts, histograms, stem-and-leaf plots, scatter plots, and box plots) that aid the description of data are introduced in this section.

3.2.1   Frequency Tables

A *one-way frequency table* shows the results of the tabulation of observations at each level of a variable. In Table 3.2, we show one-way tabulations of sex and race for the 40 patients shown in Table 3.1. Three quarters of the patients are males, and over 87 percent of the patients are whites.

Table 3.2   Frequencies of sex and race for 40 patients in DIG40.

Sex	Number of Patients	Percentage	Race	Number of Patients	Percentage
Male	30	75.0	White	35	87.5
Female	10	25.0	Nonwhite	5	12.5
Total	40	100.0	Total	40	100.0

The variables used in frequency tables may be nominal, ordinal, or continuous. When continuous variables are used in tables, their values are often grouped into categories. For example, age is often categorized into 10-year intervals. Table 3.3 shows the frequencies of age groups for the 40 patients in Table 3.1. More than one half of the patients are 60 and over. Note that the sum of percents should add up to 100 percent, although a small allowance is made for rounding. It is also worth noting that the title of the table should contain sufficient information to allow the reader to understand the table.

Table 3.3   Frequency of age groups for 40 patients in DIG40.

Age Groups	Number of Patients	Percentage
Under 40	3	7.5
40–49	6	15.0
50–59	8	20.0
60–69	11	27.5
70–79	12	30.0
Total	40	100.0

Table 3.4   Cross-tabulation of body mass index and sex for 40 patients in DIG40 with column percentages in parentheses.

Body Mass Index	Sex Male	Sex Female	Total
Under 18.5 (underweight)	1 (3.3%)	0 (0.0%)	1 (2.5%)
18.5–24.9 (normal)	10 (33.3%)	2 (20.0%)	12 (30.0%)
25.0–29.9 (overweight)	14 (46.7%)	6 (60.0%)	20 (50.0%)
30.0 & over (obese)	5 (16.7%)	2 (20.0%)	7 (17.5%)
Total	30	10	40

Two-way frequency tables, formed by the *cross-tabulation* of two variables, are usually more interesting than one-way tables because they show the relationship between the variables. Table 3.4 shows the relationship between sex and body mass index where BMI has been grouped into underweight (BMI < 18.5), normal (18.5 ≤ BMI < 25), overweight (25 ≤ BMI < 30), and obese (BMI ≥ 30). The body mass index is calculated as weight in kilograms divided by height in meters squared. There are higher percentages of females in the overweight and obese categories than those found for males, but these calculations are based on very small sample sizes.

In forming groups from continuous variables, we should not allow the data to guide us. We should use our knowledge of the subject matter, and not use the data, in determining the groupings. If we use the data to guide us, it is easy to obtain apparent differences that are not real but only artifacts. When we encounter categories with no or few observations, we can reduce the number of categories by combining or collapsing these categories into the adjacent categories. For example, in Table 3.4 the number of obesity levels can be reduced to 3 by combining the underweight and normal categories. There is no need to subdivide the overweight category, even though one-half of observations are in this category. Computer packages can be used to categorize continuous variables (recoding) and to tabulate the data in one- or two-way tables (see **Program Note 3.1** on the website).

There are several ways of displaying the data in a tabular format. In Tables 3.2, 3.3, and 3.4 we showed both numbers and percentages, but it is not necessary to show both in a summary table for presentation in journal articles. Table 3.5 presents basic patient characteristics for 200 patients from the DIG200 data set. Note that the total number (n) relevant to the percentages of each variable is presented at the top of the column and percentages alone are presented, leaving out the frequencies. The frequencies can be calculated from the percentages and the total number.

Table 3.5 Basic patient characteristics at baseline in the Digoxin clinical trial based on 200 patents in DIG200.

Characteristics		Percentage (n = 200)
Sex	Male	73.0
	Female	27.0
Race	White	86.5
	Nonwhite	13.5
Age	Under 40	3.5
	40–49	11.5
	50–59	25.0
	60–69	33.0
	70 & over	26.0
Body mass index	Underweight (< 18.5)	1.5
	Normal (18.5–24.9)	37.5
	Overweight (25–29.9)	42.5
	Obese (≥ 30)	18.5

Other data besides frequencies can be presented in a tabular format. For example, Table 3.6 shows the health expenditures of three nations as a percentage of gross domestic products (GDP) over time (NCHS 2004, Table 115). Health expenditures as a percentage of GDP are increasing much more rapidly in the United States than either Canada or United Kingdom.

3.2.2 Line Graphs

A *line graph* can be used to show the value of a variable over time. The values of the variable are given on the vertical axis, and the horizontal axis is the time variable. Figure 3.1 shows three line graphs for the data shown in Table 3.6. These line graphs also show the rapid increase in health expenditures in the United States compared with those of two other counties with national health plans. The trends are immediately clear in the line graphs, whereas one has to study Table 3.6 before the same trends are recognized.

Table 3.6 Health expenditures as a percentage of gross domestic product over time.

Year	Canada	United Kingdom	United States
1960	5.4	3.9	5.1
1965	5.6	4.1	6.0
1970	7.0	4.5	7.0
1975	7.0	5.5	8.4
1980	7.1	5.6	8.8
1985	8.0	6.0	10.6
1990	9.0	6.0	12.0
1995	9.2	7.0	13.4
2000	9.2	7.3	13.3

Source: National Center for Health Statistics, 2004, Table 115

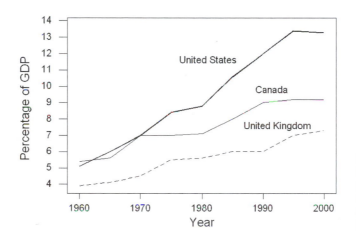

Figure 3.1 Line graph: Health expenditures as percentage of GDP for Canada, United Kingdom, and United States.

It is possible to give different impressions about the data by shortening or lengthening the horizontal and vertical axes or by including only a portion of an axis. In creating and studying line graphs, one must be aware of the scales used for horizontal and vertical axes. For example, with numbers that are extremely variable over time, a logarithmic transformation (discussed later) of the variable on the vertical axis is frequently used to allow the line graph to fit on a page.

Example 3.1

It is well accepted that blood pressure varies from day to day or even minute to minute (Armitage and Rose 1966). We present the following data on systolic blood pressure measurements for three patients taken three times a day over a three-day period in two different ways in Figure 3.2:

Patient	Day 1			Day 2			Day 3		
	8am	2pm	8pm	8am	2pm	8pm	8am	2pm	8pm
1	110	140	100	115	130	110	105	137	105
2	112	138	105	105	133	120	110	128	100
3	105	135	120	110	130	105	115	135	110

In the top graph we show the change in a patient's systolic blood pressure over the three time points for each day without connecting between days. From the line graph, we notice that the individual under study has peaks in his systolic blood pressure, and the peaks occur consistently at the same time point, giving us reason to believe that there may be a circadian rhythm in blood pressure.

Depending on the time of day when the blood pressure is measured, the patient's hypertension status may be defined differently because most cutoff points for stages of hypertension are based on fixed values that ignore the time of day. In the bottom graph the lines are connected between days, with the recognition that the time interval between days is twice as large as the measurement intervals during the day. The general trend shown in the top graph remains, but the consistency between days is less evident. Another measurement at 2am could have established the consistency between days.

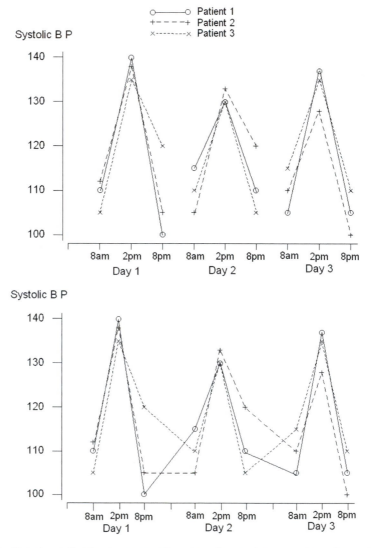

Figure 3.2 Plot of systolic blood pressure taken three times a day over a three-day period for three patients.

Example 3.2

It is possible to represent different variables in the same figure, as Figure 3.3 shows. The right vertical axis is used for lead emissions and the left vertical axis for sulfur oxide emissions. Both pollutants are decreasing, but the decrease in lead emissions is quite dramatic, from approximately 200×10^3 metric tons in 1970 to only about 8×10^3 metric tons in 1988. During this same period, sulfur oxide emissions decreased from about 20×10^6 metric tons to 21×10^6 metric tons. The decrease in the lead emissions is partially related to the use of unleaded gasoline, which was introduced during the 1970s.

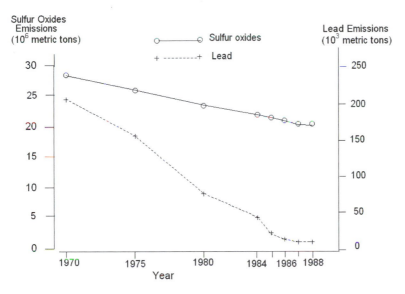

Figure 3.3   Line graph of sulfur oxides and lead emissions in the United States.
Source: National Center for Health Statistics, 1991, Table 64

3.2.3   Bar Charts

A bar chart provides a picture of data that could also be reasonably displayed in tabular format. Bar charts can be created for nominal, ordinal, or continuous data, although they are most frequently used with nominal data. If used with continuous data, the chart could be called a histogram instead of a bar chart. The bar chart can show the number or proportion of people by levels of a nominal or ordinal variable.

Example 3.3

The actual enrollment of individuals in health maintenance organizations (HMOs) in the United States was 9.1 million in 1980, 33.0 million in 1990, and 80.9 million in 2000 (NCHS 2004, Table 134). This information is displayed in Figure 3.4 using a bar chart. The numbers of people enrolled in HMOs in the United States is shown by year (ordinal variable). This bar chart makes it very clear that there has been explosive growth in HMO enrollment. The actual numbers document this growth, but it is more dramatic in the visual presentation.

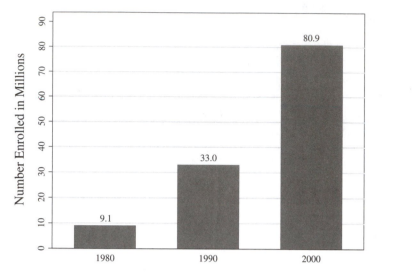

Figure 3.4 Bar chart of the number of persons (in millions) enrolled in Health Maintenance Organizations by year.
Source: National Center for Health Statistics (NCHS), 2004, Table 134

In bar charts, the length of the bar shows the number of observations or the value of the variable of interest for each level of the nominal or ordinal variable. The widths of the bar are the same for all the levels of the nominal or ordinal variable, and the width has no meaning. The levels of the nominal or ordinal variable are usually separated by several spaces that make it easier to view the data. The bars are usually presented vertically, although they could also be presented horizontally.

Bar charts can also be used to present more complicated data. The tabulated data in two- or three-way tables can be presented in bar chart format. For instance, the data in a 2 × 5 table (e.g., the status of diabetes — yes or no — by five age groups) can be presented by five bars with the length of each bar representing the proportion of people in the age group with diabetes, as shown in Figure 3.5.

When both variables in a two-way table have more than two levels each, we can use a segmented bar chart. Example 3.4 illustrates the presentation of data in a 3 × 4 table using a segmented bar chart. Data in a three-way table can be presented by a clustered

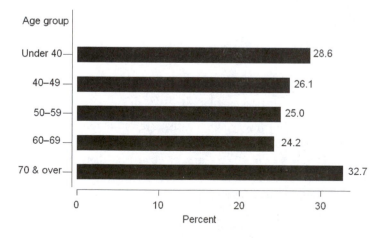

Figure 3.5 Bar chart showing proportion of people in each age group with diabetes, DIG200.

bar chart. Example 3.5 shows a presentation of data in a $2 \times 3 \times 4$ table using a clustered bar chart.

Example 3.4

To examine the relationship between obesity and age, DIG200 data are tabulated in a 3×4 table:

Obesity level	Age Group (column percent in parentheses)			
	Under 50	**50–59**	**60–69**	**70 & over**
Normal or underweight (BMI < 25)	11 (36.6)	22 (42.3)	26 (39.4)	19 (36.5)
Overweight (25 ≤ BMI < 30)	11 (36.6)	23 (44.2)	30 (45.5)	21 (40.4)
Obese (BMI ≥ 30)	8 (26.7)	7 (13.5)	10 (15.2)	12 (23.1)
Total	30	52	66	52

The data in this table are presented in Figure 3.6 using two types of segmented bar charts. The first segmented bar chart is based on frequencies (top figure), and the second segmented bar chart is based on percentages (bottom figure). The top figure shows that nearly two-thirds of obese patients are in the 60 and over age groups. The bottom figure shows that the obesity is more prevalent in the under 50 age group.

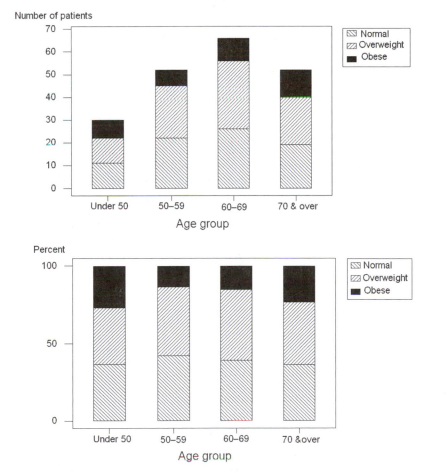

Figure 3.6   Segmented bar charts for levels of obesity by age group, DIG200 (the normal category includes underweight as well as normal)

Example 3.5

To examine how the prevalence of diabetes differs by the level of obesity and age, the DIG200 data are tabulated in a $2 \times 3 \times 4$ table. The results are presented in Figure 3.7 using a clustered bar chart. Three bars depicting the percent of diabetes in three levels of obesity are clustered in each of the age categories. It is interesting to note that the level of obesity is closely associated with the prevalence of diabetes in all age groups except for the 70 and over age group.

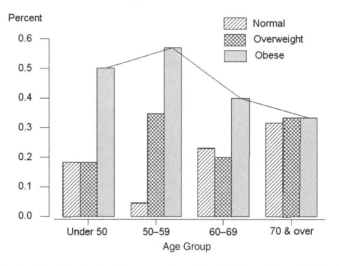

Figure 3.7 Clustered bar charts showing proportion of people in each level of obesity (the normal category includes underweight as well as normal) and age group who have diabetes.

It is often possible for "graphs to conceal more than they reveal" by making comparisons across groups less evident (van Belle 2002). To highlight that individuals categorized as obese have a higher percentage of diabetes across all age categories with the exception of the 70 and over age group, we may introduce a line graph as shown in Figure 3.7. Careful attention should be paid when constructing graphical presentations of data, and possibly several methods should be considered when exploring data in order to find the graph that best captures the data's structure.

Many computer packages are available for creating bar charts (see **Program Note 3.2** on the website).

3.2.4 Histograms

As we said earlier, a histogram is similar to a bar chart but is used with interval/ratio variables. The values are grouped into intervals (often called bins or classes) that are usually of equal width. Rectangles are drawn above each interval, and the area of rectangle represents the number of observations in that interval. If all the intervals are of equal width, then the height of the interval, as well as its area, represents the frequency of the interval. In contrast to bar charts, there are no spaces between the rectangles unless there are no observations in some interval.

Table 3.7 Frequency of individual systolic blood pressures (mmHg): DIG200.

Value	Freq.	Value	Freq.	Value	Freq.	Value	Freq.	Value	Freq.	Value	Freq.
85	1	105	1	116	8	128	3	138	1	150	12
90	5	106	2	118	5	130	23	139	2	152	3
95	2	108	2	120	25	131	1	140	26	155	1
96	1	110	16	122	4	132	2	142	1	160	3
100	14	112	1	124	4	134	1	144	3	162	1
102	1	114	5	125	3	135	2	145	1	165	1
104	2	115	2	126	1	136	1	148	1	170	5

We demonstrate here the construction of a histogram for the data on systolic blood pressure values from patients in the DIG200. Before creating the histogram, however, we create a one-way table that will facilitate the creation of the histogram. Table 3.7 gives the frequency of systolic blood pressure values (SBP) for each individual in the DIG200. Note that there are 199 observations because one individual in the placebo group has missing information on her systolic blood pressure.

After inspecting the data, you should notice that a large proportion of the blood pressure values appear to end in zero — 137 out of 199, actually. All the values are even numbers, with the exception of 17 observations, and 15 values that end in 5. This suggests that the person who recorded the blood pressure values may have had a preference for numbers ending in zero. This type of finding is not unusual in blood pressure studies; however, despite this possible digit preference, we are going to create some histograms based on these values shown in Table 3.7.

The following questions must be answered before we can draw the histograms for these data:

1. How many intervals should there be?
2. How large should the intervals be?
3. Where should the intervals be located?

Tarter and Kronmal (1976) discuss these questions in some depth. There are no hard and fast answers to these questions; only guidelines are provided.

The number of intervals is related to the number of observations. Generally 5 to 15 intervals would be used, with a smaller number of intervals used for smaller sample sizes. There is a trade-off between many small intervals, which allow for greater detail with few observations in any category, and a few large intervals, with little detail and many observations in the categories.

One method of determining the number of intervals is suggested by Sturges and elaborated by Scott (1979). The suggested formula is $(\log_2 n + 1)$, where n is the number of observations, to calculate the number of intervals required to construct a histogram. Therefore, the width of the interval can be calculated using the expression $(x_{max} - x_{min})/(\log_2 n + 1)$. Since there are 199 observations in Table 3.7, we need to find the value of $\log_2 199 + 1$. This value is 8.64, and we round it up to 9, meaning that 9 intervals should be used to construct the histogram.

We refer the reader to Appendix A for information on logarithms and how to calculate logarithms with different bases. The graph shown here also gives some feel for the shape of the logarithmic curve with 2 as the base. Briefly, $\log_2 199$ can be calculated dividing

$\log_{10}199$ by $\log_{10}2$, which is 7.64. The base 10 logarithm is available on most calculators or computer software. Alternatively, the value of $\log_2 199$ can be read from the graph. The dotted line in the graph shows that the value of $\log_2 199$ is about 7.6.

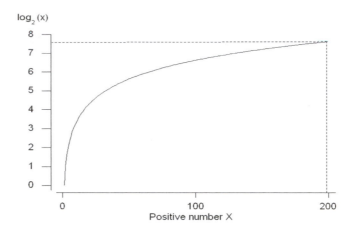

Table 3.8 illustrates the 9 intervals, and the interval width can be calculated using the expression $(x_{max} - x_{min})/(\log_2 n + 1)$. Since $(170 - 85)/8.64 = (85)/8.64 = 9.84$, we round the interval width to 10 mmHg. Notice in Table 3.8 that the notation [85–95) means all values from 85 to 95 but not including 95. Here we use the bracket ([ ) to indicate that the value should be included in the interval, whereas the parenthesis ( )) means up to the value but not including it. We have started the intervals with the value of 85, although we could have also begun the first interval with the value of 80.

This is a reasonable approach unless there are some relatively large or small values. In this case, exclude these unusual values from the difference calculation and adjust the minimum and maximum values accordingly. The location of the intervals is also arbitrary. Most researchers either begin the interval with a rounded number or have the midpoint of the interval be a round number. The computer packages create histograms using the procedures similar to the preceding approach with options to accommodate the users' request (see **Program Note 3.3** on the website).

Figure 3.8 displays the histogram for the data in Table 3.8.

Table 3.8 Intervals of histogram suggested by Sturges for the systolic blood pressure data in Table 3.7.

Class (Bin)	Class Width (Bin Width)	Frequency	Relative Frequency	Cumulative Relative Frequency	Cumulative Frequency
1	[85–95)	6	3.02	3.02	6
2	[95–105)	20	10.05	13.07	26
3	[105–115)	27	13.57	26.63	53
4	[115–125)	48	24.12	50.75	101
5	[125–135)	34	17.09	67.84	135
6	[135–145)	36	18.09	85.93	171
7	[145–155)	17	8.54	94.47	188
8	[155–165)	5	2.51	96.98	193
9	[165–175)	6	3.02	100.00	199
	Total	199	100.00		

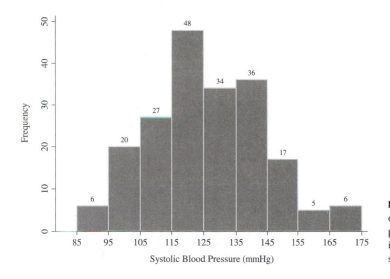

Figure 3.8 Histogram of 199 systolic blood pressure values using 9 intervals of size 10 starting at 85.

Example 3.6

Create histograms to compare the distributions of systolic blood pressures between individuals under 60 years of age and those 60 and over using the DIG200 data set. We begin by displaying the number of observations, the minimum value, and the maximum value for each of the age groups.

Under 60 years of age: $n = 81$, minimum = 90 mmHg, maximum = 170 mmHg
60 years and over: $n = 118$, minimum = 85 mmHg, maximum = 170 mmHg

We use Sturges' rule to determine the number of intervals that should be used to construct each histogram. The suggested number of intervals are:

Under 60 years of age: $(170 - 90)/(\log_2 81 + 1) = 10.9$ or 11 intervals
60 years and over: $(170 - 85)/(\log_2 118 + 1) = 10.8$ or 11 intervals

The same number of intervals is indicated. Even when different numbers of intervals were indicated, it will be better to keep the number of intervals the same for a better comparison.

Figure 3.9 presents two histograms for these groups. The first histogram displays the SBP of patients under 60 years of age and the second histogram for the 60 years and over group.

Notice that in this case the relative frequencies are used rather than frequencies mainly because the histograms are to be compared and the two groups have an unequal number of observations as just shown (i.e., there are 81 patients under 60 years of age and 118 who are 60 years and over). Relative frequencies allow for comparisons between two or more groups even if the groups do not have the same number of subjects. It is obvious, for subjects 60 years and over, that the highest

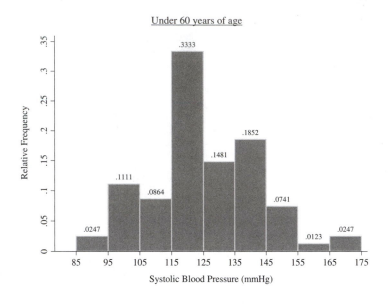

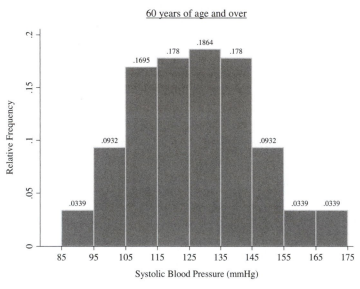

Figure 3.9 Histograms for systolic blood pressure distributions by age group.

percentages of systolic blood pressure readings fall in the intervals between 105 and 145 mmHg. Subjects under 60 years of age have a third of their systolic blood pressure observations in the interval between 115 and 125 mmHg. After comparing the two histograms, it is easy to see that the older age group has a higher concentration of subjects with systolic blood pressure values above 135 mmHg, an observation that was clearly expected.

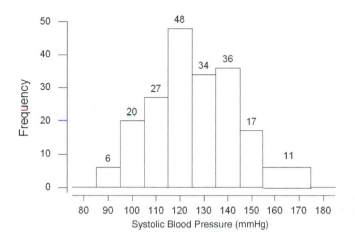

Figure 3.10 Histogram for systolic blood pressure with uneven intervals, DIG200.

It is possible for histograms constructed from the same data to have different shapes. The shapes of the histogram depend on the number of intervals used and how the boundaries are set. These differences in constructing the histogram may lead to different impressions about the data. However, histograms say basically the same thing about the distribution of the sample data even though their shapes are different.

Equal size intervals are used in most histograms. In case the use of unequal size intervals is desired, we must make some adjustments. Since the area of the rectangle for a category in a histogram reflects the frequency of the category, we need to adjust the height of an uneven size interval to keep the area at the same size. For example, assume we are interested in determining the number of subjects with SBP 155 mmHg and higher. We can collapse the last two intervals of the histograms presented in Figure 3.8 into one large interval that is twice as wide as the previous intervals. The histogram with the combined category is presented in Figure 3.10. Note that the frequency for the combined interval is 11, but the height of this interval is 5.5, one-half of the combined frequency. We divided the height by 2 to reflect the fact that the width of this last interval is twice as wide as the other intervals.

3.2.5   Stem-and-Leaf Plots

The *stem-and-leaf plot* looks similar to a histogram except that the stem-and-leaf plot shows the data values instead of using bars to represent the height of an interval. The stem-and-leaf plot is used for a relatively small dataset, while the histogram is used for a large dataset. Considering the systolic blood pressure readings of the 40 patients from the DIG40 data set, the stem contains the tens units and the leaves would be the ones units.

```
  4    10 | 0045
  9    11 | 05666
 16    12 | 0002488
 (8)   13 | 00000048
 16    14 | 000000044
  7    15 | 00002
  2    16 |
  2    17 | 00
```

Notice that a stem-and-leaf plot looks like a histogram except we know the values of all the observations, and histograms don't group data in the same way. The first column shows a cumulative count of all the observations from the top and from the bottom to the interval in which the median value is found. The median is the value such that 50 percent of the values are less than it, and 50 percent are greater than it. The number of observations in the interval containing the median is shown in parentheses. The second column is the stem, and the subsequent columns contain the leaves. For example, in the first row we read a stem of 10 and leaves of 0, 0, 4, and 5. Since the stem represents units of 10 and the leaf unit is 1, these four numbers are 100, 100, 104, and 105. The second row has a stem of 11, and there are 5 leaves referring to the systolic blood pressure values of 110, 115, 116, 116, and 116. Note that the first number in the second row is 9, which is the cumulative count of observations in the first two rows. There are 7 values in the third row, and the cumulative count is now 16. The median is the fourth row, and its value is 130. The method of determining the median is discussed later.

Example 3.7

Here is a stem-and-leaf plot to compare SBP (mmHg) readings of the following males and females in the DIG40 data set:

Males:	100	104	105	110	116	116	116	120	120	128	128	130	130	130	134	138	140
	140	140	140	140	140	140	144	150	150	150	152	170	170				
Females:	100	115	120	122	124	130	130	130	144	150							

Females	Stem	Males
0	10	045
5	11	0666
420	12	0088
000	13	00048
4	14	00000004
0	15	0002
	16	
	17	00

By displaying a two-sided stem-and-leaf plot, a comparison of the distributions of systolic blood pressures between males and females can be made. The comparison shows that female SBPs tend to be lower than male SBPs. The male observations have two extreme values occurring at 170 mmHg even though most of the male readings are concentrated at 140 mmHg.

A nice characteristic of the data that can be seen from histograms or stem-and-leaf plots is whether or not the data are symmetrically distributed. Data are *symmetrically distributed* when the distribution above the median matches the distribution below the median. Data could also come from a skewed or asymmetric distribution. Data from a skewed distribution typically have extreme values in one end of the distribution but no extreme values in the other end of the distribution. When there is a long tail to the right, or to the bottom if the data are presented sideways, data are said to be *positively skewed*. If there are some extremely small values without corresponding extremely large values, the distribution is said to be *negatively skewed*.

Example 3.8

A stem-and-leaf plot for the ages of patients in the DIG40 data set is

$$
\begin{array}{rl}
3 & 3 \mid 148 \\
9 & 4 \mid 455679 \\
17 & 5 \mid 00055889 \\
(11) & 6 \mid 01133455568 \\
12 & 7 \mid 00134566789 \\
1 & 8 \mid 3
\end{array}
$$

Notice that the data appears to be slightly asymmetric as the observations below the row containing the median are not grouped as tightly as those above it. In this case, we would consider the distribution of ages to be negatively skewed.

3.2.6 Dot Plots

A *dot plot* displays the distribution of a continuous variable. Consider Example 3.9 following where we want to compare the distribution of the continuous variable, systolic blood pressure, across a nominal variable such as age grouped into two categories — under 60 years of age and 60 years and over. These plots give a visual comparison of the center of the observations as well as providing some idea about how the observations vary. Like stem-and-leaf plots, dot plots are used for a relatively small data set.

Example 3.9

Dot plots comparing SBP across the age groups of "<60" and "≥60" are shown in Figure 3.11.

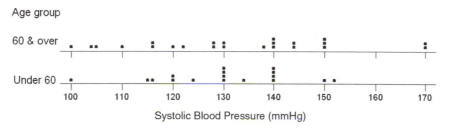

Figure 3.11 Dot plots for systolic blood pressure by age group, DIG40.

The dot plots allow us to see the data in its entirety. From the graphs, we see that the largest systolic blood pressure observation in the 60 and over group is considerably larger than the corresponding largest value in the under 60 years of age group. Also notice that dots are stacked up for observations with the same measurement value. For example, the stacked dots make it clear that there are two observations with the systolic blood pressure reading of 170 mmHg.

3.2.7   Scatter Plots

The two-dimensional *scatter plot* is analogous to the two-way frequency table in that it facilitates the examination of the relation between two variables. Unlike the two-way table, the two-dimensional scatter plot is most effectively used when the variables are continuous. Just as it is possible to have higher dimensional frequency tables, it is possible to have higher dimensional scatter plots, but they become more difficult to comprehend.

The scatter plot pictorially represents the relation between two continuous variables. In a scatter plot, a plotted point represents the values of two variables for an individual. We examine the relationship between serum creatinine levels and systolic blood pressure for 40 patients in the DIG40 data set (Table 3.1) using a scatter plot. Let us look at the top scatter plot in Figure 3.12. Each circle represents a patient's serum creatinine and systolic blood pressure values. For example, the circle in the upper left-hand corner of the plot represents the second patient (ID = 2312) in Table 3.1 with serum creatinine of 2.682 mg/dL and SBP of 104 mmHg. Overall, the scatter plot does not appear to show any relationship at all. There is a positive association between the variables when larger (smaller) values on one variable appear with larger (smaller) values of the other variable.

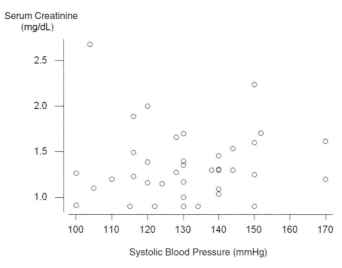

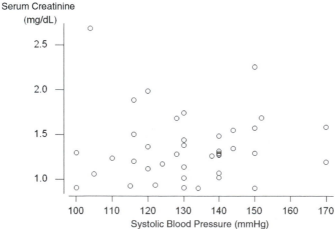

Figure 3.12 Scatter plot of serum creatinine versus systolic blood pressure for 40 patients with and without jittering, DIG40.

The association would be negative if individuals with large values of one variable tended to have small values of the other variable and conversely.

It is possible that several patients have the identical values of both variables. A careful examination of the data in Table 3.1 shows that three patients (ID = 4787, 1954, 2439) have the identical serum creatinine of 1.307 mg/dL and SBP of 140 mmHg. They are represented by one circle in the top scatter plot but by overlapping circles in the bottom scatter plot. In the bottom scatter plot a *jittering* (a very small random value) is added to the values of serum creatinine variable. If the jittering is performed for both variables, then the relative distances between circles could be slightly shifted in one or both directions.

Scatter plots are most effective for small to moderate sample sizes. When there are many variables such as in the DIG40 data set, a scatter plot matrix can be useful in displaying multiple two-way scatter plots (see Figure 3.13). From the plots we can see that there is a tendency for a very slight positive relationship between age and serum creatinine level and a slight negative relationship between serum creatinine and body mass index. There is no visual evidence of a relationship between other variables. Computer packages can be used to create stem-and-leaf plots and scatter plots (see **Program Note 3.4** on the website).

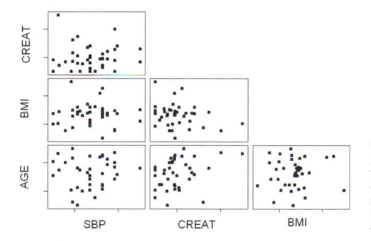

Figure 3.13   Scatter plot matrix examining the interrelationship among systolic blood pressure, creatinine, body mass index, and age, DIG40.

This completes the presentation of the pictorial tools in common use with the exception of the box plot, which is shown later in this chapter. The following material introduces the more frequently used statistics that aid us in describing and summarizing data.

3.3   Measures of Central Tendency

Simple descriptive statistics can be useful in data editing as well as in aiding our understanding of the data. The *minimum* and the *maximum* values of a variable are useful statistics when editing the data. Are the observed minimum and maximum values reasonable or even possible? For the patient's systolic blood pressure readings shown in Table 3.9, the minimum reading is 100 mmHg and the maximum is 170 mmHg. These values are somewhat unusual given that the average systolic blood pressure is approxi-

Table 3.9 Systolic blood pressure reading in ascending order, DIG40.

100	100	104	105	110	115	116	116
116	120	120	120	122	124	128	128
130	130	130	130	130	130	134	138
140	140	140	140	140	140	140	144
144	150	150	150	150	152	170	170

mately 131.4 mmHg, but they are not impossible. We will consider other ways of identifying unusual values in later sections.

3.3.1 Mean, Median, and Mode

In terms of describing data, people usually think of the average value or arithmetic mean. For example, the average systolic blood pressure was useful in determining whether or not the maximum and minimum values were reasonable. There are three frequently used measures of central tendency: the *mean*, the *median*, and the *mode*.

The *sample mean* ($\bar{x}$) is the sum of all the observed values of a variable divided by the number of observations. The *median* is defined to be the middle value — that is, the value such that 50 percent of the observed values fall above it and 50 percent fall below it. It can also be called the 50th percentile, where the *i*th percentile represents the value such that *i* percent of the observations are less than it. The mode is the most frequently occurring value.

Example 3.10

Calculate the mean systolic blood pressure reading using 40 patients in the DIG40 data set presented in Table 3.9.

The average or arithmetic mean is

$$\frac{100 + 100 + 104 + \cdots + 170}{40} = \frac{5256}{40} = 131.4 \, \text{mmHg}.$$

We can also represent the mean succinctly using symbols. We shall use upper-case X as the symbol for the variable under study — in this case, the SBP for patients in the DIG40 data set. We use lower-case x, with subscripts to distinguish each patient's systolic blood pressure, to represent the observed value of the variable. For example, the first patient's SBP is represented by x_1 and its value is 100 mmHg. The second patient's systolic blood pressure is x_2 and its value is also 100 mmHg. In the same way, x_3 is 104 mmHg, . . . , and x_{40} is 170 mmHg. Then the sum of the SBP can be represented by

$$x_1 + x_2 + x_3 + \cdots + x_{40} = \sum_{i=1}^{40} x_i.$$

The symbol Σ means summation. The value of i beneath Σ gives the subscript of the first x_i to be included in the summation process. The value above Σ gives the subscript of the last x_i to be included in the summation. The value of i increases in steps of 1 from the beginning value to the ending value. Thus, all the observations with subscripts ranging from the beginning value to the ending value are included in the sum. The formula for the sample mean variable, $\bar{x}$ (pronounced x-bar), is

$$\bar{x} = \frac{\sum\limits_{i=1}^{n} x_i}{n}$$

or more specifically in the case of this example,

$$\bar{x} = \frac{\sum\limits_{i=1}^{40} x_i}{n} = \frac{(100 + 100 + 104 + \cdots + 170)}{40} = 131.4 \,\text{mmHg.}$$

If we have the data for the entire population, not for just a sample of observations from the population, the mean is denoted by the Greek letter μ (pronounced "mu"). Values that come from samples are *statistics*, and values that come from the population are *parameters*. For example, the sample statistic $\bar{x}$ is an estimator of the population parameter μ. The population mean is defined as

$$\mu = \frac{\sum\limits_{i=1}^{N} x_i}{N}$$

where N is the population size.

In calculating the median, it is useful to have the data sorted from the lowest to the highest value as that assists in finding the middle value. Table 3.9 shows the sorted systolic blood pressure values for the 40 patients. For a sample of size n, the sample median is the value such that half ($n/2$) of the sample values are less than it and $n/2$ are greater than it. When the sample size is odd, the sample median is the $[(n + 1)/2]$th largest value. For example, the median for a sample of size 33 is thus the 17th largest value. The value 17 comes from $(33 + 1)/2$. When sample size is even, as in the case of the data on systolic blood pressure readings presented in Table 3.9, there is no observed sample value such that one-half of the sample falls below it and one-half falls above it. By convention, we use the average of the two middle sample values as the median — that is, the average of the $(n/2)$th and $[(n/2) + 1]$th largest values.

Example 3.11

Calculate the median systolic blood pressure readings using 40 patients in the DIG40 data set presented in Table 3.9. The data are already sorted in ascending order:

$$x_1 = 100, \; x_2 = 100, \; x_3 = 104, \ldots, \; x_{40} = 170.$$

Since we have an even number of patients, identify the $(n/2)$th observation or the $(40/2) = 20$th observation and the $[(n/2) + 1]$th observation or $[(40/2) + 1] = 21$st observation. Since $x_{20} = 130$ and $x_{21} = 130$, the average of these two values is 130.

The mode is the most frequently occurring value. When all the values occur the same number of times, we usually say that there is no unique mode. When two values occur the same number of times and more than any other values, the distribution is said to be bimodal. If there are three values that occur the same number of times and more than any other value, the distribution could be called trimodal. Usually one would not go beyond trimodal in labeling a distribution.

It is not unexpected to have no unique mode when dealing with continuous data, since it is unlikely that two units have exactly the same values of a continuous variable. However, in our data set of systolic blood pressure readings present in Table 3.9, the value of 140 mmHg occurs seven times, more frequently than any other reading, and is thus the mode. Although blood pressure is a continuous variable, the measurer often has a preference for values ending in zero, resulting in multiple observations of some values.

3.3.2   Use of the Measures of Central Tendency

Now that we understand how these three measures of central tendency are defined and found, when are they used? Note that in calculating the mean, we summed the observations. Hence, we can only calculate a mean when we can perform arithmetic operations on the data. We cannot perform meaningful arithmetic operations on nominal data. Therefore, the mean should only be used when we are working with continuous data, although sometimes we find it being used with ordinal data as well. The median does not require us to sum observations, and thus it can be used with continuous and ordinal data, but it also cannot be used with nominal data. The mode can be used with all types of data because it simply says which level of the variable occurs most frequently.

The mean is affected by extreme values, whereas the median is not. Hence, if we are studying a variable such as income that has some extremely large values, that is positively skewed, the mean will reflect these large values and move away from the center of the data. The median is unaffected, and it remains at the center of the data. For data that are symmetrically distributed or approximately so, the mean and median will be the same or very close to each other.

As was just mentioned, the SBP readings ranged from 100 to 170 mmHg for the 40 observations. The sample mean was 131.4 mmHg, and the sample median was 130 mmHg. These two values do not differ very much, since the data set contains observations that are relatively extreme on both the low and high end. However, the two values of 170 mmHg have caused the mean of 131.4 mmHg to be slightly larger than the median of 130 mmHg.

3.3.3   The Geometric Mean

We use another measure of central tendency when the numbers reflect population counts that are extremely variable. For example, in a laboratory setting, the growth in the number of bacteria per area is examined over time. The number of microbes per area does not change by the same amount from one period to the next, but the change is proportional to the number of microbes that were present during the previous time period. Another way of saying this is that the growth is *multiplicative*, not additive. The areas under study may also have used different media, and the microbes may not do well in some of the media, whereas in other media the growth is explosive. Hence, we may have counts in the hundred or thousands for some of the cultures, whereas a few other cultures may have counts in the millions or billions.

The arithmetic mean would not be close to the center of the values in this situation because of the effect of the extremely large values. The median could be used in this situation. However, another measure that is used in these situations is the *geometric mean*. The sample geometric mean for n observations is the nth root of the product of the values — that is,

$$\overline{x}_g = \sqrt[n]{x_1 * x_2 * \cdots x_n}.$$

Note that since the nth root is used in its calculation, the geometric mean cannot be used when a value is negative or zero.

This definition of the geometric mean is completely analogous to the arithmetic mean. The arithmetic mean is the value such that if we add it to itself $(n - 1)$ times, it equals the sum of all the observations. It is found by summing the observations and then dividing the sum by n, the sample size. Since in the preceding situation we are dealing with data resulting from a multiplicative process, our measure of central tendency should reflect this. The geometric mean is the value such that if we multiply it by itself $(n - 1)$ times, it equals the product of all the observations. It is found by multiplying the observations and then taking the nth root of the product.

When n is 2, there is little difficulty in finding the geometric mean, since the product of the two observed values is usually not large, and we know that the second root is the square root. However, for larger values of n, the product of the observed values may become very large, and we may lose some accuracy in calculating it, even when a computer is used. Fortunately, there is another way of calculating the product of the observations that does not cause any accuracy to be lost.

We can transform the observations to a logarithmic scale. Use of the logarithmic scale provides for accurate calculation of the geometric mean. After finding the logarithm of the geometric mean, we will transform the value back to the original scale and have the value of the geometric mean. In this section, we shall use logarithms to the base 10, although other bases could be used.

Again, we refer the reader to Appendix A for more information on logarithms and how to perform logarithmic transformation. The following chart shows some idea about the relationship between positive numbers and the corresponding base 10 logarithms.

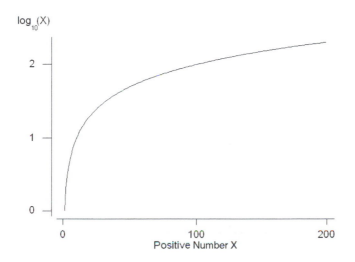

A key property of the logarithmic transformation is that the level of the mathematical operation performed on the arithmetic scale is reduced a level when the logarithmic scale is used. For example, a product on the arithmetic scale becomes a sum on the logarithmic scale. Therefore, the logarithm of the product of n values is

$$\log(x_1 * x_2 * \cdots x_n) = \sum_{i=1}^{n} \log x_i.$$

In addition, taking the nth root of a product on the arithmetic scale becomes division by n on the logarithmic scale — that is, finding the mean logarithm. In symbols, this is

$$\sqrt[n]{x_1 * x_2 * x_3 * \cdots x_n} = \frac{\sum_{i=1}^{n} \log_{10} x_i}{n} = \overline{\log_{10} x}.$$

We now have the logarithm of the geometric mean, and, to obtain the geometric mean, we must take the antilogarithm of the mean logarithm — that is,

$$\overline{x}_g = \text{antilog}\left(\overline{\log_{10} x}\right).$$

Example 3.12

Suppose that the number of microbes observed from six different areas are the following: 100, 100, 1000, 1000, 10,000, and 1,000,000. The geometric mean is found by taking the logarithm of each observation and then finding the mean logarithm. The corresponding base 10 logarithms are 2, 2, 3, 3, 4, and 6, and their mean is 3.33. The geometric mean is the antilog of 3.33, which is 2154.43. The arithmetic mean of these observations is 168,700, a much larger value than the geometric mean and also much larger than five of the six values. The usual mean does not provide a good measure of central tendency in this case. The value of the median is the average of the two middle values, 1000 and 1000, giving a median of 1000 that is of the same order of magnitude as the geometric mean.

The geometric mean has also been used in the estimation of population counts — for example, of mosquitos — through the use of capture procedures over several time points or areas. These counts can be quite variable by time or area, and hence, the geometric mean is the preferred measure of central tendency in this situation, too.

These (mean, median, mode, and geometric mean) are the more common measures of central tendency employed in the description of data. The value of central tendency, however, does not completely describe the data. For example, consider the nine systolic blood pressure readings

$$100 \quad 101 \quad 102 \quad 110 \quad 115 \quad 124 \quad 125 \quad 126 \quad 135.$$

Suppose that the four smallest observations were decreased by 10 mmHg and the four largest were increased by 10 mmHg. The values would now be the following:

90 91 92 100 115 134 135 136 145.

The means and medians of the two data sets are the same, 115 mmHg, yet the sets are very different. The sample mean of 115.3 mmHg and the sample median of 115 mmHg capture the essence of the first data set. In the second data set, however, the measures of central tendency are less informative as only one value is close to the mean and median. Therefore, some additional characteristics of the data must be used to provide for a more complete summary and description of the data and to distinguish between dissimilar data sets. The next section deals with this additional characteristic, the variability of the data.

3.4   Measures of Variability

The observations in the preceding second set of data corresponding to the systolic blood pressure of patients varied much more than those in the first set of data, but the means were the same. Hence, to provide for a more complete description of the data, we need to include a measure of its variability. A number of measures or values — the range, the interquartile range, selected percentiles, the variance, the standard deviation, and the coefficient of variation — are used to describe the variability in data.

3.4.1   Range and Percentiles

The *range* is defined as the maximum value minus the minimum value. It is simple to calculate, and it provides some idea of the spread of the data. For the patients under 60 years of age in Table 3.10, the range is the difference between 152 and 100, which is 52. In the second data set pertaining to patients 60 and over, the range is the difference between 170 and 100, which is 70.

This difference in the two ranges points to a dissimilarity between the two data sets. Although the range can be informative, the range has two major deficiencies: (1) It ignores most of the data, since only two observations are used in its definition, and (2) its value depends indirectly on sample size. The range will either remain the same or increase as more observations are added to a data set; it cannot decrease. A better measure of variability would use more of the information in the data by using more of the data points in its definition and would not be so dependent on sample size.

Percentiles, deciles, and quartiles are locations of an ordered data set that divide the data into parts. Quartiles divide the data into four equal parts. The first quartile (q1), or 25th percentile, is located such that 25 percent of the data lie below q1 and 75 percent of the data lie above q1. The second quartile (q2), or 50th percentile or median, is located

Table 3.10   Systolic blood pressure (mmHg) of patients under 60 years and 60 years and over, DIG40.

Under 60 Years					60 Years and Over				
100	115	116	120	120	100	104	105	110	116
124	130	130	130	130	116	120	122	128	128
134	140	140	140	140	130	130	138	140	140
150	152				140	144	144	150	150
					150	170	170		

such that half (50 percent) of the data lie below q2 and the other half (50 percent) of the data lie above q2. The third quartile (q3), or 75th percentile, is located such that 75 percent of the data lie below q3 and 25 percent of the data lie above q3. The *interquartile* range, the difference of the 75th and 25th percentiles (the third and first quartiles), uses more information from the data than does the range. In addition, the interquartile (or semiquartile) range can either increase or decrease as the sample size increases. The interquartile range is a measure of the spread of the middle 50 percent of the values. To find the value of the interquartile range requires that the first and third quartiles be specified, and there are several reasonable ways of calculating them. We shall use the following procedure to calculate the 25th percentile for a sample of size n:

1. If $(n + 1)/4$ is an integer, then the 25th percentile is the value of the $[(n + 1)/4]$th smallest observation.

2. If $(n + 1)/4$ is not an integer, then the 25th percentile is a value between two observations. For example, if n is 22, then $(n + 1)/4$ is $(22 + 1)/4 = 5.75$. The 25th percentile then is a value three-fourths of the way between the 5th and 6th smallest observations. To find it, we sum the 5th smallest observation and 0.75 of the difference between the 6th and 5th smallest observations.

The sample size is 40 for the systolic blood pressure data in Table 3.11. According to our procedure, we begin by sorting the data in ascending order. Next, we calculate $(40 + 1)/4$, which is 10.25. Hence the 25th percentile is a value one-fourth of the way between the 10th and 11th smallest observations. Since the 10th and 11th smallest observations have the same value of 120, the 25th percentile of the first quartile is 120 mmHg. The 75th percentile is found in the same way except that we use $3(n + 1)/4$ in place of $(n + 1)/4$. Since $3(40 + 1)/4$ yields 30.75, the 75th percentile is a value three-fourths of the way between the 30th and 31st observations. Since the 30th and 31st observations have the same value of 140, the 75th percentile, or the third quartile, is 140 mmHg. Hence, the interquartile range is $140 - 120 = 20$. Calculating the interquartile range for systolic blood pressure readings of patients under 60 years of age and 60 years and over gives the values 20 and 28, respectively. The larger interquartile range for the 60 and over age group suggests that there is more variability in the data compared to the systolic blood pressure readings for the younger age group.

The values of five selected percentiles — the 10th, 25th, 50th, 75th, and 90th — when considered together provide good descriptions of the central tendency and the spread of the data. However, when the sample size is very small, the calculation of the extreme percentiles is problematic. For example, when n is 5, it is difficult to determine how the 10th percentile should be calculated. Because of this difficulty, and also because of the instability of the extreme percentiles for small samples, we shall calculate them only when the sample size is reasonably large — say, larger than 30. The next measure of variability to be discussed is the variance, but, before considering it, we discuss the box plot because of its relation to the five percentiles.

Table 3.11 Systolic blood pressure of patients who have had a previous myocardial infarction stratified by the dose level of Digoxon treatment assigned, DIG200.

Low Dose Digoxon Treatment (0.125 mg/dL)					High Dose Digoxon Treatment (0.375 mg/dL)				
140	102	85	160	150	96	118	120	124	140
144	130	130	110	110	120	122	130	140	150

3.4.2   Box Plots

The box and whiskers plot, or just box plot, graphically gives the approximate location of the quartiles, including the median, and extreme values. The advantage of using box plots when exploring data is that several of the characteristics of the data such as outliers, symmetry features, the range, and dispersion of the data can be easily compared between different groups. The lower and upper ends (hinges) of the box mark the 25th and 75th percentiles or the locations of the first and third quartiles, while the solid band indicates the 50th percentile or the median. The whiskers represent the range of values, and the default option used in most statistical packages is to draw the whiskers out to 1.5 or 3 times the interquartile range. If the box plot is presented vertically, the area from the top edge to the bottom edge of the box represents the interquartile range.

From the systolic blood pressure data in Table 3.9, we already found the following information:

$$
\begin{aligned}
\text{minimum value} &= 100 \text{ mmHg,} \\
\text{first quartile} &= 120 \text{ mmHg,} \\
\text{median} &= 130 \text{ mmHg,} \\
\text{third quartile} &= 140 \text{ mmHg,} \\
\text{maximum value} &= 170 \text{ mmHg.}
\end{aligned}
$$

These values are plotted in a box plot in Figure 3.14.

We can use Figure 3.14 to assess the symmetry of the systolic blood pressure distribution. The box plots and histograms give us an indication of whether or not the data are skewed. For these patients, the distance from the median to the third quartile looks about the same as the corresponding distance to the first quartile. But there is a slightly longer tail to the right than to the left, indicating the distribution is slightly skewed to the right. In the Example 3.13, we go one step further by comparing the systolic blood pressures across all the age groups.

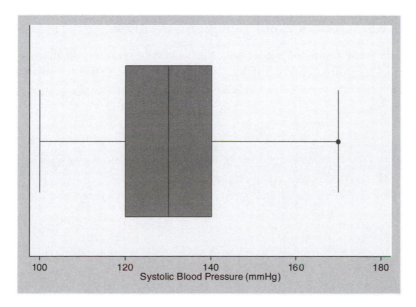

Figure 3.14   Box plot of systolic blood pressure values, DIG40.

Example 3.13

Using the data from Table 3.11, individual box and whisker plots of systolic blood pressure for the two age groups are created in Figure 3.15.

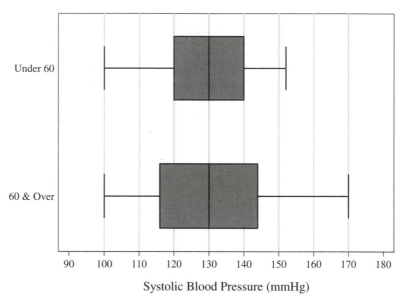

Figure 3.15 Box plots of systolic blood pressure across age groups, DIG40.

By looking at the box and whiskers plots side by side, it's possible to compare the distributions of systolic blood pressures for the two age categories. The medians are identical for both age groups. However, systolic blood pressure readings are more variable for the 60 and over group. This greater variability is shown in the larger width from the first quartile to the third quartile and through the greater range of the 60 and over group.

3.4.3 Variance and Standard Deviation

The *variance* and its square root, the *standard deviation*, are the two most frequently used measures of variability, and both use all the data in their calculations. The variance measures the variability in the data from the mean of the data. The population variance, denoted by σ^2 for a population of size N, is defined as

$$\sigma^2 = \frac{\sum_{i=1}^{N}(x_i - \mu)^2}{N}.$$

For a sample of size n, the sample variance s^2, an estimator of σ^2, is defined by

$$s^2 = \frac{\sum_{i=1}^{n}(x_i - \overline{x})^2}{n-1},$$

and the sample standard deviation is defined by

$$s = \sqrt{\frac{\sum_{i=1}^{n}(x_i - \overline{x})^2}{n-1}}.$$

The population variance could be interpreted as the average squared difference from the population mean, and the sample variance has almost the same interpretation about the sample mean.

The variance uses the sum of the squared differences from the mean divided by N, whereas the sample variance uses $n - 1$ in its denominator. Why were the squared differences chosen for use instead of the differences themselves? Perhaps the following table will clarify this. In Table 3.12 we find the systolic blood pressure readings for patients on low and high dose Digoxin treatment who have had a previous myocardial infarction.

If we consider only the 10 patients who were on high dose treatment, we can construct the information provided in Table 3.12. The sum of systolic blood pressure minus the mean must be zero since the positive differences cancel the negative differences.

Table 3.12   Differences and squared differences from the mean systolic blood pressure for 10 patients on high dose (0.375 mg/dL) Digoxin treatment who have had a previous myocardial infarction.

	SBP (mmHg)	SBP − mean	(SBP − mean)^2
	96	−30	900
	118	−8	64
	120	−6	36
	124	−2	4
	140	14	196
	120	−6	36
	122	−4	16
	130	4	16
	140	14	196
	150	24	576
Total	1260	0	2040

Additionally, why is $n - 1$ used instead of n in the denominator of the sample variance? It can be shown mathematically that the use of n results in an estimator of the population variance, which on the average slightly underestimates it. The following will give some feel for the use of $n - 1$.

In the formula for the sample variance, the population mean is estimated by the sample mean. This estimation of the population mean reduces the number of independent observations to $n - 1$ instead of n as is shown next. For example, you are told that there are three observations and that two of the values along with the sample mean are known. Can you find the value of the other observation? If you can, this means that there are only two independent observations, not three, once the sample mean is calculated. Suppose that the two values are 6 and 10 and the sample mean is 9. Since the mean of the three observations is 9, this indicates that the sum of the values is 27 and that the unknown value is $[27 - (6 + 10)] = 11$. In this sample of size three, given knowledge of the sample mean, only two of the observations are independent or free to vary. Hence, once a parameter (in this case the population mean) is estimated from the data, it reduces the number of independent observations (degrees of freedom) by one.

To account for this reduction in the number of independent observations, $n - 1$ is used in the denominator of the sample variance.

For the 10 systolic blood pressure values from patients on high dose Digoxin treatment in Table 3.12, the value of the sample variance is

$$s^2 = \frac{\sum_{i=1}^{10} (x_i - \bar{x})^2}{n - 1} = \frac{\sum_{i=1}^{10} (x_i - 126)^2}{10 - 1} = \frac{2040}{9} = \textbf{226.7},$$

and the value of the sample standard deviation is

$$s = \sqrt{\frac{\sum_{i=1}^{10} (x_i - \bar{x})^2}{n - 1}} = \sqrt{\frac{\sum_{i=1}^{10} (x_i - 126)^2}{10 - 1}} = \textbf{15.1}.$$

The sample variance and standard deviation for the 10 values from patients on low dose Digoxin treatment in Table 3.11 are 561.4 and 23.7, respectively — much larger values than the corresponding statistics for the 10 values in the high dose group. These statistics reflect the greater variation in the low dose values than in the high dose values.

The variance changes when nonconstant changes are made to all observations in the data. How does the value of the variance change when (1) a constant is added to (subtracted from) all the observations in the data set and (2) all the observations are multiplied (divided) by a constant?

The answer to the first question is that there is no change in the value of the variance, as can be seen from the following. If all the observations are increased by a constant — say, by 10 units — the mean is also increased by the same amount. Therefore, the constants will simply cancel each other out in the squared differences — that is,

$$[(x_i + 10) - (\mu + 10)]^2 = (x_i - \mu)^2$$

and thus there is no change in the sum of the squared differences or in the variance.

When all the observations are multiplied by a constant, the variance is multiplied by the square of the constant as can be seen from the following. If all the observations are multiplied by a constant — say, by 10 — the mean is also multiplied by the same amount. Therefore, in the squared differences we have

$$[(x_i * 10) - (\mu * 10)]^2 = [(x_i - \mu) * 10]^2 = (x_i - \mu)^2 * 10^2$$

and the sum of the squared differences, and thus the variance, is multiplied by the constant squared. This means that the standard deviation is multiplied by the constant. These two properties will be used in Chapter 5.

In later chapters, the variance and the standard deviation are shown to be the most appropriate measures of variation when the data come from a *normal distribution*, as knowledge of them and the mean is all that is necessary to completely describe the data. The normal distribution is the bell-shaped distribution often used in the grading of courses, and it is the most widely used distribution in statistics. The interquartile range and the five percentiles are useful statistics for characterizing the variation in data regardless of the distribution from which the data are selected, but they are not as informative as the mean and variance are when the data come from a normal distribution.

One last measure of variation is the *coefficient of variation*, defined as 100 percent times the ratio of the standard deviation to the mean. In symbols this is $(\sigma/\mu)100$ percent, and it is estimated by $(s/\bar{x})100$ percent. The coefficient of variation is a relative measure of variation, since in dividing by the mean, it directly takes the magnitude of the values into account. Large values of the coefficient suggest that the data are quite variable.

The coefficient of variation has several uses. One use is to compare the precision of different studies. If another experiment has a much smaller coefficient of variation than that in your study of the same substance, this suggests that there may be room for improvement in your study procedures. Another use is to determine whether or not there is so much variability in the data that the measure of central tendency is of little value. For example, the NCHS does not publish sample means for variables if the estimated coefficient of variation is greater than 30 percent.

Let us calculate the estimated coefficients of variation for our two sets of 10 observations in Table 3.11. For the first set, s was 23.7 and s was 15.1 in the second set. The sample mean was approximately 126 mmHg in both sets. These values lead to coefficients of variation of 18.8 percent (= [23.6946/126.1] 100 percent) and 12.0 percent in sets one and two, respectively. These values reinforce our feeling that the mean provided more useful information in the second set but was of less value in describing the data in the first set.

See **Program Note 3.5** on the website for the use of computer packages for descriptive statistics and box plots.

3.5   Rates and Ratios

Various forms of rates and ratios have been used in describing the health status and the change or growth of population. Rates and ratios are relative numbers that relate some absolute number of events to some other number such as the total population at that time. In this section we examine vital rates and population growth rates.

The rates of diseases and vital rates, which include death rates in general, infant mortality rates, feto-infant, neonatal and postneonatal mortality rates, and birth rates, are frequently used measures in public health. These rates are useful in determining the health status of a population, in monitoring the health status over time, in comparing the health status of populations, and in assessing the impact of policy changes.

For example, the infant mortality rate is often used in comparing the performance of health systems in different countries. In 2000, the United States had an infant mortality rate higher than that of 26 other nations. The U.S. rate was 6.9 infant deaths under 1 year of age per 1000 live births compared to a low rate of 3.2 for Japan. Most of the Western European nations and some Pacific Rim nations or large cities (Japan, Singapore, and Hong Kong) had lower rates than the United States. Canada's health system is often touted as a model for the United States because of its lower cost. How does Canada's infant mortality rate compare to that of the United States? Canada's infant mortality rate in 2000 was 5.3, almost 25 percent lower than the U.S. rate. The progress in reducing infant mortality has been most impressive, as can be seen from the U.S. rate for 1967 of 22.4 shown in Figure 1.1 in Chapter 1 compared to its 2000 rate of 6.9.

As can be seen from the following definition, a rate is basically a relative number multiplied by a constant. A rate is defined as the product of two parts: (1) the number of persons who have experienced the event of interest divided by the population size and (2) a standard population size. For example, according to the data compiled by the National Center for Health Statistics, there were 4,021,726 live births in an estimated population of 288,369,000 in the United States in 2002. The corresponding birth rate per 100,000 is found by taking (4,021,726/288,369,000) times 100,000, and it equals 13.9 births per 100,000 population. This is considerably lower than the corresponding rate for the United States in 1960 of 23.7 births per 100,000.

However, as is often the case with rates, there is a problem in determining the value of the denominator — that is, the 2002 U.S. population. What is meant by the 2002 population size? Is it as of January 1, July 1, December 31, or some other date? Convention is that the middle of the period (mid-2002) population size is used. An additional problem is that Census data were available for 2000 but not for 2002, which introduces some uncertainty in the value used. In this case, NCHS used an estimate of the 2002 midyear resident population based on the estimates of the U.S. Bureau of Census. The uncertainty in the value of the denominator of the rate should be of little concern given the magnitude of the numbers involved in this situation.

Vital rates are usually expressed per 1000 or per 100,000 population. As was just mentioned, infant mortality rates are expressed per 1000 live births with the exception of feto-infant mortality rates. Feto-infant mortality rates are based on the number of late fetal deaths plus infant deaths under 1 year per 1000 live births. Neonatal mortality rates are based on deaths of infants who were less than 28 days old, and postneonatal rates are based on deaths of infants between 28 and 365 days old. This split of infant deaths is useful because often the neonatal deaths may be due to genetic factors, whereas the postneonatal deaths may have more to do with the environment.

Note that as the infant mortality of 1988 rate example in Chapter 1 showed, the children whose deaths are used in the conventional method of calculating this rate may have been born in 1987, not 1988. Hence, the numerator, the number of deaths, comes from both 1987 and 1988 births, whereas the denominator is based solely on 1988 births. This should cause no problem unless something happened that caused the mortality experience or the number of births to differ greatly between the two years. One way of dealing with this possibility of a difference between the years is to combine several years of data. Often health agencies pool data over three years to provide protection against the instability of small numbers and to reduce the possible, but unlikely, effect of very different birth or mortality experiences across the years.

3.5.1 Crude and Specific Rates

Rates may be either crude or specific. *Crude rates* use the total number of events in their definition, whereas *specific rates* apply to subgroups in the population. For example, there may be age-, gender-, or race-specific birth or death rates. For an age-specific death rate, only the deaths to individuals in the specific age group are used in the numerator, and the denominator is the total number of individuals in the specific age group. Specific rates are used because they supply more information and also allow for more appropriate comparisons of groups.

For example, the crude death rate for the United States in 2002 was 847.3 per 100,000 population, and the age-specific death rates, as shown in Table 3.13, varied from 17.4 for the 5- to 14-year-old group to 14,828.3 for the 85-year-old and over group (NCHS

2004). The age-specific rates provide more information than the crude rates. For the same year the crude death rate for males was 846.6 versus 848.0 for females. There is no appreciable gender difference in the crude death rates. However, the age-specific death rates for males are higher than the female-specific rates in all age groups. Perhaps the lack of a difference between genders in the crude rate is related to differences in the age distributions. The age-specific rates by gender, shown in Table 3.13, provide a better description of the mortality experience than the crude rates. Without knowledge of the age distributions, it is difficult to conclude whether or not the age variable is responsible for the lack of a difference in the crude rates.

Table 3.13   Crude and age-specific death rates for the United States by gender in 2002.

	US Total Population	Male Population	Female Population
All ages, crude	847.3	846.6	848.0
Under 1	695.0	761.5	625.3
1–4	31.2	35.2	27.0
5–14	17.4	20.0	14.7
15–24	81.4	117.3	43.7
25–34	103.6	142.2	64.0
35–44	202.9	257.5	148.8
45–54	430.1	547.5	316.9
55–64	952.4	1,184.0	738.0
65–74	2,314.7	2,855.3	1,864.7
75–84	5,556.9	6,760.5	4,757.9
85 & over	14,828.3	16,254.5	14,209.6

Source: NCHS, 2004, Tables 1, 34, and 35 and page 442

As just shown, one problem with the use of specific rates is that they are not easily summarized. They do provide more information than the crude rate, which gives a single value for a population, but sometimes it is difficult to draw a conclusion based on the examination of the specific rates. However, because of the strong linkage between mortality and age, age often must be taken into account in the comparison of two or more populations. One way of adjusting for age or other variables while avoiding the problem of many specific rates is to use adjusted rates.

3.5.2   Adjusted Rates

Adjusted rates are weighted rates, as will be shown following. There are direct and indirect methods of adjustment; the choice of which method to use depends on what data are available. The direct method requires that we have the specific rates for each study population and a standard population. Table 3.14 provides the age-specific death rates for both male and female populations of the year 2002. The 2000 U.S. population proportions represent the standard population. The standard population provides a referent for purposes of comparison. Starting with 2001, NCHS uses the 2000 U.S. resident population as the standard for age-adjusting death rates. Prior to 2001 the 1940 U.S. population was used as the standard for age-adjusting mortality statistics. The choice of a standard population is subjective. For example, in comparing the rates between states, often the U.S. population would be used as the standard. In comparing counties of a state, the state population often would be used as the standard. For comparing rates over time, the population at a previous time point could be used as the standard. Another alternative might be to pool the populations of the areas or times under study and use the pooled population as the standard.

Table 3.14   Direct method of adjusting the 2002 U.S. male and female death rates using 2000 U.S. population as the standard.

Age	U.S. Population Proportion	Male Population		Female Population	
		Specific Rates[a]	Expected Deaths[a]	Specific Rates[a]	Expected Deaths[a]
Under 1	0.013818	761.5	10.5	625.3	8.6
1–4	0.055317	35.2	1.9	27.0	1.5
5–14	0.145565	20.0	2.9	14.7	2.1
15–24	0.138646	117.3	16.3	43.7	6.1
25–34	0.135573	142.2	19.3	64.0	8.7
35–44	0.162613	257.5	41.9	148.8	24.2
45–54	0.134834	547.5	73.8	316.9	42.7
55–64	0.087247	1,184.0	103.3	738.0	64.4
65–74	0.066037	2,855.3	188.6	1,864.7	123.1
75–84	0.044842	6,760.5	303.1	4,757.9	231.4
85 & over	0.015508	16,254.5	252.1	14,209.6	220.4
Total	1.000000		1013.7[b]		715.2[b]

[a] Per 100,000 population
[b] Age-adjusted death rate per 100,000 population
Source: NCHS, 2004, Tables 1, 34, and 35, and page 442

In performing the age adjustment in Table 3.14, the 2000 U.S. age distribution is used as the standard. The adjustment process consists of applying the male and female age-specific death rates to the standard population's age distribution and then summing the expected number of deaths over the age categories. Another way of saying this is that each age category's death rate is *weighted* by that age category's share of the standard population. The direct age-adjusted death rates for 2002 male and female populations using the U.S. as the 2000 standard population are 1013.7 and 715.2 deaths per 100,000 population, respectively. The male morality rate is about 30 percent higher than the female rate.

The indirect method is an alternative to be used when we do not have the data required for the direct method or when the specific rates may be unstable because they were based on small numbers. The indirect method requires the specific rates for the standard population and the age (or, for example, gender or race) distribution for the population to be adjusted. It is more likely that these data will be available than the age-specific death rates in the population to be adjusted. The first step in calculating the indirect age-adjusted death rate is to multiply the age-specific death rates of the standard population (the U.S.) by the corresponding age distribution of the population to be adjusted. Table 3.15 shows the calculation of indirect age-adjusted rate for American Indian or Alaskan Native male and female populations using the 2000 U.S. age-specific rates as the standard.

The observed crude death rates for American Indian/Alaskan Native male and female populations are 439.6 and 367.7 per 100,000, respectively. The male crude death rate is about 20 percent higher than the female rate. When age is taken into account, the gender difference in mortality may increase, since the average age of the female population is older than that of the male population.

In performing the indirect age standardization, the 2000 U.S. age-specific mortality rates are applied to the age distribution of the male and female populations of American Indian/Alaskan Natives. The expected death rates are created by multiplying the U.S. age-specific death rates by the proportion of people in the corresponding age groups for the male and female American Indian/Alaskan Native populations and then summing these expected numbers of deaths over the age categories. The ratio of the observed to

Table 3.15 Indirect age-adjusted death rates for the 2002 male and female populations of American Indian or Alaska Natives using the 2000 U.S. age-specific death rates as the standard.

Age	U.S. Age-Specific Rates[a] 2000	American Indian or Alaskan Native, 2002			
		Male Population		Female Population	
		Population Proportion	Expected Deaths[a]	Population Proportion	Expected Deaths[a]
All ages, crude	854.0	439.6[a]		367.7[a]	
Under 1	736.7	0.013681	10.1	0.012970	9.6
1–4	32.4	0.065798	2.1	0.063554	2.1
5–14	18.0	0.192182	3.5	0.186122	3.4
15–24	79.9	0.186971	14.9	0.175746	14.0
25–34	101.4	0.154397	15.7	0.144617	14.7
35–44	198.9	0.151792	30.2	0.154345	30.7
45–54	425.6	0.117915	50.2	0.124514	53.0
55–64	992.2	0.065798	65.3	0.070687	70.1
65–74	2,399.1	0.033225	79.7	0.038911	93.4
75–84	5,666.5	0.014332	81.2	0.020752	117.6
85 & over	15,524.4	0.003909	60.7	0.007782	120.8
Total		1.000000	413.6	1.000000	529.4
Standardized mortality ratio (SMR)		439.6/413.6 = 1.063		367.7/529.4 = 0.695	
Indirect age-adjusted death rate		854(1.063) = 907.8[a]		854(0.695) = 593.3[a]	

[a]Per 100,000 population
Source: NCHS, 2004, Tables 1, 34, and 35 and page 442

the expected death rates is the *standardized mortality ratio* (SMR). From Table 3.15, we see that the SMRs for the male and female populations are 1.063 and 0.695, respectively. The male SMR is 53 percent higher than the female SMR and the gender difference is more markedly shown, just as we expected. To find the indirect age-adjusted death rate for American Indian/Alaskan Native populations, we now multiply the crude rate for the standard population (854.0 per 100,000) by the SMRs. Thus, the indirect age-adjusted mortality rates for American Indian/Alaskan Native male and female populations are 907.8 and 593.3 per 100,000, respectively.

Both the direct and indirect age-adjustment methods can be used to adjust for more than one variable; for example, age and gender are often used together. Gender is frequently used because the mortality experiences are often quite different for females and males.

3.6   Measures of Change over Time

To understand the change in the height of a child or the growth of population over time, we may plot the data against time. We look first for an overall pattern and then for deviations from that pattern. For certain phenomena the points follow a straight line, and for other phenomena the points are nonlinear. In this section, we examine two well-known patterns of growth: linear and exponential.

3.6.1   Linear Growth

Linear growth means that a variable increases by a fixed amount at each unit of time. The height of a child or the production of food supply may take this pattern. To describe this pattern, we write a mathematical model for the straight-line growth of variable y.

$$y = a + bt.$$

In this model, b is the increment by which y changes when t increases by one unit and a is the base value of y when $t = 0$.

Example 3.14

The stature-for-age growth chart of U.S. boys is shown in Figure 3.16 (NCHS) 2006. The growth pattern exhibits a roughly linear trend between ages 2 to 15 years. For a typical child (50th percentile) a is about 34 inches (at age 2) and b is roughly 2.5

CDC Growth Charts: United States

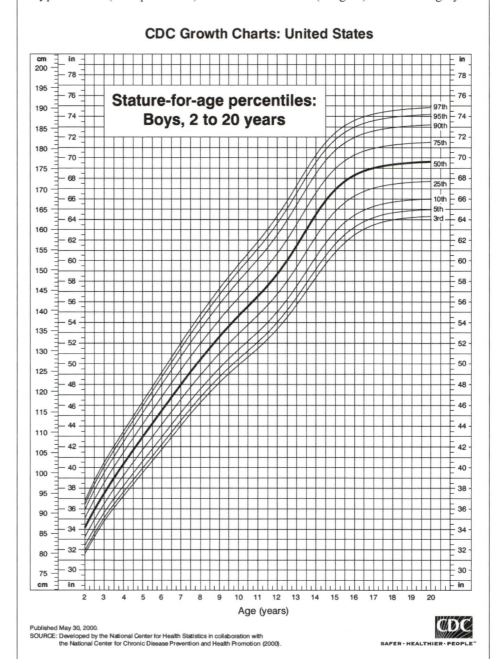

Stature-for-age percentiles: Boys, 2 to 20 years

Age (years)

Published May 30, 2000.
SOURCE: Developed by the National Center for Health Statistics in collaboration with the National Center for Chronic Disease Prevention and Health Promotion (2000).

Figure 3.16 Growth chart (stature-for-age) for U.S. boys, 2 to 20 years of age.

---

inches. From this we can tell that the stature of a 12-year-old boy would be about 59 inches [= 34 + 2.5(10)], and the chart also shows this value. The chart also shows that the stature of boys varies more as they grow older.

We will explore this linear growth model further in Chapter 13. Because no straight line usually passes exactly through all data points, we need to find a line that fits the points as well as possible. We will learn how to estimate the best fitting line from the data.

3.6.2   Geometric Growth

The population size of a community usually does not follow the linear growth model. The change in the population size over time in an area can simply be described as the number of people added or reduced between two time points. For comparison purposes, we can express the change as percent of the base population. If the time period is the same, the percent of change can be compared between populations. The percent of change from time 0 to time t in the population P is calculated by

$$\frac{P_t - P_0}{P_0}(100) = \left(\frac{P_t}{P_0} - 1\right)(100).$$

For example, the U.S. population increased from 248,709,873 in 1990 to 281,421,906 in 2000, showing a 13.15 percent increase over a 10-year period.

Percent change indicates a degree of change, but it is not yet a "rate of change." Like other vital rates, a rate of change should express change as a relative change in population size *per year*. We need to convert the percent change into an annual rate. But we cannot simply take one-tenth of the percent change (arithmetic mean) as an annual growth rate. Equal degrees of growth do not produce equal successive absolute increments because they follow the principle of compounded interest. In other words, a constant rate of growth produces larger and larger absolute increments, simply because the base of total population steadily becomes larger. Therefore, the linear growth model would not apply to population growth.

If a population is growing at an annual rate of r, then the population at time 1 would be the base plus an incremental change — that is, $(a + ar)$ or $a(1 + r)$. If the population is subject to the same constant growth rate, the population at time t will be

$$y = a(1 + r)^t.$$

Example 3.15

The geometric growth model fits well to the growth of money deposited at a bank with the interest added at the end of each year. Suppose $1000 is deposited and earns interest at an annual rate of 10 percent for 10 years. The amount in the account (y) at each anniversary date can be calculated by $y = 1000(1 + 0.1)^t$, where t ranges from 1 to 10. Figure 3.17 shows the results. The money grew more than 100 percent because the interest was compounded annually.

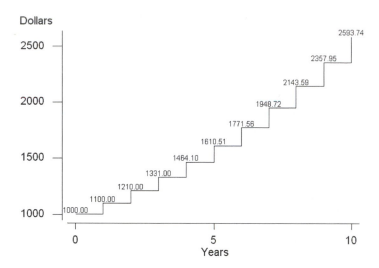

Figure 3.17 Account value over time for $1000 earning an annual interest rate of 10 percent.

If one wants to have the $1000 to be tripled over the 10-year period, then what level of annual interest rate would be required? We can solve $3000 = 1000(1 + r)^{10}$ for r as follows:

$$r + 1 = \exp\left(\frac{\ln(3)}{10}\right) = \exp\left(\frac{1.09861}{10}\right) = 1.1161.$$

One needs to find a bank that offers an annual interest rate of 11.6 percent.

3.6.3 Exponential Growth

We know that population is changing continuously as births and deaths occur throughout the year. We want to find a model that describes the growth as a continuous process. This new model is the exponential growth model and it has the following form:

$$y = ae^{rt}$$

where r is annual growth rate, e is a mathematic constant approximately equal to 2.71828, and a is the population at $t = 0$. Figure 3.18 graphically shows the exponential growth of a population of 10,000 at an annual growth rate of 5 percent over a 30-year period.

Relating to the bank interest rate example, this model assumes that the interest is compounded continuously.

Example 3.16

The U.S. population grew from 248,709,873 in 1990 to 281,421,906 in 2000. What would be an annual growth rate over the 10-year period? We can solve the following equation for r as follows:

$$281421906 = 248709873e^{10r}$$

$$\ln(281421906/248709873) = 10r$$

$$r = \left(\frac{\ln\left(281421906/248709873\right)}{10}\right) = \left(\frac{0.1236}{10}\right) = 0.01236.$$

The U.S. population grew at the annual rate of 1.24 percent.

Using the growth rate computed, we could project future size of population. Let us project the U.S. population in 2009 assuming the rate of growth remains constant.

$$y = 281421906(e^{9(0.01236)}) = 292050102$$

Over 10 million people would be added to the U.S. population in 9 years. This type of projection is acceptable for a short time period, but it should not be used for a long-range projection.

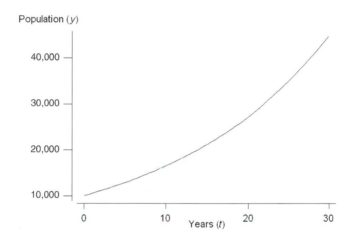

Figure 3.18 Increase of population of 10,000 at an annual rate of increase of 5 percent.

Example 3.17

(population doubling time): How long would it take to double the 2000 U.S. population assuming the annual growth rate remains constant? To answer this question, we solve the following equation for t.

$$2a = ae^{rt}, \text{ where } r = 0.01236$$

$$2 = e^{0.01236t}$$

$$\ln(2) = 0.01236t$$

$$t = \frac{\ln(2)}{0.01236} = \frac{0.69315}{0.01236} = 56.09.$$

The U.S. population will double in 56 years or in 2056.

Doubling means that $y/a = 2$ and natural logarithm of 2 is 0.69315. The solution suggests that if a population is increasing at an annual rate of 1 percent, then the population size will double in about 70 years. The time required to triple the population can be obtained by using $\ln(3)$. Similarly, the time required to increase the population by 50 percent can be obtained by using $\ln(1.5)$.

3.7 Correlation Coefficients

Earlier in the chapter, we presented a scatter plot of serum creatinine level and systolic blood pressure for 40 patients in the DIG40 data set, and we concluded that there was no appreciable association between serum creatinine and systolic blood pressure. Although this statement is informative, it is imprecise. To be more precise, a numerical value that reflects the strength of the association is needed. Correlation coefficients are statistics that reflect the strength of association.

3.7.1 Pearson Correlation Coefficient

The most widely used measure of association between two variables, X and Y, is the *Pearson correlation coefficient* denoted by ρ (rho) for the population and by r for the sample. This measure is named after Karl Pearson, a leading British statistician of the late 19th and early 20th centuries, for his role in the development of the formula for the correlation coefficient.

We want the correlation coefficient to be large, approaching +1 as a limit, as the values of the X, Y pair show an increasing tendency to be large or small together. When the values of the X, Y pair tend to be opposite in magnitude — that is, a large value of X with a small value of Y, or vice versa — the measure should be large negatively, approaching −1 as the limit. If there is no overall tendency of the values of the X, Y pair, the measure should be close to 0.

By large or small, we mean in relation to its mean. Because of the preceding requirements for the correlation coefficient, one simple function that may be of interest here is the product of $(x_i - \bar{x})$ with $(y_i - \bar{y})$. Let us focus on the sign of the differences, temporarily ignoring the magnitude. The possibilities are as follows:

$x_i - \bar{x}$	$y_i - \bar{y}$	$(x_i - \bar{x})(y_i - \bar{y})$
+	+	+
−	−	+
+	−	−
−	+	−

The product of the differences does what we want; that is, it is positive when the X, Y pairs are large or small together and negative when one variable is large and the other variable is small. By summing the product of the differences over all the sample pairs, the sum should give some indication whether there is a positive, negative, or no association in the data. If all the products are positive (negative), the sum will be a large positive (negative) value. If there is no overall tendency, the positive terms in the sum will tend to cancel out with the negative terms in the sum, driving the value of the sum toward 0.

However, the value of the sum of the products depends on the magnitude of the data. Since we want the maximum value of our measure to be 1, we must do something to remove the dependence of the measure on the magnitude of the values of the variables. If we divide the measure by something reflecting the variability in the X and Y variables, this should remove this dependence. The actual formula for r, reflecting these ideas, is

$$r = \frac{\sum_{i=1}^{n}(x_i - \overline{x})*(y_i - \overline{y})}{\sqrt{\sum_{i=1}^{n}(x_i - \overline{x})^2 * \sum_{i=1}^{n}(y_i - \overline{y})^2}}.$$

Dividing the numerator and denominator of this formula by $n-1$ enables us to rewrite the formula in terms of familiar statistics — that is,

$$r = \frac{\sum_{i=1}^{n}(x_i - \overline{x})*(y_i - \overline{y})/(n-1)}{\sqrt{s_x^2 * s_y^2}}.$$

In this version, we used the formula for the sample variance — that is, $s_x^2 = \Sigma(x_i - \overline{x})^2 / (n-1)$. The sample variance can also be expressed as $s_x^2 = \Sigma(x_i - \overline{x})(x_i - \overline{x})/n-1$. Hence, the sample variance could also be said to measure how X varies with itself. The numerator looks very similar to this, and it measures how the variables X and Y covary.

The denominator, $\sqrt{s_x^2 * s_y^2}$, standardizes r so that it varies from -1 to $+1$. For example, if $Y = X$, then the numerator becomes $\Sigma(x_i - \overline{x})^2/n-1$ — that is, s_x^2, which is the same as the denominator, and their ratio is $+1$.

For the data shown in Figure 3.12 the correlation coefficient turns out to be 0.025, confirming our earlier statement of a very slight positive relationship between serum creatinine and systolic blood pressure.

Example 3.18

We consider the following data on diastolic and systolic blood pressure readings for 12 adults.

| Systolic blood pressure: | 120 | 118 | 130 | 140 | 140 | 128 | 140 | 140 | 120 | 128 | 124 | 135 |
| Diastolic blood pressure: | 60 | 60 | 68 | 90 | 80 | 75 | 94 | 80 | 60 | 80 | 70 | 85 |

We first use a scatter plot of systolic blood pressure versus diastolic blood pressure (shown in Figure 3.19) to get a feel for the data. The jittering is added in the plot to

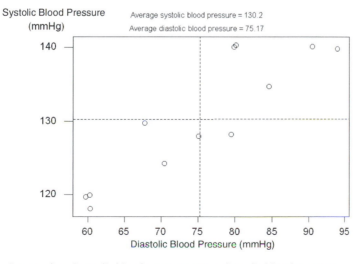

Figure 3.19   Scatter plot of systolic blood pressure versus diastolic blood pressure.

show the identical values for four adults. By adding vertical and horizontal lines showing the mean diastolic and mean systolic blood pressures, we can partition the scatter plot into four quadrants. Because most of the data cluster in the upper right and lower left quadrants, we expect that there will be a very strong correlation between these two variables.

The calculated correlation coefficient is 0.894, showing a strong positive association.

The correlation coefficient is not a general purpose measure of association, but it measures linear association — that is, the tendency of the (x_i, y_i) pairs to lie on a straight line. The following example demonstrates this point.

Example 3.19

For this example we consider the following values of Y and X:

Y:	4	1	0	1	4
X:	−2	−1	0	1	2

The sample mean of Y is 2, and the sample mean of X is 0. The pieces required to calculate r are

Y	X	$(Y-2)$	*	$(X-0)$	= product	$(Y-2)^2$	$(X-0)^2$
4	−2	2	*	−2	= −4	4	4
1	−1	−1	*	−1	= 1	1	1
0	0	−2	*	0	= 0	4	0
1	1	−1	*	1	= −1	1	1
4	2	2	*	2	= 4	4	4
Total 10	0	0		0	0	14	10

The estimated Pearson correlation coefficient, r, is then $0/\sqrt{14 * 10} = 0$. There is no linear association between Y and X. However, note that the first column (values of Y) and the last column (X^2) are the same. Hence, there is a perfect quadratic (squared) relation between Y and X that was not found by the Pearson correlation coefficient. The scatter plot in Figure 3.20 graphically shows this relationship. Connecting these points gives the parabola shape associated with a quadratic relationship.

Thus, even if r is 0, it does not mean that the two variables are unrelated; it means that there is no linear relation between the two variables. The use of a scatterplot first, followed by the calculation of r, may find the existence of a nonlinear association that could be missed when r alone is used.

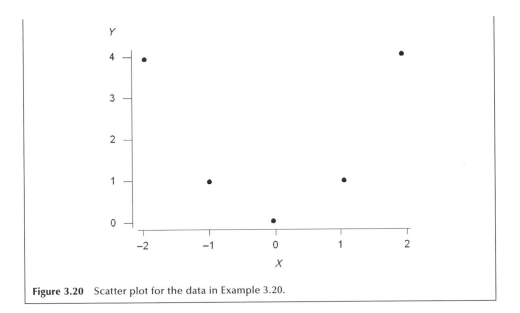

Figure 3.20   Scatter plot for the data in Example 3.20.

3.7.2   Spearman Rank Correlation Coefficient

The Pearson correlation coefficient was designed to be used jointly with normally distributed variables. However, it is used, sometimes incorrectly, with all types of data in practice. Instead of using the Pearson correlation coefficient with nonnormally distributed variables, it may be better to use a modification suggested by Spearman, an influential British psychometrician, in 1904. Spearman suggested ranking the values of Y and also ranking the values of X. These ranks are then used instead of the actual values of Y and X in the formula for the sample Pearson correlation coefficient. The result of this calculation is the sample Spearman rank correlation coefficient, denoted by r_s. In addition to being used with nonnormal continuous data, the Spearman rank correlation coefficient can also be used with ordinal data.

When ranking the data, ties (two or more subjects having exactly the same value of a variable) are likely to occur. In case of ties, the tied observations receive the same average rank. For example, if three observations of X are tied for the third smallest value, the ranks involved are 3, 4, and 5. The average of these three ranks is 4, and that is the rank that each of the three observations would be assigned. The occurrence of ties causes no problem in the calculation of the Spearman correlation coefficient when the Pearson formula is used with the ranks.

Example 3.20

Let us calculate the Spearman rank correlation coefficient for the data used in Example 3.19. The values of systolic and diastolic blood pressure values and their respective rankings are shown here. Note that there are several ties in ranking and average rankings are given.

SBP		DBP	
Value	Rank	Value	Rank
120	2.5	60	2
118	1.0	60	2
130	7.0	68	4
140	10.5	90	11
140	10.5	80	8
128	5.5	75	6
140	10.5	94	12
140	10.5	80	8
120	2.5	60	2
128	5.5	80	8
124	4.0	70	5
135	8.0	85	10

The calculated r_s is 0.866, slightly less than the Pearson correlation coefficient of 0.894.

See **Program Note 3.6** on the website for calculation of Pearson and Spearman correlation coefficients.

Conclusion

In this chapter we presented tables, graphs, and plots, as well as a few key statistics. The pictures and statistics together enable one to describe single variables and the relationship between two variables for the sample data. Although the description of the sample data and the provision of estimates of the population parameters are important, sometimes we wish to go beyond that — for example, to give a range of likely values for the population parameters or to determine whether or not it is likely that two populations under study have the same mean. Doing this requires the use of probability distributions, a topic covered in a subsequent chapter.

EXERCISES

3.1 Create a bar chart of the following data on serum cholesterol for non-Hispanic whites based on Table II-42 in *Nutrition Monitoring in the United States* (Life Sciences Research Office 1989)

Gender	Age	N	Mean Serum Cholesterol (mg/dL)[a]
Male	40–49	572	223.5
	50–59	575	228.9
	60–69	1354	226.2
	70–74	427	215.8
Female	40–49	615	218.5
	50–59	649	243.6
	60–69	1487	249.0
	70–74	533	248.3

[a]These data are from the Second National Health and Nutrition Examination Survey of noninstitutionalized persons conducted during the 1976–1980 period (NCHS 1981)

A high value of serum cholesterol is thought to be a risk factor for heart disease. The National Cholesterol Education Program (NCEP) of the National Institutes of Health in 1987 stated that the recommended value for serum cholesterol is below 200 mg/dl, and a value between 200 and 240 is considered to be the borderline. A value above 240 may indicate a problem, and NCEP recommended that a lipoprotein analysis should be performed. Based on these data, it appears that many non-Hispanic whites have serum cholesterol values that are too high, particularly women. The medical literature is also finally beginning to recognize that homocysteine is a very important risk factor for heart disease, even among people with normal levels of serum cholesterol (http://www.quackwatch.org/03HealthPromotion/homocysteine.html).

a. Give some possible reasons why non-Hispanic white males have higher mortality from heart and cerebrovascular diseases when it appears from these data that non-Hispanic white females should have the higher rates.

b. Provide a possible explanation why the serum cholesterol values for older males are lower than for the younger males and the reverse is true for females.

3.2 Create line graphs for the following expenditures for the Food Stamps Program in New York State during the 1980s.

Year	Actual Expenditures (in millions of dollars)	Inflation-adjusted Expenditures[a]
1980	745.3	745.3
1981	901.2	814.1
1982	835.7	717.3
1983	930.9	766.8
1984	904.4	709.3
1985	939.4	712.2
1986	926.5	685.3
1987	901.8	638.7
1988	909.1	613.4
1989	964.7	616.4

[a]Expenditures adjusted for inflation using the consumer price index for the Northeast Region with 1980 as the base.
Source: Division of Nutritional Sciences, 1992

What, if any, tendencies in the expenditures (both actual and inflation-adjusted) do you see? Which expenditures data do you think should be used in describing the New York State Food Stamps Program? Explain your choice.

3.3 Use line graphs to represent the short-stay hospital occupancy rates shown here.

Year	Hospital Ownership			
	Federal	Nonprofit	Proprietary	State/Local
1960	82.5	76.6	65.4	71.6
1970	77.5	80.1	72.2	73.2
1975	77.6	77.4	65.9	69.7
1980	77.8	78.2	65.2	70.7
1985	74.3	67.2	52.1	62.8
1989	71.0	68.8	51.7	64.8

Source: NCHS, 1992

Discuss the trends, if any, in these data.

3.4   Based on DIG200 data on the Web, explore prevalence of hypertension (variable name: HYPERTEN) by age, sex, and race, using appropriate descriptive tools you learned in Chapter 3. Present and discuss your findings, offering possible explanations for your findings and suggesting ways to conduct further study on the subject.

3.5   The following data on hazardous government jobs appeared as a bar chart in the "USA SNAPSHOTS" section of *USA Today* on April 30, 1992. The variable shown was the number of assaults suffered by federal officers based on 1990 FBI figures. The least number of assaults suffered were by the Internal Revenue Service (three assaults), the Bureau of Indian Affairs (five assaults), and the Postal Inspectors (six assaults). The most assaults were suffered by the Immigration and Naturalization Service with 409, followed by U.S. attorneys with 269 and the Bureau of Prisons with 185 assaults. What additional information do you need to conclude anything about which federal officers have the more hazardous — from the perspective of assaults — jobs?

3.6   A study was performed to determine which of three drugs was more effective in the treatment of a health problem. The responses of subjects who received each of three drugs (A, B, and C) were provided by Cochran (1955). The following shows the pattern of response for the 46 subjects:

	Response to		
A	**B**	**C**	**Frequency**
yes	yes	yes	6
yes	yes	no	16
yes	no	yes	2
yes	no	no	4
no	yes	yes	2
no	yes	no	4
no	no	yes	6
no	no	no	6
	Total		46

a. Give an example of a type of health problem that would be appropriate for this study.

b. Create a two-way frequency table showing the relationship between drugs A and C. Does it appear that the responses to these drugs are related?

c. Create a bar chart that shows the number of subjects with a favorable response by drug.

3.7   Using the data shown in Table 3.1, calculate the coefficient of variation for body mass index. Do you think that any measure of central tendency adequately describes these data? Explain your answer.

3.8   Lee (1980) presented survival times in months from diagnosis for 71 patients with either acute myeloblastic leukemia (AML) or acute lymphoblastic leukemia (ALL).

AML patients:
18 31 31 31 36 01 09 39 20 04 45 36 12 08 01 15 24 02 33 29 07 00 01 02 12 09 01 01 09 05 27 01 13 01 05 01 03 04 01 18 01 02 01 08 03 04 14 03 13 13 01

ALL patients:
16 25 01 22 12 12 74 01 16 09 21 09 64 35 01 07 03 01 01 22

a. Calculate the sample mean and median for both AML and ALL patients separately. Which measure do you believe is more appropriate to use with these data? Explain.

b. Create box plots, histograms, and stem-and-leaf plots to show the distributions of the survival times for AML and ALL patients. Which type of figure is more informative for these data? Which type of patient has the longer survival time after diagnosis?

c. Give examples of additional variables that are needed in order to interpret appropriately these survival times.

3.9 Is it possible to calculate the mean occupancy rate for the short-stay hospitals in 1960 given the data provided in Exercise 3.3? If it is, calculate it. If not, state why it cannot be calculated.

3.10 Provide an appropriate summarization of the following data on the results of inspections of food establishments (e.g., food processing plants, food warehouses, and grocery stores) conducted by the Division of Food Inspection Services of the New York State Department of Agriculture and Markets.

	Number Inspected		**Approximate Number Failed**	
Year	**Upstate**	**NYC & LI**[a]	**Upstate**	**NYC & LI**
1980	19,599	23,676	2,548	5,209
1982	17,183	22,767	3,093	6,830
1984	13,731	18,677	2,746	6,350
1986	10,915	15,948	2,292	6,379
1988	13,614	15,070	3,267	6,179
1990	12,609	16,285	3,026	6,677

[a]New York City and Long Island
Source: Division of Nutritional Sciences, 1992

Do you think that there were more or fewer cases of foodborne illness in New York State in 1990 than in 1980?

3.11 Diagnosis Related Groups (DRGs) are used in the payment for the health care of Medicare-funded patients. In the creation of the DRGs, suppose that the lengths of stay for 50 patients in one of the proposed groups were the following:

1	1 2 2 2	2 2 2 3 3	3 4 4 4 5	5 5 5 6 6
6	7 7 8 8	8 9 9 10 12	13 15 15 17 17	18 19 19 20 23
26 29 31 34 36	43 49 52 67 96			

Calculate the mean, standard deviation, coefficient of variation, and the five key percentiles for these data. Are these data skewed? Do the patients in this DRG appear to have homogeneous lengths of stay? Which measures, if any, should be used in the description of these data? Explain your answer.

3.12 The following data represent bacteria counts measured in water with levels of 0, 1, and 3% sodium chloride. The counts are the number per milliliter.

Level of Sodium Chloride	Counts
0%	10^7, 10^6, 10^8, 10^9, 10^8, 10^{10}
1%	10^4, 10^4, 10^5, 10^6
3%	10^3, 10^4, 10^4, 10^3, 10^5

    a. Calculate the mean and the coefficient of variation for these data.

    b. Calculate the median and the geometric mean.

    c. Comment on which measure of central tendency is appropriate for these data.

3.13 Of the estimated 1,488,939 male residents of Harris County, Texas, in 1986, there were 8,672 deaths. Of the 1,453,611 female residents, there were 6,913 deaths. The estimated 1986 U.S. population was approximately 48.7 percent male and 51.3 percent female.

    a. Calculate the crude death rate and the sex-specific death rates for Harris County in 1986.

    b. Do you believe that a sex-adjusted death rate will be very different from the crude death rate? Provide the reason for your belief.

    c. Calculate a sex-adjusted death rate for Harris County in 1986.

3.14 The Pearson correlation coefficient between age and creatinine for the data in Table 3.1 was 0.319. This suggests a modest linear relation between these two variables.

    a. Create a scatter plot of protein per age and creatinine. Is there a linear relationship? Are there any observations that clearly deviate from the linear trend?

    b. Calculate the Pearson correlation coefficient, ignoring the one or two observations that are considered to be outliers. Which measure of correlation do you think best characterizes the strength of the relation?

3.15 The U.S. population (in 1000) in 1980 and 2000 are shown below by ethnic groups:

Ethnic Groups	1980	2000
White	194,811	229,086
Black/African American	25,531	36,594
American Indian/Alaskan Native	1,420	2,984
Asian/Pacific Islander	3,729	11,757

Source: U.S. Bureau of the Census, 2000

    a. Calculate the annual growth rate, and show which group grew the fastest.

    b. Project the population in 2015 by ethnic group, assuming the growth rate remains constant over time.

    c. When will the 2000 population be doubled if the growth rate remains constant?

REFERENCES

Armitage, P., and G. A. Rose. "The Variability of Measurements of Casual Blood Pressure." *Clinical Science* 30:325–335, 1966.

Cochran, W. G. "The Comparison of Percentages in Matched Samples," *Biometrika* 37:356–366, 1955.

The Digitalis Investigative Group. "The Effect of Digoxin on Mortality and Morbidity in Patients with Heart Failure." *New England Journal of Medicine* 336:525–533, 1997.

Division of Nutritional Sciences, Cornell University in Cooperation with the Nutrition Surveillance Program of the New York State Department of Health. *New York State Nutrition: State of the State, 1992*. NYSDH, New York, 1992, Tables 2.5 and 3.2.

Kennedy, G. *Invitation to Statistics*. Blackwell, 1984, pp. 43–47.

Lee, E. T. *Statistical Methods for Survival Data Analysis*. Belmont, CA: Wadsworth, 1980.

Life Sciences Research Office, Federation of American Societies for Experimental Biology. *Nutrition Monitoring in the United States: An Update Report on Nutrition Monitoring*. DHHS Publ. No. (PHS) 89-1255. U.S. Department of Agriculture and the U.S. Department of Health and Human Services, Public Health Service, Washington, U.S. Government Printing Office, 1989.

National Center for Health Statistics. McDowell, A., A. Engel, J. T. Massey, and K. Maurer. "Plan and Operation of the Second National Health and Nutrition Examination Survey, 1976–80," *Vital and Health Statistics*, Ser. 1, No. 15, DHHS Publ. No. (PHS) 81-1317. Public Health Service, Washington, U.S. Government Printing Office, 1981.

National Center for Health Statistics. *Health, United States, 1990*. Hyattsville, MD: DHHS Pub. No. 91-1232. Public Health Service, 1991.

National Center for Health Statistics. *Health, United States, 1991 and Prevention Profile*. DHHS Publ. No. 92-1232. Public Health Service, Hyattsville, MD, 1992.

National Center for Health Statistics. *Health, United States, 2004 with Chartbook on Trends in the Health of Americans*. Hyattsville, MD: DHHS Pub. No. 2004-1232, 2004.

National Center for Health Statistics. www.cdc.gov/nchs/data/nhanes/growthcharts/set1/chart07.pdf, 2006.

Scott, D. W. "On Optimal and Data-Based Histograms." *Biometrika* 66:605–610, 1979.

Tarter, M. E. and R. A. Kronmal, "An Introduction to the Implementation and Theory of Nonparametric Density Estimation." *American Statistician* 30:105–112, 1976.

U.S. Bureau of the Census. *The 2000 Census of Population and Housing*. Summary Tape File 1A.

van Belle, G. *Statistical Rules of Thumb*. John Wiley, 2002, pp. 162–167.

Probability and Life Tables

4

Chapter Outline

As was mentioned in Chapter 3, often we want to do more than simply analyze or summarize the data in graphs or statistics. For example, we may want to determine whether two drugs or treatments are equally effective and safe or whether the age-adjusted death rates for two areas are the same. To answer these questions, we require knowledge of *probability*, the topic of this chapter.

4.1 A Definition of Probability

We have all encountered the use of probability — in the weather forecast, for example. The forecast usually involves an estimate of the probability of rain, as in the statement that "the probability of rain tomorrow is 20 percent." As its use in the weather forecast demonstrates, *probability* is a numerical assessment of the likelihood of the occurrence of an outcome of a random variable. In the weather forecast, weather is the random variable and rain is one of its possible outcomes.

Before considering the numerical assessment of likelihood, we should consider random variables. There are both discrete and continuous random variables. A *discrete* (nominal, categorical or ordinal) *random variable* is a quantity that reflects an attribute or characteristic that takes on different values with specified probabilities. A *continuous* (interval or ratio) *random variable* is a quantity that reflects an attribute or characteristic that falls within an interval with specified probabilities.

Hypertension status is a discrete random variable when the values or levels of this variable are defined as its presence (can be defined as systolic blood pressure greater than 140 mmHg, diastolic blood pressure greater than 90 mmHg, or taking antihypertensive medication) or absence. Other examples of discrete random variables include racial status, the number of children in a family, and type of health insurance. Examples of continuous random variables include height, blood pressure, and the amount of lead emissions as they are usually measured.

Table 4.1   Percent of population in selected racial groups: United States, 2000.

Race	Number	Percent
Total	281,421,906	100.0
White	211,460,626	75.1
Black or African American	34,658,190	12.3
Asian and Pacific Islander	10,641,833	3.8
American Indian and Alaskan Native	2,475,956	0.9
Some other races	15,359,073	5.5
Two or more races	6,826,228	2.4

Source: U.S. Bureau of the Census, 2000

We shall define probability of the occurrence of an outcome or interval of a random variable as its relative frequency in an infinite number of trials or in a population. A probability is a population *parameter*. An observed proportion (relative frequency) from a sample is a *statistic* that can be used to estimate a probability. We shall use the data in Table 4.1 to demonstrate the calculation of the probability of different racial categories in the United States in 2000. As shown in Table 4.1, there are four major racial groups used in the U.S. Population Census and a fifth category that combines all other races. Those who claimed two or more races are in the sixth category.

The probability of a person selected at random being white was 0.751 (= 211460626 /281421906), or 75.1 percent. The corresponding probabilities of being black, Asian and Pacific Islander, American Indian, and some other races were 0.123, 0.038, 0.009, and 0.055, respectively. Finally, the probability of a person claiming two or more races was 0.024. These six probabilities sum to 1.000 or 100.0 percent, as shown in Table 4.1.

Since a probability is the number of occurrences of an outcome divided by the total number of occurrences of all possible outcomes of the variable under study, this means that a probability cannot be larger than 1.00 or 100 percent in value. By the same reasoning, a probability cannot be smaller than 0.00 or 0 percent in value. Therefore, the only valid values for probabilities range from 0 to 1 or 0 to 100 percent. Additionally, use of the relative frequency definition means that the sum of the probabilities of all the possible outcomes of a random variable must be 1.00 or 100 percent. If a probability falls outside the 0 to 1 range, or if the sum of the probabilities of all the possible outcomes of a variable do not sum to 1 (with allowance for rounding), a mistake has been made.

For many variables in the health field, the probability of an outcome is estimated from a large number of observations and may change over time. For example, the probabilities of the different racial groups in the United States in 2020 will be different from the 2000 probabilities. As an additional example of changing probabilities, the estimates of the age-adjusted probabilities of obese persons (body mass index greater than or equal to 30) among U.S. adults (ages 20–74 years) increased from 0.151 in 1976–1980 to 0.233 in 1988–1994 and to 0.311 in 1999–2002 (NCHS 2004). This change in the values of a probability contrasts with the lack of change in the probabilities associated with physical phenomena, such as tossing a coin or a pair of dice. For example, when a fair coin is tossed, the probability of a head is assumed to be 0.5 or 50 percent, and it does not change.

The listing of the probabilities of all possible outcomes of a discrete variable is its *probability distribution*. For example, the probability distribution of the racial composition of the U.S. population in 2000 is shown in the last column of Table 4.1. More will be said about probability distributions and their use in the next chapter.

4.2   Rules for Calculating Probabilities

A few basic rules govern the calculation of probabilities of compound outcomes or events, and we will use the data in Table 4.2 to explain them. Entries in Table 4.2 are the number of live births by birth weight and the trimester in which prenatal care was begun. For example, the entry in the third row and the second column, 3271, is the number of live births to women who had begun their prenatal care during their second trimester and whose babies' birth weights were greater than 7.7 lb.

4.2.1   Addition Rule for Probabilities

The data in Table 4.2 can be used to determine whether or not there is a relation between the timing of the beginning of prenatal care and birth weight. However, before examining this issue, let us calculate a few additional probabilities. For example, the probability of a woman in Harris County, Texas, in 1986 having a low birth weight baby (less than or equal to 5.5 lb) was 0.069 (= 3541/51473). This value is very close to the 1986 value of 0.068 for the United States (NCHS 1992). Let us consider a slightly more complex example. The probability of late prenatal (third trimester) or no prenatal care is simply the sum of their individual probabilities, that is, 2337/51473 + 1695/51473 which is 0.078 (= 4032/51473). This value is slightly greater than the corresponding 1986 U.S. value of 0.060 (NCHS 1992). In these calculations of probabilities, we are considering births in Harris County, Texas, in 1986 as our population. If the intended population were Texas or the United States, then the preceding values would be sample estimates — that is, observed proportions — of the probabilities. However, a sample consisting of births in Harris County should not be used to draw inferences about births in Texas or the United States because the Harris County births are likely not to be representative of either of these two larger units.

So far, these probabilities have focused on row or column totals (marginal totals), not on the numbers in the interior of the table (cell entries). Entries in the interior of the table deal with the intersection of outcomes or events. For example, the outcome of a woman having a live birth of less than or equal to 5.5 lb and having begun her prenatal care during the first trimester is the intersection of those two individual outcomes. The

Table 4.2   Number of live births by birth weight and trimester of first prenatal care: Harris County, Texas, 1986 (excluding 1,180 births with unknown birth weight or trimester of first prenatal care).

Birth Weight	Trimester Prenatal Care Began				Total
	1st	2nd	3rd	No Care	
≤5.5 lb; ~2,500 g	2,412	754	141	234	3,541
5.6–7.7 lb; ~2,500–3,500 g	20,274	5,480	1,458	1,014	28,226
>7.7 lb; ~3,500 g	15,250	3,271	738	447	19,706
Total	37,936	9,505	2,337	1,695	51,473

Source: Harris County Health Department, 1990, Table 1.S

probability of this intersection — that is, of these two outcomes occurring together — is easily found to be 0.047 (= 2412/51473).

We just found the probability of a baby weighing less than or equal to 5.5 lb by using the row total of 3541 and dividing it by the grand total of 51,473. Note that we can also express this probability in terms of the probability of the intersection of a birth weight of less than or equal to 5.5 lb with each of the prenatal care levels — that is,

$$\Pr\{\leq 5.5\,\mathrm{lb}\} = \Pr\{\leq 5.5\,\mathrm{lb}\ \&\ \mathrm{1st\ trim.}\} + \Pr\{\leq 5.5\,\mathrm{lb}\ \&\ \mathrm{2nd\ trim.}\}$$
$$+ \Pr\{\leq 5.5\,\mathrm{lb}\ \&\ \mathrm{3rd\ trim.}\} + \Pr\{\leq 5.5\,\mathrm{lb}\ \&\ \mathrm{no\ care}\}$$
$$= \frac{2412}{51473} + \frac{754}{51473} + \frac{141}{51473} + \frac{234}{51473} = \frac{3541}{51473}.$$

This can be expressed in symbols. Let A represent the outcome of a birth weight less than or equal to 5.5 lb and B_i, $i = 1$ to 4, represent the four prenatal care levels. The symbol $\cap$ is used to indicate the intersection (to be read as "and") of two individual outcomes. Then we have

$$\Pr\{A\} = \Pr\{A \cap B_1\} + \Pr\{A \cap B_2\} + \Pr\{A \cap B_3\} + \Pr\{A \cap B_4\}$$

which, using the summation symbol, is

$$\Pr\{A\} = \sum_i \Pr\{A \cap B_i\}. \tag{4.1}$$

Suppose now that we want to find for a woman who had a live birth the probability that either the birth weight was 5.5 lb or less or the woman began her prenatal care during the first trimester. It is tempting to add the two individual probabilities — of a birth weight less than or equal to 5.5 lb and of prenatal care beginning during the first trimester — as we had done previously. However, if we added the entries in the first row (birth weights less than or equal to 5.5 lb) to those in the first column (prenatal care begun during the first trimester), the entry in the intersection of the first row and column would be included twice. Therefore, we have to subtract this intersection from the sum of the two individual probabilities to obtain the correct answer. The calculation is

$$\Pr\{\leq 5.5\,\mathrm{lb\ or\ 1st\ trim.}\} = \Pr\{\leq 5.5\} + \Pr\{\mathrm{1st\ trim.}\} - \Pr\{\leq 5.5\,\mathrm{lb\ and\ 1st\ trim.}\}$$
$$= \frac{3541 + 37936 - 2412}{51473} = 0.759.$$

This can be succinctly stated in symbols. Let A represent the outcome of live births of 5.5 lb or less and B represent the outcome of the initiation of prenatal care during the first trimester. An additional symbol $\cup$ is used to indicate the union (to be read as "or") of two individual outcomes. The intersection of these two outcomes is represented by $A \cap B$. In symbols, the rule is

$$\Pr\{A \cup B\} = \Pr\{A\} + \Pr\{B\} - \Pr\{A \cap B\}. \tag{4.2}$$

This rule also was used in the earlier example of late or no prenatal care, but, in that case, the outcomes were disjoint — that is, there was no overlap or intersection. Hence, the probability of the intersection was zero.

As the sum of the probabilities of all possible outcomes is one, if there are only two possible outcomes — say, A and not A (represented by $\bar{A}$) — we also have the following relationship:

$$\Pr\{A\} = 1 - \Pr\{\bar{A}\}. \tag{4.3}$$

4.2.2   Conditional Probabilities

Suppose we change the wording slightly in the preceding example. Based on the data in Table 4.2, we now want to find the probability of a woman having a live birth of less than or equal to 5.5 lb (event A) *conditional on or given* that her prenatal care was begun during the first trimester (event B). The word *conditional* limits our view in that we now focus on the 37,936 women who began their prenatal care during the first trimester. Thus, the probability of a woman having a live birth weighing less than or equal to 5.5 lb, given that she began her prenatal care during the first trimester, is 0.064 (= 2412 /37936). Dividing both the numerator and denominator of this calculation by 51473 (the total number of women) does not change the value of 0.064, but it allows us to define this *conditional probability* (the probability of A conditional on the occurrence of B) in terms of other probabilities. The numerator divided by the total number of women (2412 /51473) is the probability of the intersection of A and B, and the denominator divided by the total number of women (37936/51473) is the probability of B. In symbols, this is expressed as

$$\Pr\{A|B\} = \frac{\Pr\{A \cap B\}}{\Pr\{B\}} \tag{4.4}$$

where $\Pr\{A \mid B\}$ represents the probability of A given that B has occurred.

Conditional probabilities often are of greater interest than the unconditional probabilities we have been dealing with as will be shown following. Before doing that, note that we can use the conditional probability formula to find the probability of the intersection — that is,

$$\Pr\{A \cap B\} = \Pr\{A \mid B\} \cdot \Pr\{B\}. \tag{4.5}$$

Thus, if we know the probability of A conditional on the occurrence of B, and we also know the probability of B, we can find the probability of the intersection of A and B. Note that we can also express the probability of the intersection as

$$\Pr\{A \cap B\} = \Pr\{B \mid A\} \cdot \Pr\{A\}. \tag{4.6}$$

Table 4.3 repeats the data from Table 4.2 along with three different sets of probabilities. The first set of probabilities (row R) is conditional on the birth weight; that is, it uses the row totals as the denominators in the calculations. The second set (row C) is conditional on the trimester that prenatal care was begun; that is, it uses the column totals in the denominator. The third set of probabilities (row U) is the unconditional set — that is, those based on the total of 51,473 live births. The probabilities in the Total column are the probabilities of the different birth weight categories; that is, the probability distribution of the birth weight variable and those beneath the Total row are the probabilities of the different trimester categories — that is, the probability distribution of the prenatal care variable. As just mentioned, these probabilities are based on the population of births in Harris County, Texas, in 1986.

Table 4.3   Number and probabilities of live births by trimester of first prenatal care and birth weight: Harris County, Texas, 1986.

Birth Weight		Trimester Prenatal Care Began				Total
		1st	2nd	3rd	No Care	
≤5.5 lb; ~2,500 g		2,412	754	141	234	3,541
	R[a]	0.681	0.213	0.040	0.066	
	C	**0.064**	**0.079**	**0.060**	**0.138**	**0.069**
	U	0.047	0.015	0.003	0.005	
5.6–7.7 lb; ~2,500–3,500 g		20,274	5,480	1,458	1,014	28,226
	R	0.718	0.194	0.052	0.036	
	C	**0.534**	**0.577**	**0.624**	**0.598**	**0.548**
	U	0.394	0.106	0.028	0.020	
>7.7 lb; ~3,500 g		15,250	3,271	738	447	19,706
	R	0.774	0.166	0.037	0.023	
	C	**0.402**	**0.344**	**0.316**	**0.264**	**0.383**
	U	0.296	0.064	0.014	0.009	
Total		37,936	9,505	2,337	1,695	51,473
	R	0.737	0.185	0.045	0.033	1.000

[a]R, row; C, column; and U, unconditional

Let us consider the entries in the row 1, column 1 cell. The first two entries below the frequency of the cell are conditional probabilities. The value 0.681 (= 2412/3541) is the probability based on the row total; that is, it is the probability of a woman having begun her prenatal care during the first trimester given that the baby's birth weight was less than or equal to 5.5 lb. The value 0.064 (= 2412/37936) is the probability based on the column total; that is, it is the probability of a birth weight of less than or equal to 5.5 lb given that the woman had begun her prenatal care during the first trimester. The last value, 0.047 (= 2412/51473), is the unconditional probability; that is, it is based on the grand total of 51,473 live births. It is the probability of the intersection of a birth weight less than or equal to 5.5 lb with the prenatal care having been begun during the first trimester.

As Table 4.3 shows, at least three different probabilities, or observed proportions if the data are a sample, can be calculated for the entries in the two-way table. The choice of which probability (row, column, or unconditional) to use depends on the purpose of the investigation. In this case, the data may have been tabulated to determine whether or not the timing of the initiation of the prenatal care had any effect on the birth weight of the infant. If this is the purpose of the study, the column-based probabilities may be the more appropriate to use and report. The column-based calculations give the probabilities of the different birth weight categories conditional on when the prenatal care was begun. The row-based calculations give the probability of the trimester prenatal care was initiated given the birth weight category. However, these row-based probabilities are of no interest because the birth weight cannot affect the timing of the prenatal care. The unconditional probabilities are less informative in this situation, as they also reflect the row and column totals. For example, compare the unconditional probabilities in the first and third columns in the first row — 0.047 and 0.003. Even though we have seen that there is little difference in the corresponding column-based probabilities of 0.064 and 0.060, these unconditional values are very different. The value of 0.047 is larger mainly because there are 37,936 live births in the first column compared to only 2337 live births in the third column. However, the unconditional probabilities may be useful in planning and allocating resources for maternal and child health services programs.

Using the column-based values, women who began their prenatal care during the first trimester had a probability of a low birth weight baby of 0.064. This value is compared to 0.079, the probability of a low birth weight baby for those who began their prenatal care during their second trimester, to 0.060 for those who began their prenatal care during the third trimester, and to 0.138 for those who received no prenatal care. There is little difference in the probabilities of a low birth weight baby among women who received prenatal care. However, the probability of a low birth weight baby is about twice as large for women who received no prenatal care compared to women who received prenatal care. The effect of prenatal care is most clearly evident in the probability of having a baby with a birth weight of greater than 7.7 lb. In this category, the probabilities are 0.402, 0.344, 0.316, and 0.264 for the first, second, and third trimesters and no prenatal care, respectively.

Based on the trend in the probabilities of a birth weight of greater than 7.7 lb, one might conclude that there is an effect of prenatal care. However, to do so is inappropriate without further information. First, although these births can be viewed as constituting a population — that is, all the live births in Harris County in 1986 — they could also be viewed as a sample in time, one year selected from many, or in place, one county selected from many. From the perspective that these births are a sample, there is sampling variation to be taken into account, and this will be covered in Chapter 11. Second, and more important, these data do not represent a true experiment. Chapter 6 presents more on experiments, but, briefly, the women were not randomly assigned to the different prenatal care groups — that is, to the first, second, or third trimester groups or to the no prenatal care group. Thus, the women in these groups may differ on variables related to birth weight — for example, smoking habits, amount of weight gained, and dietary behavior. Without further examination of these other factors, it is wrong to conclude that the variation in the probabilities of birth weights is due to the time when prenatal care was begun.

Example 4.1

Suppose that a couple has two children and one of them is a boy. What is the probability that both children are boys? For a couple with two children, there are four possible outcomes: boy and boy, boy and girl, girl and boy, girl and girl. If one of the two children is a boy, then there are three possible outcomes, excluding the (girl and girl) outcome. Therefore, the probability of having two boys is 1/3 (one of three possible outcomes). Applying the conditional probability rule, Equation (4.4), we can calculate this probability by $(1/4) / (1 - 1/4)$.

4.2.3 Independent Events

Suppose we were satisfied that there are no additional factors of interest in the examination of prenatal care and birth weight and only the data in Table 4.2 were to be used to determine whether or not there was a relation between when prenatal care was initiated and birth weight. Row C in Table 4.3 shows the column-based probabilities — that is, those conditional on which trimester care was begun or whether care was received — and these are the probabilities to be used in the study.

If there was no relationship between the prenatal care variable and the birth weight variable — that is, if these two variables were *independent* — what values should the column-based probabilities have? If these variables are independent, this means that the birth weight probability distribution is the same in each of the columns. The last column in Table 4.3 gives the birth weight probability distribution, and this is the distribution that will be in each of the columns if the birth weight and prenatal care variables are independent. Table 4.4 shows the birth weight probability distribution for the situation when these two variables are independent.

Table 4.4   Probabilities conditional on trimester under the assumption of independence of birth weight level and trimester of first prenatal care: Harris County, Texas, 1986.

| Birth Weight | Trimester Prenatal Care Began | | | | |
	1st	2nd	3rd	No Care	Total
≤ 5.5 lb; ~2500 g	0.069	0.069	0.069	0.069	0.069
5.6–7.7 lb; ~2500–3500 g	0.548	0.548	0.548	0.548	0.548
> 7.7 lb; ~3500 g	0.383	0.383	0.383	0.383	0.383
Total	1.000	1.000	1.000	1.000	1.000

The entries in Table 4.4 are conditional probabilities — for example, of a birth weight less than or equal to 5.5 lb (A) given that prenatal care began during the first trimester (B) under the assumption of independence. Hence, under the assumption of independence of A and B, the probability of A given B is equal to the probability of A. In symbols, this is

$$\Pr\{A \mid B\} = P\{A\}$$

when A and B are independent. Combining this formula with the formula for the probability of the intersection — that is,

$$\Pr\{A \cap B\} = \Pr\{A \mid B\} \cdot \Pr\{B\}$$

yields

$$\Pr\{A \cap B\} = \Pr\{A\} \cdot \Pr\{B\}$$

when A and B are independent.

Example 4.2

For a couple with one child, there are two possible outcomes: a boy or a girl. It is assumed that the probability of a girl is the same as the probability of a boy — that is, 0.5. For a couple with two children, there are four possible outcomes, as seen in Example 4.1. The probability of each outcome is 0.25 (one out of the four possible outcomes). We have to realize that the probability of having one boy and one girl is 0.5, accounting for two of four possible outcomes. However, U.S. vital statistics consistently show that about 105 boys are born per 100 girls (a sex ratio at birth of 105; Mathews and Hamilton 2005), which suggests that the probability of having a boy is 0.51 and the probability of having a girl is 0.49. If we use these values of the case of two children, the probabilities of the four possible outcomes are 0.26 (= 0.51 [0.51]), 0.25 (= 0.51 [1 − 0.51]), 0.25 (= [1 − 0.51] 0.51), and 0.24 (= [1 − 0.51] [1 − 0.51]), respectively. The four probabilities add up to 1. The reason for multiplying two probabilities will become clearer in Example 4.3.

Example 4.3

Since the gender of the second child is independent of that of the first child, the probability of two boys in row, based on vital statistics, is 0.51(0.51) = 0.26, as shown in Example 4.2. When considering diseases, it is unlikely that the disease status of one person is independent of that of another person for many infectious diseases. However, it is likely that the disease status of one person is independent of that of another for many chronic diseases. For example, let π be the probability that a person has Alzheimer's disease. One person's Alzheimer's status should be independent of another's status. Therefore, the probability of two persons having Alzheimer's disease is the product of the probabilities of either having the disease — that is, $\pi \cdot \pi$.

Establishing the dependence (a relation exists) or independence (no relation) of variables is what much of health research is about. For example, in the disease context, is disease status related to some variable? If there is a relation (dependency), the variable is said to be a risk factor for the disease. The identification of risk factors leads to strategies for preventing or reducing the occurrence of the disease.

Example 4.4

Let us apply these definitions of probabilities to the example used for the randomized response technique in Chapter 2. In Figure 2.2 there were 12 yes responses among 36 individuals to whom the randomized response technique was administered. We can denote as the probability of yes, $\Pr(Y) = 12/36 = 1/3$. We know this observed probability is a combination of probabilities under two circumstances — that is, $\Pr(\underline{H}ead \ and \ \underline{D}runken \ driving) + \Pr(\underline{T}ail \ and \ \underline{B}orn \ in \ September \ or \ October)$. In symbols, this relationship is expressed as

$$\Pr\{Y\} = \Pr\{H \cap D\} + \Pr\{T \cap B\}.$$

The two probabilities of intersection in the right hand side of equation can be expressed in terms of conditional probabilities, applying Equation (4.6) as follows:

$$\Pr\{Y\} = \Pr\{D \mid H\} \cdot \Pr\{H\} + \Pr\{B \mid T\} \cdot \Pr\{T\}.$$

We know that $\Pr(H) = \frac{1}{2}$ and $\Pr(T) = 1 − \Pr(H) = \frac{1}{2}$. $\Pr(D \mid H)$ is unknown quantity, and we want to estimate this conditional probability. $\Pr(B \mid T)$ is known — that is, 2 months out of 12 months, $2/12 = 1/6$.

If we solve the above equation for the unknown probability, $\Pr(D \mid H)$, then we have

$$\Pr(D|H) = \frac{\Pr(Y) - \Pr(B|T) \cdot \Pr(T)}{\Pr(H)} = \frac{\left(\frac{1}{3}\right) - \left(\frac{1}{6}\right)\left(\frac{1}{2}\right)}{\left(\frac{1}{2}\right)} = \left(\frac{2}{3}\right) - \left(\frac{1}{6}\right) = \left(\frac{1}{2}\right).$$

The result is 50 percent, which is the same as in Figure 2.2.

4.3   Definitions from Epidemiology

There are many quantities used in epidemiology that are defined in terms of probabilities, particularly conditional probabilities. Several of these useful quantities are defined in this section and used in the next section to illustrate Bayes' rule.

4.3.1   Rates and Probabilities

Various rates and relative numbers are used in epidemiology. A rate is generally interpreted as a probability — as a measure of likelihood that a specified event occurs to a specified population. *Prevalence* of a disease is the probability of having the disease. It is the number of people with the disease divided by the number of people in the defined population. The observed proportion of those with the disease in a sample is the sample estimate of prevalence. When the midyear population is used for the denominator, it is possible that the numerator contains persons not included in the denominator. For example, persons with the disease that move into the area in the second half of the year are not counted in the denominator, but they are counted in the numerator. When prevalence or other quantities use midyear population or person-years lived values, they are not really probabilities or proportions, although this distinction usually is unimportant. However, this distinction is important when estimating the probability of dying from the age-specific death rate as will be discussed later in conjunction with the life table.

Incidence of a disease is the probability that a person without the disease will develop the disease during some specified interval of time. It is the number of new cases of the disease that occur during the specified time interval divided by the number of people in the population who do not already have the disease.

Prevalence provides an idea of the current magnitude of the disease problem, whereas incidence informs as to whether or not the disease problem is getting worse.

Example 4.5

We consider the incidence and prevalence rates of AIDS based on the data from *Health, United States, 2004* (NCHS 2004, Tables 1 and 52). By the end of 2002, 829,998 cases of AIDS had been reported to the Centers for Disease Control and Prevention, and of those, 42,478 cases were reported in 2002. The estimated U.S. population as of July 1, 2002, was 288,369,000. Based on the cases reported in 2002 and the estimated midyear population of 2002, the 2002 incidence rate of AIDS in the United States was 0.00014730 (42478/288369000) or 14.7 per 100,000 population. Since the rate is low, the rate is expressed as the number of cases per 100,000.

Based on the preceding data, it is difficult to estimate the prevalence rate because there is no information on the number of individuals with AIDS who had died prior to 2002. The AIDS death rate was reported starting in 1987. It steadily increased from 5.6 per 100,000 to 16.2 per 100,000 in 1995 and steadily declined to 4.9 per 100,000 in 2002. Based on an average death rate of AIDS for the last two and half decades, it is roughly estimated that about 80 percent of those diagnosed prior to 2002 had died by the end of 2001. Thus, of the 874,230 reported cases, we are assuming that 630,016 (0.8{874230 − 42479}) had died, leaving 199,982 persons with AIDS in 2002. The prevalence rate of AIDS then was 0.00069349 (= 199982/288369000) or 69.3 per 100,000 population.

A *specific rate* of a disease is the disease rate for people with a specified character- istic, such as age, race, sex, occupation, and so on. It is the conditional probability of a person having the disease given that the person has the characteristic. For example, an age-specific death rate is a death rate conditional on a specified age group, as seen in Chapter 3.

4.3.2   Sensitivity, Specificity, and Predicted Value Positive and Negative

Laboratory test results are part of the diagnostic process for determining if a patient has some disease. Unfortunately in many cases, a positive test result — that is, the existence of an unusual value — does not guarantee that a patient has the disease. Nor does a negative test result, the existence of a typical value, guarantee the absence of the disease. To provide some information on the accuracy of testing procedures, their devel- opers use two conditional probabilities: sensitivity and specificity.

The *sensitivity* of a test (symptom) is the probability that there was a positive result (the symptom was present) given that the person has the disease. The *specificity* of a test (symptom) is the probability that there was a negative result (the symptom was absent) given that the person does not have the disease. Note that one minus sensitivity is the false negative rate, and one minus specificity is the false positive rate. Thus, large values of sensitivity and specificity imply small false negative and false positive rates.

Sensitivity and specificity are probabilities of the test result conditional on the disease status. These are values that the developer of the test has estimated during extensive testing in hospitals and clinics. However, as a potential patient, we are more interested in the probability of disease status conditional on the test result. Names given to two conditional probabilities that address the patient's concerns are predicted value positive and predicted value negative. *Predicted value positive* is the probability of disease given a positive test result, and *predicted value negative* is the probability of no disease given a negative test result.

These four quantities can be expressed succinctly in symbols. Let T^+ represent a positive test result and T^- represent a negative result. The presence of disease is indicated by D^+ and its absence is indicated by D^-. Then these four quantities can be expressed as conditional probabilities as follows:

Sensitivity	$\Pr\{T^+ \mid D^+\}$
Specificity	$\Pr\{T^- \mid D^-\}$
Predicted value positive	$\Pr\{D^+ \mid T^+\}$
Predicted value negative	$\Pr\{D^- \mid T^-\}$

All four of these probabilities should be large for a screening test to be useful to the screener and to the screenee. Discussions of these and related issues are plentiful in the epidemiologic literature (Weiss 1986).

It is possible to estimate these probabilities. One way is to select a large sample of the population and subject the sample to a screening or diagnostic test as well as to a standard clinical evaluation. The standard clinical evaluation is assumed to provide the true disease status. Then the sample persons can be classified into one of the four cells in the 2 by 2 table in Table 4.5. For example, hypertension status is first screened by the sphygmomanometer in the community and by a comprehensive clinical evaluation in the clinic. Or persons are screened for mental disorders first by the DIS (Diagnostic Interview Schedule) and then by a comprehensive psychiatric evaluation. The results from a two-stage diagnostic procedure would look like Table 4.5.

Table 4.5 Disease status by test results for a large sample from the population.

Disease Status	Test Result		Total
	Positive	Negative	
Presence	a	b	$a + b$
Absence	c	d	$c + d$
Total	$a + c$	$b + d$	$a + b + c + d$

Sensitivity is estimated by $a/(a + b)$, specificity is estimated by $d/(c + d)$, predicted value positive is estimated by $a/(a + c)$, and predicted value negative is estimated by $d/(b + d)$. Similarly, the false positive rate is estimated by $c/(c + d)$ and the false negative rate by $b/(a + b)$.

For many diseases of interest, the prevalence is so low that there would be few persons with the disease in the sample. This means that the estimates of sensitivity and the predicted value positive would be problematic. Therefore, some alternate sample design must be used to estimate these conditional probabilities. When a large number of people are screened by a test in a community and a sample of persons with positive test results and those with negative test results are subjected to clinical evaluations, the predicted value positive and the predicted value negative can be directly calculated from the results of clinical evaluations, and sensitivity and specificity can be indirectly estimated. Conversely, when sensitivity and specificity are directly estimated by applying the test to persons with the disease and persons without the disease in the clinic setting, the predicted value positive and the predicted value negative can be indirectly estimated if the prevalence rate of disease is known. These indirect estimation procedures are explained in the next section.

4.3.3   Receiver Operating Characteristic Plot

In evaluating a diagnostic test for a certain disease, we need to consider relative importance of sensitivity and specificity. For incidence, if the disease in question is likely to lead to death and the preferred treatment has few side effects, then it will be more important to make sensitivity as large as possible. On the other hand, if the disease is not too serious and the known treatment has considerable side effects, then more weight might be given to specificity. The cost of the treatment given to those with positive test results could also come into consideration. In many situations, we need to consider both sensitivity and specificity. But sensitivity and specificity are relative to how we define the status of disease. Different cut-off points in the definition of the condition would give different results.

Here we illustrate how the sensitivity and specificity of a test change with respect to the cut-off point chosen for indicating a positive test result. Let us consider the case of using the serum calcium level as a test for detect hyperparathyroidism (Lundgren et al. 1997). The following data show the level of serum calcium and the status of hyperparathyroidism:

		Serum Calcium Levels mg/dL					
		8 mg/dL	**9 mg/dL**	**10 mg/dL**	**11 mg/dL**	**12 mg/dL**	**Total**
Disease Status	Negative	40	7	4	2	1	54
	Positive	2	3	5	8	17	35
	Total	42	10	9	10	18	89

If we consider 9 mg/dL or more as positive test result, the data can be summarized as follows:

		Serum Calcium Levels mg/dL		
		9 mg/dL or more	**Less than 9 mg/dL**	**Total**
Disease Status	Negative	14	**40**	54
	Positive	33	2	35
	Total	47	42	89

From this summary, the estimated sensitivity is 0.94 (= 33/35) and the specificity is 0.74 (= 40/54). As the cut-point changes, the sensitivity and specificity of the diagnostic test also change. As we increase the cut-point for serum calcium levels, the sensitivity of the test decreases and the specificity increases as shown here.

Cut-Point	Sensitivity	Specificity
<8 mg/dL | 8 mg/dL	35 / 35 = 1.00	0 / 54 = 0.00
8 mg/dL | 9 mg/dL	33 / 35 = 0.94	40 / 54 = 0.74
9 mg/dL | 10 mg/dL	30 / 35 = 0.86	47 / 54 = 0.87
10 mg/dL | 11 mg/dL	25 / 35 = 0.71	51 / 54 = 0.94
11 mg/dL | 12 mg/dL	17 / 35 = 0.48	53 / 54 = 0.98
12 mg/dL | >12 mg/dL	0 / 35 = 0.00	54 / 54 = 1.00

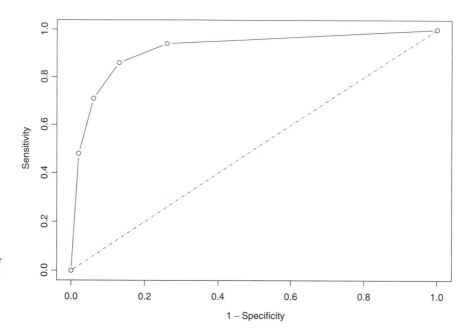

Figure 4.1 The receiver operating characteristic plot for serum calcium values and hyperparathyroidism.

We generally use the Receiver Operating Characteristic (ROC) plot to examine the tradeoff between sensitivity and specificity. This is a plot of sensitivity versus 1 − specificity. Figure 4.1 shows the ROC plot for the preceding data. By looking at the curve relative to a 45-degree line, we notice that as the curve extends farther away from the line, the accuracy of the diagnostic test improves, and as the curve draws nearer to the 45-degree line, the diagnostic test's accuracy becomes worse. Therefore, we can consider the area under the ROC curve as a measure of a diagnostic test's discrimination or the test's ability to correctly classify individuals with and without the disease. An excellent test would have an area under the curve of nearly 1.00, while a poor test would have an area under the curve of nearly 0.50.

4.4 Bayes' Theorem

We wish to find the predicted value positive and predicted value negative using the known values for disease prevalence, sensitivity, and specificity. Let us focus on predicted value positive — that is, $\Pr\{D^+ \mid T^+\}$ — and see how it can be expressed in terms of sensitivity, $\Pr\{T^+ \mid D^+\}$, specificity, $\Pr\{T^- \mid D^-\}$, and disease prevalence, $\Pr\{D^+\}$.

We begin with the definition of the predicted value positive:

$$\Pr\{D^+ \mid T^+\} = \frac{\Pr\{D^+ \cap T^+\}}{\Pr\{T^+\}}.$$

Applying Equation (4.6), $\Pr\{A \cap B\} = \Pr\{B \mid A\} \cdot \Pr\{A\}$, the probability of the intersection of D^+ and T^+ can also be expressed as

$$\Pr\{D^+ \cap T^+\} = \Pr\{T^+ \mid D^+\}\,\Pr\{D^+\}.$$

On substitution of this expression for the probability of the intersection in the definition of the predicted value positive, we have

$$\Pr\{D^+|T^+\} = \frac{\Pr\{T^+|D^+\} \cdot \Pr\{D^+\}}{\Pr\{T^+\}} \qquad (4.7)$$

which shows that predicted value positive can be obtained by dividing the product of sensitivity and prevalence by $\Pr\{T^+\}$.

Applying Equation (4.1), $\Pr\{T^+\}$ can be expressed as the sum of the probabilities of the intersection of T^+ with the two possible outcomes of the disease status: D^+ and D^-; that is,

$$\Pr\{T^+\} = \Pr\{T^+ \cap D^+\} + \Pr\{T^+ \cap D^-\}.$$

Applying Equation (4.5) to the two probabilities of intersections, we now have

$$\Pr\{T^+\} = \Pr\{T^+ \mid D^+\}\Pr\{D^+\} + \Pr\{T^+ \mid D^-\}\Pr\{D^-\}.$$

Substituting this expression in Equation (4.7), the predictive value positive is

$$\Pr\{D^+|T^+\} = \frac{\Pr\{T^+|D^+\} \cdot \Pr\{D^+\}}{\Pr\{T^+|D^+\} \cdot \Pr\{D^+\} + \Pr\{T^+|D^-\} \cdot \Pr\{D^-\}}. \qquad (4.8)$$

Note that the numerator and the first component of the denominator is the product of sensitivity and disease prevalence. The second component of the denominator is the product of $(1 - \text{specificity})$ and $(1 - \text{disease prevalence})$. Predicted value negative follows immediately, and it is

$$\Pr\{D^-|T^-\} = \frac{\Pr\{T^-|D^-\} \cdot \Pr\{D^-\}}{\Pr\{T^-|D^-\} \cdot \Pr\{D^-\} + \Pr\{T^-|D^+\} \cdot \Pr\{D^+\}}.$$

These two formulas are special cases of the theorem discovered by Reverend Thomas Bayes (1702–1761). In terms of the events A and B_i, Bayes' theorem is

$$\Pr\{B_i|A\} = \frac{\Pr\{A|B_i\} \cdot \Pr\{B_i\}}{\sum_i \Pr\{A|B_i\} \cdot \Pr\{B_i\}}.$$

Example 4.6

Consider the use of the count of blood vessels in breast tumors. A high density of blood vessels indicates a patient who is at high risk of having cancer spread to other organs (Weidner et al. 1992). Use of the count of blood vessels appears to be worthwhile in women with very small tumors and no lymph node involvement — the node-negative case. Suppose that during the development stage of this procedure, its sensitivity was estimated to be 0.85; that is, of the women who had cancer spread to other organs, 85 percent of them had a high count of blood vessels in their breast tumors. The specificity of the test was estimated to be 0.90; that is, of the women for whom there was no spread of cancer, 90 percent of them had a low count of blood vessels in their tumors. Assume that the prevalence of cancers spread from breast cancers is 0.02. Given these assumed values, what is the predicted value positive (PVP) of counting the number of blood vessels in the small tumors?

Using Equation (4.8),

$$\Pr\{D^+|T^+\} = \frac{\Pr\{T^+|D^+\} \cdot \Pr\{D^+\}}{\Pr\{T^+|D^+\} \cdot \Pr\{D^+\} + \Pr\{T^+|D^-\} \cdot \Pr\{D^-\}},$$

the answer is

$$PVP = \frac{prevalence \cdot sensitivity}{[prevalence \cdot sensitivity] + [(1 - prevalence) \cdot (1 - specificity)]}$$

$$= \frac{(0.02) \cdot (0.85)}{(0.02) \cdot (0.85) + (1 - 0.02) \cdot (1 - 0.90)} = \frac{0.017}{0.115} = 0.148.$$

Using the preceding assumed values for sensitivity, specificity, and prevalence, there is approximately a 15 percent chance of having cancer spread from a small breast tumor given a high density of blood vessels in the tumor. This value may be too low for the test to be useful. If the true values for specificity or prevalence are higher than the values just assumed, then the PVP will also be higher. For example, if the prevalence is 0.04 instead of 0.02, then the PVP is 0.262 instead of 0.148.

Example 4.7

Let us recast the question in Example 4.6 using frequencies instead of probabilities. Suppose that 20 out of every 1000 women with breast tumors have cancer spread to other organs (the prevalence of cancer spread is 0.02). Of these 20 women with cancer spread, 17 will have a high count of blood vessels (sensitivity of the test was estimated to be 0.85). Of the remaining 980 women without cancer spread, 882 will have a low count of blood vessels in their tumors (specificity of the test was estimated to be 0.90). Then what percent of women with a high density of blood vessels do actually have cancer spread (predicted value positive)?

This question can be answered easily without using the Bayes' formula. Looking at the frequencies just stated, the total number of women with high-density blood vessels is the sum of 17 from those with cancer spread and 98 (980 minus 882) from those without cancer spread. The sum is 115. Of these, 17 saw their cancers spread. Therefore, the predicted value positive is 0.148 (= 17/115), which is the same value obtained by the formula in Example 4.5. You can see it more clearly in the following 2 by 2 table:

	Blood Vessel Count		
Cancer Spread	**High**	**Low**	**Total**
Yes	(17)**	3	(20)*
No	98	(882)***	980
Total	115	885	1000

*Prevalence rate of 0.02 Predicted value positive: 17/115 = 0.148
**Sensitivity of 0.85   Predicted value negative: 882/885 = 0.997
***Specificity of 0.90

This example demonstrates that Bayes' theorem is to enhance and expedite our reasoning rather than to be memorized blindly.

4.5   Probability in Sampling

Sampling means selecting a few units from all the possible observational units in the population. To infer from the sample to the population, we need to know the probability of selection. A sample selected with unknown probability of selection cannot be linked appropriately to the population from which the sample was drawn. A sample drawn with known probability of selection is called a probability sample. We examine the simplest probability sample that assigns an equal probability of selection to every unit of observation in the population. More complex sample selection designs will be discussed in Chapter 6.

4.5.1   Sampling with Replacement

A sample that allows duplicate selections is called a sample with replacement. Allowance of duplicate selection implies that sample selections are independent — each selection is not dependent on previous selections. To understand the probability of selection in a sample with replacement, let us consider the case of selecting three units from a population of four units (A, B, C, and D). There are 64 (= 4^3) ways of selecting such samples as listed in Table 4.6.

Table 4.6   Possible samples of drawing 3 from (A, B, C and D) with replacement.

AAA ACA	BAA BCA*	CAA CCA	DAA DCA*
AAB ACB*	BAB BCB	CAB* CCB	DAB* DCB*
AAC ACC	BAC* BCC	CAC CCC	DAC* DCC
AAD ACD*	BAC* BCD*	CAD* CCD	DAD DCD
ABA ADA	BBA BDA*	CBA* CDA*	DBA* DDA
ABB ADB*	BBB BDB	CBB CDB*	DBB DDB
ABC* ADC*	BBC BDC*	CBC CDC	DBC* DDC
ABD* ADD	BBD BDD	CBD* CDD	DBD DDD

*Samples without duplications

Since selections are independent, the probability of selecting each of these samples is 1/64 (= $[1/4]^3$). As shown in the table, the total number of samples without duplications is 24($4 \times 3 \times 2$); that is, there are 4 ways to fill the first position of the sample, 3 ways to the second position, and 2 ways to fill the third position. The probability of selection with replacement samples that do not contain duplication is 0.375 (= 24/64). The probability of obtaining samples with duplications is 0.625 (= 1 − 0.375).

When selecting n units from N units in the population, there are N^n possible samples with replacement. Of these, $N(N − 1) \ldots (N − n + 1)$ samples contain no duplications. Then the probability of obtaining with replacement samples that contain duplications is

$$\Pr(duplications) = 1 - \frac{N(N-1)\cdots(N-n+1)}{N^n}. \tag{4.9}$$

Example 4.8

How likely is it that at least two students in a class of 23 will share the same birthday? The chance may be better than we might expect. If we assume that the birthdays of 23 students are independent and that each day out of 365 days in a year, eliminating February 29, is equally likely to be a student's birthday, the situation is equivalent to selection of a random sample of 23 days from the 365 days using the sampling with replacement procedure. The probability can be calculated using Equation (4.9) — that is,

$$1 - \frac{365(364)\cdots(343)}{365^{23}} = 0.507.$$

The calculation may require the use of a computer (the SAS program is available on the website). If the size of class increases to 50, the probability increases to 0.97.

Let us consider the probability of selecting a particular unit. In the list of samples in Table 4.6, unit A appears 16 times in the first position of the sample, 16 times in the second position, and 16 times in the third position. Then the probability of A being selected into the sample is $[3(16) / 4^3] = (48 / 64) = (3 / 4)$. Thus, in general, the selection probability of a unit is n / N.

4.5.2   Sampling without Replacement

A sample that does not allow duplications is called a sample without replacement. In sampling without replacement, a selection of a unit is no longer independent because the selection is conditional on the unit being not selected in a previous draw. In this sampling, once a subject is selected, it is removed from the population, and the number of units in the population is decreased by one unit. Does this decrease in the denominator as a unit is selected invalidate the equal probability of selection for subsequent units? The following example addresses this.

Suppose that a class has 30 students, and a random sample of 5 students is to be selected without allowing duplicate selections. The probability of selection for the first draw will be $1 / 30$, and that for the student selected second will be $1 / 29$, since one student was already selected. This line of thinking seems to suggest that random sampling without replacement is not an equal probability sampling model. Is anything wrong in our thinking?

We have to realize that the selection probability of $1 / 29$ for the second draw is a conditional probability. The student selected in the second draw is available for selection only if the student were not selected in the first draw. The probability of not being selected in the first draw is $29 / 30$. Thus, the event of being selected during the second draw is the intersection of the events of not being selected during the first draw and being selected during the second draw. Applying $\{\Pr\{A \cap B\} = \Pr\{A \mid B\} \cdot \Pr\{B\}$, the probability of this intersection is $(1 / 29)(29 / 30)$, which yields $1 / 30$. The same argument can be made for subsequent draws, as shown in Table 4.7.

Table 4.7   Calculation of inclusion probabilities in drawing an SRS of 5 from 30 without replacement.

Order of Draw	Conditional Probability (1)	Probability Not Selected in Previous Draws (2)	Product of (1) & (2)
1	1/30	1	1/30
2	1/29	29/30	1/30
3	1/28	$(29/30)(28/29) = 28/30$	1/30
4	1/27	$(29/30)(28/29)(27/28) = 27/30$	1/30
5	1/26	$(29/30)(28/29)(27/28)(26/27) = 26/30$	1/30

The demonstration in Table 4.7 indicates that the probability of being selected in any draw is 1 / 30, and hence the equal probability of selection also holds for sampling without replacement. Now we can state that the probability for a particular student to be included in the sample will be 5 / 30, since the student can be drawn in any one of the five draws. Thus, in general, the selection probability of a unit without replacement is n / N, the same as in the case of replacement sampling.

A sampling procedure that assigns n / N chance of being selected into the sample to every unit in the population is called *simple random sampling*, regardless of whether sampling is done with or without replacement. We usually use sampling without replacement. The distinction between sampling with and without replacement is moot when selecting a sample from large populations because the chance of selecting a unit more than once would be very small. The statement that each of the possible samples is equally likely implies that each unit in the population has the same probability of being included in the sample as demonstrated in this and the previous section.

4.6   Estimating Probabilities by Simulation

Our approach to finding probabilities has been to enumerate all possible outcomes and to base the calculation of probabilities on this enumeration. This approach works well with simple phenomena, but it is difficult to use with complex events. Another way of assessing probabilities is to simulate the random phenomenon by using repeated sampling. With the wide availability of microcomputers, the simulation approach has become a powerful tool to approach many statistical problems.

Example 4.9

Let us reconsider the question posed in Example 4.8. In a class of 30, what will be the chance of finding at least 2 students sharing the same birthday? It should be higher than the 50 percent that we found among 23 students in Example 4.8. Let us find an answer by simulation. We need to make the same assumptions as in Example 4.8. Selecting 30 days from 365 days using the sampling procedure, we can use the random number table in Appendix B. For example, we can read 30 three-digit numbers between 1 and 365 from the table and check to see if any duplicate numbers are selected. We can repeat the operation many times and see how many of the trials produced duplicates. Since this manual simulation would require considerable time, we can use a computer program (see **Program Note 4.1** on the website). The results of 10 simulations are shown in Table 4.8.

Table 4.8   Simulation to find the probability of common birthdays among 30 students.

Student	1	2*	3*	4*	5*	6*	7	8*	9*	10*
					Simulations					
1	4	2	3	44	8	3	7	5	8	12
2	10	30	10*	52	21	4	47	7	18	19
3	21	46	10*	72	24	22	48	7	27	31
4	47	67	15	85	76	23	54	18	45	48
5	48	97	23	106	91	27	80	23	50	65
6	64	100	26	116	100	42	82	37	66	80
7	65	105	35	120	113	57	93	54	90	82
8	78	106	41	123	124	64	119	59	91	103
9	93	106	53	132	143*	72	123	64	94	116
10	95	109	73	143	143*	104	137	89	97	169
11	101	133	78	151	147	107	138	109	104	175
12	115	140	86	180	150	119	140	120	132	182
13	154	145	87	181	155	132	162	138	149	193
14	165	158	163	188	166	152	179	143	153	195
15	167	191	166	208	172	167	185	173	180	208
16	185	209*	176	231	200	210	191	201	187	217
17	193	209*	186	248	205	229	199	209*	188	247
18	220	220	200	249	241	230	203	209*	189	249
19	232	223	209	255	243	233	213	215	193	261
20	242	229	220	259*	248	236	232	223	196	262*
21	257	241	251	259*	250	253	238	224	242	262*
22	282	249	260	267	263	307	252	231	250	305
23	284	268	264	270	281	321	259	239	324	307
24	285	286	265	285	283	326	267	259	333	309
25	288	317	283	286	307	327	272	274	338	321
26	299	323	295	288	310	334	287	335	354	326
27	309	335*	297	296	311	336	295	342	360*	328
28	346	335*	300	310	326	343*	308	352	360*	330
29	347	336	352	327	335	343*	313	357	360*	347
30	357	356	355	352	336	362	363	358	360*	356

Eight of these 10 trials have duplicates, which suggests that there is an 80 percent probability of finding at least one common birthday among 30 students. Not shown are the results of 10 additional trials in which 6 of the 10 had duplicates. Combining these two sets of 10 trials, the probability of finding common birthdays among 30 students is estimated to be 70 percent (= [8 + 6] / 20). Using $\Pr(duplications) = 1 - \dfrac{N(N-1)\cdots(N-n+1)}{N^n}$ we get 70.6 percent. Using 20 replicates is usually not enough to have a lot of confidence in the estimate; we usually would like to have at least hundreds of replicates.

Let us consider another example.

Example 4.10

Population and family planning program planners in Asian countries have been dealing with the effects of the preference for a son on population growth. If all couples continue to have children until they have two sons, what is the average number of children they would have? To build a probability model for this situation, we assume that genders of successive children are independent and the chance of a son is 1 / 2. To simulate the number of children a couple has, we select single digits from the random number table, considering odd numbers as boys and even numbers as girls. Random numbers are read until the second odd number is encountered, and the number of values required to obtain two odd values is noted. Table 4.9 shows the results for 20 trials (couples).

The average number of children based on this very small simulation is estimated to be 4.25 (= 85 / 20). Additional trials would provide an estimate closer to the true value of four children.

Table 4.9   Simulation of child-bearing until the second son is born.

Trial	Digits	No. of Digits	Trial	Digits	No. of Digits
1	19	2	11	37	2
2	2239	4	12	367	3
3	503	3	13	6471	4
4	4057	4	14	509	3
5	56287	5	15	940001	6
6	13	2	16	927	3
7	96409	5	17	277	3
8	125	3	18	544264882425	12
9	31	2	19	3629	4
10	425448285	9	20	045467	6
				Total number of digits	85
				Average = 85/20 = 4.25	

4.7   Probability and the Life Table

Perhaps the oldest probability model that has been applied to a problem related to health is the life table. The basic idea was conceived by John Graunt (1620–1674), and the first life table, published in 1693, was constructed by Edmund Halley (1656–1742). Later Daniel Bernoulli (1700–1782) extended the model to determine how many years would be added to the average life span if smallpox were eliminated as a cause of death. Now the life table is used in a variety of fields — for example, in life insurance calculations, in clinical research, and in the analysis of processes involving attrition, aging and wearing out of industrial products.

We are presenting the life table here to show an additional application of the probability rules described above. Table 4.10 is the abridged life table for the total U.S. population in 2002. It is based on information from all death certificates filed in the 50 states and the District of Columbia. It is called an abridged life table because it uses age-groupings instead of single years of age. If single years of age are used, it is called a complete life table. Prior to 1997, a complete life table was construed only for a census year and for all off-census years abridged life tables were constructed. Beginning with 1997 mortality data, a complete life table was constructed every year, and abridged tables are derived from the complete tables. Previously, the annual life tables were closed at age 85, but they have been extended to age 100 based on old-age mortality data from the Medicare program. Other types of life tables are available from the National Center for Health Statistics. A brief history and sources for life tables for the United States can be found in Appendix C.

Table 4.10   Abridged life table for the total U.S. population, 2002.

Age	Probability of Dying Between Ages x and x + n $_nq_x$	Number Surviving to Age x l_x	Number Dying Between Ages x and x + n $_nd_x$	Person-Years Lived Between Ages x and x + n $_nL_x$	Total Number of Person-Years Lived Above Age x T_x	Expectation of Life at Age x e_x
0–1	0.006971	100,000	697	99,389	7,725,787	77.3
1–5	0.001238	99,303	123	396,921	7,626,399	76.8
5–10	0.000759	99,180	75	495,706	7,229,477	72.9
10–15	0.000980	99,105	97	495,311	6,733,771	67.9
15–20	0.003386	99,008	335	494,345	6,238,460	63.0
20–25	0.004747	98,672	468	492,189	5,744,116	58.2
25–30	0.004722	98,204	464	489,871	5,251,927	53.5
30–35	0.005572	97,740	545	487,395	4,762,056	48.7
35–40	0.007996	97,196	777	484,164	4,274,661	44.0
40–45	0.012066	96,419	1,163	479,362	3,790,497	39.3
45–50	0.017765	95,255	1,692	472,292	3,311,135	34.8
50–55	0.025380	93,563	2,375	462,186	2,838,843	30.3
55–60	0.038135	91,188	3,478	447,838	2,376,658	26.1
60–65	0.058187	87,711	5,104	426,603	1,928,820	22.0
65–70	0.088029	82,607	7,272	395,866	1,502,217	18.2
70–75	0.133076	75,335	10,025	352,791	1,106,350	14.7
75–80	0.201067	65,310	13,132	294,954	753,560	11.5
80–85	0.304230	52,178	15,874	222,013	458,606	8.8
85–90	0.447667	36,304	16,252	140,041	236,593	6.5
90–95	0.599618	20,052	12,024	67,822	96,552	4.8
95–100 ...	0.739020	8028	5933	23,056	28,730	3.6
100+	1.000000	2095	2095	5675	5675	2.7

Source: Arias, 2004

One use of the life table is to summarize the life experience of the population. A direct way of creating a life table is to follow a large cohort — say, 100,000 infants born on the same day — until the last member of this cohort dies. For each person the exact length of life can be obtained by counting the number of days elapsed from the date of birth. This yields 100,000 observations of the length of life. The random variable is the length of life in years or even in days. We can display the distribution of this random variable and calculate the mean, the median, the first and third quartiles, and the minimum and maximum. Since most people die at older ages, we expect that the distribution is skewed to the left, and hence the median length of life is larger than the mean length of life. The mean length of life is the life expectancy. We can tabulate the data using the age intervals 0–1, 1–5, 5–10, 10–15, . . . , 95–100, and 100 or over. All the intervals are the same length — five years — except for the first two and the last interval. The first interval is of a special interest, since quite a few infants die within it. From this tabulation, we can also calculate the relative frequency distribution by dividing the frequencies by 100,000. These relative frequencies give the probability of dying in each age interval. This probability distribution can be used to answer many practical questions regarding life expectancy. For instance, what is a 20-year-old person's probability of surviving to the retirement age of 65?

However, acquiring such data poses a problem. It would take more than 100 years to collect it. Moreover, information obtained from such data may be of some historical interest but not useful in answering current life expectancy questions, since current life expectancy may be different from that of earlier times. To solve this problem, we have to find ways to use current mortality information to construct a life table. The logical current mortality data for this purpose are the age-specific death rates. For the time being, we assume that age-specific death rates measure the probability of dying in each age interval. Note that these rates are conditional probabilities. The death rate for the 5- to 10-year-old age group is computed on the condition that its members survived the previous age intervals.

As presented in Chapter 3, the age-specific death rate is calculated by dividing the number of deaths in a particular age group by the midyear population in that age group. This is not exactly a proportion, whereas a probability is. Therefore, the first step in constructing a life table is to convert the age-specific death rates to the form of a probability. One possible conversion is based on the assumption that the deaths were occurring evenly throughout the interval. Under this assumption, we expect that one-half of the deaths occurred during the first half of the interval. Thus, the number of persons at the beginning of an interval is the sum of the midyear population and one-half of the deaths that occurred during the interval. Then the conditional probability of dying during the interval is the number of deaths divided by the number of persons at the beginning of the interval. Actual conversions use more complicated procedures for different age groups, but we are not concerned about these details.

4.7.1   The First Four Columns in the Life Table

With this background, we are now ready to examine Table 4.10. The first column shows the age intervals between two exact ages. For instance, 5–10 indicates the five-year interval between the fifth and tenth birthdays. This age grouping is slightly different from those of under 5, 5–9, 10–14, and so on used in the Census publications. In the life table, age is considered a continuous variable, whereas in the Census, counting of people by age (ignoring the fractional year) is emphasized.

The second column shows the proportion of the persons alive at the beginning of the interval who will die before reaching the end of the interval. It is labeled as $_nq_x$, where the first subscript on the left denotes the length of the interval and the second subscript on the right denotes the exact age at the beginning of the interval. The first entry in the second column, $_1q_0$, is 0.006971, which is the probability of newborn infants dying during the first year of life. The second entry is $_4q_1$, which equals 0.001238. It is the conditional probability of dying during the interval between ages 1 and 5, provided the child survived the first year of life. The rest of the entries in this column are conditional probabilities of dying in a given interval for those who survived the preceding intervals. These conditional probabilities are estimated from the current age-specific death rates. Note that the last entry of column 2 is 1.000000, indicating everybody dies sometime after age 100.

Thus, we have a series of conditional probabilities of dying. Given these conditional probabilities of dying, we can also find the conditional probabilities of surviving. The probability of surviving the first year of life will be

$$(1 - {_1q_0}) = 1 - 0.006971 = 0.993029.$$

Likewise, the conditional probability of surviving the interval between exact ages 1 and 5, provided infants had survived the first year of life will be

$$(1 - {_4q_1}) = 1 - 0.001238 = 0.998762.$$

Surviving the first five years of life is the intersection of surviving the 0–1 interval and the 1–5 interval. The probability of this intersection can be obtained as the product of the probability of surviving the 0–1 interval and the conditional probability of surviving the 1–5 interval given survival during the 0–1 interval — that is,

$$\Pr\{\text{surviving the intervals 0–1 and 1–5}\} = (1 - {}_1q_0)(1 - {}_4q_1)$$

$$= (1 - 0.006971)(1 - 0.001238) = (0.993029)(0.998762) = 0.991800.$$

Similarly, the probability of surviving the first 10 years of life, the first three intervals, will be

$$(1 - {}_1q_0)(1 - {}_4q_1)(1 - {}_5q_5).$$

Using this approach, we can calculate the survival probabilities from birth to the beginning of any subsequent age intervals. These survival probabilities are reflected in the third column, the number alive, l_x, at the beginning of the interval which begins at x years of age, out of a cohort of 100,000. Note that the entries in this column may differ slightly from the product of the survival probabilities and 100,000 because, although only four digits to the right of the decimal point are shown in the second column, more digits are used in the calculations. The first entry in this column is l_0, called the *radix*, is the size of the birth cohort. The second entry, the number alive at the beginning of the interval beginning at 1 year of age, l_1, is found by taking the product of the number alive at the beginning of the previous interval and the probability of surviving that interval — that is,

$$l_1 = l_0(1 - {}_1q_0) = l_0 - (l_0 - {}_1q_0) = l_0 - {}_1d_0.$$

This quantity, l_1, is equivalent to taking the number alive at the beginning of the previous period minus the number that died during that period, ${}_1d_0$. The numbers that died during each interval are shown in the fourth column, which is labeled as ${}_nd_x$.

The number who died during the four-year age interval from 1 to 5 is ${}_4d_1$. This is found by taking the product of the number alive at the beginning of this interval, l_1, and the probability of dying during the interval, ${}_4q_1$ — that is, ${}_4d_1 = l_1({}_4q_1)$. The number alive at the beginning of the interval of 5 to 10 years of age, l_5, can be found by subtracting the number who died during the previous age interval, ${}_4d_1$, from the number alive at the beginning of the previous interval, l_1 — that is, $l_5 = l_1 - {}_4d_1$. Repeating this operation, the rest of the entries in the third and fourth columns can be obtained. The fourth column can also be obtained directly from the third column. For example,

$${}_1d_0 = l_0 - l_1, \quad {}_4d_1 = l_1 - l_5, \text{ etc.}$$

Note that the last entry of the third column is the same as the last entry in the fourth column because all the survivors at age 100 will die subsequently. Note further that the l_x value in each row is a cumulative total of ${}_nd_x$ values in that and all subsequent rows.

Dividing the entries in the third and fourth columns by 100,000, we obtain the probabilities of surviving from birth to the beginning of the current interval and dying during the current interval, respectively. Note that the entries in the fourth column sum to 100,000, meaning that the probability of dying sums to one. As we expected, the distribution is negatively skewed, with the larger probabilities of dying at older ages.

4.7.2   Some Uses of the Life Table

Before looking at expected values in the life table, we wish to show how the first four columns, particularly the third column, can be used to answer some questions regarding life chances.

Example 4.11

What is the probability of surviving from one age to a subsequent age, say from age 5 to age 20? This is a conditional probability, conditional on the survival to age 5. The intersection of the events of surviving to age 20 and surviving to age 5 is surviving to age 20. Thus, the probability of this intersection is the probability of surviving from birth to age 20. This is the number alive at the beginning of the 20–25 interval divided by the number alive at the beginning — that is, l_{20}/l_0. The probability of surviving from birth to age 5 is l_5/l_0. Therefore, the conditional survival probability from age 5 to age 20 is found by dividing the probability of the intersection by the probability of surviving to age 5 — that is,

$$\left(\frac{l_{20}}{l_0}\right)\Big/\left(\frac{l_5}{l_0}\right)=\left(\frac{l_{20}}{l_5}\right)=\frac{98672}{99180}=0.994878.$$

The survival probabilities from any age to an older age can be calculated in a similar fashion.

We know the conditional probability of dying in any single interval. However, we may be interested in the probability of dying during a period formed by the first two or more consecutive intervals.

Example 4.12

What is the probability of dying during the first 5 years of life? This probability can be found by subtracting the probability of surviving the first 5 years from 1 — that is,

$$1-(1-{}_1q_0)(1-{}_4q_1)=1-\left(\frac{l_1}{l_0}\right)\left(\frac{l_5}{l_1}\right)=1-\frac{l_5}{l_0}.$$
$$=1-0.99180=0.0820$$

This is simply 1 minus the ratio of the number alive at the beginning of the final interval of interest and 100,000.

Example 4.13

A similar question relates to the probability of dying during a period formed by two or more consecutive intervals given that one had already survived several intervals. For example, what is the probability that a 30-year-old person will die between the ages of 50 and 60? This conditional probability is found by dividing the probability of the intersection of the event of dying between the ages of 50 and 60 and the event of surviving until 30 by the probability of the event of surviving until 30 years of age. The intersection of dying between 50 and 60 and surviving until 30 is dying between 50 and 60. The probability of dying between 50 and 60 is the number of persons dying, l_{50} minus l_{60}, divided by the total number, l_0. The probability of surviving until age 30 is simply l_{30} divided by l_0. Therefore, the probability of dying between 50 and 60 given survival until 30 is

$$\left(\frac{l_{50}-l_{60}}{l_0}\right)\Big/\left(\frac{l_{30}}{l_0}\right)=\frac{l_{50}-l_{50}}{l_{30}}=\frac{93563-87711}{97740}=0.059873.$$

Example 4.14

Another slightly more complicated question concerns the joint survival of persons. Suppose that a 40-year-old person has a 5-year-old child. What will be the probability that both the parent and child survive 25 more years until the parent's retirement? If we assume that the survival of the parent and that of the child are independent, we can calculate the desired probability by multiplying the individual survival probabilities. Applying the rule for the probability of surviving from one age to a subsequent age from the first question, this is

$$\left(\frac{l_{65}}{l_{40}}\right)\cdot\left(\frac{l_{30}}{l_5}\right)=\frac{82607}{96419}\cdot\frac{97740}{99180}=0.856750\cdot0.985481=0.844311.$$

The probability that the parent will die but the child will survive during the 25 years is

$$\left(1-\frac{l_{65}}{l_{40}}\right)\cdot\left(\frac{l_{30}}{l_5}\right)=(1-0.856750)\cdot0.985481=0.141170.$$

The probability that the parent will survive but the child will die during the 25 years is

$$\left(\frac{l_{65}}{l_{40}}\right)\cdot\left(1-\frac{l_{30}}{l_5}\right)=0.856750\cdot(1-0.985481)=0.012439.$$

The probability that both the parent and the child will die during the 25 years is

$$\left(1-\frac{l_{65}}{l_{40}}\right)\cdot\left(1-\frac{l_{30}}{l_5}\right)=(1-0.856750)\cdot(1-0.985481)=0.002080.$$

These four probabilities sum to 1 because those four events represent all the possible outcomes in considering the life and death of two persons.

4.7.3   Expected Values in the Life Table

The last three columns contain the information for various expected values in the life table. The fifth column of the life table, denoted by $_nL_x$, shows the person-years lived during each interval. For instance, the first entry in the fifth column is 99,389, which is the total number of person-years of life contributed by 100,000 infants during the first year of life. This value consists of 99,303 years contributed by the infants that survived the full year. The remaining 86 (= 99389 − 99303) person-years came from 697 infants who died during the year. The value of 86 years is based on actual mortality data coupled with mathematical smoothing. It cannot be found from the first four columns in the table. The value of 86 years is much less than 348.5 expected if the deaths had been distributed uniformly during the year. This value also suggests that most of the deaths occurred during the first half of the interval. The second entry of the fifth column is much larger than the first entry, mainly reflecting that the length of the second interval is greater than the length of the first interval. Each person surviving this second interval contributed 4 person years of life.

The fifth column is often labeled as the "stationary population in the age interval." The label of stationary population is based on a model of the long-term process of birth and death. If we assume 100,000 infants are born every year for 100 years, with each birth cohort subject to the same probabilities of dying specified in the second column of the life table, then we expect that there will be 100,000 people dying at the indicated ages every year. This means that the number of people in each age group will be the numbers shown in the fifth column. This hypothetical population will maintain the same size, since the number of births is the same as the number of deaths and it also keeps the same age distribution. That is, the size and structure of population is invariant, and hence this is called a stationary population.

The sixth column of the life table, denoted by T_x, shows cumulative totals of $_nL_x$ values starting from the last age interval. The T_x value in each interval indicates the number of person years remaining in that and all subsequent age intervals. For example, the T_{95} value of 28,730 is the sum of $_5L_{95}$ (= 23056) and $_\infty L_{100}$ (= 5675).

The last column of the life table, denoted by e_x, shows the life expectancies at various ages, which are calculated by $e_x = T_x/l_x$. The first entry of the last column is the life expectancy for newborn infants, and all subsequent entries are conditional life expectancies. Conditional life expectancies are more useful information than the expectancies figured for newborn infants. For instance, those who survived to age 100 are expected to live 2.7 years more ($e_{100} = 2.7$), the last entry of the last column, whereas newborn infants are expected to live 0.06 years beyond age 100 ($T_{100} / l_0 = 5675/100000 = 0.06$).

Example 4.15

Based on T_x values, more complicated conditional life expectancies can be calculated. For instance, suppose that a 30-year-old person was killed in an industrial accident and had been expected to retire at age 65 if still alive. For how many years of unearned income should that person's heirs be compensated? The family may request a compensation for 35 years. However, based on the life table, the company argues for a smaller number of years. The total number of years of life remaining during the interval from 30 to 65 is T_{30} minus T_{65}, and there are l_{30} persons remaining at age 30 to live those years. Therefore, the average number of years of life remaining is found by

$$\frac{T_{30} - T_{65}}{l_{30}} = \frac{4762056 - 1502217}{97740} = 33.4.$$

Example 4.16

The notion of stationary population can be used to make certain inferences for population planning and manpower planning. The birth rate of the stationary population can be obtained by dividing 100,000 by the total years of life lived by the stationary population, or

$$\frac{l_0}{T_0} = \frac{100000}{7725787} = \frac{1}{77.6} = 0.013$$

or 13 per 1000 population. The death rate should be the same. But note that the birth rate equals the reciprocal of the life expectancy at birth ($1/e_0$). In other words, the birth rate (replacement rate) and death rate (attrition rate) are entirely determined by the life expectancy under the stationary population assumption.

4.7.4 Other Expected Values in the Life Table

The most widely used figures from the life table are life expectancies. These are average values. As discussed in Chapter 3, the mean value may not represent the distribution appropriately in some circumstances. Let us find the median length of life at birth. To find the median, the second quartile, we must find the value such that 50 percent of the radix falls below it. By examining column 3 in the life table, we find that 52,178 persons are alive at the beginning of the age interval 80–85, whereas only 36,304 are alive at the beginning of the interval 85–90. Since 50,000 is between 52,178 and 36,304, we know that the median is somewhere between 80 and 85 years of age. If we assume that the 15,874 deaths are uniformly distributed over this age interval, we can find the median by interpolation. We add a proportion of the five years, the length of the interval, to the age at the beginning of the interval, 80 years. The proportion is the ratio of the difference between 52,178 and 50,000 to the 15,304 deaths that occurred in the interval. The calculation is

$$median = 80 + 5 \cdot \left(\frac{52178 - 50000}{15874} \right) = 80.69.$$

As expected, the mean is smaller than the median. Perhaps, it is more enlightening to know that one-half of a birth cohort will live to age 81 than to know that an average length of life is about 77 years.

The corresponding calculations for the first and third quartiles are

$$Q_1 = 70 + 5 \cdot \left(\frac{75335 - 75000}{10025} \right) = 70.17$$

$$Q_3 = 85 + 5 \cdot \left(\frac{36304 - 25000}{16252} \right) = 88.48.$$

Conclusion

Probability has been defined as the relative frequency of an event in an infinite number of trials or in a population. Its use has been demonstrated in a number of examples, and a number of rules for the calculation of probabilities have been presented. The use of probabilities and the rules for calculating probabilities have been applied to the life table, a basic tool in public health research.

Now that we have an understanding of probability, we shall examine particular probability distributions in the next chapter.

EXERCISES

4.1 Choose the most appropriate answer.

a. If you get 10 straight heads in tossing a fair coin, a tail is _____ on the next toss.

___ more likely

___ less likely

___ neither more likely nor less likely

b. In the U.S. life table, the distribution of the length of life (or age at death) is ___.

___ skewed to the left

___ skewed to the right

___ symmetric

c. A test with high sensitivity is very good at _____.

___ screening out patients who do not have the disease.

___ detecting patients with the disease.

___ determining the probability of the disease.

d. In the U.S. life table the life expectancy (mean) is _____ the median length of life.

___ the same as

___ greater than

___ less than

e. $_4q_1$ is called a _____ because an infant cannot die in this interval unless it survived the first year of life.

___ personal probability

___ marginal probability

___ conditional probability

f. In the life table, the mean length of life for those who died during ages 0–1 is _____.

___ about 1/2 year

___ more than 1/2 year

___ less than 1/2 year

4.2 The following table gives estimates of the probabilities that a randomly chosen adult in the United States falls into each of six gender-by-education categories (based on relative frequencies from the NHANES II, NCHS 1982). The three education categories used are (1) less than 12 years, (2) high school graduate, and (3) more than high school graduation.

Categories of Education			
Gender	(1)	(2)	(3)
Female	0.166	0.194	0.164
Male	0.149	0.140	0.187

    a. What is the estimate of the probability that an adult is a high school graduate (categories 2 and 3)?

    b. What is the estimate of the probability that an adult is a female?

    c. From the NHANES II data, it is also estimated that the probability that a female is taking a vitamin supplement is 0.426. What is the estimate of the probability that the adult is a female and taking a vitamin supplement?

    d. From the NHANES II, it is also estimated that the probability of adults taking a vitamin supplement is 0.372. What is the estimate of the probability that a male is taking a vitamin supplement?

4.3 Suppose that the failure rate (failing to detect smoke when smoke is present) for a brand of smoke detector is 1 in 2000. For safety, two of these smoke detectors are installed in a laboratory.

    a. What is the probability that smoke is not detected in the laboratory when smoke is present in the laboratory?

    b. What is the probability that both detectors sound an alarm when smoke is present in the laboratory?

    c. What is the probability that one of the detectors sounds the alarm and the other fails to sound the alarm when smoke is present in the laboratory?

4.4 Suppose that the probability of conception for a married woman in any month is 0.2. What is the probability of conception in two months?

4.5 A new contraceptive device is said to have only a 1 in 100 chance of failure. Assume that the probability of conception for a given month, without using any contraceptive, is 20 percent. What is the probability of having at least one unwanted pregnancy if a woman were to use this device for 10 years? [Hint: This would be the complement of the probability of avoiding pregnancy for 10 years or 120 months. The probability of conception for any month with the use of the new contraceptive device would $0.2 * (1 - 0.99)$. This and related issues are examined by Keyfitz 1971.]

4.6 In a community, 5500 adults were screened for hypertension by the use of a standard sphygmomanometer, and 640 were found to have a diastolic blood pressure of 90 mmHg or higher. A random sample of 100 adults from those with diastolic blood pressure of 90 mmHg or higher and another random sample of 100 adults from those with blood pressure less than 90 mmHg were subjected to more intensive clinical evaluation for hypertension, and 73 and 13 of the respective samples were confirmed as being hypertensive.

    a. What is an estimate of the probability that an adult having blood pressure greater than or equal to 90 at the initial screening will actually be hypertensive (predicted value positive)?

    b. What is an estimate of the probability that an adult having blood pressure less than 90 at the initial screening will not actually be hypertensive (predicted value negative)?

c. What is an estimate of the probability that an adult in this community is truly hypertensive (prevalence rate of hypertension)?

d. What is an estimate of the probability that a hypertensive person will be found to have blood pressure greater than or equal to 90 at the initial screening (sensitivity)?

e. What is an estimate of the probability that a person without hypertension will have blood pressure less than 90 at the initial screening (specificity)?

4.7 What is the average number of children per family if every couple were to have children until a son is born? Simulate using the random number table in Appendix B or a random number generator in any statistical software.

4.8 Calculate the following probabilities from the 2002 U.S. Abridged Life Table.

a. What is the probability that a 35-year-old person will survive to retirement at age 65?

b. What is the probability that a 20-year-old person will die between ages 55 and 65?

4.10 Calculate the following expected values from the 2002 U.S. Abridged Life Table.

a. How many years is a newborn expected to live before his fifth birthday?

b. How many years is a 20-year-old person expected to live after retirement at age 65? Repeat the calculation for a 60-year-old person. How would you explain the difference?

4.11 Suppose that a couple wants to have children until they have a girl or until they have four children.

a. What is the probability that they have at least two boys?

b. What is the expected number of children?

4.12 The following are tallies of the first digits of the 50 states' populations in the 2000 U.S. Census:

Digit	1	2	3	4	5	6	7	8	9	Total
Frequencies	14	6	4	7	6	5	3	3	2	50

a. Why do you think digit 1 appears most frequently and digit 9 least frequently?

b. Tabulate the first digits of numerical data that appeared on the front page of today's newspaper, and see whether your findings conform to Benford's law (Hill 1999) [$Pr(\text{first significant digit} = d) = \log_{10}(1 + 1/d)$, $d = 1, 2, \ldots 9$].

4.13 About 1 percent of women have breast cancer. A cancer screening method can detect 80 percent of genuine cancers with a false alarm rate of 10 percent. What is the probability that women producing a positive test result really have breast cancer?

4.14 Suppose that a factory hires 500 men at age 25 and 200 women at age 25 each year. The factory maintains the fixed number of workforce. From the 2002 life tables, the following values are available: For men: $l_{25} = 97746$; $l_{65} = 78556$; $e_{25} = 51.0$; $e_{65} = 16.6$. For women: $l_{20} = 98922$; $l_{65} = 86680$; $e_{20} = 60.7$; $e_{65} = 19.5$.

a. What would be the expected number of retirees at age 65?

b. What would be the expected number of total employees?

REFERENCES

Arias, E. *United States Life Tables, 2002. National Vital Statistics Reports*, Vol. 53, No. 6. Hyattsville, MD: National Center for Health Statistics, 2004.

Harris County Health Department. Mark Canfield, ed. *The Health Status of Harris County Residents: Births, Deaths and Selected Measures of Public Health, 1980–1986*, 1990.

Hill, T. P. "The Difficulty of Faking Data," *Chance* 12(3):27–31, 1999.

Keyfitz, N. "How Birth Control Affects Births," *Social Biology* 18:109–121, 1971.

Lundgren, E., J. Rastad, E. Thrufjell, G. Akerstrom, and S. Ljunghall. "Population-based screening for primary hyperparathyroidism with serum calcium and parathyroid hormone values in menopausal women," *Surgery* 121(3):287–294, 1997.

Mathews, T. J., and B. E. Hamilton. *Trend Analysis of the Sex Ratio at Birth in the United States. National Vital Statistics Reports*, Vol. 53, No. 20. Hyattsville, MD: National Center for Health Statistics, 2005.

National Center for Health Statistics, *Second National Health and Nutrition Examination Survey (NHANES II) 1982*. Tabulation of data for adults 18 years of age and over by the authors.

National Center for Health Statistics. *Health, United States, 1991 and Prevention Profile*. Hyattsville, MD: Public Health Service. DHHS Pub. No. 92–1232, 1992.

National Center for Health Statistics. *Health, United States, 2004 with Chartbook on Trends in the Health of Americans*. Hyattsville, MD: DHHS Pub. No. 2004–1232, 2004.

U.S. Bureau of the Census. The 2000 Census of Population and Housing, Summary Tape File 1A.

Weidner, N., J. Forkman, F. Pozza, P. Bevilacqua, E. N. Allred, D. H. Morre, S. Meli, and G. Gasparini. "Tumor Angiogenesis: A New Significant and Independent Prognostic Indicator in Early State Breast Carcinoma." *Journal of the National Cancer Institute* 84(24):1875–1887, December 16, 1992, issue.

Weiss, N. S. *Clinical Epidemiology: The Study of Outcome of Disease*, New York: Oxford University Press, 1986.

Probability Distributions

Chapter Outline

This chapter introduces three probability distributions: the binomial and the Poisson for discrete random variables, and the normal for continuous random variables. For a discrete random variable, its probability distribution is a listing of the probabilities of its possible outcomes or a formula for finding the probabilities. For a continuous random variable, its probability distribution is usually expressed as a formula that can be used to find the probability that the variable will fall in a specified interval. Knowledge of the probability distribution (1) allows us to summarize and describe data through the use of a few numbers and (2) helps to place results of experiments in perspective; that is, it allows us to determine whether or not the result is consistent with our ideas. We begin the presentation of probability distributions with the binomial distribution.

5.1 The Binomial Distribution

As its name suggests, the *binomial distribution* refers to random variables with two outcomes. Three examples of random variables with two outcomes are (1) smoking status — a person does or does not smoke, (2) exposure to benzene — a worker was or was not exposed to benzene in the workplace, and (3) health insurance coverage — a person does or does not have health insurance. The random variable of interest in the binomial setting is the number of occurrences of the event under study — for example, the number of adults in a sample of size n who smoke or who have been exposed to benzene or who have health insurance. For the binomial distribution to apply, the status of each subject must be independent of that of the other subjects. For example, in the hypertension question, we are assuming that each person's hypertension status is unaffected by any other person's status.

5.1.1 Binomial Probabilities

We consider a simple example to demonstrate the calculation of binomial probabilities. Suppose that four adults (labeled A, B, C, and D) have been randomly selected and asked whether or not they currently smoke. The random variable of interest in this example is the number of persons who respond yes to the question about smoking. The possible outcomes of this variable are 0, 1, 2, 3, and 4.

The outcomes (0, 1, 2, 3, and 4) translate to estimates of the proportion of persons who answer yes (0.00, 0.25, 0.50, 0.75, and 1.00, respectively). Any of these outcomes could occur when we draw a random sample of four adults. As a demonstration, let us draw 10 random samples of size 4 from a population in which the proportion of adults who smoke is assumed to be 0.25 (population parameter). We can use a random number table in performing this demonstration. Four 2-digit numbers were taken from the first 10 rows of the first page of random number tables in Appendix B. The 2-digit numbers less 25 are considered smokers. The results are shown here:

Sample	Random Number	No. of Smokers	Prop. Smokers
1	17 17 47 59	2	0.50
2	26 58 06 84	1	0.25
3	24 04 23 38	3	0.75
4	74 83 87 93	0	0.00
5	72 86 25 09	1	0.25
6	82 27 49 45	0	0.00
7	77 58 68 91	0	0.00
8	17 80 21 66	2	0.50
9	10 27 10 61	2	0.50
10	07 78 05 54	2	0.50

Three samples have no smokers (estimate of 0.00); two samples have 1 smoker (0.25); four samples have 2 smokers (0.50); one sample has 3 smokers (0.75); and no sample has 4 smokers (1.00). The sample estimates do not necessarily equal the population parameter, and the estimates can vary considerably. In practice, a single sample is selected, and in making an inference from this one sample to the population, this sample-to-sample variability must be taken into account. The probability distribution does this, as will be seen later. Now let us calculate the binomial probability distribution for a sample size of four.

Suppose that in the population, the proportion of people who would respond "yes" to this question is π, and the probability of a response of "no" is then $1 - \pi$. The probability of each of the outcomes can be found in terms of π by listing all the possible outcomes. Table 5.1 provides this listing.

Since each person is independent of all the other persons, the probability of the joint occurrence of any outcome is simply the product of the probabilities associated with each person's outcome. That is, the probability of 4 yes responses is π^4. In the same way, the probability of three yes responses is $4\pi^3(1 - \pi)$, since there are four occurrences of three yes responses. The probability of two yes responses is $6\pi^2(1 - \pi)^2$; the probability of one yes response is $4\pi(1 - \pi)^3$; and the probability of zero yes responses is $(1 - \pi)^4$. If we know the value of π, we can calculate the numerical value of these probabilities.

Suppose π is 0.25. Then the probability of each outcome is as follows:

$$\Pr\{4 \text{ yes responses}\} = 1 * (0.25)^4 * (0.75)^0 = 0.0039 = \Pr\{0 \text{ no responses}\},$$

$$\Pr\{3 \text{ yes responses}\} = 4 * (0.25)^3 * (0.75)^1 = 0.0469 = \Pr\{1 \text{ no response}\},$$

Table 5.1   Possible binomial outcomes in a sample of size of 4 and their probabilities of occurrence.

	Person				
A	B	C	D	Probability of Occurrence	
y[a]	y	y	y	$\pi * \pi * \pi * \pi$	$= \pi^4 * (1 - \pi)^0$
y	y	y	n	$\pi * \pi * \pi * (1 - \pi)$	$= \pi^3 * (1 - \pi)^1$
y	y	n	y	$\pi * \pi * (1 - \pi) * \pi$	$= \pi^3 * (1 - \pi)^1$
y	n	y	y	$\pi * (1 - \pi) * \pi * \pi$	$= \pi^3 * (1 - \pi)^1$
n	y	y	y	$(1 - \pi) * \pi * \pi * \pi$	$= \pi^3 * (1 - \pi)^1$
y	y	n	n	$\pi * \pi * (1 - \pi) * (1 - \pi)$	$= \pi^2 * (1 - \pi)^2$
y	n	y	n	$\pi * (1 - \pi) * \pi * (1 - \pi)$	$= \pi^2 * (1 - \pi)^2$
y	n	n	y	$\pi * (1 - \pi) * (1 - \pi) * \pi$	$= \pi^2 * (1 - \pi)^2$
n	y	y	n	$(1 - \pi) * \pi * \pi * (1 - \pi)$	$= \pi^2 * (1 - \pi)^2$
n	y	n	y	$(1 - \pi) * \pi * (1 - \pi) * \pi$	$= \pi^2 * (1 - \pi)^2$
n	n	y	y	$(1 - \pi) * (1 - \pi) * \pi * \pi$	$= \pi^2 * (1 - \pi)^2$
y	n	n	n	$\pi * (1 - \pi) * (1 - \pi) * (1 - \pi)$	$= \pi^1 * (1 - \pi)^3$
n	y	n	n	$(1 - \pi) * \pi * (1 - \pi) * (1 - \pi)$	$= \pi^1 * (1 - \pi)^3$
n	n	y	n	$(1 - \pi) * (1 - \pi) * \pi * (1 - \pi)$	$= \pi^1 * (1 - \pi)^3$
n	n	n	y	$(1 - \pi) * (1 - \pi) * (1 - \pi) * \pi$	$= \pi^1 * (1 - \pi)^3$
n	n	n	n	$(1 - \pi) * (1 - \pi) * (1 - \pi) * (1 - \pi)$	$= \pi^0 * (1 - \pi)^4$

[a] y indicates a yes response and n indicates a no response

$$\text{Pr \{2 yes responses\}} = 6 * (0.25)^2 * (0.75)^2 = 0.2109 = \text{Pr \{2 no responses\}},$$

$$\text{Pr \{1 yes response\}} = 4 * (0.25)^1 * (0.75)^3 = 0.4219 = \text{Pr \{3 no responses\}},$$

$$\text{Pr \{0 yes responses\}} = 1 * (0.25)^0 * (0.75)^4 = 0.3164 = \text{Pr \{4 no responses\}}.$$

The sum of these probabilities is one, as it must be, since these are all the possible outcomes. If the probabilities do not sum to one (with allowance for rounding), a mistake has been made. Figure 5.1 shows a plot of the binomial distribution for n equal to 4 and π equal to 0.25.

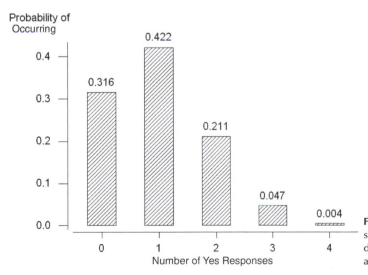

Figure 5.1   Bar chart showing the binomial distribution for $n = 4$ and $\pi = 0.25$.

Are these probabilities reasonable? Since the probability of a yes response is assumed to be 0.25 in the population, in a sample of size four, the probability of one yes response should be the largest. It is also reasonable that the probabilities of zero and two yes responses are the next largest, since these values are closest to one yes response. The probability of four yes responses is the smallest, as is to be expected. Figure 5.1 shows the rapid decrease in the probabilities as the number of yes responses moves away from the expected response of one.

In the calculation of the probabilities, there are several patterns visible. The exponent of the probability of a yes response matches the number of yes responses being considered and the exponent of the probability of a no response also matches the number of no responses being considered. The sum of the exponents is always the number of persons in the sample. These patterns are easy to capture in a formula, which eliminates the need to enumerate the possible outcomes. The formula may appear complicated, but it is really not all that difficult to use. The formula, also referred to as the *probability mass function* for the binomial distribution, is

$$\Pr\{X = x\} = \binom{n}{x} \pi^x (1 - \pi)^{n-x}$$

where $\binom{n}{x} = {}_nC_x = \dfrac{n!}{x!(n-x)!}$, $n! = n(n-1)(n-2)\cdots 1$ and $0!$ is defined to be 1. The symbol $n!$ is called n factorial, and ${}_nC_x$ is read as n combination x, which gives the number of ways that x elements can be selected from n elements without regard to order (see Appendix A for further explanations). In this formula, n is the number of persons or elements selected, and x is the value of the random variable, which goes from 0 to n. Another representation of this formula is

$$B(x; n, \pi) = \binom{n}{x} \pi^x (1 - \pi)^{n-x} = B(n - x; n, 1 - \pi)$$

where B represents binomial. The equality of $B(x; n, \pi)$ and $B(n - x; n, 1 - \pi)$ is a symbolic way of saying that the probability of x yes responses from n persons, given that π is the probability of a yes response, equals the probability of $n - x$ no responses.

The smoking situation can be used to demonstrate the use of the formula. To find the probability that $X = 3$, we have

$$\Pr(X = 3) = \binom{4}{3}(0.25)^3(0.75)^1 = \frac{4!}{3!1!}(0.015625)(0.75) = 4(0.01172) = 0.0469.$$

This is the same value we found by listing all the outcomes and the associated probabilities. There are easier ways of finding binomial probabilities, as is shown next.

There is a recursive relation between the binomial probabilities, which makes it easier to find them than to use the binomial formula for each different value of X. The relation is

$$\Pr\{X = x + 1\} = \left(\frac{n - x}{x + 1}\right)\left(\frac{\pi}{1 - \pi}\right)\Pr\{X = x\}$$

for x ranging from 0 to $n - 1$. For example, the probability that X equals 1 in terms of the probability that X equals 0 is

$$\Pr\{x=1\} = \left(\frac{4-0}{0+1}\right)\left(\frac{0.25}{0.75}\right)(0.3164) = 4\left(\frac{1}{3}\right)(0.3164) = 0.4219$$

which is the same value we calculated above.

A still easier method is to use Appendix Table B2, a table of binomial probabilities for n ranging from 2 to 20 and π beginning at 0.01 and ranging from 0.05 to 0.50 in steps of 0.05. There is no need to extend the table to values of π larger than 0.50 because $B(x; n, \pi)$ equals $B(n - x; n, 1 - \pi)$. For example, if π were 0.75 and we wanted to find the probability that $X = 1$ for $n = 4$, $B(1; 4, 0.75)$, we find $B(3; 4, 0.25)$ in Table B2 and read the value of 0.0469. These probabilities are the same because when $n = 4$ and the probability of a yes response is 0.25, the occurrence of three yes responses is the same as the occurrence of one no response when the probability of a no response is 0.75.

Another way of obtaining binomial probabilities is to use computer packages (see **Program Note 5.1** on the website). The use of computer software is particularly nice, since it does not limit the values of π to being a multiple of 0.05 and n can be much larger than 20. More will be said about how large n can be in a later section.

Table 5.2   **Probability mass (Pr{X = x}) and cumulative (Pr{X ≤ x}) distribution functions for the binomial when $n = 4$ and $\pi = 0.25$.**

x		0	1	2	3	4
Mass:	$\Pr\{X = x\}$	0.3164	0.4219	0.2109	0.0469	0.0039
Cumulative:	$\Pr\{X \le x\}$	0.3164	0.7383	0.9492	0.9961	1.0000

The probability mass function for the binomial gives $\Pr\{X = x\}$ for x ranging from 0 to n (shown in Figure 5.1). Another function that is used frequently is the *cumulative distribution function* (cdf). This function gives the probability that X is less than or equal to x for all possible values of X. Table 5.2 shows both the probability mass function and the cumulative distribution function values for the binomial when n is 4 and π is 0.25. The entries in the cumulative distribution row are simply the sum of the probabilities in the row above it, the probability mass function row, for all values of X less than or equal to the value being considered. Cumulative distribution functions all have a general shape shown in Figure 5.2. The value of the function starts with a low value and then increases over the range of the X variable. The rate of increase of the function is what varies between different distributions. All the distributions eventually reach the value of one or approach it asymptotically.

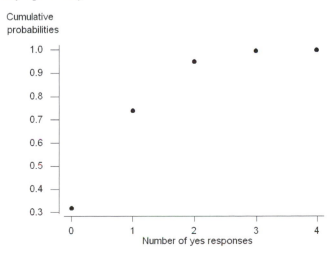

Figure 5.2   Cumulative binomial distribution for $n = 4$ and $\pi = 0.25$.

As seen above, if we know the data follow a binomial distribution, we can completely summarize the data through their two parameters, the sample size and the population proportion or an estimate of it. The sample estimate of the population proportion is the number of occurrences of the event in the sample divided by the sample size.

5.1.2 Mean and Variance of the Binomial Distribution

We can now calculate the *mean* and *variance* of the binomial distribution. The mean is found by summing the product of each outcome by its probability of occurrence — that is,

$$\mu = \sum_{x=0}^{n} x \cdot \Pr\{X = x\}.$$

This appears to be different from the calculation of the sample mean in Chapter 3, but it is really the same because in Chapter 3 all the observations had the same probability of occurrence, $1/N$. Thus, the formula for the population mean could be reexpressed as

$$\sum_{i=1}^{N} x_i \cdot \left(\frac{1}{N}\right) = \sum_{i=1}^{N} x_i \cdot \Pr\{x_i\}.$$

The mean of the binomial variable — that is, the mean number of yes responses out of n responses when n is 4 and π is 0.25 is

$$0 \cdot (0.3164) + 1 \cdot (0.4219) + 2 \cdot (0.2109) + 3 \cdot (0.0469) + 4 \cdot (0.0039) = 1.00$$
$$= n\pi$$

or in general for the binomial distribution,

$$\mu = n\pi.$$

The expression of the binomial mean as $n\pi$ makes sense, since, if the probability of occurrence of an event is π, then in a sample of size n, we would expect $n\pi$ occurrences of the event.

The variance of the binomial variable, the number of yes responses, can also be expressed conveniently in terms of π. From Chapter 3, the population variance was expressed as

$$\sigma^2 = \sum_{i=1}^{N} (x_i - \mu)/N.$$

In terms of the binomial, the X variable takes on the values from 0 to n, and we again replace the N in the divisor by the probability that X is equal to x. Thus, the formula becomes

$$\sigma^2 = \sum_{x=0}^{n} (x - n\pi)^2 \Pr\{X = x\}$$

which, with further algebraic manipulation, simplifies to

$$\sigma^2 = n\pi(1 - \pi).$$

When n is 4 and π is 0.25, the variance is then $4(0.25)(1 - 0.25)$, which is 0.75.

There is often interest in the variance of the proportion of yes responses — that is, in the variance of the number of yes responses divided by the sample size. This is the variance of the number of yes responses divided by a constant. From Chapter 3, we know that this is the variance of the number of yes responses divided by the square of the constant. Thus, the variance of a proportion is

$$\text{Var}(prop.) = \frac{n\pi(1 - \pi)}{n^2} = \frac{\pi(1 - \pi)}{n}.$$

Example 5.1

Use of the Binomial Distribution: Let us consider a larger example now. In 1990, cesarean section (c-section) deliveries represented 23.5 percent of all deliveries in the United States, a tremendous increase since 1960 when the rate was only 5.5 percent. Concern has been expressed, for example, by the Public Citizen Health Research Group (1992) in its June 1992 health letter, reporting that many unnecessary c-section deliveries are performed. Public Citizen believes unnecessary c-sections waste resources and increase maternal risks without achieving sufficient concomitant improvement in maternal and infant health. It is in this context that administrators at a local hospital are concerned, as they believe that their hospital's c-section rate is even higher than the national average. Suppose as a first step in determining if this belief is correct, we select a random sample of deliveries from the hospital. Of the 62 delivery records pulled for 1990, we found 22 c-sections. Does this large proportion of c-section deliveries, 35.5 percent (= 22/62), mean that this hospital's rate is higher than the national average? The sample proportion of 35.5 percent is certainly larger than 23.5 percent, but our question refers to the population of deliveries in the hospital in 1990, not the sample. As we just saw, we cannot infer immediately from this sample without taking sample-to-sample variability into account. This is a situation where the binomial distribution can be used to address the question about the population based on the sample.

To put the sample rate into perspective, we need to first answer a question: How likely is a rate of 35.5 percent or higher in our sample if the rate of c-section deliveries is really 23.5 percent? Note that the question includes rates higher than 35.5 percent. We must include them because if the sum of their probabilities is large, we cannot conclude that a rate of 35.5 percent is inconsistent with the national rate regardless of how unlikely the rate of 35.5 percent is.

We can use the cdf for the binomial to find the answer to this question. The cdf enables us to find the probability that a variable is less than a given value — in this case, less than the result we observed in our sample. Then we can subtract that probability from one to find how likely it is to obtain a rate as large or larger than our sample rate. It turns out to be 0.0224 (see **Program Note 5.1** on the website). This means that the probability of 22 or more c-section deliveries is 0.0224. The probability of having 22 or more c-sections is very small. It is unlikely that this hospital's c-section rate is the same as the national average, and, in fact, it appears to be higher. Further investigation is required to determine why the rate may be higher.

5.1.3 Shapes of the Binomial Distribution

The binomial distribution has two parameters, the sample size and the population proportion, that affect its appearance. So far we have seen the distribution of one binomial — Figure 5.1 — which had a sample size of 4 and a population proportion of 0.25. Figure 5.3 examines the effect of population proportion on the shape of the binomial distribution for a sample size of 10.

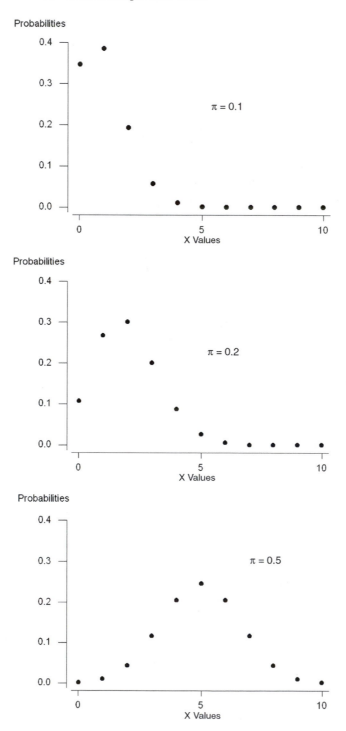

Figure 5.3 Binomial probabilities for $n = 10$ and $\pi = 0.1$, 0.2, and 0.5.

The plots in Figure 5.3 would look like bar charts if a perpendicular line were drawn from the horizontal axis to the points above each outcome. In the first plot with π equal to 0.10, the shape is quite asymmetric with only a few of the outcomes having probabilities very different from zero. This plot has a long tail to the right. In the second plot with π equal to 0.20, the plot is less asymmetric.

The third binomial distribution, with π equal to 0.50, has a mean of 5 ($= n\pi$). The plot is symmetric about its mean of 5, and it has the familiar bell shape. Since π is 0.50, it is as likely to have one less occurrence as one more occurrence — that is, four occurrences of the event of interest are as likely as six occurrences, three as likely as seven and so on, and the plot reflects this.

This completes the introduction to the binomial, although we shall say more about it later. The next section introduces the Poisson distribution, another widely used distribution.

5.2   The Poisson Distribution

The *Poisson distribution* is named for its discoverer, Siméon-Denis Poisson, a French mathematician from the late 18th and early 19th centuries. He is said to have once remarked that life is good for only two things: doing mathematics and teaching it (Boyer 1985). The Poisson distribution is similar to the binomial in that it is also used with counts or the number of events. The Poisson is particularly useful when the events occur infrequently. It has been applied in the epidemiologic study of many forms of cancer and other rare diseases over time. It has also been applied to the study of the number of elements in a small space when a large number of these small spaces are spread at random over a much larger space — for example, in the study of bacterial colonies on an agar plate.

Even though the Poisson and binomial distributions both are used with counts, the situations for their applications differ. The binomial is used when a sample of size n is selected and the number of events and nonevents are determined from this sample. The Poisson is used when events occur at random in time or space, and the number of these events is noted. In the Poisson situation, *no* sample of size n has been selected.

5.2.1   Poisson Probabilities

The Poisson distribution arises from either of two models. In one model — quantities, for example — bacteria are assumed to be distributed at random in some medium with a uniform density of λ(lambda) per unit area. The number of bacteria colonies found in a sample area of size A follows the Poisson distribution with a parameter μ equal to the product of λ and A.

In terms of the model over time, we assume that the probability of one event in a short interval of length t_1 is proportional to t_1 — that is, Pr{exactly one event} is approximately λt_1. Another assumption is that t_1 is so short that the probability of more than one event during this interval is almost zero. We also assume that what happens in one time interval is independent of the happenings in another interval. Finally, we assume that λ is constant over time. Given these assumptions, the number of occurrences of the event in a time interval of length t follows the Poisson distribution with parameter μ, where μ is the product of λ and t.

The Poisson probability mass function is

$$\Pr(X = x) = \frac{e^{-\mu}\mu^x}{x!} \qquad \text{for } x = 0, 1, 2 \ldots \ldots$$

where e is a constant approximately equal to 2.71828 and μ is the parameter of the Poisson distribution. Usually μ is unknown and we must estimate it from the sample data. Before considering an example, we shall demonstrate in Table 5.3 the use of the probability mass function for the Poisson distribution to calculate the probabilities when $\mu = 1$ and $\mu = 2$. These probabilities are not difficult to calculate, particularly when μ is an integer. There is also a recursive relation between the probability that $X = x + 1$ and the probability that $X = x$ that simplifies the calculations:

$$\Pr\{X = x + 1\} = \left(\frac{\mu}{x+1}\right)\Pr(X = x)$$

for x beginning at a value of 0. For example, for $\mu = 2$,

$$\Pr\{X = 3\} = (2/3)\ \Pr\{X = 2\} = (2/3)\ 0.2707 = 0.1804$$

which is the value shown in Table 5.3.

Table 5.3 Calculation of poisson probabilities, $\Pr\{X = x\} = e^{-\mu}\mu^x/x!$, for $\mu = 1$ and 2.

x	$\mu = 1$ e^{-1} * $1^x/x! = \Pr\{X = x\}$		$\mu = 2$ e^{-2} * $2^x/x! = \Pr\{X = x\}$	
0	0.3679 *	1/1 = 0.3679	0.1353 *	1/1 = 0.1353
1	0.3679 *	1/1 = 0.3679	0.1353 *	2/1 = 0.2707
2	0.3679 *	1/2 = 0.1839	0.1353 *	4/2 = 0.2707
3	0.3679 *	1/6 = 0.0613	0.1353 *	8/6 = 0.1804
4	0.3679 *	1/24 = 0.0153	0.1353 *	16/24 = 0.0902
5	0.3679 *	1/120 = 0.0031	0.1353 *	32/120 = 0.0361
6	0.3679 *	1/720 = 0.0005	0.1353 *	64/720 = 0.0120
7	0.3679 *	1/5040 = 0.0001	0.1353 *	128/5040 = 0.0034
8			0.1353 *	256/40320 = 0.0009
9			0.1353 *	512/362880 = 0.0002
		1.0000		0.9999

These probabilities are also found in Appendix Table B3, which gives the Poisson probabilities for values of μ beginning at 0.2 and increasing in increments of 0.2 up to 2.0, then in increments of 0.5 up to 7, and in increments of 1 up to 17. Computer software can provide the Poisson probabilities for other values of μ (see **Program Note 5.1** on the website). Note that the Poisson distribution is totally determined by specifying the value of its one parameter, μ. The plots in Figure 5.4 show the shape of the Poisson probability mass and cumulative distribution functions with $\mu = 2$.

The shape of the Poisson probability mass function with μ equal to 2 (the top plot in Figure 5.4) is similar to the binomial mass function for a sample of size 10 and π equal

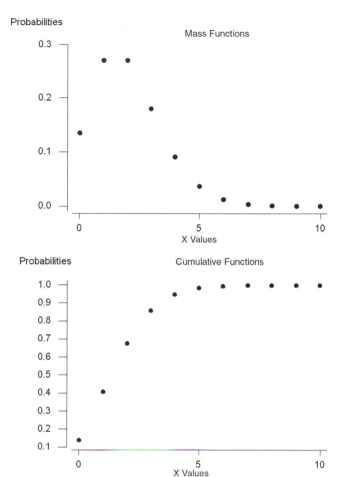

Figure 5.4   Poisson ($\mu =$ 2) probability mass and cumulative distribution functions.

to 0.2 just shown. The cdf (the bottom plot in Figure 5.4) has the same general shape as that shown in the preceding binomial example, but the shape is easier to see here, since there are more values for the X variable shown on the horizontal axis.

5.2.2   Mean and Variance of the Poisson Distribution

As just discussed, the mean is found by summing the products of each outcome by its probability of occurrence. For the Poisson distribution with parameter $\mu = 1$ (see Table 5.3), the mean is

$$mean = \sum_{x=0} x \Pr\{X = x\}$$

$$= 0(0.3679) + 1(0.3679) + 2(0.1839) + 3(0.0613)$$
$$+ 4(0.0153) + 5(0.0031) + 6(0.0005) + 7(0.0001) = 1.0000.$$

The mean of the Poisson distribution is the same as μ, which is also the parameter of the Poisson distribution. It turns out that the *variance* of the Poisson distribution is also μ.

5.2.3 Finding Poisson Probabilities

A famous chemist and statistician, W. S. Gosset, worked for the Guinness Brewery in Dublin at the turn of the 20th century. Because Gosset did not wish his competitor breweries to learn of the potential application of his work for a brewery, he published his research under the pseudonym of Student. As part of his work, he studied the distribution of yeast cells over 400 squares of a hemacytometer, an instrument for the counting of cells (Student 1907). One of the four data sets he obtained is shown in Table 5.4.

Table 5.4 Observed frequency of yeast cells in 400 squares.

	X						
	0	1	2	3	4	5	6
Frequency	103	143	98	42	8	4	2
Proportion	0.258	0.358	0.245	0.105	0.020	0.010	0.005
Poisson Probability	0.267	0.352	0.233	0.103	0.034	0.009	0.002

Do these data follow a Poisson distribution? As we just said, the Poisson distribution is determined by the mean value that is unknown in this case. We can use the sample mean to estimate the population mean μ. The sample mean is the sum of all the observations divided by the number of observations — in this case, 400. The sum of the number of cells is

$$103(0) + 143(1) + 98(2) + 42(3) + 8(4) + 4(5) + 2(6) = 529.$$

The sample mean is then 529/400 = 1.3225. Thus, we can calculate the Poisson probabilities using the value of 1.3225 for the mean. Since the value of 1.3225 for μ is not in Appendix Table B3, we must use some other means of obtaining the probabilities. We can calculate them using the recursive relation just shown. We begin by finding the probability of squares with zero cells, $e^{-1.3225}$, which is 0.2665. The other probabilities are found from this value. Computer packages can be used to calculate Poisson probabilities (see **Program Note 5.1** on the website). The results of calculation are shown in the third row of Table 5.4. Based on the visual agreement of the actual and theoretical proportions (from the Poisson), we cannot rule out the Poisson distribution as the distribution of the cell counts. The Poisson distribution agreed quite well for three of the four replications of the 400 cells that Gosset performed.

One reason for interest in the distribution of data is that knowledge of the distribution can be used in future occurrences of this situation. If future data do not follow the previously observed distribution, this can alert us to a change in the process for generating the data. It could also indicate, for example, that the blood cell counts of a patient under study differ from those expected in a healthy population or that there are more occurrences of some disease than was expected assuming that the disease occurrence follows a Poisson distribution with parameter μ. If there are more cases of the disease, it may indicate that there is some common source of infection — for example, some exposure in the workplace or in the environment.

A method of visual inspection of whether or not the data could come from a Poisson distribution is the *Poissonness plot*, presented by Hoaglin (1980). The rationale for the plot is based on the Poisson probability mass distribution formula. If the data could come from a Poisson distribution, then a plot of the sum of the natural logarithm of the frequency of x and the natural logarithm of $x!$ against the value of x should be a straight line. Using a computer package (see **Program Note 5.2** on the website) with the data in Table 5.4, a Poissonness plot is created, as shown in Figure 5.5.

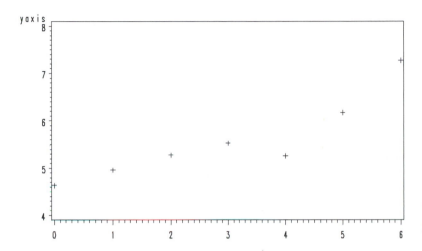

Figure 5.5   Poissonness plot for Gosset's data in Table 5.4.

The plot appears to be approximately a straight line, with the exception of a dip for $x = 4$. In Table 5.4, we see that the biggest discrepancy between the actual and theoretical proportions occurred when $x = 4$, confirmed by the Poissonness plot.

Example 5.2

Use of the Poisson Distribution: In 1986, there were 18 cases of pertussis reported in Harris County, Texas, from its estimated 1986 population of 2,942,550. The reported national rate of pertussis was 1.2 cases per 100,000 population (Harris County Health Department 1990). Do the Harris County data appear to be consistent with the national rate?

The data are inconsistent if there are too many or too few cases of pertussis compared to the national rate. This concern about both too few as well as too many adds a complication lacking in the binomial example in which we were concerned only about too many occurrences. Our method of answering the question is as follows.

First calculate the pertussis rate in Harris County. If the rate is above the national rate, find the probability of at least as many cases occurring as were observed. If the rate is below the national rate, find the probability of the observed number of cases or fewer occurring. To account for both too few as well as too many in our calculations, we double the calculated probability. Is the resultant probability large? If it is large, there is no evidence that the data are inconsistent with the national rate. If it is small, it is unlikely that the data are consistent with the national rate.

The rate of pertussis in Harris County was 0.61 cases per 100,000 population, less than the national rate. Therefore, we shall calculate the probability of 18 or fewer cases given the national rate of 1.2 cases per 100,000 population. The rate of 1.2 per 100,000 is multiplied by 29.4255 (the Harris County population of 2,942,550 divided by 100,000) to obtain the Poisson parameter for Harris County of 35.31. This value exceeds those listed in Table B3. Therefore, we can either find the probability of zero cases and use the recursive formula shown above or use the computer. Using a computer package (see **Program Note 5.1** on the website), the probability of 18 or fewer cases is found to be 0.001. Multiplying this value by 2 to account for the upper tail of the distribution gives a probability of 0.002, a very small value. It is therefore doubtful, since the probability is only 0.002, that the national rate of pertussis applies to Harris County.

This completes the introduction to the binomial and Poisson distributions. The following section introduces the normal probability distribution for continuous random variables.

5.3   The Normal Distribution

The *normal distribution* is also sometimes referred to as the *Gaussian distribution* after the German mathematician Carl Gauss (1777–1855). Gauss, perhaps the greatest mathematician who ever lived, demonstrated the importance of the normal distribution in describing errors in astronomical observations (published in 1809), and today it is the most widely used probability distribution in statistics. Recently, historians discovered that an American mine engineer, Adrian, used the similar distribution for random errors of measurements (published in 1808) (Stigler 1980). The normal distribution is so widely used because (1) it occurs naturally in many situations, (2) the sample means of many nonnormal distributions tend to follow it, and (3) it can serve as a good approximation to some nonnormal distributions.

5.3.1   Normal Probabilities

As we just mentioned, the probability distribution for a continuous random variable is usually expressed as a formula that can be used to find the probability that the continuous variable is within a specified interval. This differs from the probability distribution of a discrete variable that gives the probability of each possible outcome.

One reason why an interval is used with a continuous variable instead of considering each possible outcome is that there is really no interest in each distinct outcome. For example, when someone expresses an interest in knowing the probability that a male 45 to 54 years old weighs 160 pounds, exactly 160.000000000 . . . pounds is not what is intended. What the person intends is related to the precision of the scale used, and the person may actually mean 159.5 to 160.5 pounds. With a less precise scale, 160 pounds may mean a value between 155 and 165 pounds. Hence, the probability distribution of continuous random variables focuses on intervals rather than on exact values.

The probability density function (pdf) for a continuous random variable X is a formula that allows one to find the probability of X being in an interval. Just as the

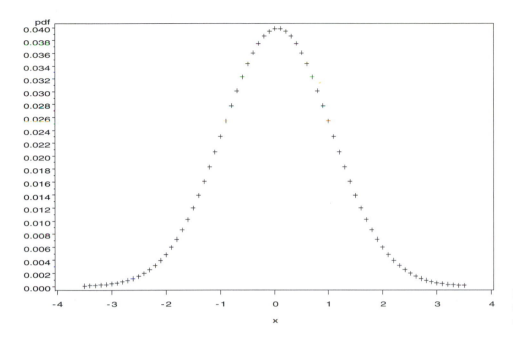

Figure 5.6   Pdf of standard normal distribution.

probability mass function for a discrete random variable could be graphed, the probability density function can also be graphed. Its graph is a curve such that the area under the curve sums to one, and the area between two points, x_1 and x_2, is equal to the probability that the random variable X is between x_1 and x_2.

The normal probability density function is

$$f(x) = \frac{1}{\sqrt{2\pi\sigma^2}} e^{-(x-\mu)^2/2\sigma^2}, \quad -\infty < x < \infty$$

where μ is the mean and σ is the standard deviation of the normal distribution, and π is a constant approximately equal to 3.14159. The normal probability density function is bell-shaped, as can be seen from Figure 5.6. It shows the standard normal density function — that is, the normal pdf with a mean of zero and a standard deviation of one — over the range of −3.5 to plus 3.5. The area under the curve is one and the probability of X being between any two points is equal to the area under the curve between those two points.

Figure 5.7 shows the effect of increasing σ from one to two on the normal pdf. The area under both of these curves again is one, and both curves are bell-shaped. The standard normal distribution has smaller variability, evidenced by more of the area being closer to zero, as it must, since its standard deviation is 50 percent of that of the other normal distribution. There is more area, or a greater probability of occurrence, under the second curve associated with values farther from the mean of zero than under the standard normal curve. The effect of increasing the standard deviation is to flatten the curve of the pdf, with a concomitant increase in the probability of more extreme values of X.

In Figure 5.8, two additional normal probability density functions are presented to show the effect of changing the mean. Increasing the mean by 3 units has simply shifted the entire pdf curve 3 units to the right. Hence, changing the mean shifts the curve to

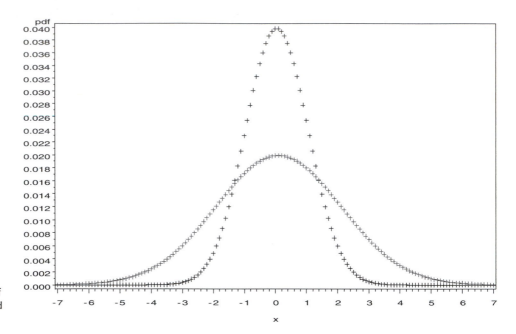

Figure 5.7 Normal pdf of $N(0, 1)$, in black, and $N(0, 2)$, in gray.

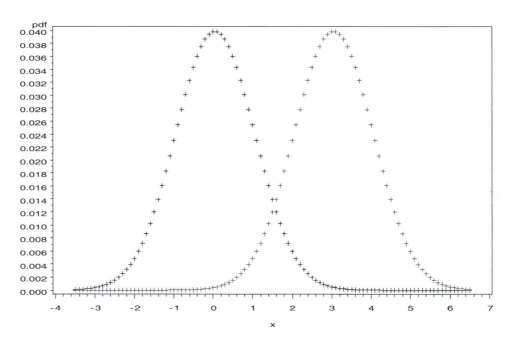

Figure 5.8 Normal pdf of $N(0, 1)$, in black, and $N(3, 1)$, in gray.

the right or left and changing the standard deviation increases or decreases the spread of the distribution.

5.3.2 Transforming to the Standard Normal Distribution

As can be seen from the normal pdf formula and the plots, there are two parameters, the mean and the standard deviation, that determine the location and spread of the normal curve. Hence, there are many normal distributions, just as there are many binomial and Poisson distributions. However, it is not necessary to have many pages of normal tables for each different normal distribution because all the normal distributions can be transformed to the standard normal distribution. Thus, only one normal table is needed, not many different ones.

Consider data from a normal distribution with a mean of μ and a standard deviation of σ. We wish to transform these data to the standard normal distribution that has a mean of zero and a standard deviation of one. The transformation has two steps. The first step is to subtract the mean, μ, from all the observations. In symbols, let y_i be equal to $(x_i - \mu)$. Then the mean of Y is μ_y, which equals

$$\mu_y = \sum \frac{x_i - \mu}{N} = \frac{\sum x_i - N\mu}{N} = \frac{N\mu - N\mu}{N} = 0.$$

The second step is to divide y_i by its standard deviation. Since we have subtracted a constant from the observations of X, the variance and standard deviation of Y is the same as that of X, as was shown in Chapter 3. That is, the standard deviation of Y is also σ. In symbols, let z_i be equal to y_i/σ. What are the mean and standard deviation of Z? The mean is still zero but the standard deviation of Z is one. This is due to the second property of the variance shown in Chapter 3 — namely, when all the observations are divided by a constant, the standard deviation is also divided by that constant. Therefore, the standard deviation of Z is found by dividing σ, the standard deviation of Y, by the constant σ. The value of this ratio is one.

Therefore, any variable, X, that follows a normal distribution with a mean of μ and a standard deviation of σ can be transformed to the standard normal distribution by subtracting μ from all the observations and dividing all the observed deviations by σ. The variable Z, defined as $(X - \mu)/\sigma$, follows the standard normal distribution. A symbol for indicating that a variable follows a particular distribution or is "distributed as" is the asymptote, $\sim$. For example, $Z \sim N(0, 1)$ means that Z follows a normal distribution with a mean of zero and a standard deviation of one. The observed value of a variable from a standard normal distribution tells how many standard deviations that value is from its mean of zero.

5.3.3   Calculation of Normal Probabilities

The cumulative distribution function of the standard normal distribution, denoted by $\Phi(z)$, represents the probability that the standard normal variable Z is less than or equal to the value z — that is, $\Pr\{Z \le z\}$. Table B4 presents the values of $\Phi(z)$ for values of z ranging from -3.79 to 3.79 in steps of 0.01. The table shows that the value of 0.9999 at $z = 3.79$, meaning that the probability of Z less than 3.79 is practically 1.0000. It also means that the area under the curve of pdf function shown in Figure 5.6 is 1.0000, a requirement for any probability distribution.

Figure 5.9 shows the cumulative distribution function for the standard normal distribution. The vertical axis gives the values of the probabilities corresponding to the values of z shown along the horizontal axis. The curve gradually increases from a probability of 0.0 for values of z around -3 to a probability of 0.5 when z is zero (as marked in Figure 5.9) and on to probabilities close to 1.0 for values of z of 3 or larger. We can calculate various probabilities associated with a normal distribution using its cdf without directly resorting to its pdf.

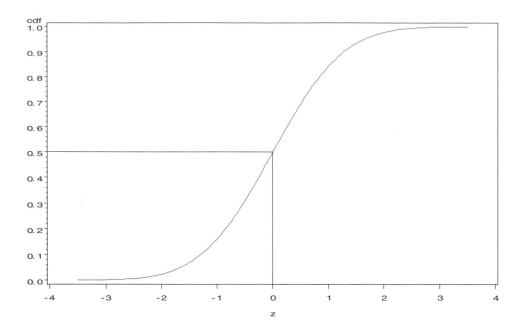

Figure 5.9   Cdf of the standard normal distribution.

Example 5.3

Probability of Being Greater than a Value: Suppose we wish to find the probability that an adult female will have a diastolic blood pressure value greater than 95 mmHg given that X, the diastolic blood pressure for adult females, follows the $N(80, 10)$ distribution. Since the values in Table B4 are for variables that follow the $N(0, 1)$ distribution, we first must transform the value of 95 to its corresponding Z value. To do this, we subtract the mean of 80 and divide by the standard deviation of 10. The z value of 95 mmHg, therefore, is

$$z = \frac{95 - 80}{10} = \frac{15}{10} = 1.5.$$

Thus, the value of the Z variable corresponding to 95 mmHg is 1.5, which means that the diastolic blood pressure of 95 is 1.5 standard deviations above its mean of 80. We now want the probability that Z is greater than 1.5. Using Table B4, look for 1.5 under the z heading and then go across the columns until reaching the .00 column. The probability of a standard normal variable being less than 1.5 is 0.9332. Thus, the probability of being greater than 1.5 is 0.0668 (= 1 − 0.9332).

Example 5.4

Calculation of the Value of the ith Percentile: Table B4 can be used to answer a slightly different question as well. Suppose that we wish to find the 95th percentile of the diastolic blood pressure variable for adult females — that is, the value such that 95 percent of adult females had a diastolic blood pressure less than it. We look in the body of the table until we find 0.9500. We find the corresponding value in the z column, and transform that value to the $N(80, 10)$ distribution. Examination of

Table B4 shows the value of 0.9495 when z is 1.64 and 0.9505 for a z of 1.65. There is no value of 0.9500 in the table. Since 0.9500 is exactly half way between 0.9495 and 0.9505, we shall use the value of 1.645 for the corresponding z. We now must transform this value to the $N(80, 10)$ distribution. This is easy to do since we know the relation between Z and X.

As $Z = (X - \mu)/\sigma$, on multiplication of both sides of the equation by σ, we have $\sigma Z = X - \mu$. If we add μ to both sides of the equation, we have $\sigma Z + \mu = X$. Therefore, we must multiply the value of 1.645 by 10, the value of σ, and add 80, the value of μ, to it to find the value of the 95th percentile. This value is 96.45 (= 16.45 + 80) mmHg.

This calculation can also be performed by computer packages (see **Program Note 5.3** on the website).

The percentiles of the standard normal distribution are used frequently, and, therefore, a shorthand notation for them has been developed. The ith percentile for the standard normal distribution is written as z_i — for example, $z_{0.95}$ is 1.645. From Table B4, we also see that $z_{0.90}$ is approximately 1.28 and $z_{0.975}$ is 1.96. By the symmetry of the normal distribution, we also know that $z_{0.10}$ is −1.28, $z_{0.05}$ is −1.645 and $z_{0.025}$ is −1.96.

The percentiles in theory could also be obtained from the graph of the cdf for the standard normal shown in Figure 5.9. For example, if the 90th percentile was desired, find the value of 0.90 on the vertical axis and draw a line parallel to the horizontal axis from it to the graph. Next, drop a line parallel to the vertical axis from that point down to the horizontal axis. The point where the line intersects the horizontal axis is the 90th percentile of the standard normal distribution.

Example 5.5

Probability Calculation for an Interval: Suppose that we wished to find the proportion of women whose diastolic blood pressure was between 75 and 90 mmHg. The first step in finding the proportion of women whose diastolic blood pressure is in this interval is to convert the values of 75 and 90 mmHg to the $N(0, 1)$ distribution. The value of 75 is transformed to an $N(0, 1)$ value by subtracting μ and dividing by σ — that is, $(75 - 80)/10$, which is −0.5, and 90 is converted to 1.0. We therefore must find the area under the standard normal curve between −0.5 and 1.0. Figure 5.10 aids our understanding of what is wanted. It also provides us with an idea of the probability's value. If the numerical value is not consistent with our idea of the value, perhaps we misused Appendix Table B4. From Figure 5.10 the area under the curve between $z = -0.5$ and $z = 1.0$ appears to be roughly $\frac{1}{2}$ of the total area.

One way of finding the area between −0.5 and 1.0 is to find the area under the curve less than or equal to 1.0 and to subtract from it the area under the curve less than or equal to −0.5. In symbols, this is

$$\Pr\{-0.5 < Z < 1.0\} = \Pr\{Z < 1.0\} - \Pr\{Z < -0.5\}.$$

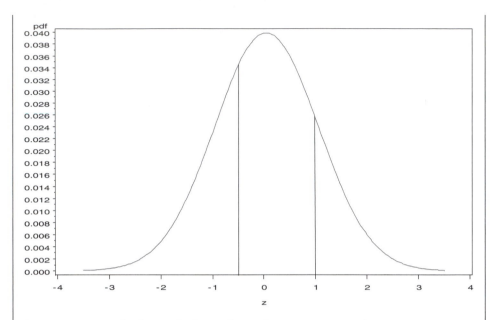

Figure 5.10   Area under the standard normal curve between $z = -0.5$ and $z = 1.0$.

From Table B4, we find that the area under the standard normal pdf curve less than or equal to 1.0 is 0.8413. The probability of a value less than or equal to -0.5 is 0.3085. Thus, the proportion of women whose diastolic blood pressure is between 75 and 90 mmHg is 0.5328 (= 0.8413 − 0.3085). Computer packages can be used to perform this calculation (see **Program Note 5.3** on the website).

5.3.4   The Normal Probability Plot

The normal probability plot provides a way of visually determining whether or not data might be normally distributed. This plot is based on the cdf of the standard normal distribution. Special graph paper, called normal probability paper, is used in the plotting of the points. The vertical axis of normal probability paper shows the values of the cdf of the standard normal. Table B4 shows the cdf values corresponding to z values of -3.79 to 3.79 in steps of 0.01, and it is not difficult to discover that that the increase in the cdf's value is not constant per a constant increase in z. It is more clearly shown in Figure 5.9. The vertical axis reflects this with very small changes in values of the cdf initially, then larger changes in the cdf's values in the middle of plot, and finally very small changes in the cdf's value. Numbers along the horizontal axis are in their natural units.

If a variable, X, is normally distributed, the plot of its cdf against X should be a straight line on normal probability paper. If the plot is not a straight line, then it suggests that X is not normally distributed. Since we do not know the distribution of X, we approximate its cdf in the following fashion.

We first sort the observed values of X from lowest to highest. Next we assign ranks to the observations from 1 for the lowest to n (the sample size) for the highest value. The ranks are divided by n and this gives an estimate of the cdf. This sample estimate is often called the *empirical distribution function*.

The points, determined by the values of the sample estimate of the cdf and the corresponding values of x, are plotted on normal probability paper. In practice, the ranks divided by the sample size are not used as the estimate of the cdf. Instead, the transformation, $(\text{rank} - 0.375)/(n + 0.25)$, is frequently used. One reason for this transformation is that the estimate of the cdf for the largest observation is now a value less than one, whereas the use of the ranks divided by n always results in a sample cdf value of one for the largest observation. A value less than one is desirable because it is highly unlikely that the selected sample actually contains the largest value in the population.

Example 5.6

We consider a small data set for vitamin A values from 33 boys shown in Table 5.5 and examine whether the data are normally distributed. An alternative to normal probability paper is the use of a computer (see **Program Note 5.4** on the website). Applying the probability plot option in a computer package to vitamin A data, Figure 5.11 is produced. The straight line helps to discern whether or not the data deviate from the normal distribution. The points in the plot do not appear to fall along a straight line. Therefore, it is doubtful that the vitamin A variable follows a normal distribution, a conclusion that we had previously reached in the discussion of symmetry in Chapter 3.

Table 5.5   Values of vitamin A, their ranks, and transformed ranks, $n = 33$.

Vit. A (IUs)	Rank	Trans.[a] Rank	Vit. A (IUs)	Rank	Trans. Rank	Vit. A (IUs)	Rank	Trans. Rank
820	1	0.0188	3747	12	0.3496	6754	23	0.6805
964	2	0.0489	4248	13	0.3797	6761	24	0.7105
1379	3	0.0789	4288	14	0.4098	8034	25	0.7406
1459	4	0.1090	4315	15	0.4398	8516	26	0.7707
1704	5	0.1391	4450	16	0.4699	8631	27	0.8008
1826	6	0.1692	4535	17	0.5000	8675	28	0.8308
1921	7	0.1992	4876	18	0.5301	9490	29	0.8609
2246	8	0.2293	5242	19	0.5602	9710	30	0.8910
2284	9	0.2594	5703	20	0.5902	10451	31	0.9211
2671	10	0.2895	5874	21	0.6203	12493	32	0.9511
2687	11	0.3195	6202	22	0.6504	12812	33	0.9812

Source: From dietary records of 33 boys[7]
[a]Transformed by $(\text{rank} - 0.375)/(n + 0.25)$

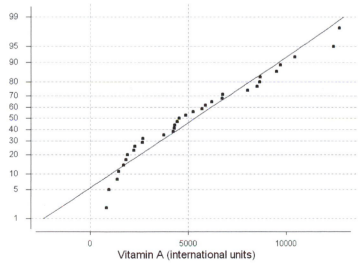

Figure 5.11   Normal probability plot of vitamin A.

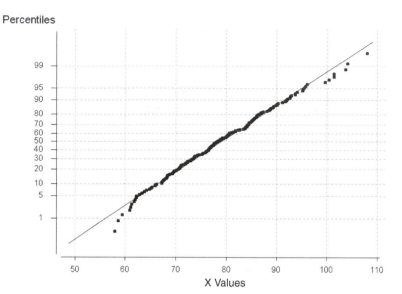

Figure 5.12 Probability plot of 200 observations from N (80, 10).

Let us now examine data from a normal distribution and see what its normality probability plot looks like. The example in Figure 5.12 uses 200 observations generated from an $N(80, 10)$ distribution. The plot looks like a straight line, but there are many points with the same normal scores. The points appear to fall mostly on a straight line as they should. The smallest observed value of X is slightly larger than expected if the data were perfectly normally distributed, but this deviation is relatively slight. Hence, based on this visual inspection, these data could come from a normal distribution.

It is difficult to determine visually whether or not data follow a normal distribution for small sample sizes unless the data deviate substantially from a normal distribution. As the sample size increases, one can have more confidence in the visual determination.

5.4   The Central Limit Theorem

As was just mentioned, one of the main reasons for the widespread use of the normal distribution is that the sample means of many nonnormal distributions tend to follow the normal distribution as the sample size increases. The formal statement of this is called the *central limit theorem*. Basically, for random samples of size n from some distribution with mean μ and standard deviation σ, the distribution of $\bar{x}$, the sample mean, is approximately $N(\mu, \sigma/\sqrt{n})$. This theorem applies for any distribution as long as μ and σ are defined. The approximation to normality improves as n increases.

The proof of this theorem is beyond the scope of this book and also unnecessary for our understanding. We shall, however, demonstrate that it holds for a very nonnormal distribution, the Poisson distribution with mean one.

Example 5.7

As seen Figure 5.4, the Poisson distribution with a mean of 1 is very nonnormal in appearance. The following demonstration consists of drawing a large number of samples — say, 100 — from this distribution, calculating the mean for each sample, and examining the sampling distribution of the sample means. We shall do this for samples of size 5, 10, and 20. Figure 5.13 shows three boxplots for each of these sample sizes. All three means are around 1, and the variances of the means are decreasing as the sample size increases.

As was just stated, the mean of the means should be 1, and the standard deviation of the means is the standard deviation divided by the square root of the sample size. It was also stated that the distribution of means should approach a normal distribution when the sample size is large. Figure 5.14 examines the case for $n = 20$. The mean is 1.003, which is very close to 1. The standard deviation is 0.2058, which is close to $0.2236(=1/\sqrt{20}$). The probability plot lines up around the straight line, suggesting that the distribution of the sample means does not differ substantially from normal distribution.

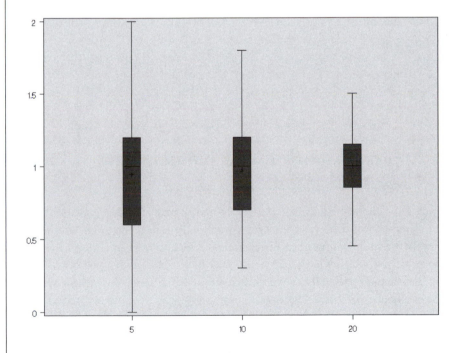

Figure 5.13   Boxplot of 100 sample means from Poisson ($\mu = 1$) for $n = 5$, 10, and 20.

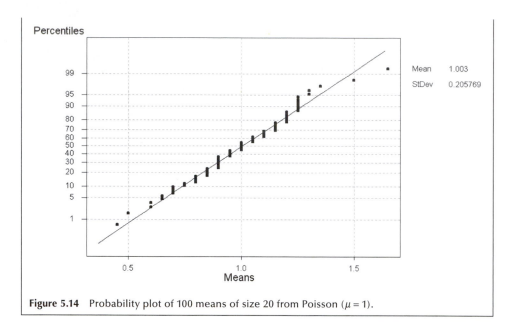

Figure 5.14 Probability plot of 100 means of size 20 from Poisson ($\mu = 1$).

Besides showing that the central limit theorem holds for one very nonnormal distributions, this demonstration also showed the effect of sample size on the estimate of the population mean. This example reinforces the idea that the mean from a very small sample may not be close to the population mean and does not warrant the use of the normal distribution. The idea of central limit theorem and sampling distribution plays a key role in referring from the sample to the population as will be discussed in subsequent chapters.

5.5 Approximations to the Binomial and Poisson Distributions

As we just said, another reason for the use of the normal distribution is that, under certain conditions, it provides a good approximation to some other distributions — in particular the binomial and Poisson distributions. This was more important in the past when there was not such a widespread availability of computer packages for calculating binomial and Poisson probabilities for parameter values far exceeding those shown in tables in most textbooks. However, it is still important today as computer packages have limitations in their ability to calculate binomial probabilities for large sample sizes or for extremely large values of the Poisson parameter. In the following sections, we show the use of the normal distribution as an approximation to the binomial and Poisson distributions.

5.5.1 Normal Approximation to the Binomial Distribution

In the plots of the binomial probability mass functions, we saw that as the binomial proportion approached 0.5, the plot began to look like the normal distribution (see Figure 5.3). This was true for sample sizes even as small as 10. Therefore, it is not surprising that the normal distribution can sometimes serve as a good approximation to the bino-

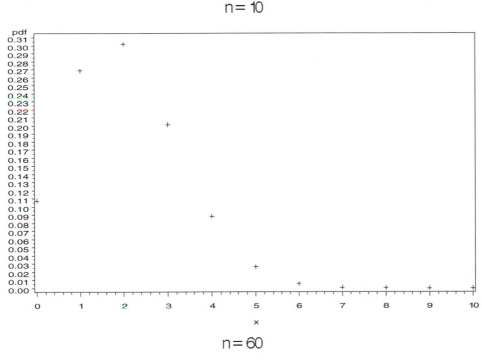

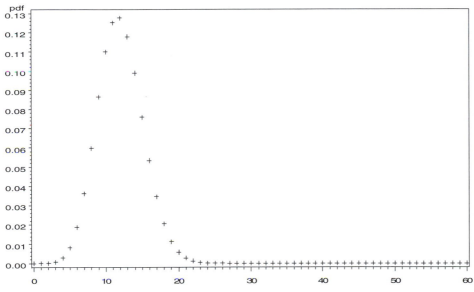

Figure 5.15 Binomial mass functions for $\pi = 0.2$ when $n = 10$ and $n = 60$.

mial distribution. Figure 5.15 demonstrates the effect of *n* on a binomial distribution, suggesting why we used the modifier *sometimes* in the preceding sentence.

Both plots in Figure 5.15 are based on $\pi = 0.2$. The first plot for $n = 10$ is skewed, and the normal approximation is not warranted. But the second plot for $n = 60$ is symmetric, and the normal distribution should provide a reasonable approximation here.

The central limit theorem provides a rationale for why the normal distribution can provide a good approximation to the binomial. In the binomial setting, there are two

Table 5.6 Sample size required for the normal distribution to serve as a good approximation to the binomial distribution as a function of the binomial proportion π.

π	0.05	0.10	0.15	0.20	0.25	0.30	0.35	0.40	0.45	0.50
n	440	180	100	60	43	32	23	15	11	10
Difference[a]	0.0041	0.0048	0.0054	0.0059	0.0059	0.0057	0.0059	0.0060	0.0049	0.0027
Mean diff.[b]	0.0010	0.0012	0.0013	0.0016	0.0016	0.0016	0.0016	0.0017	0.0016	0.0013

[a]Maximum difference between binomial probability and normal approximation
[b]Mean of absolute value of difference between binomial probability and normal approximation for all nonzero probabilities

outcomes — for example, disease and no disease. Let us assign the numbers 1 and 0 to the outcomes of disease and no disease, respectively. The sum of these numbers over the entire sample is the number of diseased persons in the sample. The mean, then, is simply the number of diseased sample persons divided by the sample size. And according to the central limit theorem, the sample mean should approximately follow a normal distribution as n increases. But if the sum of values divided by a constant approximately follows a normal distribution, the sum of the values itself also approximately follows a normal distribution. The sum of the values in this case is the binomial variable, and, hence, it also approximately follows the normal distribution.

Unfortunately, there is not a consensus as to when the normal approximation can be used — that is, when n is large enough for the central limit theorem to apply. This issue has been examined in a number of recent articles (Blyth and Still 1983; Samuels and Lu 1992; Schader and Schmid 1989). Based on work by Samuels and Lu (1992) and on some calculations we performed, Table 5.6 shows our recommendations for the size of the sample required as a function of π for the normal distribution to serve as a good approximation to the binomial distribution. Use of these sample sizes guarantees that the maximum difference between the binomial probability and its normal approximation is less than or equal to 0.0060 and that the average difference is less than 0.0017.

The mean and variance to be used in the normal approximation to the binomial are the mean and variance of the binomial, $n\pi$ and $n\pi(1 - \pi)$, respectively. Since we are using a continuous distribution to approximate a discrete distribution, we have to take this into account. We do this by using an interval to represent the integer. For example, the interval of 5.5 to 6.5 would be used with the continuous variable in place of the discrete variable value of 6. This adjustment is called the *correction for continuity.*

Example 5.8

We use the normal approximation to the binomial for the c-section deliveries example in Example 5.1. We wanted to find the probability of 22 or more c-section deliveries in a sample of 62 deliveries. The values of the binomial mean and variance, assuming that π is 0.235, are 14.57 (= 62 * 0.235) and 11.146 (= 62 * 0.235 * 0.765), respectively. The standard deviation of the binomial is then 3.339. Finding the probability of 22 or more c-sections for the discrete binomial variable is approximately equivalent to finding the probability that a normal variable with a mean of 14.57 and a standard deviation of 3.339 is greater than 21.5.

Before using the normal approximation, we must first check to see if the sample size of 62 is large enough. From Table 5.6, we see that since the assumed value of π

is between 0.20 and 0.25, our sample size is large enough. Therefore, it is okay to use the normal approximation to the binomial.

To find the probability of being greater than 21.5, we convert 21.5 to a standard normal value by subtracting the mean and dividing by the standard deviation. The corresponding z value is 2.075 (= [21.5 − 14.57]/3.339). Looking in Table B4, we find the probability of a standard normal variable being less than 2.075 is about 0.9810. Subtracting this value from one gives the value of 0.0190, very close to the exact binomial value of 0.0224 found in Example 5.1.

Example 5.9

According to data reported in Table 65 of *Health, United States, 1991* (NCHS 1992), 14.0% of high school seniors admitted that they used marijuana during the 30 days previous to a survey conducted in 1990. If this percentage applies to all seniors in high school, what is the probability that in a survey of 140 seniors, the number reporting use of marijuana will be between 15 and 25? We want to use the normal approximation to the binomial, but we must first check our sample size with Table 5.7. Since a sample of size 100 is required for a binomial proportion of 0.15, our sample of 140 for an assumed binomial proportion of 0.14 is large enough to use the normal approximation.

The mean of the binomial is 19.6 and the variance is 16.856 (= 140 ∗ 0.14 ∗ 0.86). Thus, the standard deviation is 4.106. These values are used in converting the values of 15 and 25 to z scores. Taking the continuity correction into account means that interval is really from 14.5 to 25.5.

We convert 14.5 and 25.5 to z scores by subtracting the mean of 19.6 and dividing by the standard deviation of 4.106. The z scores are −1.24 (= [14.5 − 19.6]/4.106) and 1.44 (= [25.5 − 19.6]/4.106). To find the probability of being between −1.24 and 1.44, we will first find the probability of being less than 1.44. From that, we will subtract the probability of being less than −1.24. This subtraction yields the probability of being in the interval.

These probabilities are found from Table B4 in the following manner. First, we read down the z column until we find the value of 1.44. We go across to the .00 column and read the value of 0.9251; this is the probability of a standard normal value being less than 1.44. The probability of being less than −1.24 is 0.1075. Subtracting 0.1075 from 0.9251 yields 0.8176. This is the probability that, out of a sample of 140, between 15 to 25 high school seniors would admit to using marijuana during the 30 days previous to the question being asked.

5.5.2   Normal Approximation to the Poisson Distribution

Since the Poisson tables do not show every possible value of the parameter μ, and since the tables and computer packages do not provide probabilities for extremely large values of μ, it is useful to be able to approximate the Poisson distribution. As can be seen from the preceding plots, the Poisson distribution does not look like a normal distribution for

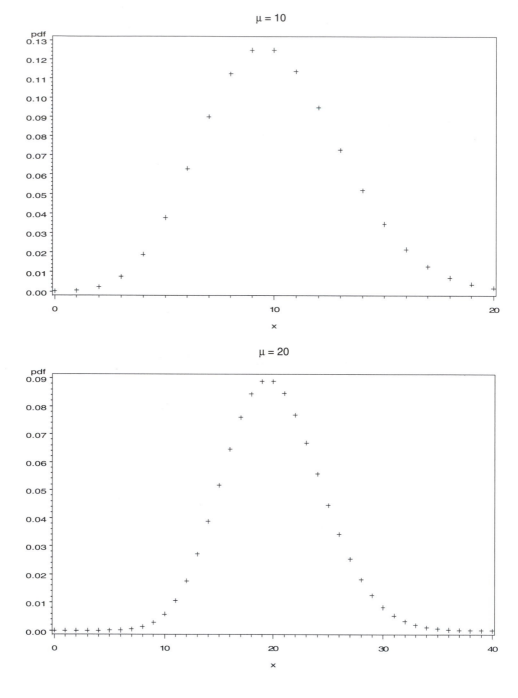

Figure 5.16   Poisson mass distributions for $\mu = 10$ and $\mu = 20$.

small values of μ. However, as the two plots in Figure 5.16 show, the Poisson does resemble the normal distribution for large values of μ. The first plot shows the probability mass function for the Poisson with a mean of 10 and the second plot shows the probability mass function for the Poisson distribution with a mean of 20.

As can be seen from these plots, the normal distribution should be a reasonable approximation to the Poisson distribution for values of μ greater than 10. The normal approximation to the Poisson uses the mean and variance from the Poisson distribution for the normal mean and variance.

Example 5.10

We use the preceding pertussis example to demonstrate the normal approximation to the Poisson distribution. In the pertussis example, we wanted to find the probability of 18 or fewer cases of pertussis, given that the mean of the Poisson distribution was 35.31. This value, 35.31, will be used for the mean of the normal and its square root, 5.942, for the standard deviation of the normal. Since we are using a continuous distribution to approximate a discrete one, we must use the continuity correction. Therefore, we want to find the probability of values less than 18.5. To do this, we convert 18.5 to a z value by subtracting the mean of 35.31 and dividing by the standard deviation of 5.942. The z value is -2.829. The probability of a Z variable being less than -2.829 or -2.83 is found from Table B4 to be 0.0023, close to the exact value of 0.001 given above.

Conclusion

Three of the more useful probability distributions — the binomial, the Poisson, and the normal — were introduced in this chapter. Examples of their use in describing data were provided. The examples also suggested that the distributions could be used to examine whether or not the data came from a particular population or some other population. This use will be explored in more depth in subsequent chapters on interval estimation and hypothesis testing.

EXERCISES

5.1 According to data from NHANES II (NCHS 1992), 26.8 percent of persons 20–74 years of age had high serum cholesterol values (greater than or equal to 240 mg/dL).

   a. In a sample of 20 persons ages 20–74, what is the probability that 8 or more persons had high serum cholesterol? Use Table B2 to approximate this value first and then provide a more accurate answer.

   b. How many persons out of the 20 would be required to have high cholesterol before you would think that the population from which your sample was drawn differs from the U.S. population of persons ages 20–74?

   c. In a sample of 200 persons ages 20–74, what is the probability that 80 or more persons had high serum cholesterol?

5.2 Based on reports from state health departments, there were 10.33 cases of tuberculosis per 100,000 population in the United States in 1990 (NCHS 1992). What is the probability of a health department, in a county of 50,000, observing 10 or more cases in 1990 if the U.S. rate held in the county? What is the probability of fewer than 3 cases if the U.S. rate held in the county?

5.3 Assume that systolic blood pressure for 5-year-old boys is normally distributed with a mean of 94 mmHg and a standard deviation of 11 mmHg. What is the probability of a 5-year-old boy having a blood pressure less than 70 mmHg? What is the probability that the blood pressure of a 5-year-old boy will be between 80 and 100 mmHg?

5.4 Less than 10 percet of the U.S. population is hospitalized in a typical year. However, the per capita hospital expenditure in the United States is generally large — for example, in 1990, it was approximately $975. Do you think that the expenditure for hospital care (at the person level) follows a normal distribution? Explain your answer.

5.5 In Harris County, Texas, in 1986, there were 173 cases of Hepatitis A in a population of 2,942,550 (HCHD 1990). The corresponding rate for the United States was 10.0 per 100,000 population. What is the probability of a rate as low as or lower than the Harris County rate if the U.S. rate held in Harris County?

5.6 Approximately 6.5 percent of women ages 30–49 were iron deficient based on data from NHANES II (LSRO 1989). In a sample of 30 women ages 30–49, 6 were found to be iron deficient. Is this result so extreme that you would want to investigate why the percentage is so high?

5.7 Based on data from the Hispanic Health and Nutrition Examination Survey (HHANES) (LSRO 1989), the mean serum cholesterol for Mexican-American males ages 20 to 74 was 203 mg/dL. The standard deviation was approximately 44 mg/dL. Assume that serum cholesterol follows a normal distribution. What is the probability that a Mexican-American male in the 20–74 age range has a serum cholesterol value greater than 240 mg/dL?

5.8 In 1988, 71% of 15- to 44-year-old U.S. women who have ever been married have used some form of contraception (NCHS 1992). What is the probability that, in a sample of 200 women in these childbearing years, fewer than 120 of them have used some form of contraception?

5.9 In ecology, the frequency distribution of the number of plants of a particular species in a square area is of interest. Skellam (1952) presented data on the number of plants of Plantago major present in squares of 100 square centimeters laid down in grassland. There were 400 squares and the numbers of plants in the squares are as follows:

Plants per Square	0	1	2	3	4	5	6	7
Frequency	235	81	43	18	9	6	4	4

Create a Poissonness plot to examine whether or not these data follow the Poisson distribution.

5.10 The Bruce treadmill test is used to assess exercise capacity in children and adults. Cumming, Everatt, and Hastman (1978) studied the distribution of the Bruce treadmill test endurance times in normal children. The mean endurance time for a sample of 36 girls 4–5 years old was 9.5 minutes with a standard deviation of 1.86 minutes. If we assume that these are the true population mean and standard deviation, and if we also assume that the endurance times follow a normal distribution, what is the probability of observing a 4-year-old girl with an endurance time of less than 7 minutes? The 36 values shown here are based on summary statistics from the research by Cumming et al. Do you believe that these data are normally distributed? Explain your answer.

Hypothetical Endurance Times in Minutes for 36 Girls 4 to 5 Years of Age											
5.3	6.5	7.0	7.2	7.5	8.0	8.0	8.0	8.0	8.2	8.5	8.5
8.8	8.8	8.9	9.0	9.0	9.0	9.0	9.5	9.8	9.8	10.0	10.0
10.6	10.8	11.0	11.2	11.2	11.3	11.5	11.5	12.2	12.4	12.7	13.3

5.11 Seventy-nine firefighters were exposed to burning polyvinyl chloride (PVC) in a warehouse fire in Plainfield, New Jersey, on March 20, 1985. A study was conducted in an attempt to determine whether or not there were short- and long-term respiratory effects of the PVC (Markowitz 1989). At the long-term follow-up visit at 22 months after the exposure, 64 firefighters who had been exposed during the fire and 22 firefighters who were not exposed reported on the presence of various respiratory conditions. Eleven of the PVC exposed firefighters had moderate to severe shortness of breath compared to only 1 of the nonexposed firefighters.

What is the probability of finding 11 or more of the 64 exposed firefighters reporting moderate to severe shortness of breath if the rate of moderate to severe shortness of breath is 1 case per 22 persons? What are two possible confounding variables in this study that could affect the interpretation of the results?

REFERENCES

Blyth, C. R., and H. A. Still. "Binomial Confidence Intervals." *Journal of the American Statistical Association* 78:108–116, 1983.

Boyer, C. B. *A History of Mathematics.* Princeton University Press, 1985, p. 569.

Cumming, G. R., D. Everatt, and L. Hastman. "Bruce Treadmill Test in Children: Normal Values in a Clinic Population." *The American Journal of CARDIOLOGY* 41:69–75, 1978.

Harris County Health Department (HCHD), Mark Canfield, editor. *The Health Status of Harris County Residents: Births, Deaths and Selected Measures of Public Health, 1980–86,* 1990.

Hoaglin, D. C. "A Poissonness Plot." *The American Statistician* 34:146–149, 1980.

Life Sciences Research Office (LSRO), Federation of American Societies for Experimental Biology: *Nutrition Monitoring in the United States — An Update Report on Nutrition Monitoring.* Prepared for the U.S. Department of Agriculture and the U.S. Department of Health and Human Services. DHHS Pub. No. (PHS) 89–1255, 1989.

Markowitz, J. S. "Self-Reported Short- and Long-Term Respiratory Effects among PVC-Exposed Firefighters." *Archives of Environmental Health* 44:30–33, 1989.

McPherson, R. S., M. Z. Nichaman, H. W. Kohl, D. B. Reed, and D. R. Labarthe. "Intake and food sources of dietary fat among schoolchildren in The Woodlands, Texas." *Pediatrics* **88**(4): 520–526, 1990.

National Center for Health Statistics. *Health, United States, 1991 and Prevention Profile.* Hyattsville, MD: Public Health Service, DHHS Pub. No. 92–1232, 1992, Tables 50 and 70.

Public Citizen Health Research Group. "Unnecessary Cesarean Sections: Halting a National Epidemic." *Public Citizen Health Research Group Health Letter* 8(6):1–6, 1992.

Samuels, M. L., and T. C. Lu. "Sample Size Requirements for the Back-of-the-Envelope Binomial Confidence Interval." *The American Statistician* 46:228–231, 1992.

Schader, M., and F. Schmid. "Two Rules of Thumb for the Approximation of the Binomial Distribution by the Normal Distribution." *The American Statistician* 43:23–24, 1989.

Skellam, J. G. "Studies in Statistical Ecology." *Biometrika* 39:346–362, 1952.

Stigler, S. M. *American Contributions to Mathematical Statistics in the Nineteenth Century.* New York: Arno Press, 1980.

Student. "On the Error of Counting with a Haemacytometer." *Biometrika* 5:351–360, 1907.

Study Designs

6

In meeting a set of data we must first check the credentials of the data: what the data represent and how the data were collected. In Chapter 2 we discussed the linkage between concepts and numbers — that is, what the data represent. As far as data collection is concerned, there are two basic methods used to obtain data: the sample survey and the designed experiment. In this chapter we examine these two basic methods and some variations of them.

6.1   Design: Putting Chance to Work

In collecting sample data we should try to avoid all potential causes of bias. When conducting an experiment, we should try to eliminate the effect of potential confounding factors. Strangely, adhering to these ideas involves the use of a chance mechanism. Let us explore why and how a chance mechanism plays a role in designing surveys and experiments.

A smart shopper is conscious of the possible variability in the quality of fruit between the top and the bottom of the fruit basket. The smart shopper looks at pieces of fruit throughout the basket, even though it is more convenient to look only at the pieces on top, before making a purchase. In the same way, a researcher is aware of the possible variability among observational units in the population. A good researcher takes steps to ensure that the process for selecting units from the population deals with this possible variability. The failure to take adequate steps would introduce a selection bias. Selecting a sample of units because of convenience also poses a problem for a researcher just as it did for the shopper. For example, the opinions of people interviewed during lunchtime on downtown street corners, although convenient to obtain, usually are not representative of the residents of the city. Those who never go to the center of the city during lunchtime are not represented in the sample, and they may have different opinions from those who go to the city center.

We are familiar with the use of a chance mechanism to remove possible biases. For example, to start a football game, a coin toss — a chance mechanism — is used in deciding which team receives the opening kickoff. The use of a chance mechanism is also involved in selecting a sample in an attempt to avoid biases. One method of drawing a fair sample is to place numbered slips of paper in a bowl, mix them up thoroughly, and then have a neutral party pick out the slips. This chance mechanism sounds fair but may not be satisfactory, as shown the following example.

Example 6.1

In 1970, Selective Service officials used a chance mechanism, a lottery, to determine who would be drafted for military service. Officials put slips of paper representing birthdates into cylindrical capsules, one birthdate per capsule, and then placed the capsules into a box. The January birthdates were put into the box first and pushed to one side, and then the February capsules were placed in the box and pushed to the side of the box with the January capsules, and then so on with March. The box was then closed, shaken several times, carried up three flights of stairs, and carried back down to the room, where the capsules were poured into a bowl. A public figure then selected the capsules to determine the order of drafting men. Figure 6.1 shows the lottery results (Fienberg 1971). It appears that the process did not work as intended, since the months at the end of the year, which were put into the container last and were not mixed thoroughly, have much smaller lottery numbers than the earlier months.

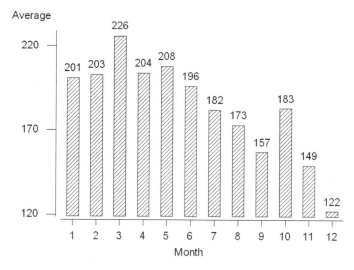

Figure 6.1 Average lottery number by month from the 1970 draft lottery.

A better way of selecting a fair sample is using random numbers that were used to estimate probabilities in Chapter 4. The random numbers can be described as the sequence of numbers we get when we draw balls numbered 0, 1, . . . 9 from an urn, replacing the ball drawn, thoroughly remixing the balls, and then drawing another ball. This process is repeated several times.

The first random numbers were produced by Tippett in 1927. It is said that Tippett obtained the numbers from the figures of areas of parishes given in the British census returns, and omitted the first two and last two digits in each figure of area. The truncated numbers were arranged in sets of four in eight columns. This 26-page book containing

Table 6.1   The first 10 rows from page 14 of Tippett's random numbers.

7816	6572	0802	6314	0702	4369	9728	0198
3204	9243	4935	8200	3623	4869	6938	7481
2976	3413	2841	4241	2424	1985	9313	2322
8303	9822	5888	2410	1158	2729	6443	2943
5556	8526	6166	8231	2438	8455	4618	4445
2635	7900	3370	9160	1620	3882	7757	4950
3211	4919	7306	4916	7677	8733	9974	6732
2748	6198	7164	4148	7086	2888	8519	1620
7477	0111	1630	2404	2979	7991	9683	5125
5379	7076	2694	2927	4399	5519	8106	8501

41,600 digits became the best-seller among technical books. Table 6.1 shows the first 10 rows of page 14 of his random number table and contains 320 digits. The appearance of each digit is random in the sense that we cannot predict the appearance of a particular digit based on the previous sequence of digits. Despite this uncertainty, we can expect that each digit is equally likely to appear. From Table 6.1, the frequencies of each digit from 0 to 9 are 27, 33, 39, 32, 36, 22, 31, 31, 35, and 34, which slightly deviates from the expect frequency (10 percent of 320).

Now random numbers are generated by computer algorithms, and most statistical software packages contain a random number generator. Table B1 in Appendix B show 1000 random digits generated from MINITAB. Considerable research is still devoted to random number generation. No definition of random numbers exists except for a vague description that they do not follow any particular pattern. The use of random numbers helps reduce the possibility of selection bias in surveys and also helps reduce the possible effect of confounders when designing experiments.

6.2   Sample Surveys and Experiments

There are many similarities as well as some differences between sample surveys and experiments. We learn the characteristics of some population from sample surveys. The sample survey focuses on the selection of individuals from the population. We discover the effect of applying a stimulus to subjects from experiments. The experimental design focuses on the formation of comparison groups that allow conclusions about the effect of the stimulus to be drawn.

As emphasized in Chapter 4, a probability sample is a carefully drawn blueprint or design as is an experiment. The blueprint or design of a survey or an experiment is based on both statistical and substantive considerations. An experiment is different from a sample survey in that the experimenter actively intervenes with the experimental subjects through the assignment of the subjects to groups, whereas the survey researcher passively observes or records responses of the survey subjects. Experiments and surveys often have different goals as well.

In a survey, the primary goal is to describe the population, and a secondary goal is to investigate the association between variables. In a survey, variables are usually not referred to as independent or dependent because all the variables can be viewed as being response variables. The survey researcher usually has not manipulated the levels of any of the variables as the experimenter does.

The goal in an experiment is to determine whether or not there is an association between the independent or predictor variables with the dependent or response variable. The different groups to which the subjects are assigned usually represent the levels of the independent variable(s). *Independent* and *dependent* were chosen as names for the variable types because it was thought that the response variable depended on the levels of the predictor variables. To determine whether or not there is an association, the experimenter assigns subjects to different levels of one or more variables — for example, to different doses of some medication. The effects of the different levels — the different doses — are found by measuring the values of an outcome variable — for example, change in blood pressure. An association exists if there is relationship between the change in blood pressure values and the dosage levels. Let us examine first how surveys are designed and then consider the basic principles of experimental design.

6.3 Sampling and Sample Designs

Sampling means selecting a few units from all the possible observational units in the population. For practical purposes, any data set is a sample. Even if a complete census is attempted, there are missing observations. This means that we must pay attention to the intended as well as the unintended sampling when evaluating a sample. This also suggests that we cannot evaluate a sample by looking at the sample itself, but we need to know what sampling method was used and how well it was executed. We are interested in the process of selection as well as the sample obtained.

Sampling is used extensively today for many reasons. In many situations a sample produces more accurate information about the population than that provided by a census. Two reasons for obtaining more accurate information from a sample are the following. As was just mentioned, a census often turns out to be incomplete, and the impact of the missing information is most often unknown. Additionally, in obtaining a sample, fewer interviewers are required, and it is likely that they will be better trained than the huge team of interviewers required when conducting a census.

Even more pragmatically, collecting data from a sample is cheaper and faster than attempting a complete census. In addition, in many situations a census is impractical or even impossible. The following three examples will illustrate situations in which sampling was used and reasons for the use of samples.

Example 6.2

Even in the U.S. Population Census, many data items are collected from a sample of households. In the 2000 Census, for example, only a few basic demographic data items — gender, age, race, and marital status — were asked from each individual in all households in the short form of the questionnaire. Many questions about socio-economic characteristics such as education, income, and occupation are included in the long form of the questionnaire that was distributed to about 17 percent of U.S. households. In small towns, a larger proportion of households received the long form to ensure reliable estimates. Conversely, in large cities, proportionately fewer households received the long form. Use of sampling not only reduced the cost of the census, but also shortened the data collection burden and time.

Example 6.3

Pharmaceutical companies routinely sample a small fraction of their products to examine the quality and the chemical contents. On the basis of this examination, a decision is made whether to accept the entire lot and ship them or reject the lot and change the manufacturing process. In this case the sample is destroyed to check the quality and, therefore, a company cannot afford to inspect the entire lot.

Example 6.4

Health departments of large urban areas monitor ambient air quality. Since the health department cannot afford to monitor the air everywhere in its coverage area, sample sites are selected and the values of several different pollutants are continuously recorded.

6.3.1   Sampling Frame

Before performing any sampling, it is important to define clearly the population of interest. Similarly, when we are given a set of data, we need to know what group the sample represents — that is, to know from what population the data were collected. The definition of population is often implicit and assumed to be known, but we should ask what the population was before using the data or accepting the information. When we read an election poll, we should know whether the population was all adults or all registered voters to interpret the results appropriately. In practice, the population is defined by specifying the *sampling frame*, the list of units from which the sample was selected. Ideally, the sampling frame must include all units of the defined population. But as we shall see, it is often difficult to obtain the sampling frame and we need to rely on a variety of alternative approaches.

The failure to include all units contained in the defined population in the sampling frame leads to selecting a biased sample. A biased sample is not representative of the population. The average of a variable obtained from a biased sample is likely to be consistently different from the corresponding value in the population. *Selection bias* is the consistent divergence of a sample value (*statistic*) from the corresponding population value (*parameter*) due to an improper selection process. Even with a complete sampling frame, selection bias can occur if proper selection rules were not followed. Two basic sources of selection bias are the use of an incomplete sampling frame and the use of improper selection procedures. The following example illustrates the importance of the sampling frame.

Example 6.5

The *Report of the Second Task Force on Blood Pressure Control in Children* (1987) provides an example of the possibility of selection bias in data. This Task Force used existing data from several studies, only one of which could be considered representative of the U.S. noninstitutionalized population. In this convenience sample, over 70 percent of the data came from Texas, Louisiana, and South Carolina, with little data from the Northeast or the West. Data from England were also used for newborns and children up to three years of age. The representativeness of these data for use in the creation of blood pressure standards for U.S. children is questionable. Unlike the *Literary Digest* survey in which the errors in the sampling were shown to lead to a wrong conclusion, it is not clear that the blood pressure standards are wrong. All we can point to is the use of convenience sampling, and with it, the likely introduction of selection bias by the Second Task Force.

Example 6.6

Telephone surveys may provide another example of the sampling frame failing to include all the members of the target population. If the target population is all the resident households in a geographical area, a survey conducted using the telephone will miss a portion of the resident households. Even though more than 90 percent of the households in the U.S. have telephones, the percentage varies with race and socioeconomic status. The telephone directory was used frequently in the past as the sampling frame, but it excluded households without telephones as well as households with unlisted numbers. A technique called *random digit dialing* (RDD) has been developed to deal with the unlisted number problem in an efficient manner (Waksberg 1978). As the name implies, telephone numbers are basically selected at random from the prefixes — the first 3 digits — thought to contain residential numbers, instead of being selected from a telephone directory. But the concern about the possible selection bias due to missing households without telephones and people who do not have a stable place of residence remains.

In order to avoid or minimize selection bias, every sample needs to be selected based on a carefully drawn sample design. The design defines the population the sample is supposed to represent, identifies the sampling frame from which the sample is to be selected, and specifies the procedural rules for selecting units. The sample data are then evaluated based on the sample design and the way the design was actually executed.

6.3.2   Importance of Probability Sampling

Any sample selected using a random mechanism that results in known chances of selection of the observational units is called a random or a probability sample. This definition requires only that the chances of selection are known. It does not require that the chances of the observational units being selected into the sample are equal. Knowledge of the chance of selection is the basis for the statistical inference from the sample to the population. A sample selected with unknown chances of selection cannot be linked appropriately to the population from which the sample was drawn. This point was explained in Chapter 4. Various sampling designs are discussed in the following sections starting with simple random sampling.

6.3.3   Simple Random Sampling

The simplest probability sample is a *simple random sample* (SRS). In an SRS, each unit in the sampling frame has the same chance of being included in the sample as any other unit. The use of an SRS removes the possibility of any bias, conscious or unconscious, on the part of the researcher in selecting the sample from the sampling frame. An SRS is drawn by the use of a random number table or random numbers generated by a computer. If the population is relatively small, we can number all units sequentially. Next we locate a starting point in the random number table, Table B1 in Appendix B. We then begin reading random numbers in some systematic fashion — for example, across a row or down a column or diagonal — but the direction of reading should be decided ahead of time. The units in the sampling frame whose unique numbers match the random numbers that have been read are selected into the sample.

Example 6.7

Suppose that we have 50 students in a classroom and they are sequentially labeled from 00 to 49 by row starting at the left end of the first row. We wish to select an SRS of 10 students. We decide to use the left-hand corner of line 1 of Table B1 as our starting point, and we will go across the row. By reading the two-digit numbers from the first row of the random digit table, the following 10 numbers are obtained:

$$17, \underline{17}, \underline{47}, 59, \underline{08}, \underline{43}, \underline{30}, 67, 70, 61$$

Since four numbers are greater than 49, they cannot be used, and we must draw additional numbers until we have 10 random numbers smaller than 50. In addition, the number 17 occurred twice. Since there is no good practical reason for including the same element twice in the sample, we should draw another number that has not been selected previously. We usually sample without replacement, as mentioned in Chapter 4. The next five valid numbers are 07, 44, 48, 36, and 47. Since the number 47 is already used, the next valid number 24 is drawn. The students whose labels match the 10 valid numbers drawn are selected as the sample.

Example 6.8

One way of dealing with the problem of drawing invalid numbers is to subtract 50 from values greater than or equal to 50 in the first set of 10 random numbers. For example, 59, 67, 70, and 61 become 09, 17, 20, and 11. We now select the students with labels 09, 17, 20, and 11. This procedure is based on the premise that each student is represented by two numbers differing by 50 in value. For example, the first student will be selected if either 00 or 50 were read, the second would be selected if either 01 or 51 were read, and so on until the last student would be selected if 49 or 99 were read. Note that even with the subtraction of 50, we again have another 17. We would still have to draw two more valid random numbers: 25 and 02 (obtained by subtracting 50 from 75 and 52) to have 10 distinct values.

In using the procedure in Example 6.8, each unit (student) in the sampling frame had the same number (two) of labels associated with it. If there are 30 students in a class, we can label them in three cycles, 1 through 30, 31 through 60, and 61 through 90, but we cannot assign 91 through 99 and 00 to any student. If we assigned these last 10 values to some of the students, some students would have three labels associated with them, whereas other students would have four labels. The students would have unequal chances of being selected. By not using the last 10 values, each student has three labels (numbers). The first student is assigned the numbers 01, 31, and 61, and the second student is assigned the numbers 02, 32, and 62, and so on for the other students.

In Examples 6.7 and 6.8, we used two-digit random numbers because we could not provide distinct labels for all 50 students with only a single digit. The number of digits to be used is dependent on the size of the population under consideration. For example, when we have 570 units in the population, we need to use three digits. A population that contains 7870 units would require four-digit random numbers.

The SRS design is modified to accommodate other theoretical and practical considerations. The common practical methods for selecting a sample include systematic sampling, stratified random sampling, single-stage cluster sampling, multistage cluster sampling, PPS (probability proportional to size) sampling, and other controlled selection procedures. These more practical designs deviate from SRS in two important ways. First, the inclusion probabilities for the elements (also the joint inclusion probabilities for sets for the elements) may be unequal. Second, the sampling unit(s) can be different from the population element of interest. These departures complicate the usual methods of estimation and variance calculation and, if no adjustments are made, can lead to a bias in estimation and statistical tests. We will consider these departures in detail, using several specific sampling designs, and examine their implications for survey analysis.

Computer packages can be used to draw random samples (see **Program Note 6.1** on the website).

6.3.4 Systematic Sampling

Systematic sampling is commonly used as an alternative to SRS because of its simplicity. It selects every kth element after a random start. Its procedural tasks are simple, and the process can easily be checked, whereas it is difficult to verify SRS by examining the results. It is often used in the final stage of multistage sampling when the field worker is instructed to select a predetermined proportion of units from the listing of dwellings in a street block. The systematic sampling procedure assigns each element in a population the same probability of being selected. This assures that the sample mean will be an unbiased estimate of the population mean when the number of elements in the population (N) is equal to k times the number of elements in the sample (n). If N is not exactly nk, then the equal probability is not guaranteed, although this problem can be ignored when N is large. When N is not exactly nk, we can use the *circular systematic sampling scheme*. In this scheme, the random starting point is selected between 1 and N (any element can be the starting point) and every kth element is selected assuming that the frame is circular (i.e., the end of list is connected to the beginning of the list).

Example 6.9

Suppose that we are taking a 1-in-4 systematic sample from a population of 11 elements: A, B, C, D, E, F, G, H, I, J, and K. Four possible samples can be drawn using the ordinary systematic sampling scheme and 11 possible samples using the circular systematic sampling. The possible samples and their selection probabilities using the ordinary systematic sampling and circular systematic sampling are shown in Table 6.2.

Ordinary systematic sampling does not guarantee equal probability sampling. For example, here the fourth sample has a different selection probability. Under the circular systematic sampling, each element can be a starting point and equal probability sampling is guaranteed in this scheme.

Table 6.2   Possible samples and selection probabilities taking 1-in-4 systematic samples from $N = 11$, using two different selection schemes.

Ordinary Systematic Sampling				Circular Systematic Sampling				
Samples			Selection Probability		Samples			Selection Probability
1.	A	E	I	3/11	1.	A	E	I	3/11
2.	B	F	J	3/11	2.	B	F	J	3/11
3.	C	G	K	3/11	3.	C	G	K	3/11
4.	D	H		2/11	4	D	H	A	3/11
					5.	E	I	B	3/11
					6.	F	J	C	3/11
					7.	G	K	D	3/11
					8.	H	A	E	3/11
					9.	I	B	F	3/11
					10.	J	C	G	3/11
					11.	K	D	H	3/11

Systematic sampling is convenient to use, but it can give an unrealistic estimate when the elements in the frame are listed in a cyclical manner with respect to a survey variable and the selection interval coincides with the listing cycle. For example, if one selects every 40th patient coming to a clinic and the average daily patient load is about 40, then the resulting sample would contain only those who came to the clinic at a certain time of the day. Such a sample may not be representative of the clinic patients. Moreover, even when the listing is randomly ordered, unlike SRS, different sets of elements may have unequal inclusion probabilities. For example, the probability of including both the ith and $(i + k)$th element is $1/k$ in a systematic sample, whereas the probability of including both the ith and $(i + k + 1)$th element is zero. This situation complicates the variance calculation.

Another way of viewing systematic sampling is that it is equivalent to selecting one cluster from k systematically formed clusters of n elements each. The sampling variance (between clusters) cannot be estimated from the one cluster selected. Thus, variance estimation from a systematic sample requires special strategies.

A modification to overcome these problems with systematic sampling is the so-called *repeated systematic sampling*. Instead of taking a systematic sample in one pass through the list, several smaller systematic samples are selected going down the list several times with a new starting point in each pass. This procedure not only guards against possible periodicity in the frame but also allows variance estimation directly from the data. The variance of an estimate from all subsamples can be estimated from the variability of the separate estimates from each subsample.

6.3.5   Stratified Random Sampling

Stratification is often used in complex sample designs. In a *stratified random sample* design, the units in the sampling frame are first divided into groups, called *strata*, and a separate SRS is taken in each stratum to form the total sample. The strata are formed to keep similar units together — for example, a female stratum and a male stratum. In this design, units need not have equal chances of being selected and some strata may be deliberately oversampled. For example, in the first National Health and Nutrition Examination Survey (NHANES I), the elderly, persons in poverty areas, and women of childbearing age were oversampled to provide sufficient numbers of these groups for in-depth analysis (NCHS 1973). If an SRS had been used, it is likely that too few people in these groups would have been selected to allow any in-depth analysis of these groups.

Another advantage of stratification is that it can reduce the variability of sample statistics over that of an SRS, thus reducing the sample size required for analysis. This reduction in variability occurs when the units in a stratum are similar, but there is variation across strata. Another way of saying this is that the reduction occurs when the variable used to form the strata is related to the variable being measured. Let us consider a small example that illustrates this point.

In this example, we wish to estimate the average weight of persons in the population. The population contains six persons: three females and three males. The weights of the females in the population are 110, 120, and 130 pounds, and the weights of the males are 160, 170, and 180 pounds. We shall form our estimate of the population average weight by taking a sample of size two without replacement.

If we use an SRS, the smallest possible estimate is 115 pounds (= [110 + 120]/2), and the largest possible estimate is 175 (= [170 + 180]/2). As an alternative, we could use a stratified random sample where the strata are formed based on gender. If one person is randomly selected from each stratum, the smallest estimate is 135 pounds (= [110 + 160]/2), and the largest estimate is 155 pounds (= [130 + 180]/2). The estimates from the stratified sample approach have less variation — that is, have greater precision than the SRS approach in this case.

The formulation of the strata requires that information on the stratification variables be available for the elements in the sampling frame. When such information is not available, stratification is not possible, but we still can take advantage of stratification by using the poststratification method. For example, stratification by race is usually desirable in social surveys but the racial identification is often not available in the sampling frame. In this case we can attempt to take race into account in the analysis after the sample is selected. Chapter 15 will provide further discussion on this topic.

6.3.6   Cluster Sampling

Most of the methods of statistical analysis assume that the data were collected using an SRS. However, when we attempt to use an SRS in the collection of data, we often encounter difficulties. Suppose we wanted an SRS of 500 adults from a large city. First, a sampling frame is not readily available. Developing a list of all adults in the city is very costly and should be considered impractical. Even though we are able to select an SRS of 500 adults from a reasonably complete list, it would be expensive to send interviewers to sample persons scattered all over the city. A solution to these practical difficulties is to sample people based on geographical areas — for example, census tracts. Most survey agencies and researchers use a *multistage cluster sample design* in this situation. First, a random sample of census tracts is selected and then neighborhood blocks within each selected tract are randomly selected. Within the selected neighborhood blocks a list of households can be prepared and a sample of households can be selected systematically from the list — say, every third household. Finally, within each of the selected households, an adult may be randomly chosen. In this example, the census tracks, neighborhood blocks, and households are the clusters used as the sampling units.

Cluster sampling is widely used but it complicates statistical estimation and analysis, since the sampling method deviates from SRS. For example, an SRS of unequal-sized clusters leads to the elements in the smaller clusters being more likely to be in the sample than those in the larger clusters. Such complications are handled either by using a special selection method or by a special analytical method. We will discuss these methods in Chapter 15.

6.3.7   Problems Due to Unintended Sampling

In analyzing data it is imperative to understand the sample design, as well as how the design was actually executed in the field. Deviations from the intended sample design are reflected in the data. Even in a well-designed survey, it is usually not possible to collect data from all the units sampled because there is almost always some nonresponse. Hence, the respondents, a subset of the sampled persons, are self-selected from the sampled persons through some procedure which is usually unknown to the designer of the study. Since the respondents are no longer a random sample of the study population, there is concern that the data may be unusable because of nonresponse bias.

If the percentage of nonresponse is small — say, less than 5 to 10 percent — there is usually little concern because the bias, if any, is also likely to be small. If the nonresponse is on the order of 20 to 30 percent, the possibility of a substantial bias exists. For example, assume that we wish to estimate the proportion of people without health insurance in our community. We select an SRS and find that 20 percent of the respondents were without health insurance. However, 1/4 of those selected to be in the sample did not respond. If we knew the proportion of those without health insurance among the nonrespondents, it would be easy to combine this value with that of the respondents to obtain the total sample estimate. The proportions of those without health insurance among the respondents and nonrespondents would be weighted by the corresponding proportion of respondents and nonrespondents in the sample.

For example, if none of these nonrespondents had health insurance, the total sample estimate would be 40 percent (= {20% × 0.75} + {100% × 0.25}), twice as large as the rate for the respondents only. If all of the nonrespondents had health insurance, then the total sample estimate becomes 15 percent (= {20% × 0.75} + {0% × 0.25}). Hence, although 20 percent of the respondents were without health insurance, the total sample estimate can range from 15 to 40 percent when 1/4 of the sample are nonrespondents.

For nonresponse bias to occur, the nonrespondents must differ from the respondents with regard to the variable of interest. In the preceding example, it may be that many of the nonrespondents were unemployed homeless whereas few of the respondents were unemployed or homeless. In this case, the respondents and nonrespondents would likely differ with regard to health insurance coverage. If they do differ, there would be a large nonresponse bias. With larger percentages of nonresponse, the likelihood of a substantial nonresponse bias is very high, and this makes the use of the data questionable. Unfortunately, many large surveys have a high percentage of nonresponse or do not mention the level of nonresponse. Data from these surveys are problematic.

Example 6.10

An example of a survey with poor response is the Nationwide Food Consumption Survey conducted in 1987–1988 for the U.S. Department of Agriculture. This survey, conducted once per decade, was to be the basis for policy decisions regarding food assistance programs. However, only about one-third of the persons who were selected for the sample participated, and, hence, the sample may not be representative of the U.S. population. An independent expert panel and the Government Accounting Office of the U.S. Congress have concluded that information from this survey may be unusable (Government Accounting Office 1991).

There is no easy solution to the nonresponse problem. The best approach is a preventive one — that is, to exert every effort to obtain a high response rate. Even if you are unable to contact the sample person, perhaps a neighbor or family member can provide some basic demographic data about the person. If a sample person refuses to participate, again try to obtain some basic data about the person. If possible, try to obtain some information about the main topic of interest in the survey. The basic demographic data can be used to compare the respondents and nonrespondents. Even if there are no differences between the two groups on the demographic variables, that does not necessarily guarantee the absence of nonresponse bias. However, it does eliminate the demographic variables as a cause of the potential nonresponse bias. If there is a difference, it may be possible to take those differences into account and create an adjusted estimator. The following calculations show one of many possible adjustment methods.

Suppose we found that there was a difference in the gender distribution between the respondents and nonrespondents. Sixty percent of the respondents were females and 40 percent were males, whereas 30 percent of the nonrespondents were females and 70 percent were males. If there were no difference in the proportions of females and males with health insurance, this difference in the gender distribution between the respondents and nonrespondents would be no problem. However, for this example, assume there was a difference. In the respondent group, 30 percent of the females were without health insurance compared to only 5 percent of the males. Figure 6.2 is a display of these percentages and of the calculations involved in creating an adjusted rate.

The corresponding percentages with health insurance are unknown for the nonrespondent group. However, if we assume that the female and male respondents' percentages with health insurance hold in the nonrespondent group, we can obtain an

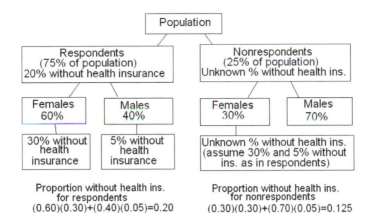

Figure 6.2 Display of the percentages for the health insurance example and calculation of the adjusted rate.

adjusted rate. The percentage of those without health insurance in the nonrespondent group under this assumption is found by weighting the proportions of females and males without health insurance by their proportions in the nonrespondent group — that is, {30% × 0.3} + {70% × 0.05}, which is 12.5 percent. We then use this value for the proportion of nonrespondents without health insurance and combine it with the proportion of respondents without health insurance to obtain a sex-adjusted estimate of the proportion of our community without health insurance. This adjusted estimated is 18.1 percent (= {75% × 0.20} + {25% × 0.125}).

The adjusted rate does not differ much from the rate for the respondents only. However, this adjusted rate was based on the assumption that the proportions of females and males without health insurance were the same for respondents and nonrespondents. If this assumption is false, which we cannot easily check, this adjusted estimate then is incorrect. Whatever method of adjustment is employed, an assumption similar to the above must be made at some stage in the adjustment process (Kalton 1983). Our message is to prevent nonresponse from occurring or to keep its rate of occurrence small.

The discussion so far has focused on *unit nonresponse* — that is, the observational unit did not participate in the survey. There is also *item nonresponse*, in which the sample person did not provide the requested information for some of the items in the survey. Just as there are no easy answers to unit nonresponse, item nonresponse or missing data also is a source of difficulty for the data analyst. Again, if the percentage of item nonresponse is small — say, less than 5 to 10 percent — it probably will not have much of an effect on the data analysis. In this case, the observations with the missing values may be deleted from the analysis. As the percentage of missing data increases, there is increasing concern about the representativeness of the sample persons remaining in the analysis. Because of this concern, statisticians have developed methods for *imputing* or creating values for the missing data (Kalton 1983). By imputing values, it is no longer necessary to delete the sample persons with the missing data from the analysis. The imputation methods range from the very simple to the complex, depending on the amount of auxiliary data available.

As an example, suppose that in a survey to estimate the per capita expenditure for health care, we decided to substitute the respondents' sample average for those with a missing value on this variable. That is a reasonable imputation. However, since age is highly related to health care expenditures, if we know the age of the sample persons, a better imputation would be to use the average expenditure from respondents in the same age group. There are other variables that could be used with age that would be even better than using age alone — for example, the combination of age and health insurance status. The sample mean from the respondents in the same age and health insurance group should be an even better estimate of the missing value than the mean from the age group or the overall mean. In using any imputation method, we must remember that the number of observations is really the number of sample persons with no missing data for the analysis performed, not the number of sample persons. We must also realize that we are assuming that the mean of the group is a reasonable value to substitute for the missing value. Using the mean smoothes the data and likely reduces variability.

Other more complicated procedures are also available. However, none of these procedures guarantee that the value substituted for the missing data is correct. It is possible that the use of imputation procedures can lead to wrong conclusions being drawn from the data. Again, the best procedure for dealing with missing data is preventive — that is, make every effort to avoid missing data in the data collection process.

6.4   Designed Experiments

Designed experiments have been used in biostatistics in the evaluation of (1) the efficacy and safety of drugs or medical procedures, (2) the effectiveness and cost of different health care delivery systems, and (3) the effect of exposure to possible carcinogens. In the following, we present the principles underlying such experiments. Limitations of experiments and ethical issues related to experiments, especially when applied to humans, are also raised. Let us consider a couple of examples to illustrate the essential points in the experimental design.

Example 6.11

The Hypertension Detection and Follow-up Program (HDFP 1979) was a community-based, clinical trial conducted in the early 1970s by the National Heart, Lung and Blood Institute (NHLBI) with cooperation of 14 clinical centers and other supporting groups. The purpose of the trial was to assess the effectiveness of treating hypertension, a major risk factor for several different forms of heart disease. For this trial, it was decided that the major outcome variable would be total mortality.

At the time of designing the HDFP trial, results of a Veterans Administration (VA) Cooperative Study were known. This study had already demonstrated the effectiveness of antihypertensive drugs in reducing morbidity and mortality due to hypertension among middle-aged men with sustained elevated blood pressure. However, the VA study included only a subset of the entire community. Applicability of its findings to those with undetected hypertension in the community, to women and to minority persons was uncertain. Therefore, it was decided to perform a study, the HDFP study, in the general community. Instead of including only people who knew that they had high blood pressure, subjects were recruited by screening people in the community.

In this clinical trial, antihypertensive therapy was the *independent or predictor variable* and the mortality rate was the *dependent or response variable*. To determine the effectiveness of the antihypertensive therapy, a comparison group was required. Thus, the study was intended to have a *treatment group* — those who received the therapy — and a *control group* — those who did not receive the therapy. However, this classic experimental design could not be used. Since the antihypertensive therapy was already known to be effective, it could not ethically be withheld from the control group. Recognizing this, the HDFP investigators decided to compare a systematic antihypertensive therapy given to those in the treatment group (Stepped Care) to the therapy received from their usual sources of care for those in the control group (Regular Care). As a result, no one was denied treatment.

Example 6.12

In Chapter 3 we introduced a data set from the Digitalis Investigation Group trial. The primary objective of the DIG trial was to determine the effect of digoxin as the cause of mortality in patients with clinical heart failure who were in sinus rhythm and whose ejection fraction was ≤0.45. A total of 302 clinical centers in the United States and Canada enrolled 7788 patients between February 1991 and September 1993 and follow-up continued until December 1995 (DIG 1995).

Eligible patients were recruited and randomized to either digoxin or placebo (dummy pill) using a random block size method (to be explained later) within each clinical center; 3889 to digoxin and 3899 to placebo. This large sample size was required to detect a 12 percent reduction in mortality by treatment and to take non-compliance into account. The trial was double blinded (both investigators and patients were not informed about the group assignment). We discuss these essential feathers of an experimental design in this section.

6.4.1 Comparison Groups and Randomization

A simple experiment may be conducted without any comparison group. For example, a newly developed AIDS education course was taught to a class of ninth graders in a high school for a semester. The level of knowledge regarding AIDS was tested before and after the course to assess the effect of the course on students' knowledge. The difference in test scores between the pre- and posttests would be taken as the effect of the instructional program. However, it may be inappropriate to attribute the change in scores to the instructional program. The change may be entirely or partially due to some influence outside the AIDS course — for example, mass media coverage of AIDS-related information. Therefore, we have to realize that when this simple experimental design is used, the outside influence, if any, is mixed with the effect of the course and it is not possible to separate them.

Thus, in studying the effect of an independent variable on a dependent variable, we have to be aware of the possible influence of an extraneous variable(s) on the dependent variable. When the effects of the independent variable and the extraneous variable cannot be separated, the variables are said to be *confounded*. In observational studies such as sample surveys, all variables are confounded with one another and the analytical task is to untangle the comingled influence of many variables that are measured at the same time. In experimental studies, the effects of extraneous variables are separated from the effect of the independent variable by adopting an appropriate design.

The basic tool for separating the influence of extraneous variables from that of the independent variable is the use of comparison groups. For example, giving the treatment to one of two equivalent groups of subjects and withholding it from the other group means that the observed difference in the outcome variable between the two groups can be attributed to the effect of the treatment. In this design, any extraneous variables would presumably influence both groups equally, and, thus, the difference between the two groups would not be influenced by the extraneous variables. The key to the successful use of this design is that the groups being compared are really equivalent before the experiment begins.

Matching is one method that is used in an attempt to make groups equivalent. For example, subjects are often matched on age, gender, race, and other characteristics, and then one member of each matched pair receives the treatment and the other does not. However, it is difficult to match subjects on many variables, and also, the researcher may not know all the important variables that should be used in the matching process. A method for dealing with these difficulties with matching is the use of *randomization*.

Randomization is the random assignment of subjects to groups. By using randomization, the researcher is attempting to (1) eliminate intentional or nonintentional selection bias — for example, the assignment of healthier subjects to the treatment group and sicker subjects to the control group; and (2) remove the effect of any extraneous variables. With large samples, the random assignment of subjects to groups should cause the distributions of the extraneous variables to be equivalent in each group, thus removing their effects.

6.4.2   Random Assignment

One way of randomly assigning subjects to groups is the use of the random sampling without replacement procedure discussed in the earlier section.

Example 6.13

Consider the case of randomly assigning 50 subjects to two groups. An SRS (without replacement) of 25 from the 50 sequentially numbered subjects is selected using a computer package (see **Program Note 6.1** on the website):

2	4	5	6	11	12	16	17	18	20
21	25	16	27	30	31	32	33	35	36
40	41	44	47	48					

These subjects are assigned to the treatment group and the remaining 25 subjects form the control group. In many randomized experiments, subjects are assigned to the groups sequentially as soon as subjects are identified, as in the HDFP trial. In

---

that case, the preceding results can be put into the following sequence of letters T (treatment group) and C (control group) that can be used to show the assignment:

C	T	C	T	T	T	C	C	C	C
T	T	C	C	C	T	T	T	C	T
T	C	C	C	T	T	T	C	C	T
T	T	T	C	T	T	C	C	C	T
T	C	C	T	C	C	T	T	C	C

If one were to assign 60 subjects to three groups, the first random sample of 20 will be assigned to the first group, the second random sample of 20 to the second group, and the remaining 20 subjects to the third group.

The method of random allocation illustrated in Example 6.13 poses some problem when the subjects are to be randomized in sequence as they are recruited in a clinical trial because the total number of eligible patients is not known in advance. As a result, it is difficult to balance the size of comparison groups. For example, the sequence of letters T and C in Example 6.13 works fine if 50 eligible patients can be recruited in a clinical center. But if only 10 eligible patients are available, then there are 4 Ts and 6 Cs in the first 10 letters in the sequence, making the size of comparison groups unbalanced.

An alternative to the preceding method is a *random block size method* (or random permuted blocks method). This is the randomization method used in the DIG trial described in Example 6.12. We illustrate an example for the blocks of size 4. We can list all the different possible sequences of allocations of four successive patients containing two Ts and two Cs as follows:

1. T T C C
2. T C T C
3. T C C T
4. C T T C
5. C T C T
6. C C T T

Blocks are then chosen at random by selecting random numbers between 1 and 6. This could be done, for example, with a fair die or by using Table B1, ignoring the digits 7, 8, 9, and 0. The first five eligible random digits from the first row of Table B1 are 1, 1, 4, 5, and 4. By choosing the previous corresponding blocks, we have the following sequence of allocations:

T, T, C, C,   T, T, C, C,   C, T, T, C,   C, T, C, T,   C, T, T, C.

It may be possible for an investigator to discover the pattern when the block size is small. To alleviate this problem, the block size is often changed in as allocation proceeds. It will be difficult to discover a pattern of sequence when the block size is 10 or more.

6.4.3   Sample Size

The random assignment of subjects to groups does not guarantee the equivalence of the distributions of the extraneous variables in the groups. There must be a sufficiently large number of subjects in each group for randomization to have a high probability of causing the distributions of the extraneous variables to be similar across groups. As discussed earlier, use of larger random samples decreases the sample-to-sample variability and increases our confidence that the sample estimates are closer to the population parameters. In the same way, a greater number of subjects in the treatment and control groups increase our confidence that the two groups are equivalent with respect to all extraneous factors.

To make this point clearer, consider the following example. A sample of 10 adults is taken from the Second National Health and Nutrition Examination Survey (NHANES II) data file, and 5 of the 10 persons are randomly assigned to the treatment group, and the other 5 are assigned to the control group. The two groups are compared with respect to five characteristics. The same procedure is repeated for sample sizes of 40, 60, and 100, and the results are shown in Table 6.3.

Table 6.3   Comparison of treatment and control groups for different group sizes.

Characteristics[a]	Treatment	Control	Treatment	Control
	$(n_1 = 5)$	$(n_2 = 5)$	$(n_1 = 20)$	$(n_2 = 20)$
Percent male	60	20	60	35
Percent black	0	20	5	20
Mean years of education	12.6	11.2	12.9	13.0
Mean age	38.8	41.6	40.7	34.0
Percent smokers	60	40	27	23
	$(n_1 = 30)$	$(n_2 = 30)$	$(n_1 = 50)$	$(n_2 = 50)$
Percent male	43	50	42	44
Percent black	17	10	16	16
Mean years of education	12.7	12.9	11.7	12.5
Mean age	39.7	40.2	42.1	42.5
Percent smokers	32	35	34	34

[a]Observations are weighted using the NHANES II sampling weights.

The treatment and control groups are not very similar when n is 10. As the sample size increases, the treatment and control groups become more similar. When n is 100, the two groups are very similar. It appears that at least 30 to 50 persons are needed in each of the treatment and control groups for them to be reasonably similar. The sample size considerations will be discussed further in Chapters 7 and 9.

In the HDFP clinical trial shown in Example 6.11, over 10,000 hypertensive persons were screened through community surveys and included in the study. These subjects were randomly assigned to either the Stepped Care or Regular Care groups. Because of this random assignment and the large number of subjects included in the trial, the Stepped Care and the Regular Care groups were very similar with respect to many important characteristics at the beginning of the trial. Table 6.4 is a demonstration of the similarities. The randomization and the sufficiently large sample size also give us confidence that these two groups were equivalent with respect to other characteristics that are not listed in Table 6.4.

Table 6.4 Comparison of Stepped Care and Regular Care participants by selected characteristics at entry to the hypertension detection and follow-up program.

Characteristics (Number of Participants)	Stepped Care (5485)	Regular Care (5455)
Mean age in years	50.8	50.8
Percent black men	19.4	19.9
Percent black women	24.5	24.8
Mean systolic blood pressure, mmHg	159.0	158.5
Mean diastolic blood pressure, mmHg	101.1	101.1
Mean pulse beats/minute	81.7	82.2
Mean serum cholesterol, mg/dL	235.0	235.4
Mean plasma glucose, mg/dL	178.5	178.9
Percent smoking >10 cigarettes/day	25.6	26.2
Percent with history of stroke	2.5	2.5
Percent with history of myocardial infarction	5.1	5.2
Percent with history diabetes	6.6	7.5
Percent taking antihypertension medication	26.3	25.7

Source: HDFP, 1979

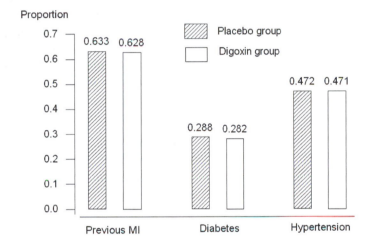

Figure 6.3 Proportions of patients with previous myocardial infarction, diabetes, and hypertension for placebo and digoxin groups in the DIG trial. *Source:* Digitalis Investigation Group, 1995

The DIG trial shown in Example 6.12 also used a large sample size recruited by clinical centers in the United States and Canada. The recruited patients were assigned to either placebo or digoxin treatment by a random block size method just described. As shown in Figure 6.3, medical history of the subjects with respect to myocardial infarction, diabetes, and hypertension is about the same, providing assurance that these and other clinical conditions would not pose as confounding factors for the comparison of two experimental groups.

The sample size required for an experiment depends on three factors: (1) the amount of variation among the experimental subjects, (2) the magnitude of the effect to be detected, and (3) the level of confidence associated with the study. When the experimental subjects are similar, a smaller sample size can be used than when the subjects differ. For example, a laboratory experiment using genetically engineered mice does not require as large a sample size as the same experiment using mice trapped in the wild. There is less likelihood of extraneous variables existing in the study using the genetically engineered mice. Hence, a smaller sample should be acceptable, since there is less need to control for extraneous variables. The fact that the sample size for the experiment depends on the size of the effect to be detected is not surprising. Since it should be more difficult to detect a small effect of the independent variable than a large effect, the sample size must reflect this. This is one of the reasons that the HDFP trial and DIG trial used a large sample size. Both trials attempt to detect a relatively small difference between the treatment and control groups. The relation between the sample size and the confidence associated with the study will be explored further in Chapters 7 and 8.

6.4.4 Single- and Double-Blind Experiments

So far we have been concerned with the statistical aspects of the design of an experiment. This means the use of comparison groups, the random assignment of subjects to the groups, and the need for an adequate number of subjects in the groups. An additional concern is the possible bias that can be introduced in an experiment. Let us consider some possible sources of bias and possible ways to avoid them.

In drug trials, particularly in those involving a placebo, the subjects are often *blinded* — that is, they are not informed whether they have received the active medication or a placebo. This is done because knowledge of which treatment has been provided may affect the subject's response. For example, those assigned to the control group may lose interest, whereas those receiving the active medication, because of expectations of a positive result, may react more positively. Studies in which the treatment providers know but the subjects are unaware of the group assignment are called *single-blind* experiments.

In most drug trials, both the subjects and the treatment providers are unaware of the group assignment. The treatment providers are blinded because they also have expectations about the reaction to the treatment. These expectations may affect how the experimenter measures or interprets the results of the experiment. When both the subjects and the experimenters are unaware of the group assignment, it is called a *double-blind* experiment.

Example 6.14

Let us examine one double-blind, randomized experiment conducted by a Veterans Administration research team (Goldman et al. 1988). They used the experimental design shown in Figure 6.4 to determine whether antiplatelet therapies improve saphenous vein graft patency after coronary artery bypass grafting.

In this experimental design, there are four treatment groups (four regimens of drug therapy) and a control group (placebo). Both the patients and the doctors were blinded, and only the designers of the trial, who were not directly involved in patient treatment, knew the group assignment. A total of 772 consenting patients were randomized, and postoperative treatment was started six hours after surgery and continued for one year.

As was to be expected, this experiment encountered problems in retaining subjects during the course of the experiment. The final analysis was based on 502 patients who underwent the late catheterization. These patients had a total of 1618 grafts. Of the 270 patients not included in the final analysis, 154 refused to undergo catheterization, 32 were lost to follow-up, 31 died during treatment, 42 had medical

---

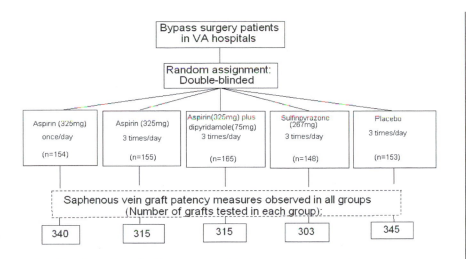

Figure 6.4 Experimental design for Veterans Administration Cooperative Study on Effect of Antiplatelet Therapy.
Source: Goldman et al., 1988

complications, and data on 11 patients were not available in the central laboratory (Goldman et al. 1989). Although we may expect that these problems are fairly evenly distributed among the groups because of the random assignment of subjects, the sample size was reduced considerably. This suggests that we needed to increase the initial sample size in anticipation of the loss of some subjects during the experiment.

There are other types of precautions that must be taken to avoid potential biases. In addition to statistical aspects, the experiment designer must provide detailed procedures for handling experimental subjects, monitoring compliance of all participants, and collecting data. For this purpose a study protocol must be developed, and the experimenter is responsible for adherence to the protocol by all participants. Similar to the problem of nonresponse in sample surveys, the integrity of experiments are often threatened by unexpected happenings such as the loss of subjects during the experiment and changes in the experimental environment. Steps must be taken to minimize such threats.

6.4.5   Blocking and Extraneous Variables

Thus far we have considered the simplest randomization, the random assignment of subjects to groups without any restriction. This design is known as a *completely randomized design*. The role of this design in experimental design is the same as that of the simple random sample design in survey sampling. As was mentioned earlier, in completely randomized designs, we attempt to remove the effects of extraneous variables by randomization. However, a reasonably large sample size is required before we can have confidence in the randomization process.

Another experimental design for eliminating the effects of extraneous variables known or thought to be related to the dependent variable uses *blocking*. Blocking means directly taking these extraneous variables into account in the design. For example, in a study of the effects of different diets on weight loss, subjects are often blocked or grouped into different initial weight categories. Within each block, the subjects are then randomly assigned to the different diets. We block based on initial weight because it is thought that weight loss may be related to the initial weight. Designs using blocking do not rely entirely on randomization to remove the effects of these important extraneous variables. Blocking guarantees that each diet has subjects with the same distribution of initial weights; randomization cannot guarantee this. Blocking in experiments is similar to stratification in sample surveys. The experimental design that uses blocks to control the effect of one extraneous variable is called a *randomized block design*. This name indicates that randomization is performed separately within each block.

Blocking is also used for administrative convenience. The VA Cooperative Study discussed in the previous section had 11 participating hospitals located throughout the United States. Since the subjects were randomized separately at each site, each participating hospital was a block. In this case, the blocking was done for administrative convenience while also controlling for the variation among hospitals.

In the previous section, we saw that the SRS design can be modified and extended as required to meet the demands of a wide variety of sampling situations. The completely randomized experimental design can similarly be expanded to accommodate many different needs in experimentation. Only one factor is considered in a completely randomized design. When two or more factors are considered, a *factorial design* can be used. For example, a clinical trial testing two different drugs simultaneously can be conducted in a 2 by 2 or 2^2 factorial design. Two levels of drug A and two levels of drug B will form four experimental groups: both drug A and B, A only, B only, and control (no drug). Study subjects are randomly assigned to the four groups. From this design we can examine the main effects of A and B as well as the interaction of two drugs.

The randomized block design just examined can also be expanded to accommodate more than one independent variable or block on more than one extraneous variable. We will discuss further these more complex experimental designs in Chapter 12.

6.4.6   Limitations of Experiments

The results of an experiment apply to the population from which the experimental subjects were selected. Sometimes this population may be very limited — for example, patients may be selected from only one hospital or from one clinic within the hospital. In situations like these, does this mean that we must perform similar experiments in many more hospitals to determine if the results can be generalized to a larger population — for example, to all patients with the condition being studied? From a statistical perspective, the answer is yes. However, if based on substantive reasons, we can argue that there is nothing unique about this hospital or clinic that should affect the experiment, then it may be possible to generalize the results to the larger population of all patients with the condition. This generalization is based on substantive reasoning, not on statistical principles.

For example, the results of the VA Cooperative Study may be valid only for male veterans. It certainly would be difficult to generalize the results to females without more information. It may be possible to generalize the results to all males who are known to have hypertension, but this requires careful scrutiny. We must know whether or not the

VA medical treatment of hypertension is comparable to that received by males in the general population. Does the fact that the men served in the military cause any difference, compared to those who were not in the military, in the effect of the medical intervention? If differences are suspected, then we should not generalize beyond the VA system.

On the other hand, the results of the HDFP should apply more widely, since the subjects were screened from random samples of residents in 14 different communities and then randomly assigned to the comparison groups. This use of accepted statistical principles of random sampling from the target population and randomizing these subjects to comparison groups makes it reasonable to generalize the results.

Another limitation of an experiment stems from its dependency on the experimental conditions (Deming 1975). Often experiments take place in a highly controlled, artificial environment, and the observed results may be confounded with these factors. Dr. Lewis Thomas's experience (Thomas 1984) is a case in point. While he was waiting to return home from Guam at the end of World War II, he conducted an experiment on several dozen rabbits left in a medical science animal house. He tested a mixed vaccine consisting of heat-killed streptococci and a homogenate of normal rabbit heart tissue, and the test produced spectacular and unequivocal results. All the rabbits that received the mixture of streptococci and heart tissue became ill and died within two weeks. The histologic sections of their hearts showed the most violent and diffuse myocarditis he had ever seen. The control rabbits injected with streptococci alone or with heart tissue alone remained healthy and showed no cardiac lesions. Upon returning to the Rockefeller Institute, he replicated the experiment using the Rockefeller stock of rabbits. He repeated the experiment over and over, but he never saw a single sick rabbit. One explanation for the spectacular results of the Guam experiment is that there may have been some type of a latent virus in the Guam rabbit colony. As Dr. Thomas said, "I had all the controls I needed. I wasn't bright enough to realize that Guam itself might be a control."

As Dr. Thomas's experience shows, we have to be careful not to deceive ourselves and extrapolate beyond our data. The experimental data consist of not only the observed difference between the treatment and control groups, but also the conditions and circumstances under which the experiment was conducted. These include the method of investigation, the time and place, and the duration of the test and other conditional factors. For example, in interpreting the results of drug trials, there is no statistical method by which to extrapolate the safety record of a drug beyond the period of the experiment, nor to a higher level of dosage, nor to other types of patients. The toxic effect of the medication may manifest itself only after a longer exposure, at higher levels of dosage or for other types of patients. Therefore, extrapolation of experimental results must be done with great care, if at all. Better than extrapolation is a replication of the study for different types of subjects under different conditions.

Implicit in the naming of experimental variables as being dependent and independent is the idea of cause and effect — that is, changes in the levels of the independent variables cause corresponding changes in the dependent variable. However, it is difficult to demonstrate a cause-and-effect relationship. It is sometimes possible to demonstrate this in very carefully designed experiments. However, in most situations in which statistics are used, positive results do not mean a cause-and-effect relationship but only the existence of an association between the dependent and independent variables.

Finally, statistical principles of experimentation can sometimes be in conflict with our cultural values and ethical standards. Experimenting, especially on human beings, can lead to many problems. If the experiment can potentially harm the subjects or impinge upon their privacy or individual rights, then serious ethical questions arise. The harm can be direct physical, psychological, or mental damage, or it may be the withholding of potential benefits. As was seen in the HDFP study, to avoid withholding the benefits of antihypertensive therapy, the study designers used the Regular Care group instead of a placebo group as the control group. When the potential direct harm is obvious, we cannot subject human beings to an experiment.

To protect human subjects from potential harm or from an invasion of privacy, an *informed consent* is required for experiments and even for interviews in sample surveys. This consent has to be voluntary. However, it is not difficult to recognize the possibility for pressuring patients to participate in a clinical trial. To prevent undue pressure being applied to patients or other potential study participants, all organizations receiving funds from the Federal government are required to have an institutional review committee (OSTP 1991). It is this committee's task to evaluate the study protocol to see if it provides adequate safeguards for the rights of the study participants.

6.5　Variations in Study Designs

As just seen, the essential characteristics of an experiment are that the investigators initially randomly assign the study subjects to the treatment and control groups (parallel comparison groups), administer the treatment, and observe what happens prospectively. Such an experimental design is hard to use in practice due to ethical and other practical reasons. The following quasi-experimental designs attempt to emulate an experimental situation, accommodating certain practical constraints.

6.5.1　The Crossover Design

The *crossover design* uses one group of experimental subjects, and each level of treatment is given at different times to each subject. The simplest crossover uses two periods and two levels of treatments. To control the effect of the order of applying two treatments, subjects are often randomly assigned group A and group B. Subjects in group A receive treatment 1 and subjects in group B receive treatment 2 in the first period. In the second period the treatments are switched. It is possible that a treatment effect in the first period can carry residual effects to the second period. In order to minimize the carryover effect, a *washout period* is often established between the two treatment periods. Some studies have a *run-in period* before the first treatment period begins, so as to wash out any residual effects of any previous medications. An obvious advantage of a crossover design is the cost saved by using a fewer subjects. This design, however, would require a long period of experimentation. The lack of a parallel comparison group and random assignment would make it difficult to distinguish the within-subject variation from the between-subject variation and to control the confounding effects.

Example 6.15

A 2×2 crossover design was used to compare ibuprofen (usual prescribed treatment) and lysine acetyl salicylate (Aspergesic, over-the-counter medicine) in the treatment of rheumatoid arthritis (Hill et al. 1990). Thirty-six patients were randomly assigned to the two equal treatment order groups at entry. After two weeks on their first treatment, patients crossed over to the other treatment, and two weeks later, the trial ended. There was no run-in or wash-out period, but the trial was double-blinded. They used the *double-dummy* procedure where each patient always received a combination of two pills, the appropriate active treatment plus a placebo that was indistinguishable from the other active treatment. The treatment periods were of equal length and relatively short. At baseline a general medical examination was carried out, and the recorded baseline values of the two groups were found to be similar. At the end of each treatment period a clinic visit was made to assess grip strength, blood pressure, and so forth. During the treatment periods, patients recorded a pain assessment score on a 1–5 scale (1 = no pain, 2 = mild pain, 3 = moderate pain, 4 = severe pain, 5 = unbearable pain). Five patients withdrew from the trial, one patient was considered noncompliant, and one patient failed to report his scores for the second period. The average pain scores for the 29 patents were analyzed.

6.5.2   The Case-Control Design

The *case-control design* identifies a set of subjects with a disease or certain condition (the cases) and another set without the disease (the controls). These two groups are compared with respect to a risk factor (the treatment). In this study the outcome is specified first, and the risk factor is assessed retrospectively. Since the subjects are not assigned to the case and control groups, the effects of confounding factors are not controlled. To overcome this problem, the two groups are matched with respect to some confounders such as age and gender. Although the two groups may be comparable in terms of age and gender, the effect of other confounders still remain. Another drawback of the case-control design is that it cannot be used to measure the incidence or prevalence of the disease because of the retrospective assessment of the risk factor.

Example 6.16

In January 1984 six cases of Legionnaires' disease were reported to the health authority in Reading, United Kingdom (Anderson 1985). All of them became ill between December 15 and 19, 1983. After a thorough investigation, the authority discovered seven unreported cases. The cases had no obvious factor in common except for visiting the Reading town center just before their illness. A case-control study was conducted to compare exposure between the 13 cases and a selected set of 36 people without the disease (the controls). Frequencies of visiting the town center just before Christmas were high for both groups, but most of the cases and fewer controls visited the Butts Center shopping mall. This suggested that the Butts Center might be a source of the legionella bacterium. In their analysis of data, they matched the cases and controls with respect to age, sex, neighborhood, and mobility status. One case had one control, another case had two controls, and the remaining

11 cases had three controls. In this investigation the case-control approach appears to be the only one possible. The investigation needed to be conducted quickly in fear of further infections. Other study designs may be impractical in this type of situation.

6.5.3 The Cohort Study Design

In the *cohort study design*, study subjects are followed up through time to record incidences of disease. The simplest approach is to select two groups of subjects at the baseline. One group consists of subjects who possess some special attribute that is thought to be a possible risk factor for a disease of interest, while the other group does not. For instance, a group of coal miners who are free of lung cancer and a group of lung cancer–free farmers are followed up several years to record the lung cancer incidences. This is a prospective study design, mimicking conditions of an experiment, and thus we can measure incidence. However, in the cohort study design the study subjects are not assigned to the groups and confounding is not controlled. Similar to the case-control study, the groups can be matched with respect to selected confounders, but the matching cannot provide protection against all possible confounders.

Example 6.17

In a cohort study of 34,387 menopausal women in Iowa, intakes of vitamin A, C, and E were assessed in 1986 (Kushi 1996). In the period up to the end of 1992, 879 of these women were newly diagnosed with breast cancer. The investigators examined the effect of vitamin use and level of intake for each vitamin on breast cancer incidence. Since women were not randomly assigned to vitamin use and level of intake groups, confounding factors were not controlled effectively. However, this study provided valuable information for further investigation. Randomization might not be possible for ethical and practical reasons.

These alternative study designs are widely used in epidemiological studies because of ethical and practical reasons. But the analysis of data from these studies requires the use of special methods and the analytical results need to be interpreted recognizing the limitations of the study design used. The data from matched studies need to be analyzed taking into account the matching. These methods will be discussed in the subsequent chapters. Data from these studies will be used to illustrate the methods of analysis in subsequent chapters.

Conclusion

In this chapter we saw how to collect data using sample surveys and designed experiments. We examined the use of a chance mechanism in drawing samples and assigning experimental subjects to comparison groups. We also presented some practical issues that cause more complicated sample designs to be used and experimental designs to be modified. Regardless of the complexity of the sample design, as long as we know the selection probability, we can infer from the sample to the population. A requirement of a good experimental design is that it reduces the chance of extraneous variables being confounded with the experimental variables. Randomization and blocking are basic tools for preventing this confounding. When these tools are used appropriately, it is possible to analyze the data to determine whether or not it is likely that an association exists between the dependent variable and the independent variables. Analysis of data from complex surveys would require special considerations, which we will discuss in Chapter 15. Experimental data are analyzed using the designs described here in Chapters 8 and 12. Even after performing the experiment appropriately, care must be used in interpreting the experimental results. We must not unduly extrapolate the findings from our experiment, but recognize that replication may be necessary for the appropriate generalization to the target population.

EXERCISES

6.1 Choose the most appropriate response from the choices listed after each question.

  a. To determine whether a given set of data is a random sample from a defined population, one must _____.
    __ analyze the data.
    __ know the procedure used to select the sample.
    __ use a mathematical proof.

  b. A simple random sample is a sample chosen in such a way that every unit in the population has a(n) _____ chance of being selected into the sample.
    __ equal
    __ unequal
    __ known

  c. In the random number table, Appendix Table B1, approximately what percent of numbers are 9 or 2?
    __ 20
    __ 10
    __ unknown

  d. Sampling with replacement from a large population gives virtually the same result as sampling without replacement.
    __ true
    __ false

  e. In a stratified random sample, the selection probability for each element within a stratum is _____.
    __ equal.
    __ unequal.
    __ unknown.

  f. A probability sample is a sample chosen in such a way that each possible sample has a(n) _____ chance of being selected.
    __ equal
    __ unequal
    __ known
    __ unknown

6.2 If a population has 2000 members in it, how would you use Table B-1, the table of random numbers, to select a simple random sample of size 25? Assume that the 2000 members in the population have been assigned numbers from 0 to 1999. Beginning with the first row in Table B1, select the 25 subjects for the sample.

6.3 In the following situations, do you consider the selected sample to be a simple random sample? Provide your reasoning for your answer.

a. A college administrator wishes to investigate students' attitudes concerning the college's health services program. A 10 percent random sample is to be selected by distributing questionnaires to students whose student ID number ends with a 5.

b. A medical researcher randomly selected five letters from the alphabet and abstracted data from the charts of patients whose surnames start with any of those five letters.

6.4 In the NHANES II, 27 percent of the target sample did not undergo the health examination. In the examined sample, the weighted estimate of the percent overweight was 25.7 percent (NCHS 1992).

a. Assuming that these data were collected via an SRS, what is the range for the percent overweight in the target sample?

b. Should any portion of the population be excluded in the measurement of overweight?

6.5 Discuss how sampling can be used in the following situations by defining (1) the population, (2) the unit from which data will be obtained, (3) the unit to be used in sampling, and (4) the sample selection procedure:

a. A student is interested in estimating the total number of words in this book.

b. A city planner is interested in estimating the proportion of passenger cars that have only one occupant during rush hours.

c. A county public health officer is interested in estimating the proportion of dogs that have been vaccinated against rabies.

6.6 For each of the following situations discuss whether or not random sampling is used appropriately and why the use of random sampling is important:

a. A doctor selected every 20th file from medical charts arranged alphabetically to estimate the percent of patients who have not had any clinic visits during the past 24 months.

b. A city public health veterinarian randomly selected 50 out of 500 street corners and designated a resident at each corner to count the number of stray dogs for one week. He multiplied the number of stray dogs counted at the 50 corners by 10 as an estimate of the number of stray dogs in the city.

c. A hospital administrator reported to the board of directors that his extensive conversations with two randomly selected technicians revealed no evidence of support for a walkout by hospital technicians this year.

6.7 An epidemiologist wishes to estimate the average length of hospitalization for cancer patients discharged from the hospitals in her region of the country. There are 500 hospitals with the number of beds ranging from 30 to 1200 in the region.

a. Discuss what difficulties the researcher might encounter in drawing a simple random sample.

b. Offer suggestions for drawing a random sample.

6.8 Discuss the advantages and disadvantages of the following sampling frames for a survey of the immunization levels of preschool children:

a. Telephone directory

b. The list of children in kindergarten

c. The list of registered voters

6.9 Discuss the interpretation of the following surveys:

a. A mail survey was conducted of 1000 U.S. executives and plant managers. After a month, 112 responses had been received. The report of the survey results stated that Japan, Germany, and South Korea were viewed as being better competitors than the U.S. in the world economy. Also one-third of the managers did not believe their own operations were making competitive improvements.

b. A weekly magazine reported that most American workers are satisfied with the amount of paid vacation they are allowed to take. This conclusion was based on the results of a telephone poll of 522 full-time employees (margin of error is plus or minus 4%; "Not sure" omitted). The question asked was "Should you have more time off or is the amount of vacation you have fair?"

More time off   33%

Current amount fair   62%

6.10 Choose the most appropriate response from the choices listed under each question:

a. Which of the following is not required in an experiment?

__ designation of independent and dependent variables

__ random selection of the subjects from the population

__ use of a control group

__ random assignment of the subjects to groups

b. The main purpose of randomization is to balance between experimental groups the effects of extraneous variables that are _____.

__ known to the researcher.

__ not known to the researcher.

__ both known and unknown to the researcher.

c. The experimental groups obtained by randomization may fail to be equivalent to each other, especially when _____.

__ the sample size is very small.

__ blocking is not used.

__ matching is not used.

d. Which, if any, of the following is an inappropriate analogy between random sampling and randomized experiments?

__ simple random sampling–completely randomized experiment

__ stratified random sampling–randomized complete block design

__ random selection–random assignment

e. A randomized experiment is intended to eliminate the effect of _____.

__ independent variable.

__ confounded extraneous variables.

__ dependent variable.

f. If the number of subjects randomly assigned to experimental groups increases, then the treatment and control groups are likely to be
_____.

___ more similar to each other.
___ less similar to each other.
___ neither of the above.

6.11 A middle school principal wants to implement a newly developed health education curriculum for 30 classes of 7th graders that are taught by 6 teachers. However, the available budget for teacher training and resource material is sufficient for implementing the new course in only half of the classes. A teacher suggests that an experiment can be designed to compare the effectiveness of the new and old curricula.

a. Design an experiment to make this comparison, explaining how you would carry out the random assignment of classes and what precautions you would take to minimize hidden bias.

b. How would you select teachers for the new curriculum?

6.12 To examine the effect of the seat belt laws on traffic accident casualties, the National Highway Traffic Safety Administration compared fatalities among those jurisdictions that were covered by seat belt laws (the Covered Group) with those jurisdictions that were not covered by seat belt laws (the Other Group). They found that among the Covered Group, 24 belt law jurisdictions, fatalities were 6.6 percent lower than the number forecasted from past trends. In the Other Group, observed fatalities were 2 percent above the forecasted level (Campbell and Campbell 1988).

a. Explain whether or not you attribute the difference between these two groups to seat belt laws.

b. Provide some possible extraneous variables that might have influenced the effect difference and explain why these variables may have had an effect.

6.13 A large-scale experiment was carried out in 1954 to test the effectiveness of the Salk poliomyelitis vaccine (Francis et al. 1955). After a considerable debate, the randomized placebo (double-blind) design was used in approximately half of the participating areas and the "observed control" design was used in the remaining areas. In the latter areas, children in the second grade were vaccinated and children in the first and third grades were considered as controls (no random assignment was used). In both areas, volunteers participated in the study, but polio cases were monitored among all children in participating areas. The following results were announced on April 12, 1955, at the University of Michigan:

Study Type and Group	Study Subjects	Polio Case Rate[a] (per 100,000)			
		Total	Paralytic	Nonparalytic	Fatal
Placebo Control Areas:					
Vaccinated	200,745	28	16	12	0
Placebo	201,229	71	57	13	2
Not inoculated[b]	338,778	46	36	11	0
Observed Control Areas:					
Vaccinated	221,998	25	17	8	0
Controls	725,173	54	46	8	2
Not inoculated[c]	123,605	44	35	9	0

[a]Based on confirmed cases
[b]Nonvolunteers in the participating areas
[c]Second graders not inoculated
Source: Reference 19, Tables 2 and 3

a. Why was it necessary to use so many subjects in this trial?

b. What extraneous variables could have been confounded with the vaccination in the observed control areas?

6.14 To test whether or not oat bran cereal diet lowers serum lipid concentrations (as compared with a corn flakes diet), an experiment was conducted (Anderson et al. 1990). In this experiment 12 men with undesirably high serum total-cholesterol concentrations were randomly assigned to one of the two diets upon admission to the metabolic ward. After completing the first diet for two weeks, the subjects were switched to the other diet for another two weeks. This is a crossover design in which each subject received both diets in sequence. Eight subjects were hospitalized in the metabolic ward for a continuous four-week period, and the remaining subjects were allowed a short leave of absence, ranging from 3 to 14 days, between diet regiments for family emergencies or holidays. The results indicated that the oat bran cereal diet compared with the corn flakes diet lowered serum total-cholesterol and serum LDL-cholesterol concentrations significantly by 5.4 percent and 8.5 percent, respectively.

a. Discuss how this crossover design is different from the two-group comparison design studied in this chapter. What are the advantages of a crossover design?

b. The nutritional effects of the first diet may persist during the administration of the second diet. Is the carryover effect effectively controlled in this experiment?

c. Discuss any other factors that may have been confounded with the type of cereal.

6.15 To determine the efficacy of six different antihypertensive drugs in lowering blood pressure, a large experiment was conducted at 15 clinics (Materson 1993). After a washout phase lasting four to eight weeks (using a placebo without informing the subjects), a total of 1292 male veterans whose diastolic blood pressure was between 95 and 109 mmHg were randomly assigned in a double-blind manner to one of the six drugs or a placebo. Each medication was prepared in three dose levels (low, medium, and high). The average age of the subjects was 59; 48 percent were black, and 71 percent were already on antihypertensive treatment at screening. All medications were started at the lowest dose, and the dose was increased every two weeks, as required, until a diastolic blood pressure of less than 90 mmHg was reached without intolerance to the drug on two consecutive visits or until the maximal drug dose was reached.

The blood pressure measurement during treatment was taken as the mean of the blood pressures recorded during the last two visits. The following table shows the number of subjects assigned, the number that withdrew during the treatment and the results on reduction in diastolic blood pressure:

Experimental Group	Number Assigned	Number Withdrawn	Reduction in diastolic BP:		
			Mean	Std	% Success*
1. Hydrochlorothiazide	188	15	10	6	57
2. Atenlol	178	16	12	6	65
3. Captopril	188	23	10	7	54
4. Clonidine	178	13	12	6	65
5. Diltiazem	185	12	14	5	75
6. Prazosin	188	29	11	7	56
7. Placebo	187	29	5	7	33
Total	1292	137			

*Proportion of patients reaching the target blood pressure (diastolic blood pressure <90 mmHg)
Source: Reference 22, Tables 2 and 3, Figure 1

a. Discuss why the patients were not informed about the use of a placebo during the initial washout period.

b. More than 10 percent of the subjects withdrew from the study during the treatment, and there were more withdrawals in some groups than in other groups. Discuss how the withdrawals may affect the experimental results.

c. Discuss how widely you can generalize the results of this experiment.

6.16 A randomized trial was conducted to test the effects of an educational program to reduce the use of psychoactive drugs in nursing homes. Six matched pairs of nursing homes were selected for this trial. The matching was based on the size of nursing home, type of ownership, and level of drug use. Professional staff and aides participated in an educational program at one randomly selected nursing home in each pair. At baseline, the drug use status was determined for all residents of the nursing homes ($n = 823$), and a blinded observer performed standardized clinical assessments of the residents who were taking psychoactive medications. After the five-month program, drug use and patient clinical status were reassessed and the educational program was found to have reduced the use of psychoactive drugs in the nursing homes (Avorn et al. 1992).

a. How would you characterize the experimental design used in this study?

b. If the effectiveness of the educational program is related to the organizational and leadership types of the nursing home staff, is the effect of this confounder effectively controlled in this study? If not, how would you modify the experimental design?

c. Obviously not all the nursing homes that could be matched were included in this study. How might this limitation affect the study findings?

d. Discuss to what extent the study findings can be extrapolated to nursing homes in other states.

REFERENCES

Anderson, J. W., D. B. Spencer, C. C. Hamilton, et al. "Oat-Bran Cereal Lowers Serum Total and LDL Cholesterol in Hypercholesterolemic Men." *American Journal of Clinical Nutrition* 52:495–499, 1990.

Anderson, P., C. Bartlett, G. Cook, and M. Woodward. "Legionnaires Disease in Reading — Possible Association with a Cooling Tower." *Community Medicine* 7:202–207, 1985.

Avorn, J., S. B. Soumerai, D. E. Everitt, et al. "A Randomized Trial of a Program to Reduce the Use of Psychoactive Drugs in Nursing Homes." *The New England Journal of Medicine* 327:168–173, 1992.

Campbell, B. J., and F. A. Campbell. "Injury Reduction and Belt Use Associated with Occupant Restraint Laws." In Graham, J. D. (ed.) *Preventing Automobile Injury: New Findings from Evaluation Research*, Chapter 2. Dover, MA: Auburn House Publishing Co., 1988.

Deming, W. E. "On Probability As a Basis For Action." *The American Statistician* 29:146–152, 1975.

Digitalis Investigation Group. "Rationale, Design, Implementation, and Baseline Characteristics of Patients in the DIG Trial: A Large, Simple, Long-Term Trial to Evaluate the Effect of Digitalis on Mortality in Heart Failure." *Controlled Clinical Trials* 17:77–97, 1995.

Fienberg, S. E. "Randomization and Social Affairs: The 1970 Draft Lottery." *Science* 171:255–261, 1971.

Francis, T., Jr., R. F. Korns, R. B. Voight, et al. "An Evaluation of the 1954 Poliomyelitis Vaccine Trials: Summary Report." *American Journal of Public Health* 45, Supplement:1–63, 1955.

Goldman, S., J. Copeland, T. Moritz, et al. "Improvement in Early Saphenous Vein Graft Patency After Coronary Artery Bypass Surgery with Antiplatelet Therapy: Results of a Veterans Administration Cooperative Study." *Circulation* 77:1324–1332, 1988.

Goldman, S., J. Copeland, T. Moritz, et al. "Saphenous Vein Graft Patency 1 Year After Coronary Artery Bypass Surgery and Effects of Antiplatelet Therapy: Results of a Veterans Administration Cooperative Study." *Circulation* 80:1190–1197, 1989.

Government Accounting Office. "Nutrition Monitoring: Mismanagement of Nutrition Survey Has Resulted in Questionable Data." GAO/RCED-91-117, 1991.

Hill, J., H. A. Bird, G. C. Fenn, C. E. Lee, M. Woodward, and V. Wright. "A Double Blind Crossover Study to Compare Lysine Acetyl Salicylate (Aspergesic) with Ibuprofen in the Treatment of Rheumatoid Arthritis." *Journal of Clinical Pharmacologic Therapeutics*, 15:205–211, 1990.

Hypertension Detection and Follow-up Program (HDFP) Cooperative Group. "Five-Year Findings of the Hypertension Detection and Follow-up Program," *Journal of the American Medical Association* 242:2562–2571, 1979.

Kalton, G. *Compensating for Missing Survey Data*. Research Report Series, Institute for Social Research, the University of Michigan, 1983.

Kushi, L. H., R. M. Fee, T. A. Sellers, W. Zheng, and A. R. Folsom. "Intake of Vitamins A, C, and E and Postmenopausal Breast Cancer. The Iowa Women's Health Study." *American Journal of Epidemiology* 144:165–174, 1996.

Materson, B. J., D. J. Reda, W. C. Cushman, et al. "Single-Drug Therapy for Hypertension in Men: A Comparison of Six Antihypertensive Agents with Placebo." *The New England Journal of Medicine* 328:914–921, 1993.

National Center for Health Statistics: Plan and Operation of the Health and Nutrition Examination Survey, United States, 1971–73. *Vital and Health Statistics*. Series 1, No. 10a. DHEW Pub. No. (HSM) 73-1310, 1973.

National Center for Health Statistics. *Health, United States, 1991 and Prevention Profile*. Hyattsville, MD: Public Health Service. DHHS Pub. No. 92-1232, 1992.

The NHLBI Task Force on Blood Pressure Control in Children. "The Report of the Second Task Force on Blood Pressure Control in Children, 1987." *Pediatrics* 79:1–25, 1987.

Office of Science and Technology Policy (OSTP). "Federal Policy for the Protection of Human Subjects: Notices and Rules." *Federal Register* 56:28003–28032, 1991.

Thomas, L. *The Youngest Science: Notes of a Medicine Watcher*. New York: Bantam Books, 1984.

Tippett, L. H. C. "Random Sampling Numbers". *Tracks of Computers*. No. 15. Ed. E. S. Pearson, Cambridge University Press, 1927.

Waksberg, J. "Sampling Methods for Random Digit Dialing." *Journal of the American Statistical Association* 73:40–46, 1978.

Interval Estimation

7

Chapter Outline

In Chapter 5 we saw that variation occurs when we use a sample instead of the entire population. For example, in the presentation of the binomial distribution, we saw that the sample estimates of the population proportion varied considerably from sample to sample. In this chapter, we present *prediction, confidence*, and *tolerance intervals*, quantities that allow us to take the variation in sample results into account in describing the data. These intervals represent specific types of *interval estimation* — the provision of limits that are likely to contain either (1) the population parameter of interest or (2) future observations of the variable. Interval estimation thus provides more information about the population parameter than the *point estimation* approach that we met in Chapter 3. In that chapter, we provided a single value as the estimate of the population parameter without giving any information about the sampling variability of the estimator. For example, knowledge of the value of the sample mean, a point estimate of the population mean, does not tell us anything about the variability of the sample mean. Interval estimation addresses this variability.

7.1 Prediction, Confidence, and Tolerance Intervals

The material in this and the following section is based on material presented by Vardeman (1992) and Walsh (1962). To understand the difference between these three intervals (prediction, confidence, and tolerance), consider the following. Dairies add vitamin D to milk for the purpose of fortification. The recommended amount of vitamin D to be added to a quart of milk is 400 IUs (10 μg). If a dairy adds too much vitamin D, perhaps over 5000 IUs, the possibility exists that a consumer will develop hypervitaminosis D — that is, vitamin D toxicity.

A *prediction interval* focuses on a single observation of the variable — for example, the amount of vitamin D in the next bottle of milk. A *confidence interval* focuses on a population parameter — for example, the mean or median amount of vitamin D per bottle in a population of bottles of milk. Thus, the prediction interval is of more interest to the consumer of the next bottle of milk, whereas the confidence interval is of more interest to the dairy. A *tolerance interval* provides limits such that there is a high level of confidence that a large proportion of values of the variable will fall within them. For example, besides being interested in the mean, the dairy owner or a regulatory agency also wants to be confident that a large proportion of the bottles' vitamin D contents are within a specified tolerance of the value of 400 IUs. We begin our treatment of these intervals with distribution-free intervals.

7.2 Distribution-Free Intervals

When the method for forming the different intervals is independent of how the data are distributed, the resultant intervals are said to be *distribution free*. Distribution-free intervals are based on the rank order of the sample values, with the following notations for rank order. The smallest of the x values is indicated by $x_{(1)}$, the second smallest by $x_{(2)}$, and so on, to the largest value that is denoted by $x_{(n)}$. The $x_{(i)}$ are called *order statistics*, since the subscripts show the order of the values.

We shall use hypothetical data showing the amount of vitamin D in 30 bottles of milk selected at random from one dairy. The values are shown in rank order in Table 7.1.

Based on this sample, $x_{(1)}$ equals 289 IUs, $x_{(2)}$ is 326 IUs and so on to $x_{(30)}$, which equals 485 IUs.

Table 7.1   Values of vitamin D (IUs) in a hypothetical sample of 30 bottles.

289	355	376	392	406	433
326	363	379	395	410	434
339	364	384	396	413	456
346	370	386	398	422	471
353	373	389	403	427	485

7.2.1   Prediction Interval

As a consumer of milk, our major concern about vitamin D is that the milk does not contain an amount of vitamin D that is toxic to us. We are not too concerned about there being too little vitamin D in the bottle. Based on the hypothetical sample of vitamin D contents in 30 bottles of milk, we can form a one-sided prediction interval — our concern focuses on the upper limit — for the amount of vitamin D in the bottle of milk that we are going to purchase.

A natural one-sided prediction interval in this case is from 0 to the maximum observed value of vitamin D (485 IUs) in the sample. The level of confidence associated with this interval, from 0 to 485 IUs, is 96.8 percent (= 30/31). This value can be found from the consideration of the order statistics and the real number line. For example, we have the line

```
|__1__|__2__|__3__|                          |__30__|__31__
 0    x(1)   x(2)   x(3)         . . . . .      x(30)
```

and there are 31 intervals along this line. The vertical marks (|) indicate the location of the order statistics along the line, and the numbers above the line between the |'s indicate the interval number. There are 31 intervals, and the next observation can fall into any one of the intervals. Of these 31 intervals, 30 have values less than the maximum value. Hence, we are 96.8 percent confident that the vitamin D content in the next bottle will be between zero and the observed maximum value.

Note that we used the word *confidence* instead of *probability* here. We use confidence because we are using the sample data as the basis of estimating the probability distribution of the vitamin D content. If we used the probability distribution of the vitamin D content instead of using its sample estimate, the empirical distribution function, we would use the word *probability*. In repeated sampling, we expect that 96.8 percent of the prediction intervals, ranging from zero to the observed maximum in each sample of size 30, would contain the next observed vitamin D content.

The use of the second largest value, $x_{(29)}$, as the upper limit of the interval results in a prediction confidence level of 93.5 percent (= 29/31). An attraction of this interval is that it provides a slightly shorter interval with a maximum of 471 IUs, but we are slightly less confident about it. Based on either of these intervals, the consumer should not be worried about purchasing a bottle that has a value of vitamin D that would cause vitamin D poisoning.

For a two-sided interval, a natural interval would be from the minimum observed value, $x_{(1)}$, to the maximum observed value, $x_{(30)}$. In this case, the two-sided interval is from 289 to 485 IUs. The confidence level associated with this prediction interval is 93.5 percent (= 29/31). Of the 31 intervals just shown, there is one below the minimum value and one above the maximum value. Hence, there are 29 chances out of 31 that the next observed value will fall between the minimum and maximum values.

With a sample size of 30, it is not possible to have a distribution-free, two-sided, 95 percent prediction interval. The smallest sample size that attains the 95 percent level is 39. When n is 39, there are 40 intervals, and 2/40 equals 0.05. This calculation shows that it is easy to determine how large a sample is required to satisfy prediction interval requirements.

7.2.2 Confidence Interval

The dairy wants to know, on average, how much vitamin D is being added to the milk. If the interval estimate for the central tendency differs much from 400 IUs, the dairy may have to change its process for adding vitamin D. One way of obtaining the interval estimate is to use a distribution-free confidence interval.

Distribution-free confidence intervals are used to provide information about population parameters — for example, the median and other percentiles. There are two approaches to finding confidence intervals for percentiles: (1) the use of order statistics and (2) the use of the normal approximation to the binomial distribution. The first approach is generally used for smaller samples, whereas the second approach is used for larger samples.

Use of Order Statistics and the Binomial Distribution: The lower and upper limits of the $(1 - \alpha)100$ percent confidence interval for the pth percentile of X are the order statistics $x_{(j)}$ and $x_{(k)}$, where the values of j and k, j less than k, are to be determined. The limits of the confidence interval for the pth percentile of X are the values $x_{(j)}$ and $x_{(k)}$ that satisfy the following inequality:

$$\Pr\{x_{(j)} < p^{th}\ percentile < x_{(k)}\} \geq 1 - \alpha$$

and this is equivalently

$$\Pr\{x_{(j)} \geq p^{th}\ percentile\} + \Pr\{x_{(k)} \leq p^{th}\ percentile\} \leq \alpha$$

If we require that both terms in the sum be less than or equal to $\alpha/2$, from the first term, we have

$$\Pr\{at\ most\ j - 1\ observations < p^{th}\ percentile\} \leq \alpha/2.$$

This is a situation with two outcomes: an observation is less than the pth percentile, or it is greater than or equal to the pth percentile. The probability that an observation is less than the pth percentile is p. The variable of interest is the number of observations, out of the n, that are less than the pth percentile. Thus, this variable follows a binomial distribution with parameters n and p. Knowing the values of n and p enables us to find the value of j because j must satisfy the following inequality:

$$\sum_{i=0}^{j-1} \frac{n!}{i!(n-i)!} p^i (1-p)^{n-i} \leq \alpha/2.$$

The inequality used to find the value of k is

$$\sum_{i=k}^{n} \frac{n!}{i!(n-i)!} p^i (1-p)^{n-i} \leq \alpha/2.$$

Putting these two inequalities together means that the binomial sum from j to $k-1$ must be greater than or equal to $1 - \alpha$. Here we have dropped the requirement that the sums of the probabilities from 0 to $j - 1$ and from k to n both must be less than $\alpha/2$. The values of j and k are found from the binomial table, Table B2, or by using a computer package such as SAS or Stata.

For example, suppose we want to find a 95 percent confidence interval for the median, the 50th percentile, for the vitamin D values from the dairy used in Table 7.1. The sample estimate of the median is the average of the 15th and 16th smallest values — that is, 390.5 IUs (= [389 + 392]/2).

To find the 95 percent confidence interval for the median in the population of bottles of milk from the selected dairy, we use the binomial distribution. For this problem we need a binomial distribution with $n = 30$ and $\pi = 0.5$, shown in Table 7.2. Since Table B2 does not have values for n larger than 20, we used SAS to obtain the distribution. The order and observations from Table 7.1 are also shown in the last two columns in Table 7.2. There may be more than one pair of values of j and k that satisfy the requirement that the sum of the binomial probabilities from j to $k-1$ is greater than or equal to $1 - \alpha$. To choose from among these pairs, we shall select the pair whose difference $(k-j)$ is the smallest. In the special case of the median, we shall require that k equals $n - j + 1$; this requirement gives the same number of observations in both tails of the distribution.

Table 7.2   Cumulative binomial distribution with $n = 30$ and $\pi = 0.5$ and sorted observations in Table 7.1.

x	Pr $(X \le x)$	No.	Observation
0	0.0000	1	289
1	0.0000	2	326
2	0.0000	3	339
3	0.0000	4	346
4	0.0000	5	353
5	0.0002	6	355
6	0.0007	7	363
7	0.0026	8	364
8	0.0081	9	370
9	<u>0.0214</u>	<u>10</u>	<u>373</u>
10	0.0494	11	376
11	0.1002	12	379
12	0.1808	13	384
13	0.2923	14	386
14	0.4278	15	389
15	0.5722	16	392
16	0.7077	17	395
17	0.8192	18	396
18	0.8998	19	398
19	0.9506	20	403
20	<u>0.9786</u>	<u>21</u>	<u>406</u>
21	0.9919	22	410
22	0.9974	23	413
23	0.9993	24	422
24	0.9998	25	427
25	1.0000	26	433
26	1.0000	27	434
27	1.0000	28	456
28	1.0000	29	471
29	1.0000	30	485
30	1.0000		

The sum of the probabilities from j to $k - 1$ must be greater than or equal to 0.95. Examination of the cumulative probabilities tells us that j is 10 and k is 21. The sum of the probabilities between 10 and 20 is 0.9572 ($= 0.9786 - 0.0214$). If j were 11 and k were 20, the sum of the probabilities between 11 and 19 is 0.9012, less than the required value of 0.95. Thus, the approximate 95 percent (really closer to 96%) confidence interval for the median is from 373 IUs ($= x_{(10)}$) to 406 IUs ($= x_{(21)}$). The use of distribution-free intervals does not necessarily provide intervals that are symmetric about the sample estimator. For example, the sample median value, 390.5 IUs, is not in the exact middle of the confidence interval.

Note that the confidence interval for the median is much narrower than the approximate 95 percent prediction interval, from 289 to 485 IUs, for a single observation. As we saw in Chapter 3, there is much less variability associated with a mean or median than with a single observation, and this is additional confirmation of that.

As we can observe from the preceding, the use of distribution-free intervals does not provide exactly 95 percent levels. The level of confidence associated with these intervals is a function of the sample size as well as which order statistics are used in the creation of the interval.

It is also possible to create one-sided confidence intervals for parameters. For example, if the goal were to create an upper one-sided confidence interval for the median, we would find the value of k such that

$$\sum_{i=k}^{n} \frac{n!}{i!(n-i)!} p^i (1-p)^{n-i} \leq \alpha$$

for a p having the value of 0.50. The upper one-sided confidence interval for the median is from 0 to $x_{(k)}$ where k's value is found from the above inequality.

***Use of the Normal Approximation to the Binomial*:** For larger sample sizes, the normal approximation to the binomial distribution can be used to find the values of j and k. The sample size must be large enough to satisfy the requirements for the use of the normal approximation. Since p is 0.50, the sample size of 30 bottles from the dairy is large enough.

As before, we want to find the value of j such that the probability of the binomial variable, Y, being less than or equal to $j - 1$ is less than or equal to $\alpha/2$ — that is,

$$\Pr\{Y \leq j - 1\} \leq \alpha/2.$$

Use of the continuity correction converts this to

$$\Pr\{Y \leq j - 0.5\} \leq \alpha/2.$$

To convert Y to the standard normal variable, we must subtract np, the estimate of the mean, and divide by $\sqrt{np(1-p)}$, the estimate of the standard error. This yields

$$\Pr\left\{\frac{Y - np}{\sqrt{np(1-p)}} \leq \frac{j - 0.5 - np}{\sqrt{np(1-p)}}\right\} \leq \frac{\alpha}{2}.$$

This can be rewritten as

$$\Pr\left\{Z \leq \frac{j - 0.5 - np}{\sqrt{np(1-p)}}\right\} \leq \frac{\alpha}{2}.$$

If we change this inequality to equality — that is, the probability is equal to $\alpha/2$ — we can find a unique value for j. The value of the term on the right side of the inequality inside the brackets is simply $z_{\alpha/2}$, and hence we can find the value of j from the equation

$$j - 0.5 - np = z_{\alpha/2}\sqrt{np(1-p)}$$

or

$$j = z_{\alpha/2}\sqrt{np(1-p)} + 0.5 + np.$$

In the preceding example, p was 0.50, n was 30, and α was 0.05. Since the value of $z_{0.025}$ is -1.96, we have

$$j = -1.96\sqrt{30(0.5)(0.5)} + 0.5 + 30(0.5)$$

or j is 10.13. To ensure that the level of the confidence interval is at least $(1 - \alpha) * 100$ percent, we must round down the value of j to the next smaller integer, 10, and we round up the value of k, found following, to the next larger integer.

The value of k is found from the equation

$$k = z_{1-\alpha/2}\sqrt{np(1-p)} + 0.5 + np$$

which yields a k equal to 20.87, which is rounded to 21. Thus, the 95 percent confidence interval is from 373 IUs ($= x_{(10)}$) to 406 IUs ($= x_{(21)}$). In this case, the binomial and the normal approximation approaches resulted in the same confidence limits.

7.2.3   Tolerance Interval

As we said before, tolerance intervals are of most interest to the dairy or to a regulatory agency. The tolerance limits are values such that we have a high level of confidence that a large proportion of the bottles have vitamin D contents located between the lower and upper tolerance limits. These upper and lower limits of the tolerance interval can be used in determining whether or not the process for adding vitamin D is under control. If the limits are too wide, the dairy may have to modify its process for adding vitamin D to the milk.

The dairy does not want to add too much vitamin D to the milk because of the possible problems for the consumer and the extra cost associated with using more vitamin D than required. At the same time, the dairy must add enough vitamin D to be in compliance with truth in advertising legislation.

As with the prediction interval, it is reasonable to use the smallest and largest observed values for the lower and upper limits of the tolerance interval, although other values could be used. We also have to specify the proportion of the population, p, that we want to include within the tolerance interval. Given the tolerance interval limits and the proportion of values to be included within it, we can calculate the confidence level, γ, associated with the interval.

In symbols, the tolerance interval limits are the order statistics $x_{(j)}$ and $x_{(k)}$ such that

$$\Pr\left[\Pr\{X \le x_{(k)}\} - \Pr\{X \le x_{(j)}\} \ge p\right] = \gamma.$$

The quantity, $\Pr\{X \le x_{(k)}\} - \Pr\{X \le x_{(j)}\}$, is the proportion of the population values contained in the tolerance interval for this sample. Let us call the above quantity W_{kj}. In symbols we then have $\Pr\{W_{kj} \ge p\} = \gamma$. The variable W_{kj} is either less than p or greater than or equal to p. This is a binomial situation, and, therefore, we can use the same approach as in the confidence interval section to find the value of γ. The value of γ can be expressed in terms of the binomial summation as

$$\gamma = \sum_{i=0}^{k-j-1} \frac{n!}{i!(n-i)!} p^i (1-p)^{n-i}.$$

If we use the minimum, $x_{(1)}$, and the maximum, $x_{(n)}$, for the limits, $k - j - 1$ becomes $n - 1 - 1$, which equals $n - 2$. It is therefore easy to find the value of this summation for i ranging from 0 to $n - 2$ because that sum is equal to 1 minus the binomial sum from $n - 1$ to n. In symbols, the value of γ is

$$1 - [p^n] - [np^{n-1}(1 - p)].$$

Suppose we want our tolerance interval to contain 95 percent of the observations. Let's calculate the confidence level associated with the tolerance interval of 289 to 485 IUs. In this case, n is 30 and p is 0.95. The value of γ is found by taking $1 - 0.95^{30}$

$- 30(0.95)^{29}(1 - 0.95)$, which equals 0.4465. There is not a high level of confidence associated with this tolerance interval. This confidence level is contrasted with the 0.935 level associated with the prediction interval. It is not surprising that the confidence level of the prediction interval is much higher than that of the tolerance interval because the prediction interval is based on the location of a single future value whereas the tolerance interval is based on the location of a large proportion of the population values.

The interval from 289 to 485 IUs is the widest interval we can have using the sample data since these are the minimum and maximum observed values. We can increase our confidence by either (1) decreasing p, the proportion of the population to be included in the tolerance interval or (2) by taking a larger sample.

Let us reduce p to 90 percent. The confidence level for this interval is increased to 0.8162, a much more reasonable value. Instead of reducing p, let us increase the sample size from 30 to 60. The confidence level associated with the increased sample size is 0.8084, also a much more reasonable value. Table 7.3 shows the sample size required to have 90, 95, and 99 percent confidence associated with tolerance intervals that have 80, 90, 95, and 99 percent coverage of the distribution, based on the use of $x_{(1)}$ and $x_{(n)}$.

Table 7.3 Sample size required for the tolerance interval to have the indicated confidence level for the specified coverage proportions based on the use of $x_{(1)}$ and $x_{(n)}$.

Coverage	Confidence Level		
Proportion	90%	95%	99%
0.80	18	22	31
0.90	38	46	64
0.95	77	93	130
0.99	388	473	662

From these calculations and the general formula for calculating, we can see the relationships between p, the values of k and j, n and γ. We can investigate the values of these quantities before we have performed the study and can modify the proposed study design if we are not satisfied with the values of p and γ.

A one-sided tolerance interval is sometimes of interest. Suppose that there was interest in the upper one-sided tolerance interval. In this case, the tolerance interval ranges from 0 to $x_{(n)}$ and the confidence associated with this interval is found by taking $1 - p^n$ — that is, one minus the binomial term calculated for i equal to n.

7.3 Confidence Intervals Based on the Normal Distribution

If the data are from a known probability distribution, knowledge of this distribution allows more informative (smaller) intervals to be constructed for the parameters of interest or for future values. We begin this presentation by showing how to create confidence intervals for a variety of population parameters, assuming that the data come from a normal distribution. The central limit theorem and the sampling distribution of statistics (e.g., sample mean) presented in Chapter 5 provide the rationale for interval estimation based on the normal distribution. Following the material on confidence intervals, we show how to use the normal distribution in the creation of prediction and tolerance intervals. We begin the confidence interval presentation with the population mean and follow it with the confidence interval for the population proportion that can also be viewed as a mean.

7.3.1   Confidence Interval for the Mean

In the preceding material, we saw how to construct a confidence interval for the population median. That confidence interval gave information to the dairy about the amount of vitamin D being added to the milk. As an alternative to the median, a confidence interval for the mean could have been used. To find a confidence interval for the mean, assuming that the data follow a specific distribution, we must know the sampling distribution of its estimator. We must also specify how confident we wish to be that the interval contains the population parameter. The sample mean is the estimator of the population mean, and the sampling distribution of the sample mean is easily found.

Since we are assuming the data follow a normal distribution, the sample mean — the average of the sample values — also follows a normal distribution. However, this assumption is not crucial. Even if the data are not normally distributed, the central limit theorem states that the sample mean, under appropriate conditions, will approximately follow a normal distribution.

To specify the normal distribution completely, we also have to provide the mean and variance of the sample mean. First we develop the confidence interval for the mean assuming population variance is known and extend it to the situation where population variance is unknown and it is estimated from the sample.

Known Variance: In Chapter 5, we saw that the mean of the sample mean was μ, the population mean, and its variance was σ^2/n. The standard deviation of the sample mean is thus $\sigma/\sqrt{n}$, and it is called the *standard error* of the sample mean $(\bar{x})$. The use of the word *error* is confusing, since no mistake has been made. However, it is the traditional term used in this context. The term *standard error* is used instead of standard deviation when we are discussing the variation in a sample statistic. The term *standard deviation* is usually reserved for discussion of the variation in the sample data themselves. Thus, the standard deviation measures the unit-to-unit variation, while the standard error measures the sample-to-sample variation.

We now address the issue of how confident we wish to be that the interval contains the population mean (μ). From the material on the normal distribution in Chapter 5, we know that

$$\Pr\{-1.96 < Z < 1.96\} = 0.95$$

where Z is the standard normal variable. In terms of the sample mean, this is

$$\Pr\left\{-1.96 < \frac{\bar{x} - \mu}{\left(\sigma/\sqrt{n}\right)} < 1.96\right\} = 0.95.$$

But we want an interval for μ, not for Z. Therefore, we must perform some algebraic manipulations to convert this to an interval for μ. First we multiply all three terms inside the braces by $\sigma/\sqrt{n}$. This yields

$$\Pr\left\{-1.96\left(\frac{\sigma}{\sqrt{n}}\right) < \bar{x} - \mu < 1.96\left(\frac{\sigma}{\sqrt{n}}\right)\right\} = 0.95.$$

We next subtract $\bar{x}$ from all the expressions inside the braces, and this gives

$$\Pr\left\{-1.96\left(\frac{\sigma}{\sqrt{n}}\right) - \bar{x} < -\mu < 1.96\left(\frac{\sigma}{\sqrt{n}}\right) - \bar{x}\right\} = 0.95.$$

This interval is about $-\mu$; to convert it to an interval about μ, we must multiply each term in the brackets by -1. Before doing this, we must be aware of the effect of multiplying an inequality by a minus number. For example, we know that 3 is less than 4. However, -3 is greater than -4, so the result of multiplying both sides of an inequality by -1 changes the direction of the inequality. Therefore, we have

$$\Pr\left\{1.96\left(\frac{\sigma}{\sqrt{n}}\right) + \bar{x} > \mu > -1.96\left(\frac{\sigma}{\sqrt{n}}\right) + \bar{x}\right\} = 0.95.$$

We reorder the terms to have the smallest of the three quantities to the left — that is,

$$\Pr\left\{\bar{x} - 1.96\left(\frac{\sigma}{\sqrt{n}}\right) < \mu < \bar{x} + 1.96\left(\frac{\sigma}{\sqrt{n}}\right)\right\} = 0.95$$

or, more generally,

$$\Pr\left\{\bar{x} - z_{1-\alpha/2}\left(\frac{\sigma}{\sqrt{n}}\right) < \mu < \bar{x} + z_{1-\alpha/2}\left(\frac{\sigma}{\sqrt{n}}\right)\right\} = 1 - \alpha.$$

The $(1 - \alpha) * 100$ percent confidence interval limits for the population mean can be expressed as

$$\bar{x} \pm z_{1-\alpha/2}\left(\frac{\sigma}{\sqrt{n}}\right).$$

The result of these manipulations is an interval for μ in terms of σ, n, 1.96 (or some other z value), and $\bar{x}$. The sample mean, $\bar{x}$, is the only one of these quantities that varies from sample to sample. However, once we draw a sample, the interval is fixed as the sample mean's value, $\bar{x}$, is known. Since the interval will either contain or not contain μ, we no longer talk about the probability of the interval containing μ.

Although we do not talk about the probability of an interval containing μ, we do know that in repeated sampling, intervals of the preceding form will contain the parameter, μ, 95 percent of the time. Thus, instead of discussing the probability of an interval containing μ, we say that we are 95 percent confident that the interval from $[\bar{x} - 1.96(\sigma/\sqrt{n})]$ to $[\bar{x} + 1.96(\sigma/\sqrt{n})]$ will contain μ. Intervals of this type are therefore called *confidence intervals*. This reason for the use of the word *confidence* is the same as that discussed in the preceding distribution-free material. The limits of the confidence interval usually have the form of the sample estimate plus or minus some distribution percentile — in this case, the normal distribution — times the standard error of the sample estimate.

Example 7.1

The 95 percent confidence interval for the mean systolic blood pressure for 200 patients can be found based on the dig200 data set introduced in Chapter 3. We assume that the standard deviation for this patient population is 20 mmHg. As the sample mean, $\bar{x}$, based on a sample size of 199 (one missing value) observations, was found to be 125.8 mmHg, the 95 percent confidence interval for the population mean ranges from $\left[125.8 - 1.96\left(20/\sqrt{199}\right)\right]$ to $\left[125.8 - 1.96\left(20/\sqrt{199}\right)\right]$ — that is, from 123.0 to 128.6 mmHg.

Table 7.4 illustrates the concept of confidence intervals. It shows the results of drawing 50 samples of size 60 from a normal distribution with a mean of 94 and a standard deviation of 11. These values are close to the mean and standard deviation of the systolic blood pressure variable for 5-year-old boys in the United States as reported by the NHLBI Task Force on Blood Pressure Control in Children (1987).

In this demonstration, 4 percent (2 out of 50 marked in the table) of the intervals did not contain the population mean, and 96 percent did. If we draw many more samples, the proportion of the intervals containing the mean will be 95 percent. This is the basis for the statement that we are 95 percent confident that the confidence interval, based on our single sample, will contain the population mean.

If we use a different value for the standard normal variable, the level of confidence changes accordingly. For example, if we had started with a value of 1.645, $z_{0.95}$, instead

Table 7.4   Simulation of 95% confidence intervals for 50 samples of $n = 60$ from the normal distribution with $\mu = 94$ and $\sigma = 11$ (standard error = 1.42).

Sample	Mean	Std	95% CI	Sample	Mean	Std	95% CI
1	94.75	10.25	(91.96, 97.54)	26	94.61	11.49	(91.82, 97.39)
2	94.85	10.86	(92.06, 97.63)	27	92.79	9.36	(90.00, 95.58)
3	94.71	10.09	(91.92, 97.50)	28	96.00	12.19	(93.22, 98.79)
4	94.03	12.27	(91.24, 96.82)	29	95.99	11.36	(93.20, 98.78)
5	93.77	10.05	(90.98, 96.56)	30	93.98	11.74	(91.19, 96.76)
6	92.54	9.32	(89.76, 95.33)	31	95.36	13.08	(92.57, 98.15)
7	93.40	12.07	(90.62, 96.19)	32	91.10	8.69	(88.31, 93.89)*
8	93.97	11.02	(91.18, 96.75)	33	93.85	12.94	(91.06, 96.63)
9	96.33	9.26	(93.54, 99.12)	34	96.01	9.63	(93.22, 98.79)
10	93.56	12.01	(90.78, 96.35)	35	95.20	8.94	(92.41, 97.99)
11	94.94	10.81	(92.15, 97.73)	36	95.64	9.41	(92.85, 98.43)
12	94.66	12.08	(91.88, 97.45)	37	94.74	10.31	(91.95, 97.53)
13	94.21	11.02	(91.42, 97.00)	38	93.52	10.30	(90.73, 96.31)
14	94.55	9.98	(91.76, 97.34)	39	92.92	10.27	(90.13, 95.71)
15	93.57	11.50	(90.79, 96.36)	40	95.08	10.07	(92.30, 97.87)
16	95.99	12.01	(93.20, 98.78)	41	93.88	10.53	(91.09, 96.66)
17	93.86	12.53	(91.08, 96.65)	42	95.38	9.98	(92.59, 98.17)
18	92.02	13.58	(89.23, 94.81)	43	94.38	11.65	(91.59, 97.17)
19	95.16	12.03	(92.38, 97.95)	44	91.55	10.63	(88.76, 94.33)
20	94.99	12.00	(92.20, 97.78)	45	95.41	12.79	(92.62, 98.20)
21	94.65	11.18	(91.86, 97.43)	46	92.40	10.57	(89.62, 95.19)
22	92.86	12.52	(90.07, 95.64)	47	96.00	11.45	(93.21, 98.78)
23	93.99	11.76	(91.20, 96.78)	48	95.39	10.56	(92.60, 98.18)
24	91.44	10.75	(88.65, 94.22)	49	97.69	10.89	(94.90, 100.47)*
25	96.07	11.89	(93.28, 98.86)	50	95.01	10.61	(92.22, 97.79)

*Does not contain 94

of 1.96, $z_{0.975}$, the confidence level would be 90 percent instead of 95 percent. The $z_{0.95}$ value is used with the 90 percent level because we want 5 percent of the values to be in each tail. The lower and upper limits for the 90 percent confidence interval for the population mean for the data in the first sample of 60 observations are 92.41 [= 94.75 − 1.645(1.42)] and 97.09 [= 94.75 + 1.645(1.42)], respectively. This interval is narrower than the corresponding 95 percent confidence interval of 91.96 to 97.54. This makes sense, since, if we wish to be more confident that the interval contains the population mean, the interval will have to be wider. The 99 percent confidence interval uses $z_{0.995}$, which is 2.576, and the corresponding interval is 91.09 [= 94.75 − 2.576(1.42)] to 98.41 [= 94.75 + 2.576(1.42)].

The fifty samples shown in Table 7.4 had sample means, based on 60 observations, ranging from a low of 91.1 to a high of 97.7. This is the amount of variation in sample means expected if the data came from the same normal population with a mean of 94 and a standard deviation of 11. The Second National Task Force on Blood Pressure Control in Children (1987) had study means ranging from 85.6 (based on 181 values) to 103.5 mmHg (based on 61 values), far outside the range just shown. These extreme values suggest that these data do not come from the same population, and this then calls into question the Task Force's combination of the data from these diverse studies.

The size of the confidence interval is also affected by the sample size that appears in the $\sigma/\sqrt{n}$ term. Since n is in the denominator, increasing n decreases the size of the confidence interval. For example, if we doubled the sample size from 60 to 120 in the preceding example, the standard error of the mean changes from $1.42(=11/\sqrt{60})$ to $1.004(=11/\sqrt{120})$. Doubling the sample size reduces the confidence interval to about 71 percent ($=1/\sqrt{2}$) of its former width. Thus, we know more about the location of the population mean, since the confidence interval is shorter as the sample size increases.

The size of the confidence interval is also a function of the value of σ, but to change σ means that we are considering a different population. However, if we are willing to consider homogeneous subgroups of the population, the value of the standard deviation for a subgroup should be less than that for the entire population. For example, instead of considering the blood pressure of 5-year-old boys, we consider the blood pressure of 5-year-old boys grouped according to height intervals. The standard deviation of systolic blood pressure in the different height subgroups should be much less than the overall standard deviation.

Another factor affecting the size of the confidence interval is whether it is a one-sided or a two-sided interval. If we are only concerned about higher blood pressure values, we could use an upper one-sided confidence interval. The lower limit would be zero, or $-\infty$ for a variable that had positive and negative values, and the upper limit is

$$\bar{x} + z_{1-\alpha}\left(\frac{\sigma}{\sqrt{n}}\right).$$

This is similar to the two-sided upper limit except for the use of $z_{1-\alpha}$ instead of $z_{1-\alpha/2}$.

Unknown Variance: When the population variance, σ^2, is unknown, it is reasonable to substitute its sample estimator, s^2, in the confidence interval calculation. There is a problem in doing this, though. Although $(\bar{x}-\mu)/(\sigma/\sqrt{n})$ follows the standard normal distribution, $(\bar{x}-\mu)/(s/\sqrt{n})$ does not. In the first expression, there is only one random variable, $\bar{x}$, whereas the second expression involves the ratio of two random variables, $\bar{x}$ and s. We need to know the probability distribution for this ratio of random variables.

Fortunately, Gosset, who we encountered in Chapter 5, already discovered the distribution of $(\bar{x} - \mu)/(s/\sqrt{n})$. The distribution is called *Student's t* — crediting Student, the pseudonym used by Gosset — or, more simply, the *t distribution*. For large values of n, sample values of s are very close to σ, and, hence, the t distribution looks very much like the standard normal. However, for small values of n, the sample values of s vary considerably, and the t and standard normal distributions have different appearances. Thus, the t distribution has one parameter, the number of independent observations used in the calculation of s. In Chapter 3, we saw that this value was $n - 1$, and we called this value the degrees of freedom. Hence, the parameter of the t distribution is the degrees of freedom associated with the calculation of the standard error. The degrees of freedom are shown as a subscript — that is, as t_{df}. For example, a t with 5 degrees of freedom is written as t_5.

Figure 7.1 shows the distributions of t_1 and t_5 compared with the standard normal distribution over the range of -3.8 to 3.8. As we can see from these plots, the t distribution with one degree of freedom, the lowest curve, is considerably flatter — that is, there is more variability than for the standard normal distribution, the top curve in the figure. This is to be expected, since the sample mean divided by the sample standard deviation is more variable than the sample mean alone. As the degrees of freedom increase, the t distributions become closer and closer to the standard normal in appearance. The tendency for the t to approach the standard normal distribution as the number of degrees of freedom increases can also be seen in Table 7.5, which shows selected percentiles for several t distributions and the standard normal distribution. A more complete t table is found in Appendix Table B5.

Now that we know the distribution of $(\bar{x} - \mu)/(s/\sqrt{n})$, we can form confidence intervals for the mean even when the population variance is unknown. The form for the confidence interval is similar to that preceding for the mean with known variance except that s replaces σ and the t distribution is used instead of the standard normal

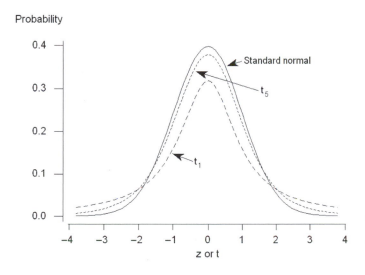

Figure 7.1
Distributions of t_1 and t_5 compared with z distribution.

Table 7.5 Selected percentiles for several t distributions and the standard normal distribution.

Distribution	Percentiles			
	0.80	0.90	0.95	0.99
t_1	1.376	3.078	6.314	31.821
t_5	0.920	1.476	2.015	3.365
t_{10}	0.879	1.372	1.813	2.764
t_{30}	0.854	1.310	1.697	2.457
t_{60}	0.848	1.296	1.671	2.390
t_{120}	0.845	1.289	1.658	2.358
Standard normal	0.842	1.282	1.645	2.326

distribution. Therefore, the lower and upper limits for the $(1 - \alpha) * 100$ percent confidence interval for the mean when the variance is unknown are $\left\{ \bar{x} - t_{n-1, 1-\alpha/2} \left(s/\sqrt{n} \right) \right\}$ and $\left\{ \bar{x} + t_{n-1, 1-\alpha/2} \left(s/\sqrt{n} \right) \right\}$, respectively.

Let us calculate the 90 percent confidence interval for the population mean of the systolic blood pressure for 5-year-old boys based on the first sample data in Table 7.4 (row 1). A 90 percent [= $(1 - \alpha) * 100$ percent] confidence interval means that α is 0.10. Based on a sample of 60 observations, the sample mean was 94.75 and the sample standard deviation was 10.25 mmHg. Thus, we need the 95th (= $1 - \alpha/2$) percentile of a t distribution with 59 degrees of freedom. However, neither Table 7.5 nor Table B5 shows the percentiles for a t distribution with 59 degrees of freedom. Based on the small changes in the t distribution for larger degrees of freedom, there should be little error if we use the 95th percentile for a t_{60} distribution. Therefore, the lower and upper limits are approximately

$$94.75 - 1.671 \left(\frac{10.25}{\sqrt{60}} \right) \quad \text{and} \quad 94.75 + 1.671 \left(\frac{10.25}{\sqrt{60}} \right)$$

or 92.54 and 96.96 mmHg, respectively.

If we use a computer package (see **Program Note 7.1** on the website) to find the 95th percentile value for a t_{59} distribution, we find its value is 1.6711. Hence, there is little error introduced in this example by using the percentiles from a t_{60} instead of a t_{59} distribution.

7.3.2 Confidence Interval for a Proportion

We are frequently exposed to the confidence interval for a proportion. Most surveys about opinions or voting intentions today report the margin of error. This quantity is simply one half the width of the 95 percent confidence interval for the proportion. Finding the confidence interval for a proportion, π, can be based on either the binomial or normal distribution. The binomial distribution is generally used for smaller samples and it provides an exact interval whereas the normal distribution is used with larger samples and provides an approximate interval. Let us examine the exact interval first.

Use of the Binomial Distribution: Suppose we wish to find a confidence interval for the proportion of restaurants that are in violation of local health ordinances. A simple random sample of 20 restaurants is selected, and, of those, four are found to have violations. The sample proportion, p, which is equal to 0.20 (= 4/20), is the point estimate of π, the population proportion. How can we use this sample information to create the $(1 - \alpha) * 100$ percent confidence interval for the population proportion?

This is a binomial situation, since there are only two outcomes for a restaurant — that is, a restaurant either does or does not have a violation. The binomial variable is the number of restaurants with a violation and we have observed its value to be 4 in this sample.

The limits of the confidence interval for the proportion are those values that make this outcome appear to be unusual. Another way of stating this is that the lower limit is the proportion for which the probability of 4 or more restaurants is equal to $\alpha/2$. Correspondingly, the upper limit is the proportion for which the probability of 4 or fewer restaurants is equal to $\alpha/2$. The two charts in Appendix Table B6 can be used to find the 95 and 99 percent confidence intervals.

Example 7.2

Suppose that we want the 95 percent confidence interval for $p = 0.20$ and $n = 20$. We use the first chart (Confidence Level 95 Percent) of Table B6, and, since the sample proportion is less than 0.50, we read across the bottom until we find the sample proportion value of 0.20. We then move up along the line corresponding to 0.20 until it intersects the first curve for a sample size of 20. Since p is less than 0.50, we read the value of the lower limit from the left vertical axis; it is slightly less than 0.06. To find the upper limit, we continue up the vertical line corresponding to 0.20 until we reach the second curve for a sample size of 20. We read the upper limit from the left vertical axis, and its value is slightly less than 0.44. The approximate 95 percent confidence limits are 0.06 and 0.44. Note that this interval is not symmetric about the point estimation. If p is greater than 0.5, we locate p across the top and read the limits from the right vertical axis.

Another method of finding the upper and lower limits of a confidence interval based on a binomial distribution is to find these values by trial and error.

Example 7.3

Suppose that we wish to find the 90 percent confidence interval for $p = 0.20$ ($x = 4$) and $n = 20$. This means that α is 0.10 and $\alpha/2$ is 0.05. We wish to find the probability of being less than or equal to 4 and being greater than or equal to 4 for different binomial proportions. For the upper limit, we can try some value above 0.20, say, 0.35 and calculate Pr $(X \leq x)$. If Pr $(X \leq x)$ is larger than $\alpha/2$, then we will try a larger value of p — say, 0.4. We can try this process until Pr $(X \leq x)$ is close enough to $\alpha/2$. For the lower limit, we try some value of p smaller than 0.20, say 0.1 and calculate Pr $(X \geq x)$, which is $1 -$ Pr $(X \leq x - 1)$. If $1 -$ Pr $(X \leq x - 1)$ is smaller than $\alpha/2$, then we try a smaller value of p — say, 0.07. Continue this process until $1 -$ Pr $(X \leq x - 1)$ is close enough to $\alpha/2$. Computers can perform this iterative process quickly. An SAS program produced the 90 percent confidence interval (0.0714, 0.4010). An option of getting a binomial confidence interval is available in most programs (see **Program Note 7.2** on the website).

Use of the Normal Approximation to the Binomial: Let us now consider the use of the normal approximation to the binomial distribution. The sample proportion, p, is the binomial variable, x, divided by a constant, the sample size. Since the normal distribution was shown in Chapter 5 to be a good approximation for the distribution of x when the sample size was large enough, it also serves as a good approximation to the distribution of p. The variance of p is expressed in terms of the population proportion, π, and it is $\pi(1 - \pi)/n$. Because π is unknown, we estimate the variance by substituting p for π in the formula.

The sample proportion can also be viewed as a mean as was discussed in Chapter 5. Therefore, the confidence interval for a proportion has the same form as that of the mean, and the limits of the interval are

$$\left(p - z_{1-\alpha/2}\sqrt{\frac{p(1-p)}{n}} + \frac{1}{2n}, \quad p + z_{1-\alpha/2}\sqrt{\frac{p(1-p)}{n}} + \frac{1}{2n} \right).$$

The $1/(2n)$ is the continuity correction term required because a continuous distribution is used to approximate a discrete distribution. For large values of n, the term has little effect and many authors drop it from the presentation of the confidence interval.

Example 7.4

The local health department is concerned about the protection of children against diphtheria, pertussis, and tetanus (DPT). To determine if there is a problem in the level of DPT immunization, the health department decides to estimate the proportion immunized by drawing a simple random sample of 150 children who are 5 years old. If the proportion of children in the community who are immunized against DPT is clearly less than 75 percent, the health department will mount a campaign to increase the immunization level. If the proportion is clearly greater than 75 percent, the health department will shift some resources from immunization to prenatal care. The department decides to use a 99 percent confidence interval for the proportion to help it reach its decision.

Based on the sample, 86 families claimed that their child was immunized, and 54 said their child was not immunized. There were 10 children for whom the immunization status could not be determined. As was mentioned in Chapter 6, there are several approaches to dealing with the unknowns. Since there are only 10 unknowns, we shall ignore them in the calculations. Thus, the value of p is 0.614 (= 86/140), much lower than the target value of 0.75. If all 10 of the children with unknown status had been immunized, then p would have been 0.640, not much different from the value of 0.614, and still much less than the target value of 0.75.

Applying the preceding formula, the 99 percent confidence interval ranges from

$$0.614 - 2.576\sqrt{\frac{0.614(0.386)}{140}} + \frac{1}{2(140)} \quad \text{to} \quad 0.614 + 2.576\sqrt{\frac{0.614(0.386)}{140}} + \frac{1}{2(140)}$$

or from 0.504 to 0.724. Since the upper limit of the 99 percent confidence interval is less than 0.75, the health department decides that it is highly unlikely that the proportion of 5-year-old children who are immunized is as large as 0.75. Therefore, the health department will mount a campaign to increase the level of DPT immunization in the community.

> If the issue facing the health department was whether or not to add resources to the immunization program, not to shift any resources away from the program, a one-sided interval could have been used. The 99 percent upper one-sided interval uses $z_{0.99}$ instead of $z_{0.995}$ in its calculation and it ranges from 0 to 0.713. This interval also does not contain 0.75. Therefore, resources should be added to the immunization program.

Let us use the normal approximation to find the confidence for the data in Example 7.3. The confidence interval for π based on $p = 0.2$ and $n = 20$ using the normal distribution is (0.0779, 0.3721) compared to (0.0714, 0.4010) based on the binomial distribution (Example 7.3). The former interval is symmetric, while the latter interval is not symmetric. The use of the normal distribution can give a negative lower limit when used with a small p and a small n. For this extreme case the binomial distribution is recommended. The charts in Table B6 suggest that the normal approximation is satisfactory for a large n and can be used even for a relatively small n when p is close to 0.5.

7.3.3   Confidence Intervals for Crude and Adjusted Rates

In Chapter 3, we presented crude, specific, and direct and indirect adjusted rates. However, we did not present any estimate for the variance or standard deviation of a rate, quantities that are necessary for the calculation of the confidence interval. Therefore, we begin this material with a section on how to estimate the variance of a rate.

Rates are usually based on the entire population. If this is the case, there is really no need to calculate their variances or confidence intervals for them. However, we often view a population rate in some year as a sample in location or time. From this perspective, there is justification for calculating variances and confidence intervals. If the value of the rate is estimated from a sample, as is often done in epidemiology, then it is important to estimate the variance and the corresponding confidence interval for the rate. If the rate is based on the occurrence of a very small number of events — for example, deaths — the rate may be unstable and it should not be used in this case. We shall say more about this later.

Variances of Crude and Adjusted Rates: The crude rate is calculated as the number of events in the population during the year divided by the midyear population. This rate is not really a proportion, but it is very similar to a proportion, and we shall treat it as if it were a proportion. The variance of a sample proportion, p, is $\pi(1 - \pi)/n$. Thus, the variance of a crude rate is approximated by the product of the rate (converted to a decimal value) and one minus the rate divided by the population total.

From the data on rates in Chapter 3, we saw that the crude death rate for American Indian/Alaskan Native males in 2002 was 439.6 per 100,000. The corresponding estimated 2002 American Indian/Alaskan Native male population was 1,535,000. Thus the estimated standard error, the square root of the variance estimate, for this crude death rate is

$$\sqrt{\frac{0.004396\,(1 - 0.004396)}{1535000}} = 0.0000534$$

or 5.3 deaths per 100,000 population.

The direct age-adjusted rate is a sum of the age-specific rates, $(sr)_i$'s, in the population under study weighted by the age distribution, w_i's, in the standard population. In symbols, this is $\Sigma[w_i(sr)_i]$, where w_i is the proportion of the standard population in the ith age group and $(sr)_i$ is the age-specific rate in the ith age category. The age-specific rate is calculated as the number of events in the age category divided by the midyear population in that age category. Again, this rate is not a proportion, but it is very similar to a proportion. We shall approximate the variance of the age-specific rates by treating them as if they were proportions. Since the w_i's are from the standard population that is usually very large and stable, we shall treat the w_i's as constants as far as the variance calculation is concerned. Since the age-specific rates are independent of one another, the variance of the direct adjusted rate, that is, the variance of this sum, is simply the sum of the individual variances

$$Var\left(\sum w_i\left(sr\right)_i\right) = \sum Var\left(w_i\left(sr\right)_i\right) = \sum w_i^2\left(\frac{\left(sr\right)_i\left(1-\left(sr\right)_i\right)}{n_i}\right)$$

where n_i is the number of persons in the ith age subgroup in the population under study.

Considering the U.S. mortality data as a sample in time, we can calculate the approximate variance of the direct age-adjusted death rate. The data to be used are the 2002 U.S. male age-specific death rates along with the U.S. male population totals and the 2000 U.S. population proportions by age from Table 3.14. Table 7.6 repeats the relevant data and shows the calculations. The entries in the last column are all quite small, less than 0.00000001, and therefore, only their sum is shown. The standard error of the direct age-adjusted mortality rate is 0.0000117 (= square root of variance). The direct age-adjusted rate was 1013.7 deaths per 100,000 population, and the standard error of the rate is 1.2 deaths per 100,000. The magnitude of the standard error here is not unusual, and it shows why the sampling variation of the adjusted rate is often ignored in studies involving large population bases.

For the indirect method, the adjusted rate can be viewed as the observed crude rate in the population under study multiplied by a ratio. The ratio is the standard population's

Table 7.6 Calculation of the approximate variance for the age-adjusted death rate by the direct method for U.S. males in 2002.

Age i	U.S. Male Age-Specific Rates $(sr)_i$	U.S. Male Population n_i	U.S. Population Proportion[a] w_i	$\Sigma[w_i^2(sr)_i(1-(sr)_i)/n_i]$
Under 1	0.007615	2,064,000	0.013818	
1–4	0.000352	7,962,000	0.055317	
5–14	0.000200	21,013,000	0.145565	
15–24	0.001173	20,821,000	0.138645	
25–34	0.001422	20,203,000	0.135573	
35–44	0.002575	22,367,000	0.162613	
45–54	0.005475	19,676,000	0.134834	
55–64	0.011840	12,784,000	0.087242	
65–74	0.028553	8,301,000	0.066037	
75–84	0.067605	5,081,000	0.044842	
85 & over	0.162545	1,390,000	0.015508	
Total		141,661,000	1.000000	1.37×10^{10}

[a]U.S. total population proportion in 2000 (the standard)

crude rate divided by the rate obtained by weighting the standard population's age-specific rates by the age distribution from the study population. This ratio is viewed as a constant in terms of approximating the variance. Hence, the approximation of the variance of the indirect adjusted rate is simply the square of the ratio multiplied by the variance of the study population's crude rate.

Using the data from Chapter 3, the standard population's (the 2000 U.S. population) crude rate was 854.0 deaths per 100,000 population. The combination of the standard population's age-specific rates with the study population's (the 2002 American Indian/Alaskan Native male) age distribution yielded 413.6 deaths per 100,000 population. The crude rate for American Indian/Alaskan Native male was 439.6 deaths per 100,000 population. Thus, the approximate standard error, the square root of the variance, of the indirect age-adjusted death rate is

$$\sqrt{\left(\frac{0.008540}{0.00413.6}\right)^2\left(\frac{0.004396(1-0.004396)}{1535000}\right)} = 0.00011025$$

or 11 per 100,000.

***Formation of the Confidence Interval*:** To form the confidence interval for a rate, we require knowledge of its sampling distribution. Since we are treating crude and specific rates as if they are proportions, the confidence intervals for these rates will be based on the normal approximation as just shown for the proportion. Therefore, the confidence interval for the population crude rate (θ) is

$$cr - z_{1-\alpha/2}\sqrt{\frac{cr(1-cr)}{n}} < \theta < cr + z_{1-\alpha/2}\sqrt{\frac{cr(1-cr)}{n}}$$

where cr is the value of the crude rate based on the observed sample.

For example, the 95 percent confidence interval for the 2002 American Indian/Alaskan Native male crude death rate is

$$0.00439.6 - 1.96(0.0000534) < \theta < 0.00439.6 + 1.96(0.0000534)$$

or from 0.004291 to 0.004501. Thus, the confidence interval for the crude death rate is from 429.1 to 450.1 deaths per 100,000 population.

The confidence intervals for the rates from the direct and indirect methods of adjustment have the same form as that of the crude rate. For example, the 95 percent confidence interval for the indirect age-adjusted death rate for 2002 American Indian/Alaskan Native male is found by taking

$$907.8 \pm 1.96(11.0) = 907.8 \pm 21.6$$

and thus the limits are from 886.2 to 929.4 deaths per 100,000 population.

***Minimum Number of Events Required for a Stable Rate*:** As we just mentioned, rates based on a small number of occurrences of the event of interest may be unstable. To deal with this instability, a health agency for a small area often will combine its mortality data over several years. By using the estimated coefficient of variation, the estimated standard error of the estimate divided by the estimate and multiplied by 100 percent, we can determine when there are too few events for the crude rate to be stable.

Recall that in Chapter 3 we said that if the coefficient of variation was large, the data had too much variability for the measure of central tendency to be very informative. Values of the coefficient of variation greater than 30 percent — others might use slightly larger or smaller values — are often considered to be large. We shall use this idea with the crude rate to determine how many events are required so that the rate is stable.

For example, the coefficient of variation for the 1986 crude mortality rate of Harris County is 0.904 percent (= [0.0000479/0.005296] * 100). This rate, less than 1 percent, is very reliable from the coefficient of variation perspective. It turns out that the coefficient of variation of the crude rate can be approximated by $(1/\sqrt{d}) * 100$ percent, where d is the number of events. For example, the total number of deaths for Harris County in 1986 was 12,152 and (1/12152) * 100 is 0.907 percent, essentially the same result as above.

Thus, we can use the approximation $(1/\sqrt{d}) * 100$ percent for the coefficient of variation. Setting the coefficient of variation to 20, 30, and 40 percent, yields 25, 12, and 7 events, respectively. If the crude rate is based on fewer than seven events, it certainly should not be reported. If we require that the coefficient of variation be less than 20 percent, there must be at least 25 occurrences of the event for the crude rate to be reported.

7.4   Confidence Interval for the Difference of Two Means and Proportions

We often wish to compare the mean or proportion from one population to that of another population. The confidence interval for the difference of two means or proportions facilitates the comparison. As will be seen the following sections, the method of constructing the confidence interval is different, depending on whether the two means or proportions are independent or not and depending on what assumptions are made.

7.4.1   Difference of Two Independent Means

Examples of comparing two independent means include the following. Is the mean change in blood pressure for men with mild to moderate hypertension the same for men taking different doses of an angiotensin-converting enzyme inhibitor? Is the mean length of stay in a psychiatric hospital equal for patients with the same diagnosis but under the care of two different psychiatrists? Given the following, there is an interest in the mean change in air pollution — specifically, in carbon monoxide — from 1991 to 1992 for neighboring states A and B. There was no change in gasoline formulation in State A, whereas State B required on January 1, 1992, that gasoline must consist of 10 percent ethanol during the November to March period.

One reason for interest in the confidence interval for the difference of two means is that it can be used to address the question of the equality of the two means. If there is no difference in the two population means, the confidence interval for their difference is likely to include zero.

Known Variances: The confidence interval for the difference of two means has the same form as that for a single mean; that is, it is the difference of the sample means plus or minus some distribution percentile multiplied by the standard error of the difference of the sample means. Let's convert these words to symbols. Suppose that we draw samples of sizes n_1 and n_2 from two independent populations. All the observations are assumed to be independent of one another — that is, the value of one observation does not affect the value of any other observation. The unknown population means are μ_1 and μ_2, the sample means are $\bar{x}_1$ and $\bar{x}_2$, and the known population variances are σ_1^2 and σ_2^2, respectively. The variances of the sample means are σ_1^2/n_1 and σ_2^2/n_2, respectively. Since the means are from two independent populations, the standard error of the difference of the sample means is the square root of the sum of the variances of the two sample means — that is,

$$\sqrt{\frac{\sigma_1^2}{n_1} + \frac{\sigma_2^2}{n_2}}.$$

The central limit theorem implies that the difference of the sample means will approximately follow the normal distribution for reasonable sample sizes. Thus, we have

$$Z = \frac{(\bar{x}_1 - \bar{x}_2) - (\mu_1 - \mu_2)}{\sqrt{\sigma_1^2/n_1 + \sigma_2^2/n_2}}.$$

Therefore, the $(1 - \alpha) * 100$ percent confidence interval for the difference of population means, $\mu_1 - \mu_2$, is

$$\left((\bar{x}_1 - \bar{x}_2) - z_{1-\alpha/2} \sqrt{\frac{\sigma_1^2}{n_1} + \frac{\sigma_2^2}{n_2}}, (\bar{x}_1 - \bar{x}_2) + z_{1-\alpha/2} \sqrt{\frac{\sigma_1^2}{n_1} + \frac{\sigma_2^2}{n_2}} \right).$$

Example 7.5

Suppose we wish to construct a 95 percent confidence interval for the effect of different doses of Ramipril, an angiotensin-converting enzyme converting inhibitor, used in treating high blood pressure. A study reported changes in diastolic blood pressure using the values at the end of a four-week run-in period as the baseline and measured blood pressure after two, four, and six weeks of treatment (Walter, Forthofer, and Witte 1987). We shall form a confidence interval for the difference in mean decreases from baseline to two weeks after treatment was begun between doses of 1.25 mg and 5 mg of Ramipril. The sample mean decreases are 10.6 ($\bar{x}_1$) and 14.9 mmHg ($\bar{x}_2$) for the 1.25 and 5 mg doses, respectively, and n_1 and n_2 are both equal to 53. Both σ_1 and σ_2 are assumed to be 9 mmHg. The 95 percent confidence interval for $\mu_1 - \mu_2$ is calculated as follows:

$$\left((10.6 - 14.9) - 1.96 \sqrt{\frac{81}{53} + \frac{81}{53}}, (10.6 - 14.9) + 1.96 + \sqrt{\frac{81}{53} + \frac{81}{53}} \right)$$

or ranging from −7.98 to −0.62. The value of 0 is not contained in this interval. Since the difference in mean decreases is negative, it appears that the 5 mg dose of Ramipril is associated with a greater decrease in diastolic blood pressure during the first two weeks of treatment when considering only these two doses.

Unknown but Equal Population Variances: If the variances are unknown but assumed to be equal, data from both samples can be combined to form an estimate of the common population variance. Use of the sample estimator of the variance calls for the use of the t instead of the normal distribution in the formation of the confidence interval. The pooled estimator of the common variance, s_p^2, is defined as

$$s_p^2 = \frac{\sum\limits_{i=1}^{n_1}(x_{1i} - \bar{x}_1)^2 + \sum\limits_{i=1}^{n_2}(x_{2i} - \bar{x}_2)^2}{n_1 + n_2 - 2}$$

and this can be rewritten as

$$s_p^2 = \frac{(n_1 - 1)s_1^2 + (n_2 - 1)s_2^2}{n_1 + n_2 - 2}.$$

The pooled estimator is a weighted average of the two sample variances, weighted by the respective degrees of freedom associated with the individual sample variances and divided by sum of the degrees of freedom associated with each of the two sample variances.

Now that we have an estimator of σ^2, we can use it in estimating the standard error of the difference of the sample means, $\bar{x}_1$ and $\bar{x}_2$. Since we are assuming that the population variances for the two groups are the same, the standard error of the difference of the sample means is

$$\sqrt{\frac{\sigma^2}{n_1} + \frac{\sigma^2}{n_2}} = \sigma\sqrt{\frac{1}{n_1} + \frac{1}{n_2}}$$

and its estimator is

$$s_p = \sqrt{\frac{1}{n_1} + \frac{1}{n_2}}.$$

The corresponding t statistic is

$$t = \frac{(\bar{x}_1 - \bar{x}_2) - (\mu_1 - \mu_2)}{s_p\sqrt{1/n_1 + 1/n_2}}$$

and the $(1 - \alpha) * 100$ percent confidence interval for $(\mu_1 - \mu_2)$ is

$$\left((\bar{x}_1 - \bar{x}_2) - t_{n-2, 1-\alpha/2}s_p\sqrt{1/n_1 + 1/n_2}, (\bar{x}_1 - \bar{x}_2) + t_{n-2, 1-\alpha/2}s_p\sqrt{1/n_1 + 1/n_2}\right)$$

where n is the sum of n_1 and n_2.

Example 7.6

Suppose that we wish to calculate the 95 percent confidence interval for the difference in the proportion of caloric intake that comes from fat for fifth- and sixth-grade boys compared to seventh- and eighth-grade boys in suburban Houston. The sample data are shown in Table 7.7. The proportion of caloric intake that comes from fat is found by converting the grams of fat to calories by multiplying by nine (9 calories result from 1 gram of fat) and then dividing by the number of calories consumed.

Table 7.7   Total fat,[a] calories, and the proportion of calories from total fat for the 33 boys.

Grades 7 and 8			Grades 5 and 6		
Total Fat	Calories	Prop. from Fat	Total Fat	Calories	Prop. from Fat
567	1,823	0.311	1,197	3,277	0.365
558	2,007	0.278	891	2,039	0.437
297	1,053	0.282	495	2,000	0.248
1,818	4,322	0.421	756	1,781	0.424
747	1,753	0.426	1,107	2,748	0.403
927	2,685	0.345	792	2,348	0.337
657	2,340	0.281	819	2,773	0.295
2,043	3,532	0.578	738	2,310	0.319
1,089	2,842	0.383	738	2,594	0.285
621	2,074	0.299	882	1,898	0.465
225	1,505	0.150	612	2,400	0.255
783	2,330	0.336	252	2,011	0.125
1,035	2,436	0.425	702	1,645	0.427
1,089	3,076	0.354	387	1,723	0.225
621	1,843	0.337			
666	2,301	0.289			
1,116	2,546	0.438			
531	1,292	0.411			
1,089	3,049	0.357			

[a]Total fat has been converted to calories by multiplying the number of grams by 9.

The sample mean for the 14 fifth- and sixth-grade boys is 0.329 compared to 0.353 for the 19 seventh- and eighth-grade boys. These values of percent of intake from fat are slightly above the recommended value of 30 percent (Life Sciences Research Office 1989). The corresponding standard deviations are 0.0895 and 0.0974, which support the assumption of equal variances.

The estimate of the pooled standard deviation is therefore

$$s_p = \sqrt{\frac{13(0.0895^2) + 18(0.0974^2)}{14 + 19 - 2}} = 0.094.$$

The estimate of the standard error of the difference of the sample means is

$$0.094\sqrt{1/14 + 1/19} = 0.033.$$

To find the confidence interval, we require $t_{31, 0.975}$. This value is not shown in Table B5, but, based on the values for 29 and 30 degrees of freedom, an approximate value for it is 2.04. Therefore, the lower and upper limits are

$$[(0.329 - 0.353) - 2.04\,(0.033)] \text{ and } [(0.329 - 0.353) + 2.04\,(0.033)]$$

or −0.092 and 0.044. Since zero is contained in the 95 percent confidence interval, there does not appear to be a difference in the mean proportions of calories that come from fat for fifth- and sixth-grade boys compared to seventh- and eighth-grade boys in suburban Houston.

Unknown and Unequal Population Variances: If the population variances are different, this poses a problem. There is a procedure for obtaining an exact confidence interval for the difference in the means when the population variances are unequal, but it is much more complex than the other methods in this book (Kendall and Stuart 1967).

Because of this complexity, most researchers use an approximate approach to the problem. The following shows one of the approximate approaches.

Since the population variances are unknown, we again use a t-like statistic. This statistic is

$$t' = \frac{(\bar{x}_1 - \bar{x}_2) - (\mu_1 - \mu_2)}{\sqrt{s_1^2/n_1 + s_2^2/n_2}}.$$

The t distribution with the degrees of freedom shown next can be used to obtain the percentiles of the t' statistic. The degrees of freedom value, df, is

$$df = \frac{\left(s_1^2/n_1 + s_2^2/n_2\right)^2}{\left(s_1^2/n_1\right)^2/(n_1 - 1) + \left(s_2^2/n_2\right)^2/(n_2 - 1)}.$$

This value for the degrees of freedom was suggested by Satterthwaite (1946). It is unlikely to be an integer, and it should be rounded to the nearest integer.

The approximate $(1 - \alpha) * 100$ percent confidence interval for the difference of two independent means when the population variances are unknown and unequal is

$$(\bar{x} - \bar{x}_2) - t_{df,1-\alpha/2} s_{\bar{x}_1 - \bar{x}_2} < (\mu_1 - \mu_2) < (\bar{x}_1 - \bar{x}_2) + t_{df,1-\alpha/2} s_{\bar{x}_1 - \bar{x}_2}$$

where the estimate of the standard error of the difference of the two sample means is

$$s_{\bar{x}_1 - \bar{x}_2} = \sqrt{\frac{s_1^2}{n_1} + \frac{s_2^2}{n_2}}.$$

Example 7.7

In Exercise 3.8, we presented survival times from Exercise Table 3.3 in Lee (1980) on 71 patients who had a diagnosis of either acute myeloblastic leukemia (AML) or acute lymphoblastic leukemia (ALL). In one part of the exercise, we asked for additional variables that should be considered before comparing the survival times of these two diagnostic groups of patients. One such variable is age. Let us examine these two groups to determine if there appears to be an age difference. If there is a difference, it must be taken into account in the interpretation of the data. To examine if there is a difference, we find the 99 percent confidence interval for the difference of the mean ages of the AML and ALL patients. Since we have no knowledge about the variation in the ages, we shall assume that the variances will be different. Table 7.8 shows the ages and survival times for these 71 patients.

The sample mean age for the AML patients, $\bar{x}_1$, is 49.86 and s_1 is 16.51 based on the sample size, n_1, of 51 patients. The sample mean, $\bar{x}_2$, for the 20 ALL patients is 36.65 years and s_2 is 17.85. This is the information needed to calculate the confidence interval. Let's first calculate the sample estimate of the standard error of the difference of the means:

$$s_{\bar{x}_1 - \bar{x}_2} = \sqrt{\frac{16.51^2}{51} + \frac{17.85^2}{20}} = 4.61.$$

We next calculate the degrees of freedom, df, to be used and we find it from

Table 7.8   Ages and survival times of the AML and ALL patients (age and survival times are in the same order).

AML Patients												
Age	20	25	26	26	27	27	28	28	31	33	33	33	34
	36	37	40	40	43	45	45	45	45	47	48	50	50
	51	52	53	53	56	57	59	59	60	60	61	61	61
	62	63	65	71	71	73	73	74	74	75	77	80	
Survival Time	18	31	31	31	36	01	09	39	20	04	45	36	12
in Months	08	01	15	24	02	33	29	07	00	01	02	12	09
	01	01	09	05	27	01	13	01	05	01	03	04	01
	18	01	02	01	08	03	04	14	03	13	13	01	

ALL Patients												
Age	18	19	21	22	26	27	28	28	28	28	34	36	37
	47	55	56	59	62	83	19						
Survival Time	16	25	01	22	12	12	74	01	16	09	21	09	64
in Months	35	01	07	03	01	01	22						

$$df = \frac{\left(16.51^2/51 + 17.85^2/20\right)^2}{\left(\dfrac{\left(16.51^2/51\right)^2}{51-1} + \dfrac{\left(17.85^2/20\right)^2}{20-1}\right)} = 32.501.$$

This value is rounded to 33. The 99.5 percentile of the t distribution with 33 degrees of freedom is about midway between the value of 2.750 (30 degrees of freedom) and 2.724 (35 degrees of freedom) in Appendix Table B5. We shall interpolate and use a value of 2.7344 for the 99.5 percentile of the t distribution with 33 degrees of freedom. Therefore, the approximate 99 percent confidence interval for the difference of the mean ages is

$$(49.86 - 36.65) - 2.7344\,(4.61) < \mu_1 - \mu_2 < (49.86 - 36.65) + 2.7344\,(4.61)$$

or

$$0.60 < \mu_1 - \mu_2 < 25.82.$$

Since zero is not contained in this confidence interval, there is an indication of a difference in the mean ages. If the survival patterns differ between patients with these two diagnoses, it may be due to a difference in the age of the patients.

How large would the confidence interval have been if we had assumed that the unknown population variances were equal? Using the approach in the previous section, the pooled estimate of the standard deviation, s_p, is

$$\sqrt{\frac{(51-1)61.51^2 + (20-1)17.85^2}{51+20-2}} = 16.89.$$

This leads to an estimate of the standard error of the difference of the two means of

$$16.89\sqrt{\frac{1}{51} + \frac{1}{20}} = 4.456.$$

Thus the confidence interval, using an approximation of 2.65 to the 99.5 percentile of the t distribution with 69 degrees of freedom, is

$$(49.86 - 36.65) - 2.65\,(4.456) < \mu_1 - \mu_2 < (49.86 - 36.65) + 2.65\,(4.456)$$

or

$$1.20 < \mu_1 - \mu_2 < 25.02.$$

This interval is slightly narrower than the preceding confidence interval found. However, both intervals lead to the same conclusion about the ages in the two diagnosis groups. For the use of a computer for this calculation, see **Program Note 7.3** on the website.

In practice, we usually know little about the magnitude of the population variances. This makes it difficult to decide which approach, equal or unequal variances, should be used. We recommend that the unequal variances approach be used in those situations when we have no knowledge about the variances and no reason to believe that they are equal. Fortunately, as we just saw, often there is little difference in the results of the two approaches. Some textbooks and computer packages recommend that we first test to see if the two population variances are equal and then decide which procedure to use. Several studies have been conducted recently and conclude that this should not be done (Gans 1991; Markowski and Markowski 1990; Moser and Stevens 1992).

7.4.2 Difference of Two Dependent Means

Dependent means occur in a variety of situations. One situation of interest occurs when there is a preintervention measurement of some intervention and a postintervention measurement. Another dependent mean situation occurs when there is a matching or pairing of subjects with similar characteristics. One subject in the pair receives one type of treatment and the other member in the pair receives another type of treatment. Measurements on the variable of interest are made on both members of the pair. In both of these situations, there is some relation between the values of the observations in a pair. For example, the preintervention measurement for a subject is likely to be correlated with the postintervention measurement on the same subject. If there is a nonzero correlation, this violates the assumption of independence of the observations. To deal with this relation (dependency), we form a new variable that is the difference of the observations in the pair. We then analyze the new variable, the difference of the paired observations.

Consider the blood pressure example just presented. Suppose that we focus on the 1.25 mg dose of Ramipril. We have a value of the subject's blood pressure at the end of a four-week run-in period and the corresponding value after two weeks of treatment for 53 subjects. There are 106 measurements, but only 53 pairs of observations and only 53 differences for analysis. The mean decrease in diastolic blood pressure after two weeks of treatment for the 53 subjects is 10.6 mmHg, and the sample standard deviation of the difference is 8.5 mmHg. The confidence interval for this difference has the form of the confidence interval for the mean from a single population. If the population variance is known, we use the normal distribution; otherwise we use the t distribution. We assumed that the population standard deviation was 9 mmHg previously, and we shall use that value here. Thus, the confidence interval will use the normal distribution — that is,

$$\overline{x}_d - z_{1-\alpha/2}\left(\frac{\sigma_d}{\sqrt{n}}\right) < \mu_d < \overline{x}_d + z_{1-\alpha/2}\left(\frac{\sigma_d}{\sqrt{n}}\right)$$

where the subscript d denotes difference.

Let us calculate the 90 percent confidence interval for the mean decrease in diastolic blood pressure. Table B4 shows that the 95th percentile of the standard normal is 1.645. Thus, the confidence interval is

$$10.6 - 1.645\left(\frac{9}{\sqrt{53}}\right) < \mu_d < 10.6 + 1.645\left(\frac{9}{\sqrt{53}}\right)$$

which gives an interval ranging from 8.57 to 12.63 mmHg. Since zero is not contained in the interval, it appears that there is a decrease from the end of the run-in period to the end of the first two weeks of treatment.

If we had ignored the relation between the pre- and postintervention values and used the approach for independent means, how would that have changed things? The mean difference between the pre- and postvalues does not change, but the standard error of the mean difference does change. We shall assume that the population variances are known and that σ_1, for the preintervention value, is 7 mmHg and σ_2 is 8 mmHg. The standard error of the differences, wrongly ignoring the correlation between the pre- and postmeasures, is then

$$\sqrt{\frac{7^2}{53} + \frac{8^2}{53}} = 1.46.$$

This is larger than the value of $9/\sqrt{53}$ (= 1.236) just found when taking the correlation into account. This larger value for the standard error of the difference (1.46 versus 1.236) makes the confidence interval larger than it would be had the correct method been used.

This experiment was to examine the dose-response relation of Ramipril. It consisted of a comparison of the changes in the pre- and postintervention blood pressure values for three different doses of Ramipril. If the purpose had been different — for example, to determine whether or not the 1.25 mg dose of Ramipril had an effect — this type of design may not have been the most appropriate. One problem with this type of design — measurement, treatment, measurement — when used to establish the existence of an effect is that we have to assume that nothing else relevant to the subjects' blood pressure values occurred during the treatment period. If this assumption is reasonable, then we can attribute the decrease to the treatment. However, if this assumption is questionable, then it is problematic to attribute the change to the treatment. In this case, the patients received a placebo — here, a capsule that looked and tasted liked the medication to be taken later — during the four-week run-in period. There was little evidence of a placebo effect, a change that occurs because the subject believes that something has been done. A placebo effect, when it occurs, is real and may reflect the power of the mind to affect disease conditions. This lack of a placebo effect here lends credibility to attributing the decrease to the medication, but it is no guarantee.

7.4.3 Difference of Two Independent Proportions

In this section, we want to find the $(1 - \alpha) * 100$ percent confidence interval for the difference of two independent proportions — that is, π_1 minus π_2. We shall assume that the sample sizes are large enough so that it is appropriate to use the normal distribution as an approximation to the distribution of p_1 minus p_2. In this case, the confidence interval for the difference of the two proportions is approximate. Its form is very similar to that for the difference of two independent means when the variances are not equal.

The variance of the difference of the two independent proportions is

$$\frac{\pi_1(1-\pi_1)}{n_1} + \frac{\pi_2(1-\pi_2)}{n_2}.$$

Since the population proportions are unknown, we shall substitute the sample proportions, p_1 and p_2, for them in the variance formula. The $(1 - \alpha) * 100$ percent confidence interval for $\pi_1 - \pi_2$ then is

$$(p_1 - p_2) - z_{1-\alpha/2}\sqrt{\frac{p_1(1-p_1)}{n_1} + \frac{p_2(1-p_2)}{n_2}} < \pi_1 - \pi_2$$
$$< (p_1 - p_2) + z_{1-\alpha/2}\sqrt{\frac{p_1(1-p_1)}{n_1} + \frac{p_2(1-p_2)}{n_2}}.$$

Because we are considering the difference of two proportions, the continuity correction terms cancel out in taking the difference.

Example 7.8

Holick et al. (1992) conducted a study of 13 milk processors in five eastern states. They found that only 12 of 42 randomly selected samples of milk that they collected contained 80 to 120 percent of the amount of vitamin D stated on the label. Suppose that 10 milk processors in the Southwest are also studied and that 21 of 50 randomly selected samples of milk contained 80 to 120 percent of the amount of vitamin D stated on the label. Construct a 99 percent confidence interval for the difference of proportions of milk that contain 80 to 120 percent of the amount of vitamin D stated on the label between these eastern and southwestern producers.

Since the sample sizes and the proportions are relatively large, the normal approximation can be used. The estimate of the standard error of the sample difference is

$$\sqrt{\frac{(12/42)(1-12/42)}{42} + \frac{(21/50)(1-21/50)}{50}} = 0.0987.$$

The value of $z_{0.995}$ is found from Table B4 to be 2.576. Therefore, the 99 percent confidence interval is

$$(0.286 - 0.420) - 2.576\,(0.0987) < \pi_1 - \pi_2 < (0.286 - 0.420) + 2.576\,(0.0987)$$

which is $(-0.388, 0.120)$.

Since zero is contained in the confidence interval, there is little indication of a difference in the proportion of milk samples with vitamin D content within the 80 to 120 percent range of the amount stated on the label between these eastern and southwestern milk producers.

7.4.4   Difference of Two Dependent Proportions

Suppose that a sample of n subjects has been selected to examine the relationship between the presences of a particular attribute at two time points for the same individuals (paired observations). The situation could also be used to examine the relationship between two different attributes for the same individuals. The sample data for these situations can be arranged as follows:

Attribute at Time		
1	**2**	**Number of Subjects**
Present	Present	a
Present	Absent	b
Absent	Present	c
Absent	Absent	d
Total		n

Then the estimated proportion of subjects with the attribute at time 1 is $p_1 = (a + b)/n$, and the estimated proportion with the attribute at time 2 is $p_2 = (a + c)/n$. The difference between the two estimated proportions is

$$p_d = p_1 - p_2 = \frac{a+b}{n} - \frac{a+c}{n} = \frac{b-c}{n}.$$

Since the two population probabilities are dependent, we cannot use the same approach for estimating the standard error of the difference that we used in the previous section. Instead of showing the steps in the derivation of the formula, we simply present the formula for the estimated standard error (Fleiss 1981).

$$Estimated\ SE\ (p_d) = \frac{1}{n}\sqrt{(b+c) - \frac{(b-c)^2}{n}}.$$

The confidence interval for the difference of two dependent proportions, $\pi_d\ (= \pi_1 - \pi_2)$, is then given by

$$p_d - z_{1-\alpha/2}SE(p_d) < \pi_d < p_d + z_{1-\alpha/2}SE(p_d).$$

Example 7.9

Suppose that 100 students took both biostatistics and epidemiology tests, and 18 failed in biostatistics ($p_1 = 0.18$) and 10 failed in epidemiology ($p_2 = 0.10$). There is an 8 percentage point difference ($p_d = 0.08$). The confidence interval for the difference of these two failure rates cannot be constructed using the method in the previous subsection because the two rates are dependent. We need additional information to assess the dependency. Nine students failed both tests ($p_{12} = 0.09$), and this reflects the dependency. The dependency between p_1 and p_2 can be seen more clearly when the data are presented in a 2 by 2 table.

Biostatistics	Epidemiology		
	Failed	Passed	Total
Failed	9 (a)	9 (b)	18
Passed	1 (c)	81 (d)	82
Total	10	90	100 (n)

The marginal totals reflect the two failure rates. The numbers in the diagonal cells (a, d) are concordant pairs of test scores (those who passed or failed both tests), and those in the off-diagonal cells (b, c) are discordant pairs (those who passed one test but failed the other). Important information for comparing the two dependent failure rates is contained in discordant pairs, as the estimated difference of the two proportions and its estimated standard error are dependent on b and c.

Using the standard error equation, we have

$$\frac{1}{100} \sqrt{(9+1) - \frac{(9-1)^2}{100}} = 0.0306.$$

Then the 95 percent confidence interval for the difference of these two dependent proportions is

$$0.08 - 1.96\,(0.0306) < \pi_d < 0.08 + 1.96\,(0.0306)$$

or (0.0200, 0.1400). This interval does not include 0, suggesting that the failure rates of these two tests are significantly different. However, this method is not recommended for small frequencies and further discussion will follow in conjunction with hypothesis testing in the next chapter.

7.5   Confidence Interval and Sample Size

One important point about the confidence interval for the population mean is that its width can be calculated before the sample is selected. The half-width of the confidence interval is

$$z_{1-\alpha/2} \sqrt{\frac{\sigma}{\sqrt{n}}}.$$

When σ and n are known, the width can be calculated. If the interval is viewed as being too wide to be informative, we can change one of the values used (z, n, or σ) in calculating the width to see if we can reduce it to an acceptable value. The two most common ways of reducing its width are by decreasing our level of confidence (reducing the z value) or by increasing the sample size (n); however, there are limits for both of these choices. Most researchers prefer to use at least the 95 percent level for the confidence interval although the use of the 90 percent level is not uncommon. To drop below the

90 percent level is usually unacceptable. Researchers may be able to increase the sample size somewhat, but the increase requires additional resources that are often limited.

Example 7.10

Suppose that we wish to estimate the mean systolic blood pressure of girls who are 120 to 130 cm (approximately 4 feet to 4 feet 3 inches) tall. We assume that the standard deviation of the systolic blood pressure variable for girls in this height group is 7 mmHg. Given this information, how large a sample is required so that the half-width of the 95 percent confidence interval is no more than 3 mmHg wide?

The half-width of the confidence interval can be equated to the specified half-width — that is

$$1.96\left(\frac{7}{\sqrt{n}}\right) = 3.$$

This equation can be solved for n, multiplying both sides by $\sqrt{n}$ and squaring both sides, which gives

$$n = \left(\frac{1.96(7)}{3}\right)^2 = 20.9.$$

Since n must be an integer, the next highest integer value, 21, is taken to be the value of n.

The formula for n, given a specified half-width, d, for the $(1 - \alpha) * 100$ percent confidence interval is

$$n = \left(\frac{z_{1-\alpha/2}\,\sigma}{d}\right)^2.$$

So far, we have been assuming that σ is known; however, in practice, we seldom know the population standard deviation. Sometimes the literature or a pilot study provides an estimate of its value that we may use for σ.

For the case of proportion, the sample size can be calculated by the following formula:

$$n = \left(\frac{z_{1-\alpha/2}\,\sqrt{\pi(1-\pi)}}{d}\right)^2.$$

In this formula π is the population proportion and $\pi(1-\pi)/n$ is the variance of binomial distribution as shown in Chapter 4. The population proportion is seldom known when calculating the sample size. Again, the literature or a pilot study may provide an estimate. In cases when we have no information for π, we can use $\pi = 0.5$. This practice is based on the fact that $\pi(1-\pi)$ is the maximum when $\pi = 0.5$ and the calculated sample size will be sufficient for any value of π.

The confidence interval for the difference between two independent means, μ_1 and μ_2, can be used to determine the sample size required when there are two equal-size experimental groups. We assume that the same known population variance is σ^2 and two equal random samples of size n are to be taken. Then the half-width of the confidence interval for the difference of two means simplifies to

$$z_{1-\alpha/2}\sqrt{\frac{\sigma^2}{n}+\frac{\sigma^2}{n}}.$$

As before, let d be the half-width of the desired confidence interval. Equating the preceding quantity to d and solving for n we have

$$n=2\left(\frac{z_{1-\alpha/2}\sigma}{d}\right)^2.$$

For the case of the difference of two independent proportions, the required sample size can be calculated by

$$n=2\left(\frac{z_{1-\alpha/2}\sqrt{\pi_1(1-\pi_1)+\pi_2(1-\pi_2)}}{d}\right)^2.$$

Example 7.11

A researcher wants to be 99 percent confident ($z = 2.567$) that the difference in the mean systolic blood pressure of boys and girls be estimated within plus and minus 2 mmHg ($d = 2$). How large a sample size does the researcher need in each group? We will assume that the sample size is large enough that the normal distribution approximation can be used. We also assumed that the standard deviation of the systolic blood pressure for boys and girls are the same, and it is 8 mmHg. The required sample size is

$$n=2\left(\frac{2.567(8)}{2}\right)^2=210.9.$$

The required sample size is 211 in each group.

In the planning of a statistical study, the determination of sample size is not as simple as the preceding example may suggest. If you want a high level of confidence and a small interval, a very large sample size is required. The difficulty lies in deciding what level of confidence to aim for within the limit imposed by available resources. The balancing of the level of confidence against availability of resources may require an iterative process until a solution is found that satisfies both requirements.

7.6   Confidence Intervals for Other Measures

We next consider confidence intervals for the variance and the Pearson correlation coefficient. Interval estimation for other measures such as the odds ratio and regression coefficient will be discussed in subsequent chapters.

7.6.1   Confidence Interval for the Variance

Besides being useful in describing the data, the variance is also frequently used in quality control situations. It is one way of stating how reliable the process under study is. For example, in Chapter 2 we presented data on the measurement of blood lead levels by different laboratories. We saw from that example that great variability in the measurements made by laboratories exists, and the variance is one way to characterize that variability. Variability within laboratories can be due to different technicians, failure to calibrate the equipment, and so forth. It is critically important that measurements of the same sample within a laboratory have variability less than or equal to a prespecified small amount. Thus, based on the sample variance for a laboratory for measuring blood lead, we wish to determine whether or not the laboratory's variance is in compliance with the standards. The confidence interval for the population variance provides one method of doing this.

To construct the confidence interval for the population variance, we need to know the sampling distribution of its estimator, the sample variance, s^2. The sampling distribution of s^2 can be examined by (1) taking a repeated random sample from a normal distribution, (2) calculating a sample variance from each sample, and (3) plotting a histogram of sample variances. When we take a repeated random sample of size 3, the distribution of sample variances looks like the black line in Figure 7.2. The distribution for df = 2 is very asymmetric with a long tail to the right, suggesting that there is tremendous variability in the sample variances. This large variation is expected as each sample variance was based on only three observations. When we increase the sample size to 6 (df = 5), the distribution of sample variances is not so asymmetric and the tail to the right is much shorter than in the first distribution. When we increase the sample size to 11 (df = 10), the distribution of sample variances is almost symmetric. We can see that the sampling distributions for the three samples sizes are very different; that is, they depend on the sample size.

It appears that the distribution of the sample variance does not match any of the probability distributions we have encountered so far. Fortunately, when the data come from a normal distribution, the distribution of the sample variance is known. The sample variance (s^2), multiplied by $(n - 1)/\sigma^2$, follows a *chi-square* (χ^2) *distribution*. Two

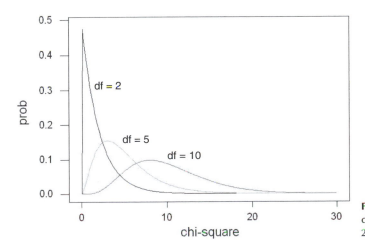

Figure 7.2 Chi-square distributions with df = 2, df = 5, and df = 10.

eminent 19th-century French mathematicians, Laplace and Bienaymé, played important roles in the development of the chi-square distribution. Karl Pearson, an important British statistician previously encountered in connection with the correlation coefficient, popularized the use of the chi-square distribution in the early 20th century. As we just saw, the distribution of the sample variance depends of the sample size, actually on the number of independent observations (degrees of freedom) used to calculate s^2. Therefore, Appendix Table B7 shows percentiles of the chi-square distribution for different values of the degrees of freedom parameter.

To create a confidence interval for the population variance, we begin with the probability statement

$$\Pr\left\{\chi^2_{n-1,\alpha/2} < \frac{(n-1)s^2}{\sigma^2} < \chi^2_{n-1,1-\alpha/2}\right\} = 1-\alpha.$$

This statement indicates that the confidence interval will be symmetric in the sense that the probability of being less than the lower limit is the same as that of being greater than the upper limit. However, the confidence limit will not be symmetric about s^2. This probability statement is in terms of s^2, however, and we want a statement about σ^2. To convert it to a statement about σ^2, we first divide all three terms in the braces by $(n-1)s^2$. This yields

$$\Pr\left\{\frac{\chi^2_{n-1,1-\alpha/2}}{(n-1)s^2} < \frac{1}{\sigma^2} < \frac{\chi^2_{n-1,\alpha/2}}{(n-1)s^2}\right\} = 1-\alpha.$$

The interval is now about $1/\sigma^2$, not σ^2. Therefore, we next take the reciprocal of all three terms, which changes the direction of the inequalities. For example, we know that 3 is greater than 2, but the reciprocal of 3, which is 1/3 or 0.333, is less than the reciprocal of 2, which is 1/2 or 0.500. Thus, we have

$$\Pr\left\{\frac{(n-1)s^2}{\chi^2_{n-1,\alpha/2}} > \sigma^2 > \frac{(n-1)s^2}{\chi^2_{n-1,1-\alpha/2}}\right\} = 1-\alpha$$

and reversing the directions of the inequalities to have the smallest term on the left, yields

$$\Pr\left\{\frac{(n-1)s^2}{\chi^2_{n-1,1-\alpha/2}} < \sigma^2 < \frac{(n-1)s^2}{\chi^2_{n-1,\alpha/2}}\right\} = 1-\alpha.$$

It is also possible to create one-sided confidence intervals for the population variance. For example, the lower one-sided confidence interval for the population variance is

$$\frac{(n-1)s^2}{\chi^2_{n-1,1-\alpha}} < \sigma^2 < \infty.$$

Example 7.12

Let's apply this formula to an example. From 1988 to 1991, eight persons in Massachusetts were identified as having vitamin D intoxication due to receiving large doses of vitamin D_3 in fortified milk (Jacobus, Holick, and Shao 1992). The problem was traced to a local dairy that had tremendous variability in the amount of vitamin D added to individual bottles of milk. Homogenized whole milk showed the greatest variability based on measurements made in April and June 1991, with a low value of less than 40 IUs and a high of 232,565 IUs of vitamin D_3 per quart. These values are contrasted with the requirement for at least 400 IUs (10 μg) to no more than 500 IUs of vitamin D per quart of milk in Massachusetts.

The Food and Drug Administration (FDA) found poor compliance with the requirement for 400 IUs of vitamin D per quart of vitamin D fortified milk in a 1988 survey (Holick et al. 1992). Based on this poor compliance, the FDA urged that the problem be corrected; otherwise it would institute a regulatory program. Suppose that compliance is defined in terms of the mean and standard error of the mean vitamin D concentration in milk. The mean concentration should be 400 IUs with a variance of less than 1600 IUs. To determine if a milk producer is in compliance, a simple random sample of milk cartons from the producer is selected and the amount of vitamin D in the milk is ascertained. It is decided that if the 90 percent lower one-sided confidence interval for the variance contains 1600 IUs, the process used by the producer to add vitamin D is said to be within the acceptable limits for variability. This is an approach for determining compliance that greatly favors the producer.

A random sample of 30 cartons is selected and the sample variance for the vitamin D in the milk is found to be 1700 IUs. The 90 percent confidence interval uses $\chi^2_{29,0.90}$, where the first subscript is the degrees of freedom parameter and the second subscript is the percentile value. The value from Table B7 is 39.09. The lower limit is found from [29(1700)]/39.09, which gives the value of 1261.3. Since the 90 percent confidence interval does contain 1600 IUs, the producer is said to be in compliance with the variability requirement. To find that a producer is not in compliance requires a sample variance to be at least 2156.5.

A key assumption in calculating the confidence interval for the population variance is that the data come from a normal distribution. If the data are from a very nonnormal distribution, the use of the preceding formula for calculating the confidence interval can be very misleading.

To find the confidence interval for the population standard deviation, we take the square root of the variance's confidence interval limits. Thus, the lower limit of the confidence interval for σ in the above example is 35.5 IUs.

7.6.2 Confidence Interval for the Pearson Correlation Coefficient

In Chapter 3, we presented ρ, the Pearson correlation coefficient, which is used in assessing the strength of the linear relation between two jointly normally distributed variables. We presented a formula for finding r, the sample Pearson correlation coefficient. We also found the correlation between systolic and diastolic blood pressures, based on the 12 adults in Example 3.18, to be 0.894, suggestive of a strong positive relation. Although this point estimate of ρ is informative, more information is provided by the interval estimate. For example, if the sampling variation of r were so large that the 95 percent confidence interval for ρ contains zero, we would not be impressed by the strength of the relation between total fat and protein.

It turns out that the sampling distribution of r is not easily characterized. However, the father of modern statistics, Ronald Fisher, showed that a transformation of r approximately follows a normal distribution. This transformation is

$$z' = 0.5[\log_e(1 + r) - \log_e(1 - r)]$$

and it provides the basis for the confidence interval for ρ. The mean of z' is $[\log_e (1 + \rho) - \log_e (1 - \rho)]$ and its standard deviation, $\sigma_{z'}$, is $1/\sqrt{(n-3)}$. Note that for convenience, $\log_e$ is often written as ln, and we shall do that following. Thus, we can employ the procedures we have just used for finding the confidence interval for the transformed value of ρ — that is,

$$z' - z_{1-\alpha/2}\sigma_{z'} < 0.5[\ln(1 + \rho) - \ln(1 - \rho)] < z' + z_{1-\alpha/2}\sigma_z.$$

There is one simplification we can make that allows us to have to take only one natural logarithm in the calculation instead of finding two natural logarithms. In the presentation of the geometric mean in Chapter 3, we saw that the sum of logarithms of two terms is the logarithm of the product of the terms — that is,

$$\ln x_1 + \ln x_2 = \ln(x_1 x_2).$$

In the same way, the difference of logarithms of two terms is the logarithm of the quotient of the terms — that is,

$$\ln x_1 - \ln x_2 = \ln\left(\frac{x_1}{x_2}\right).$$

Thus, we have the relation

$$z' = 0.5[\ln(1+r) - \ln(1-r)] = 0.5\ln\left(\frac{1+r}{1-r}\right).$$

Let us apply these formulas for finding the 95 percent confidence interval for the correlation between systolic and diastolic blood pressure for 12 adults just mentioned. Since r is 0.894, z' is

$$(0.5)\ln\left(\frac{1+0.894}{1-0.894}\right) = (0.5)\ln 17.8679 = (0.5)2.8830 = 1.4415.$$

The standard deviation of z' is $1/\sqrt{12-3}$, which is 0.3333. Thus the interval for $(0.5)\ln[(1 + \rho)/(1 - \rho)]$ is from $01.4415 - 1.96(0.3333)$ to $1.4415 + 1.96(0.3333)$ or from 0.7882 to 2.0948.

To find the confidence interval for ρ, we first perform the inverse transformation on twice the lower and upper limits of the interval just calculated. The inverse transformation of the natural logarithm, ln, is the exponential transformation. This means that

$$\exp(\ln x) = x.$$

After obtaining the exponential of twice a limit, call it a, further manipulation leads to the following equation:

$$\text{limit for } \rho = \frac{a-1}{a+1}.$$

The exponential of twice the lower limit — that is, two times 0.7882 — is the exponential of 1.5764, which is 4.83785, and this is the value used for a for the lower limit. The lower limit for ρ is

$$\frac{\exp[2(0.7882)]-1}{\exp[2(0.7882)]+1} = \frac{4.8375-1}{4.8375+1} = 0.657.$$

Similarly, the upper limit for ρ is

$$\frac{\exp[2(2.0948)]-1}{\exp[2(2.0948)]+1} = \frac{65.9926-1}{65.9926+1} = 0.970.$$

Therefore, the 95 percent confidence interval for the Pearson correlation coefficient between systolic and diastolic blood pressure in the population is from 0.657 to 0.970. The interval does not include 0. Thus, it is reasonable to conclude that there is a strong positive association between systolic and diastolic blood pressures among patients in the DIG clinical trial. These calculations are easily performed by a program (see **Program Note 7.4** on the website). The preceding material also applies to the Spearman correlation coefficient for sample sizes greater than or equal to 10.

7.7 Prediction and Tolerance Intervals Based on the Normal Distribution

As we have seen, knowledge that the data follow a specific distribution can be used effectively in the creation of confidence intervals. This knowledge can also be used in the formation of prediction and tolerance intervals, and this use is shown next.

7.7.1 Prediction Interval

The distribution-free method for forming intervals used specific observed values of the variable under study. In contrast, the formation of intervals based on the normal distribution uses the sample estimates of its parameters: the mean and standard deviation. Assuming that the data follow the normal distribution, the prediction interval is formed by taking the sample mean plus or minus some value. This form is the same as that used in the construction of the confidence interval for the population mean. However, we know that the prediction interval will be much wider than the confidence interval, since the prediction interval focuses on a single future observation.

The confidence interval for the mean, when the population variance is unknown, is

$$\bar{x} \pm t_{n-1,1-\alpha/2}\left(\frac{s}{\sqrt{n}}\right).$$

The estimated standard error of the sample mean, $s/\sqrt{n}$, can also be expressed as $\sqrt{[s^2(1/n)]}$. The variance of a future observation is the sum of the variance of an observation about the sample mean and the variance of the sample mean itself, that is, $\sigma^2 + \sigma^2/n$. Thus, the estimated standard error of a future observation is $\sqrt{[s^2(1+1/n)]}$ and the corresponding prediction interval is

$$\bar{x} \pm t_{n-1,1-\alpha/2} s \sqrt{1 + \frac{1}{n}}.$$

Let us calculate the prediction interval for the systolic blood pressure data just used in the calculation of the 90 percent confidence interval for the mean. The sample mean was 94.75 mmHg, and the sample standard deviation was 10.25 mmHg, based on a sample size of 60. The value of $t_{59,0.95}$ used in the 90 percent confidence interval was 1.671. The value of $s\sqrt{(1+1/n)}$ is $10.335 (= 10.25\sqrt{[1+1/60]})$. Therefore, the prediction interval is

$$94.75 \pm 1.671 \ (10.335)$$

and the lower and upper limits are 77.48 and 112.02 mmHg, respectively. These values are contrasted with 92.54 and 96.96 mmHg, limits of the confidence interval for the mean. Thus, as expected, the 90 percent prediction interval for a single future observation is much wider than the corresponding 90 percent confidence interval for the mean.

7.7.2   Tolerance Interval

The tolerance interval is also formed by taking the sample mean plus or minus some quantity, k, multiplied by the estimate of the standard deviation. Since the derivation of k is beyond the level of this book, we shall simply use its value found in Table B8. In symbols, the $(1 - \alpha) * 100$ percent tolerance interval containing p percent of the population based on a sample of size n is

$$\bar{x} \pm k_{n,p,1-\alpha} \ s.$$

Let us use Table B8 to find the 90 percent tolerance interval containing 95 percent of the systolic blood pressure values in the population based on the first sample of 60 observations from above. From Table B8 we find that $k_{60,0.95,0.90}$'s value is 2.248. Therefore, the tolerance interval is

$$94.75 \pm 2.248 \ (10.25)$$

which gives limits of 71.71 and 117.79. One-sided prediction and tolerance intervals based on the normal distribution are also easy to construct.

Conclusion

In this chapter, the concept of interval estimation was introduced. We presented prediction, confidence, and tolerance intervals and explained their applications. We showed how distribution-free intervals and intervals based on the normal distribution were calculated. The idea and use of confidence intervals discussed in this chapter will be explored further to introduce methods of testing statistical hypotheses in the next two chapters. Parenthetically, it is worth pointing out that the idea of confidence interval is often expressed as a margin of error in journalistic reporting, which refers to one-half of the width of a two-sided confidence interval.

We also pointed out that characteristics — for example, size — of the intervals could be examined before actually conducting the experiment. If the characteristics of the interval are satisfactory, the investigator uses the proposed sample size. If the characteristics are unsatisfactory, the design of the experiment, the topic of the next chapter, needs to be modified.

EXERCISES

7.1 Assume that the AML patients shown in Exercise 3.7 can be considered a simple random sample of all AML patients.
   a. Calculate the 95 percent confidence interval for the population mean survival time after diagnosis for AML patients.
   b. Interpret this confidence interval so that someone who knows no statistics can understand it.
   c. Calculate the approximate 95 percent confidence interval for the median survival time. Compare the intervals for the population mean and median.
   d. There are two methods for forming the tolerance interval. Use both methods to form the approximate 95 percent tolerance interval containing 90 percent of the survival times for the population of AML patients. Which method do you think is the more appropriate one to use here? Provide your rationale.

7.2 Calculate a 90 percent confidence interval for the population median length of stay based on the data from the patient sample shown in Exercise 3.10. Is it appropriate to calculate a confidence interval for the population mean based on these data? Support your answer.

7.3 Find a study from the health literature that uses confidence intervals for one of the statistics covered in this chapter. Provide a reference for the study and briefly explain how confidence intervals were used.

7.4 The following table shows the average annual fatality rate per 100,000 workers based on the 1980–1988 period by state along with the state's composite score on a scale created by the National Safe Workplace Institute (NSWI). The scale takes into account prevention and enforcement activities and compensation paid to the victim. The data are taken from the Public Citizen Health Research Group (1992).

State	Fatality[a] Rate	NSWI[b] Score	State	Fatality Rate	NSWI Score	State	Fatality Rate	NSWI Score
CT	1.9	65	SC	6.7	26	LA	11.2	31
MA	2.4	73	VT	6.8	38	NE	11.3	27
NY	2.5	76	IL	6.9	76	NV	11.5	30
RI	3.3	59	NC	7.2	47	TX	11.7	72
NJ	3.4	80	WA	7.7	55	KY	11.9	32
AZ	4.1	40	IN	7.8	47	NM	12.0	14
MN	4.3	64	ME	7.8	67	AR	12.5	11
NH	4.5	56	TN	8.1	24	UT	13.5	26
OH	4.8	55	OK	8.7	53	ND	13.8	21
MI	5.3	63	AL	9.0	25	MS	14.6	25
MO	5.3	42	KS	9.1	15	SD	14.7	25
MD	5.7	46	IA	9.2	54	WV	16.2	47
DE	5.8	40	CO	9.3	52	ID	17.2	22
HI	6.0	25	FL	9.3	48	MT	21.6	28
PA	6.1	55	VA	9.9	60	WY	29.5	12
WI	6.3	58	GA	10.3	36	AK	33.1	59
CA	6.5	81	OR	11.0	63			

[a]Average annual fatality rate per 100,000 workers based on 1980–1988 data
[b]National Safe Workplace Institute Score (116 is the maximum and a higher score is better)

During the 1980–1988 period, the National Institute of Occupational Safety and Health reported that there were 56,768 deaths in the workplace. The preceding rates are based on that number. The National Safety Council reported 105,500 deaths for the same period. Do you think that there should be any relationship between the fatality rates and the NSWI scores? If you think that there is a nonzero correlation, will it be positive or negative? Explain your reasoning. Calculate the Pearson correlation coefficient for these data. Is there any reason to calculate a confidence interval based on the correlation value you calculated? Why or why not?

7.5   There is some concern today about excessive intakes of vitamins and minerals, possibly leading to nutrient toxicity. For example, many persons take vitamin and mineral supplements. It is estimated that 35 percent of the adult U.S. population consumes vitamin C in the form of supplements (LSRO 1989). Based on survey results, among users of vitamin C supplements, the median intake was 333 percent of the recommended daily allowance. Suppose that you take a tablet that claims to contain 500 mg vitamin C. Which type of interval — prediction, confidence, or tolerance — about the vitamin C content in the tablets is of most interest to you? Explain your reasoning.

7.6   In a test of a laboratory's measurement of serum cholesterol, 15 samples containing the same known amount (190 mg/dL) of serum cholesterol are submitted for measurement as part of a larger batch of samples, one sample each day over a three-week period. Suppose that the following daily values in mg/dL for serum cholesterol for these 15 samples were reported from the laboratory:

| 180 | 190 | 197 | 199 | 210 | 187 | 192 | 199 | 214 | 237 | 188 | 197 | 208 | 220 | 239 |

Assume that the variance for the measurement of serum cholesterol is supposed to be no larger than 100 mg/dL. Construct the 95 percent confidence interval for this laboratory's variance. Does 100 mg/dL fall within the confidence interval? What might be an explanation for the pattern shown in the reported values?

7.7   The percentage of persons in the United States without health insurance in 1991 was 14.1 percent, or approximately 35.5 million persons. The following data show the percent of persons without health insurance in 1991 by state (PCHRG 1993) along with the 1990 population of the state (U.S. Bureau of the Census 1991). The District of Columbia is treated as a state in this presentation. Calculate the sample Pearson correlation coefficient between the state population total and its percent without health insurance. How can these counts be viewed as a sample? Calculate a 95 percent confidence interval for the Pearson correlation coefficient in the population. Does there appear to be a strong linear relation between these two variables? Provide at least one additional variable that may be related to the proportion without health insurance in each state and provide a rationale for your choice.

State	Population[a]	Percent without Health Insurance	State	Population	Percent without Health Insurance
New England			East South Central		
ME	1.23	11.1	KY	3.69	13.1
NH	1.11	10.1	TN	4.88	13.4
VT	0.56	12.7	AL	4.04	17.9
MA	6.02	10.9	MS	2.57	18.9
RI	1.00	10.2			
CT	3.29	7.5	West South Central		
			AR	2.35	15.7
Mid Atlantic			LA	4.22	20.7
NY	17.99	12.3	OK	3.15	18.2
NJ	7.73	10.8	TX	16.99	22.1
PA	11.88	7.8			
			Mountain		
East North Central			MT	0.80	12.7
OH	10.85	10.3	ID	1.01	17.8
IN	5.54	13.0	WY	0.45	11.3
IL	11.43	11.5	CO	3.29	10.1
MI	9.30	9.0	NM	1.52	21.5
WI	4.89	8.0	AZ	3.67	16.9
			UT	1.72	13.8
West North Central			NV	1.20	18.7
ND	0.64	7.6			
SD	0.70	9.9	Pacific		
NE	1.58	8.3	WA	4.87	10.4
KS	2.48	11.4	OR	2.84	14.2
MN	4.38	9.3	CA	29.76	18.7
IA	2.78	8.8	AK	0.55	13.2
MO	5.12	12.2	HI	1.11	7.0
South Atlantic					
DE	0.67	13.2			
MD	4.78	13.1			
VA	6.19	16.3			
WV	1.79	15.7			
FL	12.94	18.6			
NC	6.63	14.9			
SC	3.49	13.2			
GA	6.48	14.1			
DC	0.61	25.7			

[a]Population is expressed in millions

7.8 Calculate the mean state proportion of those without health insurance from data in Exercise 7.7. Is this number the same as the overall U.S. percentage? Explain how the state information can be used to obtain the overall U.S. percentage of 14.1.

7.9 Suppose you are planning a simple random sample survey to estimate the mean family out-of-pocket expenditures for health care in your community during the last year. In 1990, the approximate per capita (not per family) out-of-pocket expenditure was $525 (NCHS 1992). From previous studies in the literature, you think that the population standard deviation for family out-of-pocket expenditures is $500. You want the 90 percent confidence interval for the community mean family out-of-pocket expenditures to be no wider than $100.

    a. How many families do you require in the sample to satisfy your requirement for the width of the confidence interval for the mean?

    b. Do you believe that family out-of-pocket expenditures follow the normal distribution? Support your answer.

c. Regardless of your answer, assume that you said that the family out-of-pocket expenditures do not follow a normal distribution. Discuss why it is still appropriate to use the material based on the normal distribution in finding the confidence interval for the population mean.

d. In the conduct of the survey, how would you overcome reliance on a person's memory for out-of-pocket expenditures for health care for the past year?

7.10 In 1979, the Surgeon General's *Report on Health Promotion and Disease Prevention* and its follow-up in 1980 established health objectives for 1990. One of the objectives was that the proportion of 12- to 18-year-old adolescents who smoked should be reduced to below 6 percent (NCHS 1992). Suppose that you have monitored progress in your community toward this objective. In a survey conducted in 1983, you found that 17 of 90 12- to 18-year-old adolescents admitted that they were smokers. In your 1990 simple random sample survey, you found 11 of 85 12- to 18-year-old adolescents who admitted that they smoked.

a. Construct a 95 percent confidence interval for the proportion of smokers among 12- to 18-year-old adolescents in your community. Is 6 percent contained in the confidence interval?

b. Construct a 99 percent confidence interval for the difference in the proportion of smokers among 12- to 18-year-old adolescents from 1983 to 1990. Do you believe that there is a difference in the proportion of smokers among the 12- to 18-year-old adolescents between 1983 and 1990? Explain your answer.

c. Briefly describe how you would conduct a simple random sample of 12- to 18-year-old adolescents in your community. Do you have confidence in the response to the question about smoking? Provide the rationale for your answer. What is a method that might improve the accuracy of the response to the smoking question?

7.11 Construct the 95 percent confidence interval for the difference in the population mean survival times between the AML and ALL patients shown in Table 7.6. Since there appears to be a difference in mean ages between the AML and ALL patients, perhaps we should adjust for age. One way to do this is to calculate age-specific confidence intervals. For example, calculate the confidence interval for the difference in population mean survival times for AML and ALL patients who are less than or equal to 40 years old. Is the confidence interval for those less than or equal to 40 years of age consistent with the confidence interval which has ignored the ages? How else might we adjust for the age variable in the comparison of the AML and ALL patients?

7.12 Suppose we wish to investigate the claims of a weight loss clinic. We randomly select 20 individuals who have just entered the program, and we follow them for six weeks. The clinic claims that its members will lose on the average 10 pounds during the first six weeks of membership. The beginning weights and the weights after six weeks are shown following. Based on this sample of 20 individuals, is the clinic's claim plausible?

Person	Beginning Weight	Weight at 6 Weeks	Person	Beginning Weight	Weight at 6 Weeks
1	147	143	11	246	239
2	163	151	12	218	222
3	198	184	13	143	135
4	261	245	14	129	124
5	233	229	15	154	136
6	227	220	16	166	159
7	158	161	17	278	263
8	154	147	18	228	205
9	162	155	19	173	164
10	249	254	20	135	122

7.13 In a study of aplastic anemia patients, 16 of 41 patients on one treatment achieved complete or partial remission after three months of treatment compared to 28 of 43 patients on another treatment (Frickhofen et al. 1991). Construct a 99 percent confidence interval on the difference in proportions that achieved complete or partial remission. Does there appear to be a difference in the population proportions of the patients who would achieve complete or partial remission on these two treatments?

7.14 In 1970, Japanese American women had a fertility rate (number of live births per 1000 women ages 15–44) of 51.2, considerably lower than the rate of 87.9 for all U.S. women in this age group. Use the following data to calculate an age-adjusted fertility rate for Japanese American women and approximate the standard deviation of the age-adjusted rate.

Age	U.S. Age-Specific Fertility Rate	Number of Japanese American Women
15–19	69.6	24,964
20–24	167.8	23,435
25–29	145.1	22,093
30–34	73.3	23,055
35–39	31.7	32,935
40–44	8.6	34,044

Source: U.S. Population Census, 1970, P(2)-1G and U.S. Vital Statistics, 1970

REFERENCES

Fleiss, J. L. *Statistical Methods for Rates and Proportions*, 2nd ed. New York: Wiley, 1981.

Frickhofen, N., J. P. Kaltwasser, H. Schrezenmeier, "Treatment of Aplastic Anemia with Antilymphocyte Globulin and Methylprednisolone with or without Cyclosporine." *The New England Journal of Medicine* 324:1297–1304, 1991.

Gans, D. J. "Letter to the Editor — Preliminary Test on Variances." *The American Statistician* 45:258, 1991.

Holick, M. F., Q. Shao, W. W. Liu, "The Vitamin D Content of Fortified Milk and Infant Formula." *The New England Journal of Medicine* 326:1178–1181, 1992.

Jacobus, C. H., M. F. Holick, Q. Shao, et al. "Hypervitaminosis D Associated with Drinking Milk." *The New England Journal of Medicine* 326:1173–1177, 1992.

Kendall, M. G., and A. Stuart. *The Advanced Theory of Statistics, Volume 2, Inference and Relationship*, 2nd edition. New York: Hafner Publishing Company, 1967.

Lee, E. T. *Statistical Methods for Survival Data Analysis*. Belmont, CA: Wadsworth, 1980.

Life Sciences Research Office (LSRO), Federation of American Societies for Experimental Biology. *Nutrition Monitoring in the United States — An Update Report on Nutrition Monitoring.* Prepared for the U.S. Department of Agriculture and the U.S. Department of Health and Human Services. DHHS Pub. No. (PHS) 89-1255, 1989.

Markowski, C. A., and E. P. Markowski. "Conditions for the Effectiveness of a Preliminary Test of Variance." *The American Statistician* 44:322–326, 1990.

Moser B. K., and G. R. Stevens. "Homogeneity of Variance in the Two-Sample Means Test." *The American Statistician* 46:19–21, 1992.

National Center for Health Statistics. *Health, United States, 1991 and Prevention Profile.* Hyattsville, Maryland: Public Health Service. DHHS Pub. No. 92-1232, 1992.

The NHLBI Task Force on Blood Pressure Control in Children. "The Report of the Second Task Force on Blood Pressure Control in Children, 1987." *Pediatrics* 79:1–25, 1987.

Public Citizen Health Research Group. "Work-Related Injuries Reached Record Level Last Year." *Public Citizen Health Research Group Health Letter* 8(12):1–3, 9, 1992.

Public Citizen Health Research Group (PCHRG). "The Growing Epidemic of Uninsurance." *Public Citizen Health Research Group Health Letter* 9(1):1–2, 1993.

Satterthwaite, F. E. "An Approximate Distribution of Estimates of Variance Components." *Biometrics Bulletin* 2:110–114, 1946.

U.S. Bureau of the Census. 1970 Census of Population, Subject Reports: Japanese, Chinese, and Filipinos in the United States, P(2)-1G, Government Printing Office, Washington, D.C.; National Center for Health Statistics (1975). *U.S. Vital Statistics, 1970*, Volume I: Natality. Government Printing Office, Washington, D.C., 1973.

U.S. Bureau of the Census. 1990 Census of Population and Housing, Summary Tape File 1A on CD-ROM Technical Documentation prepared by Bureau of the Census. Washington: The Bureau, 1991.

Vardeman, S. B. "What About the Other Intervals?" *The American Statistician* 46:193–197, 1992.

Walsh, J. E. "Nonparametric Confidence Intervals and Tolerance Regions." In *Contributions to Order Statistics*, edited by A. E. Sarhan and B. G. Greenberg. New York: John Wiley & Sons, 1962.

Walter U., R. Forthofer, and P. U. Witte. "Dose-Response Relation of Angiotensin Converting Enzyme Inhibitor Ramipril in Mild to Moderate Essential Hypertension." *American Journal of Cardiology* 59:125D–132D, 1987.

Tests of Hypotheses

8

Chapter Outline

In this chapter, we formally introduce the testing of hypotheses, define key terms to help us succinctly communicate the ideas of hypothesis testing, and show how to conduct tests. The statistical ideas used in the tests of hypotheses share the same roots with those used in confidence intervals presented in the previous chapter. Therefore, we do not repeat the details on the distributions of the test statistics that we presented in Chapter 7.

8.1 Preliminaries in Tests of Hypotheses

Hypothesis testing is a way of organizing and presenting evidence that helps us reach a decision. Although the confidence interval and the test of hypothesis can be used to reach the same conclusion, their emphases are different. The confidence interval provides limits that are likely to contain the parameter. These limits can also be used to test a hypothesis, but that is not necessarily the reason why they were created. The test of hypothesis aids in reaching a decision about whether or not we believe that the hypothesized value of the parameter is correct. The use of the test of hypothesis also provides additional information about the decision that is not provided with the confidence interval. Example 8.1 illustrates the basic idea.

Example 8.1

Let us consider a situation associated with the decision to proceed with the marketing of a new drug for reducing cholesterol. This decision was reached because it is unlikely that the greater mean reduction of serum cholesterol observed in a sample of patients receiving a new drug, when compared to the reduction achieved for a sample of patients who received the standard treatment, was due to chance. Or the decision may be for the local health department to allocate more resources to an immunization campaign for childhood diseases. This decision was reached because, based on the sample proportion immunized, it is unlikely that the proportion of 5-year-old children in the community that have the required immunizations equals the targeted value of 95 percent.

There are negative outcomes associated with making a wrong decision, and these must be weighed carefully. If the decision to market the new drug is wrong — that is, it is not an improvement over the standard treatment — patients may pay more money for no additional benefit or for a treatment that does not work. However, if the decision were not to market and the drug was better, patients would lose by not having access to a better treatment, and the company would lose because it did not realize the profit from this drug. If the health department's decision to conduct an immunization campaign is wrong — that is, the proportion of 5-year-old children immunized in the community is at least 95 percent — scarce resources would be misdirected. Other needy programs would not receive additional resources. However, if the decision were not to conduct the campaign when it was needed, there would be increased risk of unnecessary disease in preschool children.

We use another example to clarify these notions and to lead into the definitions used in tests of hypotheses.

Example 8.2

Suppose two diets are proposed for losing weight. We have 12 pairs of individuals, matched on age (±5 years), sex, initial weight (±10 pounds), and level of exercise. One member of the pair is assigned at random to diet 1 and the other member is assigned to diet 2. Individuals remain on their diets for six weeks and are then reweighed. We wish to determine whether or not the diets are equivalent from a weight loss perspective. Table 8.1 shows how the data — the weight losses for those on diets 1 and 2 and the within pair difference — may be presented.

There are several ways of analyzing these data. We demonstrate a very simple approach here and other approaches will be shown later. We shall examine the proportion of pairs in which the person on diet 1 had the greater weight loss. If the diets do not differ with regards to weight loss, assuming there are no ties in weight loss, the proportion should be 0.50. Deviations from 0.50 suggest that there is a difference in the diets in terms of weight loss. If there are ties in the weight losses, the hypothesis being tested is that the proportion of pairs in which the person on diet 1 had the greater weight loss is the same as the proportion of pairs in which the person on diet 2 had the greater weight loss. Note that we have converted the hypothesis in words into something that we can deal with analytically.

Table 8.1   Weight losses (pounds) by diet for 12 pairs of individuals.

Diet	Pairs											
	1	2	3	4	5	6	7	8	9	10	11	12
1	x_1	x_2	x_3	x_4	x_5	x_6	x_7	x_8	x_9	x_{10}	x_{11}	x_{12}
2	y_1	y_2	y_3	y_4	y_5	y_6	y_7	y_8	y_9	y_{10}	y_{11}	y_{12}
Difference	d_1	d_2	d_3	d_4	d_5	d_6	d_7	d_8	d_9	d_{10}	d_{11}	d_{12}

8.1.1   Terms Used in Hypothesis Testing

The statistical terms in hypothesis testing are defined, and their underlying concepts are explained following, based on Example 8.2.

Null and Alternative Hypotheses: The hypothesis being tested is called the null hypothesis and is denoted by H_0. The *null hypothesis* is that π, the proportion of pairs in the population for which persons on diet 1 would show the greater weight loss, is 0.50. The *alternative hypothesis*, denoted by H_a or H_1, to the null hypothesis is that π is not equal to 0.50. In symbols, these hypotheses are

$$H_0: \pi = 0.50 \quad \text{and} \quad H_a: \pi \neq 0.50.$$

We either reject or fail to reject the null hypothesis. If we reject the null hypothesis, we are expressing a belief that the alternative hypothesis is true. If there are ties in the weight losses, the alternative hypothesis is that the proportion of pairs in which the person on diet 1 had the greater weight loss is not equal to the proportion of pairs in which the person on diet 2 had the greater weight loss.

Type I and Type II Errors: If we reject the null hypothesis in favor of the alternative hypothesis, there are two possible outcomes. Either we have correctly rejected the null hypothesis or we have falsely rejected it. Falsely rejecting the null hypothesis is called a Type I error. In this example, the Type I error is claiming that the proportion of pairs for which diet 1 showed the greater weight loss is not equal to 0.50 when, in fact, it is 0.50.

If we fail to reject the null hypothesis, again there are two possible outcomes. Either we have failed to reject the null hypothesis when it should have been rejected or we have correctly failed to reject the null hypothesis. Failing to reject the null hypothesis when it should have been rejected is called a Type II error. The Type II error in this example is claiming that the proportion of pairs for which diet 1 showed the greater weight loss is 0.50 when, in fact, the proportion is different from 0.50. Figure 8.1 shows these four possibilities. The probability of a Type I error is usually labeled α, and the probability of a Type II error is usually labeled β. Ideally we would like to keep both of these probabilities as small as possible, although we usually focus more on the Type I error and its probability.

Our Decision about the Null Hypothesis	Reality: Null Hypothesis Is	
	True	False
True	**Good**	**Type II Error**
False	**Type I Error**	**Good**

Figure 8.1 Possibilities associated with a test of hypothesis.

The Test Statistic: The *test statistic*, the basis for the test of hypothesis, is the number of pairs out of the 12 sample pairs for which those on diet 1 achieved the greater weight loss. Equivalently, the observed sample proportion of pairs for which those on diet 1 achieved the greater weight loss, p, could be used. The test is based on the sign of the difference and, therefore, this particular test is called the sign test. Now that we know what hypothesis is to be tested and what test statistic is to be used, we must specify the decision rule to be used.

8.1.2   Determination of the Decision Rule

The decision rule specifies which values of the test statistic (or some function of it) will cause us to reject the null hypothesis in favor of the alternative hypothesis. The decision rule is based on the probabilities of the Type I and II errors. The probabilities of Type I and Type II errors are found from consideration of the distribution of the test statistic. In this example, the test statistic follows the binomial distribution. The binomial is used because there are only two outcomes: diet 1 is better or diet 2 is better (again ignoring the possibility of a tie in weight loss). We begin by assuming that the null hypothesis is true — that is, π is 0.50. Because we know that n is 12, we know both parameters of the binomial distribution. The probability distribution of the possible outcomes is shown in the following table and in Figure 8.2.

No. of Times Diet 1 Is Better	Probability	No. of Times Diet 1 Is Better	Probability
0	0.0002	7	0.1934
1	0.0030	8	0.1208
2	0.0161	9	0.0537
3	0.0537	10	0.0161
4	0.1208	11	0.0030
5	0.1934	12	0.0002
6	0.2256		

What values of the test statistic would cause us to reject the null hypothesis that π is 0.50 in favor of the alternative hypothesis? Large deviations from six pairs for which diet 1 was better — that is, large deviations from π of 0.50 — are suggestive that the diets have different effects. Thus, either very large or very small values of the test statistic would cause us to question the null hypothesis. As we can see from Figure 8.2, it is highly unlikely to observe either very large or very small values of the test statistic if π is really 0.50.

One- and Two-Sided Tests: The test we are considering is called a *two-sided test*, since either large or small values of the test statistic cause us to question the truth of

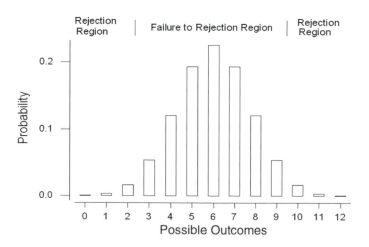

Figure 8.2 Bar chart showing the binomial probability distribution for $n = 12$ and $\pi = 0.5$.

the null hypothesis. A *one-sided test* occurs when only values in one direction cause us to question the null hypothesis. For example, if we were the developers of diet 1, we might only be interested in whether or not diet 1 was better than diet 2, not whether or not it was worse than diet 2. If this were the situation, the null hypothesis remains that π is equal to 0.50, but the alternative hypothesis becomes that π is greater than 0.50. In symbols, this is

$$H_0: \pi = 0.50 \quad \text{versus} \quad H_\text{a}: \pi > 0.50.$$

In this case, only large values of the test statistic would cause us to reject the null hypothesis in favor of the alternative hypothesis.

Use of a one-sided test makes it easier to detect departures from the null hypothesis in the indicated direction — that is, π greater than 0.50. However, the use of a one-sided test means that if the departure is in the other direction — that is, π is less than 0.50 — it won't be detected.

Calculation of the Probabilities of Type I and Type II Errors: Suppose that we decide to reject the null hypothesis whenever we observe a test statistic of 0 or 12 pairs — that is, the values of 0 and 12 form the rejection or critical region. The values from 1 to 11 then form the failure to reject or acceptance region. The probability of a Type I error, is thus the probability of observing 0 or 12 pairs in which diet 1 had the greater weight loss when π is actually 0.50. From the preceding probability mass function, we see that is 0.0004. That's great! There is almost no chance of making this error, and this is almost as small as we can make it. Of course, we could decide never to reject the null hypothesis, and, then, there would be zero probability of a Type I error. That is unrealistic, however.

We are pleased with this decision rule because it has an extremely small probability of a Type I error. However, what is the value of β, the probability of a Type II error, associated with this decision rule? To be able to calculate β, we have to be more specific about the alternative hypothesis. The preceding alternative hypothesis is quite general in that it only says π is not equal to 0.50. However, just as we used a specific value, the value 0.50, for π in calculating the probability of a Type I error, we must specify a value of π other than 0.50 to be used in calculating the probability of a Type II error. We must move from the general alternative to a specific alternative hypothesis to be able to calculate a value for β. This means that there is not merely one β associated with the decision rule; rather, there is a value of β corresponding to each alternative hypothesis.

What specific value of π should be used in the alternative hypothesis? We should have little interest in the alternative that π is 0.51 instead of the null hypothesis value of 0.50. The difference between 0.51 and 0.50 is of little practical interest. For all practical intent, if π is really 0.51, there is little difference in the diets. As the value of π departs more and more from 0.50, the ability to detect these departures becomes more important. We may not all agree at which point π differs enough from 0.50 to be important. Some may say 0.60 is different enough, whereas others may say that π must be at least 0.70 for the difference to be important. Most would certainly agree that we should reject the equality of the diets if diet 1 provides for greater weight loss in 80 percent of the pairs.

Let us assume that π is really 0.80, not 0.50, and find the value for β. The binomial distribution for an n of 12 and a proportion of 0.80 is shown below.

No. of Times Diet 1 Is Better	Probability	No. of Times Diet 1 Is Better	Probability
0	0.0000	7	0.0532
1	0.0000	8	0.1328
2	0.0000	9	0.2363
3	0.0001	10	0.2834
4	0.0005	11	0.2062
5	0.0033	12	0.0687
6	0.0155		

Type II error is failing to reject the null hypothesis when it should be rejected. Since our decision rule is to reject only when we observe a test statistic of 0 or 12, we will fail to reject for the values of 1 through 11. The probability of 1 through 11 when π is actually 0.80 is 0.9313 (= $1 - 0.0000 - 0.0687$). Therefore, use of this decision rule yields an α of 0.0004 and a β of 0.9313. The probability of the Type I error is very small, but the probability of the Type II error, corresponding to the value of 0.80 for π, is quite large.

8.1.3   Relationship of the Decision Rule, α and β

If we change our decision rule to reject the null hypothesis more often, we will increase α but decrease β — that is, there is an inverse relation between α and β. For example, if we increase the rejection region by including values 1 and 11 in addition to 0 and 12, the value of α becomes 0.0064 (= $0.0002 + 0.0030 + 0.0030 + 0.0002$). These probabilities are found from the probability distribution based on the value for π of 0.50. The new value for β, based on this expansion of the rejection region, and using 0.80 for π, is 0.7251 (= $1 - 0.2062 - 0.0687$). The probability of a Type I error remains quite small, but the probability of a Type II error is still large.

If the decision rule is to reject for values of the test statistic of 0 to 2 and 10 to 12, then α's value is increased to 0.0386 (= $2 [0.0161 + 0.0030 + 0.0002]$) and the value of β is reduced to 0.4417 (= $1 - 0.0687 - 0.2062 - 0.2834$). The probability of a Type I error is still reasonable, whereas, although the probability of Type II error is much smaller than previously, it is still quite large. However, a further change in the decision rule to include the values of the test statistic of 3 and 9 increases the value of α to 0.1460 (= $0.0386 + 2[0.0537]$), which is now becoming large.

What Are Reasonable Values for α and β? There are no absolute values that indicate that the probability of error is too large. It is a matter of personal choice, although convention suggests that an α greater than 0.10 is unacceptable. Most investigators set α to 0.05, and some set it to 0.01. There is less guidance for the choice of β. It again is a matter of personal choice. However, the implications of the Type II error play a role in how large a β can be tolerated. A value of 0.20 for β is used frequently in the literature. Investigators often ignore the Type II error because (1) the hypothesis has been framed such that the Type I error is of much greater interest than the Type II error, or (2) it is often difficult to find the value of β.

Ways to Decrease β Without Increasing α: We were in a bind when we left the example above. The value of β was too large and, if we tried to reduce it by further enlargement of the rejection region, we made α too large. One way of decreasing β without increasing α is to change the alternative hypothesis or to increase the sample size.

(1) **Changing Alternative Hypothesis:** The specific alternative hypothesis that we had used previously in calculating β was that π was equal to 0.80. We selected the value of 0.80 because if diet 1 performed better for 80 percent of the pairs, we believed this indicated a very important difference between the diets. If we are willing to change what we consider to be a very important difference, we can reduce β. For example, by increasing the value of π in the alternative hypothesis from 0.80 to 0.90, β will decrease. However, this means that we no longer consider it to be important to detect that π was really 80 percent instead of the hypothesized 50 percent. We will focus on the test's ability to detect a very large difference — that is, the difference between 0.90 and 0.50 — and not worry that the test has a small chance of detecting smaller differences.

The following shows the probability mass function for the binomial with a sample size of 12 and a proportion of 0.90.

No. of Times Diet 1 Is Better	Probability	No. of Times Diet 1 Is Better	Probability
0	0.0000	7	0.0038
1	0.0000	8	0.0213
2	0.0000	9	0.0853
3	0.0000	10	0.2301
4	0.0000	11	0.3766
5	0.0001	12	0.2824
6	0.0004		

If we again use the rejection region of 0 to 2 and 10 to 12, the probability of the Type I error is still 0.0386, since that was calculated based on π being 0.50. However, β is the probability of not rejecting that π is 0.50 when it is actually 0.90. This probability is the sum of the probabilities of the outcomes 3 through 9 in the preceding distribution, and that sum is 0.1109. Now both the values of α and β are reasonable.

By changing the alternative hypothesis, we have not changed the value of β for the alternative of π being 0.80. The β-value corresponding to a π of 0.80 and a rejection region of 0 to 2 and 10 to 12 remains 0.4417. What has changed is what we consider to be an important difference. If a lesser difference is considered to be important, the probability of the Type II error for that value of π can be calculated. Table 8.2 shows

Table 8.2   Probability of Type II error and power for specific alternative hypotheses based on a rejection region of 0 to 2 and 10 to 12.

Alternative Hypothesis	Probability of Type II Error	Power
$\pi = 0.55$	0.9507	0.0493
$\pi = 0.60$	0.9137	0.0863
$\pi = 0.65$	0.8478	0.1522
$\pi = 0.70$	0.7470	0.2530
$\pi = 0.75$	0.5778	0.4222
$\pi = 0.80$	0.4416	0.5584
$\pi = 0.85$	0.2642	0.7358
$\pi = 0.90$	0.1109	0.8891
$\pi = 0.95$	0.0195	0.9805

the values of the Type II errors for several values of the alternative hypothesis based on a rejection region of 0 to 2 and 10 to 12.

The probability of a Type II error decreases as the value of π used in the alternative hypothesis moves farther away from its value in the null hypothesis. This makes sense, since it should be easier to detect greater differences than smaller ones. As this table shows, there is a very high chance of failing to detect departures from 0.50 less than 0.30 to 0.35 in magnitude. The last column in Table 8.2 is *power*, the probability of rejecting the null hypothesis when it should be rejected — that is, when the alternative hypothesis is true. From the table we can see that power is 1 minus β. Power is often used in the literature when discussing the properties of a test statistic instead of using the probability of a Type II error. From the values in Table 8.2, it is possible to create a *power curve* — that is, to graph the values of power versus the values of π used in the alternative hypothesis. Figure 8.3 shows a portion of the power curve for values of π greater than 0.50. Statisticians use power curves to compare different test statistics.

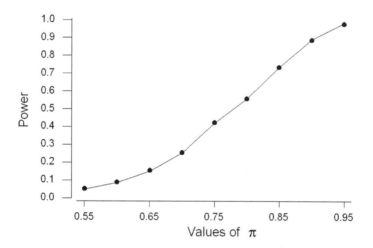

Figure 8.3 Portion of the power curve for π values greater than 0.5.

The preceding trade-off as a way of reducing β may not be very satisfactory. We still may feel that 80 percent is very different from 50 percent. As an alternative, we could increase the sample size instead of changing the alternative hypothesis.

(2) Increasing the Sample Size: None of the calculations shown so far have required the observed sample data. All these calculations are preliminary to the actual collection of the data. Therefore, if the probabilities of errors are too large, we can still change the experiment. As just mentioned, increasing the sample size is one way of decreasing the error probabilities, but doing this increases the resources required to perform the experiment. There is a trade-off between the sample size and the error probabilities.

Suppose we can afford to find and follow 15 pairs instead of the 12 pairs we initially intended to use. We still use the binomial distribution in the calculation of the error probabilities where π remains 0.50, but now n is equal to 15. The binomial probability mass function with these parameters is shown next.

No. of Times Diet 1 Is Better	Probability	No. of Times Diet 1 Is Better	Probability
0	**0.0000**	8	0.1964
1	**0.0005**	9	0.1527
2	**0.0032**	10	0.0917
3	**0.0139**	11	0.0416
4	0.0416	12	**0.0139**
5	0.0917	13	**0.0032**
6	0.1527	14	**0.0005**
7	0.1964	15	**0.0000**

Let us use a rejection region of 0 to 3 and 12 to 15. If we do this, the probability of a Type I error is 0.0352 (= 2 [0.0005 + 0.0032 + 0.0139]). The probability of a Type II error, based on the alternative that π is 0.80, uses the binomial distribution with parameters 15 and 0.80 and this probability mass function is now shown.

No. of Times Diet 1 Is Better	Probability	No. of Times Diet 1 Is Better	Probability
0	0.0000	8	**0.0139**
1	0.0000	9	**0.0430**
2	0.0000	10	**0.1031**
3	0.0000	11	**0.1876**
4	0.0000	12	0.2502
5	**0.0001**	13	0.2309
6	**0.0007**	14	0.1319
7	**0.0034**	15	0.0352

The probability of failing to reject a null hypothesis when it should be rejected — that is, of being in the acceptance region (values 4 to 11), when π is 0.80 — is 0.3518 (= 0.0001 + 0.0007 + 0.0034 + 0.0139 + 0.0430 + 0.1031 + 0.1876). The probability of a Type I error, 0.0352, is similar to its preceding value, 0.0386, when we considered this same alternative hypothesis. The probability of a Type II error has decreased from 0.4417 above when n was 12 to 0.3518 now for an n of 15. A further increase in the sample size can reduce β to a more acceptable level. For example, when n is 20, use of values 0 to 5 and 15 to 20 for the rejection region leads to an α of 0.0414 and a β of 0.1958.

8.1.4   Conducting the Test

The procedure used in conducting a test begins with a specification of the null and alternative hypotheses. In this example, they are

$$H_0:\ \pi = 0.50 \quad \text{versus} \quad H_a:\ \pi \neq 0.50.$$

We must decide on the significance level to be used in conducting the test. The *significance level* is the probability of a Type I error that we are willing to accept. We use the conventional significance level of 0.05 in this example.

Based on the preceding calculations, we have decided to increase the sample size to 20. We will reject the null hypothesis if the value of the test statistic is from 0 to 5 or from 15 to 20. Use of this sample size and decision rule keeps the probability of a Type I error less than 0.05 and also keeps β reasonably small when considering large departures from the null hypothesis. With discrete data, the probability of a Type I error usually does not equal the significance level exactly. The decision rule used with discrete

Table 8.3   Weight losses (pounds) by diet for 20 pairs of individuals.

Pairs	Diet 1	Diet 2	Difference
1	9	7	2
2	4	6	−2
3	11	9	2
4	7	12	−5
5	−4	3	−7
6	13	8	5
7	6	5	1
8	3	−1	4
9	8	14	−6
10	10	8	2
11	8	6	2
12	9	8	1
13	14	15	−1
14	11	7	4
15	5	7	−2
16	−3	4	−7
17	6	−2	8
18	7	4	3
19	13	10	3
20	9	5	4

data is chosen so that it results in a probability of a Type I error being as close as possible to and less than the desired significance level. The data are collected and shown in Table 8.3.

There are 13 pairs for which persons on diet 1 had the greater weight loss. As 13 does not fall into the rejection region of 0 to 5 or 15 to 20, we fail to reject the null hypothesis in favor of the alternative hypothesis at the 0.05 significance level. The observed result is not statistically significant.

The **p**-*value*: Another statistic often reported is the *p-value* of the test, the probability of a Type I error associated with the smallest rejection region that includes the observed value of the test statistic. Another way of stating this is that the *p*-value is the level at which the observed result would just be statistically significant. In this example, since we are conducting a two-sided test, the smallest rejection region including the observed result of 13 is the region from 0 to 7 and 13 to 20. Examination of Table B2 for an *n* of 20 and a π of 0.50 yields a probability of being in this region of 0.2632 (= 2[0.0370 + 0.0739] + 0.0414). The value of 0.0414 is the value associated with the region from 0 to 5 and 15 to 20, and to that we have added the probabilities associated with the outcomes 6, 7, 13, and 14. The *p*-value is thus 0.2632.

Some statisticians do not believe in the decision rule approach to testing hypotheses. They believe that the *p*-value provides information regardless of whether or not the hypothesis is rejected. The *p*-value tells how likely the observed result is, assuming that the null hypothesis is true. For example, these statisticians see little difference in *p*-values of 0.05001 and 0.04999, although in the first case we would fail to reject the null hypothesis at the 0.05 significance level, whereas in the second case we would reject the null hypothesis. For these statisticians, the key information to be obtained from the study is that there is roughly 1 chance in 20 that we would have obtained the observed result if the null hypothesis were true. Using the *p*-value in this way is very reasonable.

8.2   Testing Hypotheses about the Mean

Suppose we wish to analyze the Digoxin clinical trial data shown in Table 3.1. However, before performing the analyses, we wish to determine whether or not the population represented by the sample of 200 patients differs from the national adult population as far as systolic blood pressure is concerned. Therefore, we first test the hypothesis that the mean systolic blood pressure for the patients in the Digoxin clinical trial is the same as the national average.

From the calculations in Chapter 3, we know that the sample mean, based on 199 patients (one missing value), is 125.8 mmHg. Based on national data (Lee and Forthofer 2006), we take the national average to be 122.3 mmHg. The test of hypothesis about the population mean, just like the confidence interval, uses the normal distribution if the population variance is known or the t distribution if the variance is unknown. We first assume that the variance is known.

8.2.1   Known Variance

In Chapter 7, when we formed the 95 percent confidence interval for the population mean, we assumed that the population standard deviation was 20 mmHg or that the variance was 400 mmHg. We shall use that value in the test of hypothesis about the population mean. The null and alternative hypotheses are

$$H_0: \mu = \mu_0 \quad \text{and} \quad H_a: \mu \neq \mu_0$$

where μ_0 is 122.3 mmHg in this example. To be able to compare the test results with the confidence interval from Chapter 7, we conduct the test at the 0.05 significance level.

The test statistic is $z \ (= [\bar{x} - \mu_0]/[\sigma/\sqrt{n}\,])$, the standard normal statistic. If the null hypothesis is true, z will follow the standard normal distribution. The rejection region is thus defined in terms of percentiles of the standard normal distribution. For a two-sided alternative, if z is either less than or equal to $z_{\alpha/2}$ or greater than or equal to $z_{1-\alpha/2}$, we reject the null hypothesis in favor of the alternative hypothesis. In symbols, this is

$$\frac{\bar{x} - \mu_0}{\sigma/\sqrt{n}} \leq z_{\alpha/2} \quad \text{or} \quad \frac{\bar{x} - \mu_0}{\sigma/\sqrt{n}} \geq z_{1-\alpha/2}.$$

If the test statistic is not in the rejection region — that is,

$$z_{\alpha/2} < z < z_{1-\alpha/2}$$

we fail to reject the null hypothesis in favor of the alternative hypothesis. Let us calculate the test statistic for the systolic blood pressure. The z value is

$$\frac{125.8 - 122.3}{20/\sqrt{199}} = 2.47.$$

Since 2.47 falls in the rejection region — that is, it is greater than 1.96 ($= z_{1-0.025}$) or is less than −1.96 — we reject the null hypothesis. This situation is shown pictorially in Figure 8.4. The p-value for this test is the probability of observing a standard normal variable with either a value greater than 2.47 or less than −2.47. This probability is found to be 0.014.

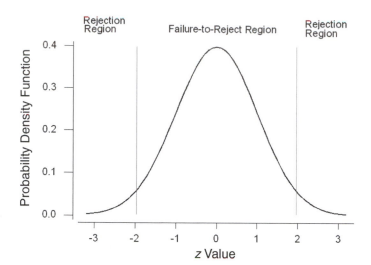

Figure 8.4
Representation of the rejection and failure-to-reject regions in terms of the standard normal statistic for $\alpha = 0.05$.

Equivalence of Confidence Intervals and Tests of Hypotheses: Recall that in Chapter 7 when we found the $(1 - \alpha) * 100$ percent confidence interval for the population mean, we started with the expression

$$\Pr\left\{-z_{1-\alpha/2} < \frac{\bar{x} - \mu}{\sigma/\sqrt{n}} < z_{1-\alpha/2}\right\} = 1 - \alpha.$$

We manipulated this expression, and we obtained the following expression:

$$\bar{x} - z_{1-\alpha/2}\left(\frac{\sigma}{\sqrt{n}}\right) < \mu < \bar{x} + z_{1-\alpha/2}\left(\frac{\sigma}{\sqrt{n}}\right).$$

If we replace μ in the middle portion of the preceding first expression by μ_0, the middle portion is the z statistic for testing the hypothesis that μ equals μ_0. Since the confidence interval was derived from this test statistic, this means that if μ_0 is contained in the confidence interval, then the corresponding z statistic must also be in the failure to reject (acceptance) region. If μ_0 is not in the confidence interval, then the z statistic is in the rejection region — that is, it is less than or equal to $-z_{1-\alpha/2}$ or greater than or equal to $z_{1-\alpha/2}$.

In this case, the hypothesized value of 122.3 mmHg is not contained in the 95 percent confidence interval for the population mean. We saw in Chapter 7 (Example 7.1) that the confidence interval ranges from 123.0 to 128.6 mmHg. Therefore, we know that the test statistic will be in the rejection region, and, hence, we will reject the null hypothesis. In addition, using the same logic, from the confidence interval, we know we would reject the null hypothesis for any μ_0 not in the range from 123.0 to 128.6 mmHg.

This same type of argument for the linkage of the test of hypothesis and the corresponding confidence interval can be used with the other tests of hypotheses presented in this chapter. Thus, the confidence interval is also very useful from a test of hypothesis perspective. However, the confidence interval does not provide the p-value of the test, also a useful statistic.

One-Sided Alternative Hypothesis: If we are concerned only when the patients have elevated blood pressure, the null and alternative hypotheses are

$$H_0: \mu = \mu_0 \quad \text{and} \quad H_a: \mu > \mu_0.$$

The test statistic does not change, but the rejection region is a one-sided region now. We reject the null hypothesis in favor of the alternative hypothesis if z is greater than or equal to $z_{1-\alpha}$, or equivalently, if $\bar{x}$ is greater than or equal to $\mu_0 + [z_{1-\alpha}(\sigma/\sqrt{n})]$. A one-sided rejection region is shown in Figure 8.5 in terms of the z test statistic.

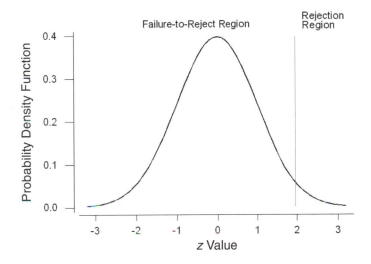

Figure 8.5
Representation of the rejection and failure-to-reject regions for a one-sided alternative of greater than the mean for $\alpha = 0.05$.

If we are concerned only when the patient's blood pressure is too low, the null and alternative hypotheses are

$$H_0: \mu = \mu_0 \quad \text{and} \quad H_a: \mu < \mu_0.$$

We now reject if z is less than or equal to z_α, or equivalently, if $\bar{x}$ is less than or equal to $\mu_0 + [z_\alpha (\sigma/\sqrt{n})]$. In this case, the rejection region in Figure 8.5 moves to the lower end of the distribution.

Power of the Test: Before collecting the data, suppose that we wanted to be confident that, if the systolic blood pressure of patients in the Digoxin clinical trial was substantially more than the national average, we could detect this higher mean blood pressure. By substantially more, we mean 3 percent or more above the national average of 122.3 mmHg. Thus, we wish to conclude that there is a difference between the study subjects and the national average if the study subjects have a population mean of 126.0 mmHg or more. The use of 3 percent is subjective and other values could be used.

The null and alternative hypotheses for this situation are

$$H_0: \mu = \mu_0 \quad \text{and} \quad H_a: \mu > \mu_0.$$

We use a significance level of 0.01. Thus, the rejection region includes all z more than or equal to $z_{0.99}$ — that is, z greater than or equal to 2.326. In terms of $\bar{x}$ the rejection region includes all values of $\bar{x}$ greater than or equal to

$$\mu_0 + z_{1-0.01}\left(\frac{\sigma}{\sqrt{n}}\right) = 122.3 + 2.326\left(\frac{20}{\sqrt{199}}\right) = 125.6.$$

Figure 8.6 shows the rejection and acceptance regions in terms of $\bar{x}$ as well as showing its distribution under the alternative hypothesis.

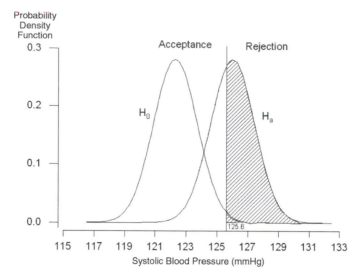

Figure 8.6 Rejection and acceptance regions of testing H_0: $\mu = 122.3$ versus H_a: $\mu = 126.0$ at the 0.01 significance level.

The shaded area provides a feel for the size of the power — the probability of rejecting the null hypothesis when it should be rejected — of the test. Power is the proportion of the area under the alternative hypothesis curve that is in the rejection region — that is, greater than or equal to 125.6 mmHg.

Let us find the power of the test and see if it agrees with our expectations about it based on Figure 8.6. Power is the probability of being in the rejection region — that is, of the sample mean being greater than or equal to 125.6 mmHg, assuming that the alternative hypothesis ($\mu = 126.0$) is true. To find this probability, we convert 125.6 to a standard normal value by subtracting the mean of 126.0 mmHg and dividing by $\sigma/\sqrt{n}$. Thus, the z value is

$$\frac{125.6 - 126.0}{20/\sqrt{199}} = -0.28.$$

The probability of a standard normal variable being greater than or equal to -0.28 is found from Table B4 to be 0.6103 ($= 1 - 0.3897$).

The power of the test is 61 percent. If this value is not large enough, there are several methods of increasing the power. One way is to increase the sample size. For example, let us increase the sample size to 600. Then the z value is

$$\frac{125.6 - 126.0}{20/\sqrt{600}} = -0.49.$$

The probability of a standard normal variable being greater than or equal to -0.49 is 0.6879 ($= 1 - 0.3121$), almost 8 percent larger than the power associated with the sample size of 199.

As was discussed earlier, another way of changing the power is to change the significance level. Let us decrease the significance level to 0.05. Doing this reduces the size of the rejection region. All values of $\bar{x}$ that are greater than or equal to

$$122.3 + 1.645\left(\frac{20}{\sqrt{199}}\right) = 124.6$$

now are in the rejection region. Using this significance level and still using a sample size of 199, the z value becomes

$$\frac{124.6 - 126.0}{20/\sqrt{199}} = -1.06.$$

The probability of a standard normal variable being greater than or equal to −1.06 is 0.855, more than 80 percent, which is often used as the desired level for power in the literature.

Another way of increasing the power is to redefine what we consider to be a substantial difference. If our emphasis were on detecting a blood pressure 5 percent more than the national average, instead of 3 percent more, we would have a higher power. As 5 percent of 122.3 mmHg is 6.1 mmHg, the null and alternative hypotheses become

$$H_0: \mu = 122.3 \quad \text{and} \quad H_a: \mu = 128.4.$$

The z statistic becomes

$$\frac{125.6 - 128.4}{20/\sqrt{199}} = -1.97$$

and the probability of a standard normal variable being more than or equal to −1.97 is 0.976. The power associated with the alternative that μ equals 126.0 has not changed, but our emphasis on what difference is important has changed. We have a much higher chance of detecting this greater difference, from 128.4 instead of from 126.0, between the null and alternative hypotheses.

Example 8.3

Let us consider another example of the calculation of power. Suppose that we have reason to suspect that the serum cholesterol level of male college students in a rural town is lower than the national average and we are planning a study to test this. The null and alternative hypotheses for the study are

$$H_0: \mu = 188 \text{ mg/dL} \quad \text{and} \quad H_a: \mu < 188 \text{ mg/dL}.$$

We use a value of 30 mg/dL for the standard deviation of serum cholesterol level for male college students.

We must choose a specific value for the mean serum cholesterol level under the alternative hypothesis. We have selected the value of 170 mg/dL, a difference of 18 mg/dL from the national mean, as an important difference that we wish to be able to detect. For this study, our initial plans call for a sample size of 50.

To find the power, we must first find the acceptance and rejection regions. Let us perform the test at the 0.01 significance level. Therefore, the rejection region consists of all values of z less than or equal to $z_{0.01}$ ($= -2.326$). In terms of the sample mean, the rejection region consists of values of $\bar{x}$ less than or equal to

$$\mu_0 + z_{0.01}\left(\frac{\sigma}{\sqrt{n}}\right) = 188 + (-2.326)\left(\frac{30}{\sqrt{50}}\right) = 178.132.$$

Figure 8.7 shows this situation.

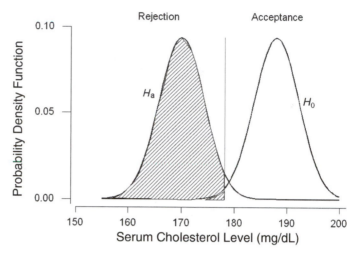

Figure 8.7 Rejection and acceptance regions for testing $H_0: \mu = 94$ versus $H_a: \mu = 100$ at the 0.05 significance level.

Once we know the boundary between the acceptance and rejection regions, we convert the boundary to a z value by subtracting the mean under the alternative hypothesis and dividing by the standard error. For this example, the z value is

$$z = \frac{178.132 - 170}{30/\sqrt{50}} = 1.92.$$

The power of the test is the probability of observing a z-statistic with a value less than or equal to 1.92. From Table B4, we find the power to be 0.9726. This value is consistent with what we would have expected based on Figure 8.7. A study with 50 male college students has an excellent chance of detecting a mean value 18 mg/dL below the national average.

The key point about power is that calculations like in Example 8.3, or like those discussed in the material on confidence intervals, should be performed before any data are collected. These calculations give some indication about whether or not it is worthwhile to conduct an experiment before the resources are actually expended.

8.2.2 Unknown Variance

If the variance is unknown, the t statistic is used in place of the z statistic — that is,

$$t = \frac{\bar{x} - \mu_0}{\left(s/\sqrt{n}\right)}$$

in the test of the null hypothesis that the mean is the particular value μ_0. The rejection region for a two-sided alternative is $(t \leq t_{n-1,\alpha/2})$ or $(t \geq t_{n-1,1-\alpha/2})$.

Suppose that we did not know the value of σ for the systolic blood pressure in the dig200 data or that we were uncomfortable in using the value of 20 mmHg for σ. Then we would substitute s for σ and use the t distribution in place of the z distribution. In this case, the value of the t statistic is

$$t = \frac{125.8 - 122.3}{18.2/\sqrt{199}} = 2.71.$$

To be consistent with the test shown above, we shall also perform this test at the 0.05 significance level. Therefore, t is compared to $t_{198,0.025}$, which is -1.972, and to $t_{198,0.975}$, which is 1.972. Since 2.71 is in the rejection region, we reject the null hypothesis in favor of the alternative. Not surprisingly, this result is very similar to that obtained when the z statistic was used. The results are similar because there was little difference between the values of s and σ, and, since the sample size is large, the critical values of the t and normal distributions are also close in value. The t test for one mean can be performed by the computer (see **Program Note 8.1** on the website).

8.3   Testing Hypotheses about the Proportion and Rates

In this section we focus on situations for which the use of the normal distribution as an approximation for the binomial distribution is appropriate. In general, these are situations in which the sample size is large.

Example 8.4

In Chapter 7 (Example 7.4) we considered the immunization level of 5-year-olds. The health department took a sample and, based on the sample, would decide whether or not to provide additional funds for an immunization campaign. In Example 7.4 we examined both the 99 percent confidence interval and a one-sided interval. Since the health department will provide additional funds if the proportion of immunization is less than 75 percent, we consider a one-sided test here, considering the following null and alternative hypotheses

$$H_0: \pi = \pi_0 = 0.75 \quad \text{and} \quad H_a: \pi < \pi_0 = 0.75.$$

The test statistic for this hypothesis is

$$z = \frac{|p - \pi_0| - 1/(2n)}{\sqrt{p(1-p)/n}}.$$

If $(p - \pi_0)$ is positive, a positive sign is assigned to z; if the difference is negative, a minus sign is assigned to z. The rejection region consists of values of z less than or equal to z_α. This framework is very similar to that used with the population mean, the only difference being the use of the continuity correction with the proportion.

The sample proportion, p, had a value of 0.614 based on a sample size of 140. Thus, the calculation of z is

$$z = \frac{|0.614 - 0.75| - 1/(2\{140\})}{\sqrt{0.614(1 - 0.614)/140}} = 3.219.$$

Since $(p - \pi_0)$ is negative, the test statistic's value is -3.219. If the test is performed at the 0.01 significance level, values of z less than or equal to -2.326 form the rejection region. Since z is less than -2.326, we reject the null hypothesis in favor of the alternative. The health department should devote more funds to an immunization effort. This conclusion agrees with that reached based on the confidence interval approach in Chapter 7.

The continuity correction can be eliminated from the calculations for relatively large sample sizes because its effect will be minimal. For example, if we had ignored the continuity correction in this example, the value of the test statistic would be -3.306, not much different from -3.219. The computer can be used to analyze these data (see **Program Note 8.2** on the website).

The same procedure applies to the test of crude and adjusted rates. Just as in Chapter 7, we treat rates as if they were proportions. This treatment allows for a simple approximation to the variance of a rate and also gives a justification for the use of the normal distribution as an approximation to the distribution of the rate. Thus, our test statistic has the same form as that used for the proportion.

Example 8.5

Suppose that we wish to test, at the 0.05 significance level, that the 2002 age-adjusted death rate for the American Indian/Alaskan Native male population, obtained by the indirect method of adjustment (using the 2002 U.S. age-specific death rates as the standard), is equal to the 2002 direct adjusted death rate for U.S. white male population of 992.9 per 100,000 (NCHS 2004). The alternative hypothesis is that the rates differ. In symbols, the null and alternative hypotheses are

$$H_0: \theta = \theta_0 = 0.009929 \quad \text{and} \quad H_a: \theta \neq \theta_0.$$

The test statistic, z, for this hypothesis is

$$(\hat{\theta} - \theta_0)/(\text{approximate standard error of } \hat{\theta})$$

where $\hat{\theta}$ is 907.8 per 100,000, the 2002 indirect age-adjusted death rate for the American Indian/Alaskan Native male population. In Chapter 7 we found the approximation to the standard error of $\hat{\theta}$ was 11 per 100,000. If this value of z is less than or equal to -1.96 ($= z_{0.025}$) or greater than or equal to 1.96 ($= z_{0.975}$), we reject the null hypothesis in favor of the alternative hypothesis. The value of z is

$$\frac{0.009078 - 0.009929}{0.00011} = -7.74.$$

Since -7.74 is in the rejection region, we reject the null hypothesis in favor of the alternative hypothesis at the 0.05 significance level. There is sufficient evidence to suggest that the indirect age-adjusted death rate for the American Indian/Alaskan Native male population is significantly different from the U.S. white male rate. The p-value for this test is obtained by taking twice the probability that a z statistic is less than or equal to -7.74; the p-value is less than 0.00001.

As we have previously discussed, this test makes sense only if we view the American Indian/Alaskan Native population data as a sample in time or place.

The tests for the crude rate and for the adjusted rate obtained by the direct method of adjustment have the same form as the preceding.

8.4   Testing Hypotheses about the Variance

In Chapter 7 we saw that $(n-1)s^2/\sigma^2$ followed the chi-square distribution with $n-1$ degrees of freedom. Therefore, we shall base the test of hypothesis about σ^2 on this statistic. The null and alternative hypotheses are

$$H_0: \sigma^2 = \sigma_0^2 \quad \text{and} \quad H_a: \sigma^2 \neq \sigma_0^2.$$

We shall define X^2 to be equal to $(n-1)\,s^2/\sigma^2$. When X^2 is greater than or equal to $\chi^2_{n-1,1-\alpha/2}$ or when X^2 is less than or equal to $\chi^2_{n-1,\alpha/2}$, we reject H_0 in favor of H_a.

For a one-sided alternative hypothesis, for example, $H_a: \sigma^2 < \sigma_0^2$, the rejection region is $X^2 \leq \chi^2_{n-1,\alpha}$. If the alternative is $H_a: \sigma^2 > \sigma_0^2$, the rejection region is $X^2 \geq \chi^2_{n-1,1-\alpha}$.

Example 8.6

Returning to the vitamin D in milk example discussed in Chapter 7, suppose we wish to test the hypothesis that the producer is in compliance with the requirement that the variance be less than 1600. We doubt that the producer is in compliance, and, therefore, we shall use the following null and alternative hypotheses:

$$H_0: \sigma^2 = 1600 \quad \text{and} \quad H_a: \sigma^2 > 1600.$$

Since this is a one-sided test, we are implicitly saying that the null hypothesis is that the population variance is less than or equal to 1600 versus the alternative that the variance is greater than 1600. We shall perform the test at the 0.10 significance level. Thus, the test statistic, X^2, which equals

$$\frac{(n-1)s^2}{\sigma_0^2}$$

is compared to $\chi^2_{29,0.90}$. If X^2 is greater than or equal to 39.09, obtained from Table B7, we reject the null hypothesis in favor of the alternative hypothesis. Using the value of 1700 for s^2 and 30 for n from Chapter 7, the value of X^2 is

$$\frac{29(1700)}{1600} = 30.81.$$

Since X^2 is not in the rejection region, we fail to reject the null hypothesis. There is not sufficient evidence to suggest that the producer is not in compliance with the variance requirement. This is the same conclusion reached when the confidence interval approach was used in Chapter 7. Figure 8.8 shows the rejection and acceptance regions for this test.

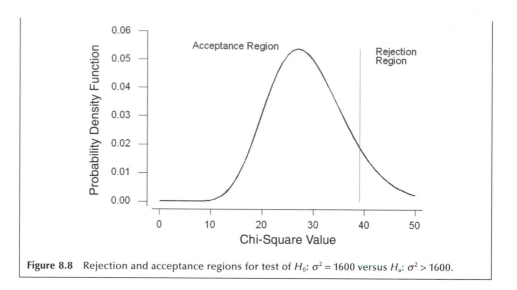

Figure 8.8   Rejection and acceptance regions for test of H_0: $\sigma^2 = 1600$ versus H_a: $\sigma^2 > 1600$.

As was mentioned in Chapter 7, the chi-square distribution begins to resemble the normal curve as the degrees of freedom becomes large. Figure 8.8 is a verification of that fact. From this figure, we also see that the p-value for the test statistic is large — approximately 0.40.

8.5   Testing Hypotheses about the Pearson Correlation Coefficient

In Chapter 7, we saw that the z' transformation, $z' = (1/2) \ln[(1 + r)/(1 - r)]$, approximately followed a normal distribution with a mean of $(1/2) \ln[(1 + \rho)/(1 - \rho)]$ and a standard error of $1/\sqrt{n-3}$. Therefore, to test the null hypothesis of H_0: $\rho = \rho_0$ versus an alternative hypothesis of H_a: $\rho \neq \rho_0$, we shall use the test statistic, λ, defined as $\lambda = (z' - z_0')\sqrt{n-3}$ where z_0' is $(1/2) \ln[(1 + \rho_0)/(1 - \rho_0)]$. If λ is less than or equal to $z_{\alpha/2}$ or greater than or equal to $z_{1-\alpha/2}$, we reject the null hypothesis in favor of the alternative hypothesis.

There is often interest as to whether or not the Pearson correlation coefficient is zero. If it is zero, then there is no linear association between the two variables. In this case, the test statistic simplifies to

$$\lambda = z'\sqrt{n-3}.$$

Example 8.7

Table 8.4 shows infant mortality rates for 1988 and total health expenditures as a percentage of gross domestic product in 1987 for selected 21 countries. It is thought that there should be some relation between these two variables. We translate these thoughts into the following null and alternative hypotheses:

$$H_0: \rho = 0.00 \quad \text{and} \quad H_a: \rho \neq 0.00.$$

Table 8.4   1988 infant mortality rates and 1987 health expenditures as a percentage of gross domestic product for selected countries.[a]

Country	1988 Infant Mortality Rate[b]	1987 Health Expenditures as Percentage of GDP
Japan	4.8	6.8
Sweden	5.8	9.0
Finland	6.1	7.4
Netherlands	6.8	8.5
Switzerland	6.8	7.7
Canada	7.2	8.6
West Germany	7.5	8.2
Denmark	7.5	6.0
France	7.8	8.6
Spain	8.1	6.0
Austria	8.1	7.1
Norway	8.3	7.5
Australia	8.7	7.1
Ireland	8.9	7.4
United Kingdom	9.0	6.1
Belgium	9.2	7.2
Italy	9.3	6.9
United States	10.0	11.2
New Zealand	10.8	6.9
Greece	11.0	5.3
Portugal	13.1	6.4

[a]Infant mortality rates are from National Center for Health Statistics, 1992, Table 25, and health expenditures are from National Center for Health Statistics, 1991, Table 104.
[b]Infant mortality rates are deaths to infants under 1 year of age per 1000 live births.

and the null hypothesis will be tested at the 0.10 significance level. The rejection region consists of values of λ that are less than or equal to -1.645 ($= z_{0.05}$) or greater than or equal to 1.645.

Applying the formula from Chapter 3 to the data in Table 8.4, we find a correlation coefficient, r, to be -0.243. From this value z' and λ can be calculated:

$$z' = 0.5\left(\ln\left(\frac{1-0.243}{1+0.243}\right)\right) = -0.248 \text{ and}$$

$$\lambda = -0.248\sqrt{21-3} = -1.052.$$

Since λ is not in the rejection regions, we fail to reject the null hypothesis. The p-value of this test is 0.29 (twice of probability $[z \le -1.052]$). The correlation is not significantly different from zero. Let us check whether we can draw the same conclusion from the corresponding confidence interval. Using the method discussed in Chapter 7, a 90 percent confidence interval is $(-0.562, 0.139)$ in which zero is included. The computer can be used to perform this test (see **Program Note 8.3** on the website), and most programs give correlation coefficients and their associated p-values.

This procedure can be used with the Spearman correlation coefficient for sample sizes greater than or equal to 10.

8.6   Testing Hypotheses about the Difference of Two Means

When we test hypotheses about the difference of two means, we need to first check whether the two means come from independent samples or from a single sample. When the two means are calculated from the sample, they are dependent. The test procedures used are different depending on whether the means are independent or dependent.

8.6.1   Difference of Two Independent Means

We begin with the consideration of independent means under various assumptions. The first test assumes that the variances are known, followed by the assumption that the variances are unknown but equal and then unknown and unequal. After these sections, we consider the difference of two dependent means.

Known Variances: The null hypothesis of interest for the difference of two independent means is

$$H_0: \mu_1 - \mu_2 = \Delta_0$$

where Δ_0 is the hypothesized difference of the two means. Usually Δ_0 is zero — that is, we are testing that the means have the same value. The alternative hypothesis could be either

$$H_a: \mu_1 - \mu_2 \neq \Delta_0$$

or that the difference is greater (less) than Δ_0. Regardless of the alternative hypothesis, when the variances are known, the test statistic is

$$z = \frac{(\bar{x}_1 - \bar{x}_2) - \Delta_0}{\sqrt{\sigma_1^2/n_1 + \sigma_2^2/n_2}}.$$

The rejection region for the two-sided alternative includes values of z less than or equal to $z_{\alpha/2}$ or greater than or equal to $z_{1-\alpha/2}$. The rejection region for the greater than alternative includes values of z greater than or equal to $z_{1-\alpha}$ and the rejection region for the less than alternative includes values of z less than or equal to z_α.

Example 8.8

We return to the Ramipril example from Chapter 7 and test the hypothesis that μ_1, the mean decrease in diastolic blood pressure associated with the 1.25 mg dose, is the same as μ_2, the mean decrease for the 5 mg dose. In practice, we should not initially focus on only two of the three doses; all three doses should be considered together at the start of the analysis. However, at this stage, we do not know how to analyze three means at one time — the topic of the next chapter. Therefore, we are temporarily ignoring the existence of the third dose (2.5 mg) of Ramipril that was used in the actual experiment.

As we expect that the higher dose of medication will have the greater effect, the null and alternative hypotheses are

$$H_0: \mu_1 - \mu_2 = 0 \quad \text{and} \quad H_a: \mu_1 - \mu_2 < 0.$$

We perform the test at the 0.05 significance level; thus, if the test statistic is less than -1.645 ($= z_{0.05}$), we shall reject the null hypothesis in favor of the alternative hypothesis. The sample mean decreases, $\bar{x}_1$ and $\bar{x}_2$, are 10.6 and 14.9 mm Hg, respectively, and both sample means are based on 53 observations. Both σ_1 and σ_2 are assumed to be 9 mmHg. Therefore, the value of z, the test statistic, is

$$z = \frac{(10.6 - 14.9) - 0}{\sqrt{81/53 + 81/53}} = -2.46.$$

Since the test statistic is less than -1.645, we reject the null hypothesis in favor of the alternative hypothesis. There appears to be a difference in the effects of the two doses of Ramipril with the higher dose being associated with the greater mean decrease in diastolic blood pressure at the 0.05 significance level.

Unknown but Equal Population Variances: The null and alternative hypotheses are the same as in the preceding section. However, the test statistic for the difference of two independent means, when the variances are unknown, changes to

$$t = \frac{(\bar{x}_1 - \bar{x}_2) - \Delta_0}{s_p \sqrt{1/n_1 + 1/n_2}}.$$

For a two-sided alternative hypothesis, the rejection region includes values of t less than or equal to $t_{n-2, \alpha/2}$ or greater than or equal to $t_{n-2, 1-\alpha/2}$, where n is the sum of n_1 and n_2.

Example 8.9

Let us test, at the 0.05 significance level, the hypothesis that there is no difference in the population mean proportions of total calories coming from fat for fifth- and sixth-grade boys and seventh- and eighth-grade boys. The alternative hypothesis is that there is a difference — that is, that Δ_0 is not zero. The rejection region includes values of t less than or equal to -2.04 ($= t_{31, 0.025}$) or greater than or equal to 2.04.

From Chapter 7, we know that $\bar{x}_1$, the sample mean proportion for the 14 fifth- and sixth-grade boys, is 0.329, and the corresponding value, $\bar{x}_2$, for the 19 seventh- and eighth-grade boys is 0.353. The value of s_p is 0.094. Therefore, the test statistic's value is

$$t = \frac{(0.329 - 0.353) - 0}{0.094\sqrt{1/14 + 1/19}} = -0.727.$$

Since t is not in the rejection region, we fail to reject the null hypothesis. There does not appear to be a difference in the proportion of calories coming from fat at the 0.01 significance level.

The computer can be used to perform this test (see **Program Note 8.4** on the website).

Unknown and Unequal Population Variances: The test statistic for testing the null hypothesis of a specified difference in the population means — that is,

$$H_0: \mu_1 - \mu_2 = \Delta_0$$

assuming that the population variances are unequal, is given by

$$t' = \frac{(\bar{x}_1 - \bar{x}_2) - \Delta_0}{\sqrt{s_1^2/n_1 + s_2^2/n_2}}.$$

The statistic t' approximately follows the t distribution with degrees of freedom, df, given by

$$df = \frac{\left(s_1^2/n_1 + s_2^2/n_2\right)^2}{\left[\left(s_1^2/n_1\right)^2/(n_1-1) + \left(s_2^2/n_2\right)^2/(n_2-1)\right]}.$$

For a two-sided alternative, if t' is less than or equal to $t_{df,\alpha/2}$ or greater than or equal to $t_{df,1-\alpha/2}$, we reject the null hypothesis in favor of the alternative hypothesis. If the alternative hypothesis is

$$H_a: \mu_1 - \mu_2 < \Delta_0,$$

the rejection region consists of values for t' of less than or equal to $t_{df,\alpha}$. If the alternative hypothesis is

$$H_a: \mu_1 - \mu_2 > \Delta_0,$$

the rejection region consists of values for t' of greater than or equal to $t_{df,1-\alpha}$.

Example 8.10

In Chapter 7, we examined the mean ages of the AML and ALL patients. Suppose we will consider that no difference in the population mean ages exists if the mean age of AML patients minus the mean age of ALL patients is less than or equal to 5 years. Thus, the null and alternative hypotheses are

$$H_0: \mu_1 - \mu_2 = 5 \text{ and } H_a: \mu_1 - \mu_2 > 5.$$

We shall perform this test at the 0.01 significance level, which means that we shall reject the null hypothesis in favor of the alternative hypothesis if t' is greater than or equal to 2.446 ($= t_{33,0.99}$). The degrees of freedom of 33 for the t value is obtained by

$$df = \frac{\left(16.51^2/51 + 17.85^2/20\right)^2}{\left(\dfrac{\left(16.51^2/51\right)^2}{51-1} + \dfrac{\left(17.85^2/20\right)^2}{20-1}\right)} = 32.501.$$

Using the values for the sample means, standard deviations, and sample sizes from Chapter 7, we calculate t' to be

$$t' = \frac{(49.86 - 36.65) - 5}{\sqrt{16.51^2/51 + 17.85^2/20}} = 1.781.$$

Since t' is less than 2.446, we fail to reject the null hypothesis. There is not sufficient evidence to conclude that the difference in ages is greater than 5 years. Usually one would test the hypothesis of no difference instead of a difference of 5 years. However, by testing the difference of 5 years, we were able to demonstrate the calculations for a nonzero Δ_0. The computer can be used to perform this test (see **Program Note 8.5** on the website).

As we emphasized in Chapter 7, we seldom know much about the magnitude of the two variances. Therefore, in those situations in which we know little about the variances and have no reason to believe that they are equal, we recommend that the unequal variances assumption should be used.

8.6.2   Difference of Two Dependent Means

The test to be used in this section is the paired t test, one of the more well-known and widely used tests in statistics. The null hypothesis to be tested is that the mean difference of the paired observations has a specified value — that is,

$$H_0 : \mu_d = \mu_{d0}$$

where μ_{d0} is usually zero. The test statistic is

$$t_d = \frac{\bar{x}_d - \mu_{d0}}{s_d \sqrt{n}}.$$

The rejection region for a two-sided alternative hypothesis includes values of t_d less than or equal to $t_{df,\alpha/2}$ or greater than or equal to $t_{df,1-\alpha/2}$. The rejection region for the alternative of less than includes values of t_d that are less than or equal to $t_{df,\alpha}$, and the rejection region for the alternative of greater than includes values of t_d that are greater than or equal to $t_{df,1-\alpha}$.

Example 8.11

We use this method to examine the effect of the 1.25 mg level of Ramipril. We shall analyze the first six weeks of observation of the subjects — four weeks of run-in followed by two weeks of treatment. The null hypothesis is that the mean difference in diastolic blood pressure between the value at the end of the run-in period and the value at the end of the first treatment period is zero. The alternative hypothesis of interest is that there is an effect — that is, that the mean difference is greater than zero. In symbols, the hypotheses are

$$H_0 : \mu_d = 0 \text{ and } H_a : \mu_d > 0.$$

We perform the test at the 0.10 significance level. Thus, the rejection region includes values of t_d that are greater than or equal to 1.298 ($= t_{52,0.90}$), using 52 degrees of freedom because there were 53 pairs of observations that are being analyzed.

From Chapter 7, we find that the sample mean difference in diastolic blood pressure after the two weeks of treatment was 10.6 mmHg for the 53 subjects. The sample standard deviation of the differences was 8.5 mmHg. Based on these data, we can calculate the value of t_d, and it is

$$t_d = \frac{(10.6 - 0)}{8.5\sqrt{53}} = 9.08.$$

Since t_d is greater than 1.298, we reject the null hypothesis in favor of the alternative hypothesis. It appears that there is a difference between the value of diastolic blood pressure at the end of the run-in period and the treatment period with the blood pressure at the end of the treatment period being significantly less than that at the end of the run-in period. Note that we only said that there was a difference, but we did not attribute the difference to the medication.

As we have discussed in Chapters 6 and 7, drawing any conclusion from this type of study design is very difficult. There are two concerns — the presence of extraneous factors and reversion to the mean — associated with this design. Without some control group, it is difficult to attribute any effects that are observed in the study group to the intervention because of the possibility of extraneous factors. In a tightly controlled experiment, the researcher may be able to remove all extraneous factors, but it is difficult. The presence of a control group is also useful in providing an estimate of the reversion-to-the-mean effect if such an effect exists. Thus, we are suggesting that the paired t test should be used with great caution — that is, in only those situations for which we believe that there are no extraneous factors and no reversion-to-the-mean effect. In other cases, we would randomly assign study subjects either to the control group or to the intervention group and compare the differences of the pre- and post-measures for both groups.

If we are comfortable with the use of the paired t test, it can easily be performed by the computer (see **Program Note 8.6** on the website).

8.7   Testing Hypotheses about the Difference of Two Proportions

As in the comparison of two means, when we test hypotheses about the difference of two proportions, we need to first check whether the two proportions come from independent samples or from a single sample. When two proportions come from a single sample or paired observations, they are dependent. The test procedures used are different depending on whether the proportions are independent or dependent.

8.7.1   Difference of Two Independent Proportions

As in Chapter 7, we are considering the case of two independent proportions. The null hypothesis is

$$H_0: \pi_1 - \pi_2 = \Delta_0$$

where Δ_0 usually is taken to be zero. The test statistic for this hypothesis, assuming that the sample sizes are large enough for the use of the normal approximation to the binomial to be appropriate, is

$$z_{\pi d} = \frac{(p_1 - p_2) - \Delta_0}{\sqrt{p_1(1 - p_1)/n_1 + p_2(1 - p_2)/n_2}}.$$

The rejection region for a two-sided alternative includes values of $z_{\pi d}$ that are less than or equal to $z_{\alpha/2}$ or greater than or equal to $z_{1-\alpha/2}$. If the alternative is less than, the rejection region consists of values of $z_{\pi d}$ that are less than or equal to z_α; if the alternative is greater than, the rejection region consists of values of $z_{\pi d}$ that are greater than or equal to $z_{1-\alpha}$.

Example 8.12

We test the hypothesis, at the 0.01 significance level, that there is no difference in the proportions of milk that contain 80 to 120 percent of the amount of vitamin D stated on the label between the eastern and southwestern milk producers. The alternative hypothesis is that there is a difference. From Chapter 7, we find the values of p_1 and p_2 are 0.286 and 0.420, respectively. Thus, the test statistic is

$$z_{\pi d} = \frac{(0.286 - 0.420) - 0}{\sqrt{0.286(1 - 0.286)/42 + 0.420(1 - 0.420)/50}} = -1.358.$$

Since $z_{\pi d}$ is not in the rejection region, we fail to reject the null hypothesis. The computer can be used to perform this test (see **Program Note 8.7** on the website).

8.7.2   Difference of Two Dependent Proportions

In Chapter 7 we discussed the confidence interval for the difference between two dependent proportions, π_d. Recall that the proportions of a particular attribute at two time points for the same individuals are not independent and the sample data for these situations is arranged as follows:

Attribute at Time		
1	**2**	**Number of Subjects**
Present	Present	a
Present	Absent	b
Absent	Present	c
Absent	Absent	d
Total		n

The estimated proportion of subjects with the attribute at time 1 is $p_1 = (a + b)/n$, and the estimated proportion with the attribute at time 2 is $p_2 = (a + c)/n$. The difference between the two estimated proportions is

$$p_d = p_1 - p_2 = (a + b)/n - (a + c)/n = (b - c)/n.$$

Here we want to test the difference of two dependent proportions. The null and alternative hypotheses in this situation are

$$H_o: \pi_d = \Delta_0 \text{ versus } H_a: \pi_d \neq \Delta_0.$$

The test statistic, assuming the sample size is large enough for the normal approximation to the binomial to be appropriate, is

$$z_{\pi d} = \frac{p_d - \Delta_0 - 1/n}{\text{Estimated } SE(p_d)}$$

where $1/n$ is the continuity correction suggested by Edwards (1948) and Δ_0 is zero in most situations. Expression of the estimated standard error for the difference of two dependent proportions was given in Chapter 7, and we repeat it here:

Then the test statistic becomes

$$z_{\pi d} = \frac{(b-c)/n - 1/n}{(1/n)\sqrt{(b+c)-(b-c)^2/n}} = \frac{b-c-1}{\sqrt{(b+c)-(b-c)^2/n}}.$$

This test is valid when the average of the discordant cell frequencies ($[b + c]/2$) is 5 or more. When it is less than 5, a binomial test is recommended instead of the z-test. The binomial test can be done by restricting our attention to the $(b + c)$ pairs. Under the null hypothesis the difference of the proportions conveniently follows a binomial distribution with $\pi = 0.5$ and sample size of $(b + c)$.

Example 8.13

We use the same data from the biostatistics and epidemiology test in Example 7.9 and the data were tabulated in a 2 by 2 table.

	Epidemiology		
Biostatistics	**Failed**	**Passed**	**Total**
Failed	**9** (a)	**9** (b)	**18**
Passed	**1** (c)	**81** (d)	**82**
Total	**10**	**90**	**100** (n)

Let us test the null hypothesis that there is no difference between the failure rates in biostatistics (18 percent) and epidemiology (10 percent) against the two-sided alternative hypothesis that they are different. We use a significance level of 0.05. The test statistic, $z_{\pi d}$, is

$$z_{\pi d} = \frac{9 - 1 - 1}{\sqrt{(9+1)-(9-1)^2/100}} = 2.289.$$

Since 2.289 is larger than 1.96, we reject the null hypothesis, suggesting that the two failure rates are significantly different. This result is consistent with the conclusion based on a 95 percent confidence interval shown in Example 7.9. The p-value of this test is 0.022. If we conducted this test at the 0.01 significance level, we could not reject the null hypothesis.

8.8 Tests of Hypotheses and Sample Size

We considered the sample size issue in the context of confidence interval in Chapter 7. We now consider the sample size in the context of hypothesis testing. As seen in testing hypotheses, the decision rule is based on the probabilities of Type I and Type II errors and increasing the sample size is one way of decreasing the error probabilities. Thus, specification of Type I and Type II errors leads to the determination of required sample size. We consider the sample size issue for three situations below.

Testing a Single Mean: The z-values specifying α for a μ_0 (null hypothesis and β for a μ_1 (alternative hypothesis) leads to the determination of n. The z value specifying the upper α percentage point of the normal distribution is

$$z_\alpha = \frac{\bar{x} - \mu_0}{\sigma/\sqrt{n}}$$

and the z value specifying the lower β percentage point is

$$z_\beta = \frac{\bar{x} - \mu_1}{\sigma/\sqrt{n}}.$$

Solving these two equations for n, eliminating $\bar{x}$, we get

$$n = \left[\frac{(z_\alpha - z_\beta)\sigma}{\mu_1 - \mu_0} \right]^2.$$

Note that the above formula includes the population standard deviation σ. Thus, in addition to specification of two error levels, we must have some idea of the underlying variability of a variable under consideration. The formula suggests that a larger sample size is required when the alternative hypothesis (μ_1) is closer to the null hypothesis (μ_0). While z_α is different depending on whether the test is specified as one-sided or two-sided, z_β always refers to one side of the normal curve. Also note that z_β carries the opposite sign of z_α.

Example 8.14

Let us consider an example. A researcher wants to determine a required sample size for a study to test whether male patients who do not exercise have elevated serum uric acid vales. The serum uric acid levels of males are known to be distributed normally with mean = 5.4 mg/100 mL and standard deviation of 1. The investigator wants to perform a one-sided test at the 5 percent significance level. The null hypothesis is that the population mean is 5.4. This indicates $\mu_0 = 5.4$ and $z_\alpha = 1.645$. He further specifies that if the true difference is as much as 0.4 mg/100 mL, he wishes to risk 10 percent chance of failing to reject the null hypothesis. This indicates that $\mu_1 = 5.8$ and $z_\beta = -0.28$. Then the required sample is

$$n = \left[\frac{(1.645 + 1.282)(1)}{5.8 - 5.4} \right]^2 = 53.5 \approx 54.$$

Testing a Single Proportion: As in the case of proportion, the specification of α for a π_0 and β for a π_1 would lead to the following two equations, ignoring the continuity correction:

$$z_\alpha = \frac{p - \pi_0}{\sqrt{\pi_0(1-\pi_0)/n}} \quad \text{and} \quad z_\beta = \frac{p - \pi_1}{\sqrt{\pi_1(1-\pi_1)/n}}.$$

Solving for n, eliminating p, we get

$$n = \left[\frac{z_\alpha \sqrt{\pi_0(1-\pi_0)} - z_\beta \sqrt{\pi_1(1-\pi_1)}}{\pi_1 - \pi_0}\right]^2.$$

Example 8.15

Consider the planning of a survey to find out how smoking behavior changed while students were in college. A comprehensive survey four years ago found that 30 percent of freshmen smoked. The investigator wants to know how many seniors to be sampled this year. He wants to perform a two-tailed test at the 0.05 level. This suggests that $z_\alpha = 1.96$. The null hypothesis is $\pi_0 = 0.3$. He also states that if the proportion is changed as much as 5 percentage points, then he wishes to risk 10 percent chance of failing to reject the null hypothesis. This indicates $\pi_1 = 0.35$ and $z_\beta = -1.28$ (as mentioned earlier, one-tailed z value is used for z_β). Then the required sample size is

$$n = \left[\frac{1.96\sqrt{(.3)(.7)} - (-1.282)\sqrt{(.35)(.65)}}{.35 - .3}\right]^2 = 910.48 \approx 911.$$

Testing the Difference of Two Means: The required sample size for testing the difference of two independent means can also be determined in a similar manner. We assume equal variance in two groups ($\sigma_1^2 = \sigma_2^2$) and an equal division of the sample size between the two groups ($n_1 = n_2$). Specifying α error for the null hypothesis ($\mu_1 - \mu_2 = \Delta_0$), we have

$$z_\alpha = \frac{(\bar{x}_1 - \bar{x}_2) - \Delta_0}{\sqrt{\sigma_1^2/n_1 + \sigma_2^2/n_2}} = \frac{(\bar{x}_1 - \bar{x}_2) - \Delta_0}{\sqrt{2\sigma_1^2/n_1}}.$$

Specifying β error for the alternative hypothesis ($\mu_1 - \mu_2 = \Delta_1$), we have

$$z_\beta = \frac{(\bar{x}_1 - \bar{x}_2) - \Delta_1}{\sqrt{2\sigma_1^2/n_1}}.$$

By solving these two equations for n_1, eliminating $(\bar{x}_1 - \bar{x}_2)$, we get

$$n_1 = 2\left[\frac{(z_\alpha - z_\beta)\sigma_1}{\Delta_1 - \Delta_0}\right]^2.$$

Since Δ_0 is zero in most application, the denominator is usually Δ_1, the value specified in the alternative hypothesis. Note that n_1 is the sample size in each group and the total sample size for the study is $2n_1$.

Example 8.16

Let us consider a case of designing a clinical nutritional study of special diet regimen to lower blood pressure among hypertensive adult males (diastolic blood pressure over 90 mmHg). The investigator expects to demonstrate that the new diet would reduce diastolic blood pressure by 4 mmHg in three months. He is willing to risk a

Type I error of 5 percent and a Type II error of 10 percent for a one-sided test. NHANES III data show that the mean and standard deviation of diastolic blood pressure among hypertensive males are 95.4 mmHg and 5.6 mmHg. The required sample size in each group can be calculated by

$$n_1 = 2\left(\frac{5.6(1.645+1.282)}{4}\right)^2 = 33.5 \approx 34.$$

The proposed study would need a total of 68 subjects, allocated randomly and equally between the treatment and the control group.

As discussed in Chapter 7, the determination of sample size for a study is not as simple as the preceding example may suggest. In practice, we seldom know the population standard deviation, and we need to obtain an estimate of its value from the literature or from a pilot study. Setting the error levels low may lead to a very large sample size that can not possibly be carried out. The balancing of the error levels against availability of resources may require an iterative process until a satisfactory solution is found.

8.9   Statistical and Practical Significance

We must not confuse statistical significance with practical significance. For example, in the diet study discussed earlier, if we had a large enough sample, an observed value for p of 0.51 could be significantly different from the null hypothesis value of 0.50. However, this finding would be of little practical use. For a result to be important, it should be both statistically and practically significant. The test determines statistical significance, but the investigator must determine whether or not the observed difference is large enough to be practically significant.

When reporting the results of a study, many researchers have simply indicated whether or not the result was statistically significant and/or given only the p-value associated with the test statistic. This is useful information, but it is more informative to include the confidence interval for the parameter as well.

In conclusion, while hypothesis tests are useful to check whether observed results are attributable to chance, estimates of effects should also be considered before conclusions are drawn.

Conclusion

In this chapter we have introduced hypothesis testing and the associated terminology. A key point is that the calculation of the probabilities of errors should be conducted before the study is performed. By doing this, we can determine whether or not the study, as designed, can deliver answers to the question of interest with reasonable error levels. We also added to the material on confidence intervals that was presented in Chapter 7, demonstrating the equivalence of the confidence intervals to the test of hypothesis. We showed how to test hypotheses about the more common parameters used with normally distributed data and how to calculate power for a test of hypothesis about the mean when the population variance was known. In addition, we presented statistics to be used in the tests of hypotheses about the difference of two means and two proportions. This latter material prepares us for the analysis of variance, in which we extend the test of hypothesis to comparing two or more means. Finally, we pointed out that statistical significance must not be confused with practical significance.

EXERCISES

8.1   In the diet study with a sample size of 20 pairs, suppose that we used a rejection region of 0 to 4 and 16 to 20. The null and alternative hypotheses are the same as in the chapter, and we are still interested in the specific alternative that π is 0.80. What are the values of α and β based on this decision rule? What is the power of the test for this specific alternative? We again observed 13 pairs favoring diet 1. What is the p-value for this result?

8.2   Suppose that the null and alternative hypotheses in the diet study were

$$H_0: \pi = 0.50 \text{ versus } H_a: \pi > 0.50.$$

Conduct the test at the 0.05 significance level. What is the decision rule that you will use? What are the probabilities of Type I and Type II errors for a sample size of 20 pairs and the specific alternative that π is 0.80?

8.3   What specific alternative value for π do you think indicates an important difference in the diet study? Provide an example of another study for which the binomial distribution could be used. What value would you use for the specific alternative for π in your study? What is the rationale for your choice for π in this new study?

8.4   Complete Table 8.2 by providing the values of power for π ranging from 0.05 to 0.50 in increments of 0.05. Graph the values of the power function versus the values of π for π ranging from 0.05 to 0.95. This graph is the power curve of the binomial test using the critical region of 0 to 2 and 10 to 12. What is the value of power when π is 0.50? Is there a specific name for this value? Describe the shape of the power curve. Discuss why the power curve, when the null hypothesis is π is equal to 0.50, must have this shape.

8.5   Frickhofen et al. (1991) performed a study on the effect of using cyclosporine in addition to antilymphocyte globulin and methylprednisolone in the treatment of aplastic anemia patients. There was a sample of 43 patients that received the cyclosporine in addition to the other treatment. Assume that the use of antilymphocyte globulin and methylprednisolone without cyclosporine results in complete or partial remission in 40 percent of aplastic anemia patients at the end of three months of treatment. We wish to determine if the use of cyclosporine can increase significantly the percentage of patients with complete or partial remission. What are the appropriate null and alternative hypotheses? Assume that the test is to be performed at the 0.01 significance level. What is the decision rule to be used? What is the probability of a Type II error based on the sample size of 43 and your decision rule? Twenty-eight patients achieved complete or partial remission at the end of three months. Is this a statistically significant result at the 0.01 level? What is the p-value of the test?

8.6   In a recent study, Hall (1989) examined the pulmonary functioning of 135 male Caucasian asbestos product workers. An earlier study had suggested that the development of clinical manifestations of the exposure to asbestos required a minimum of 20 years. Therefore, Hall partitioned his data set into two groups, one with less than 20 years of exposure to asbestos and the other with 20 or more years of exposure. Two of the variables used to examine pulmonary function are the forced vital capacity (FVC) measured in liters and the percent of the predicted FVC value where the prediction is based on age, height, sex, and race. Age is a particularly important variable to consider, since there is a strong positive correlation between FVC and age. The sample means and standard deviations of FVC and percent of the predicted FVC for each of the two groups are as follows:

| | Length of Exposure | | | |
| | <20 Years ($n = 66$) | | 20 Years ($n = 69$) | |
Variable	Mean	S.D.	Mean	S.D.
FVC (L)	5.19	0.78	4.27	0.63
% Predicted. FVC	104	9.7	45	12.8

Choose the more appropriate of these two variables to use in a test of whether or not there is a difference in the means of the two exposure groups. Perform the test at the 0.05 significance level. Explain your choice for which variable to use and also your choice of a one- or two-sided alternative hypothesis. What assumption did you make about the population variances? Does this study support the idea that there is a difference between those with less than 20 years of exposure and those with 20 or more years of exposure? What is the p-value of the test? What, if any, other variable should be taken into account in the analysis?

8.7 Kirklin et al. (1981) performed a study of infants less than 3 months old who underwent open heart surgery. There were 175 infants in their study based on data from 1967 to 1980. It was suggested that the survival probabilities improved over time. To examine this, the data were broken into two time periods. Test the hypothesis that there is a difference in the survival probabilities over these two time periods versus the alternative hypothesis of no difference over time at the 0.01 significance level. Use the following hypothetical data, based on data presented in the study.

Date	Probability of Survival	Sample Size
Jan. 1967 to Dec. 1973	0.46	66
Jan. 1974 to July 1980	0.64	109

Provide possible reasons why there might be a difference in the survival probabilities over time.

8.8 Data from the National Institute of Occupational Safety and Health for the 1980–88 period were used to obtain estimates of the annual workplace fatality rates by state (PCHRG 1992). The average annual state rates over the nine-year period are given in Exercise 7.4. There is tremendous variability in the rates, ranging from a low of 1.9 to a high of 33.1 deaths per 100,000 workers. Provide some possible reasons for this variability. For the state of your residence, test the hypothesis of no difference in the crude workplace fatality rate and the national average of 7.2 per 100,000 workers. Exercise 7.7 gives the population total for your state. Perform this test against a two-sided alternative at the 0.05 significance level. What is the *p*-value of the test? Provide possible reasons why there is or is not evidence of a difference between your state and the national average.

8.9   In the study by Reisin et al. (1978) one of the goals was to observe the effect of weight loss without salt restriction on blood pressure. We shall focus on one of the intervention groups, the group that was on a weight reduction program and given no medication. The program consisted of a strict diet with caloric intake reduced to about 50 percent of the usual adult intake for a two-month period. Before examining the data for an effect on blood pressure, it is necessary to determine whether or not the diet worked. The summary weight data for the sample of 24 patients was a mean reduction of 8.8 kilograms, and the standard deviation of the weight changes was 4.3 kilograms. This is a paired-*t* test situation for this single group. However, there was also a control group that was not part of the weight reduction effort. During the period when the study group lost an average of 8.8 kilograms, the control group showed an average decrease of only 0.7 kilograms. The results from the control group increase our confidence in the use of the paired-*t* test here. Test the null hypothesis of no weight reduction versus the appropriate one-sided alternative hypothesis at the 0.01 significance level. Did the weight reduction program work?

8.10   There have been a number of drug recalls during 1993 because of the failure of the drugs to meet dissolution specifications, content uniformity specifications, or because of subpotency (PCHRG 1993). Three products from the Parke-Davis Division of the Warner-Lambert Company were recalled. One of the products, Tedral, did not meet either the dissolution or content uniformity specifications.

Suppose that the content uniformity specification as expressed in terms of the variance. For example, say that the variance of the amount of phenobarbital in tablets was supposed to be less than or equal to 0.015 grams^2. We selected a sample of 30 tablets and found the sample standard deviation of phenobarbital to be 0.14 grams. Test the appropriate hypothesis to determine, at the 0.10 level, whether there is compliance with the content uniformity specification for the amount of phenobarbital in the tablets.

8.11   In Chapter 7, using data from Table 7.6, we saw that there was a statistically significant (at the 0.01 level) difference in the mean ages of the AML and ALL patients. The difference in ages is important, particularly if the length of survival is strongly related to age. Calculate the sample Pearson correlation coefficient between age and length of survival based on all the patients in Table 7.6. Then test the null hypothesis, at the 0.05 level, that the population correlation coefficient is −0.30 versus the alternative hypothesis that the correlation is less (more negative) than −0.30. Here we are using −0.30 or more negative values to indicate a strong inverse correlation. Based on your analysis, is it necessary to control for the effect of age in the comparison of the length of survival of the AML and ALL patients?

8.12   In Exercise 7.10, we examined progress towards the Surgeon General's goal of reducing the proportion of 12- to 18-year-old adolescents who smoked to below 6 percent for a hypothetical community. We found that in 1990, of the 12- to

18-year-olds in the sample, 11 of 85 admitted that they smoked. Test the hypothesis that the hypothetical community has already attained the Surgeon General's goal at the 0.05 significance level. Should you use a one- or two-sided alternative hypothesis? Explain your reasoning.

8.13 Opponents of a national health system argue that it will lead to rationing of services, something that is viewed as being unacceptable to people in the United States. To determine how people in the United States really felt about rationing of services, the American Board of Family Practice had a survey conducted and some of the results are reported by Potter and Porter (1989). One question asked whether or not people would approve of rationing medical attention in the case of a terminal illness. Suppose that we have decided that there is substantial support for rationing if the proportion of the population who would approve of rationing in this case is 40 percent. In the sample of 1007 Americans, 34 percent supported rationing in the case of terminal illness. Test the hypothesis that the population proportion equals 40 percent versus the alternative hypothesis that it is less than 40 percent. Use the 0.01 significance level. It is interesting to note that 43 percent of the physicians surveyed supported rationing in this situation.

8.14 Anderson et al. (1990) performed a study on the effects of oat bran on serum cholesterol for males with high or borderline high values of serum cholesterol. High values of serum cholesterol are greater than or equal to 240 mg/dL (6.20 mmol/L). We wish to use the data from the study to determine whether or not there is a linear relation between body mass index and serum cholesterol. The body mass index is defined as weight (in kilograms) divided by the square of height (in meters). The data are

Body Mass Index	Serum Cholesterol	Body Mass Index	Serum Cholesterol
29.0	7.29	26.3	8.04
21.6	8.43	21.8	7.96
27.2	5.43	24.8	5.77
25.2	6.96	24.5	6.23
25.1	6.65	23.5	6.26
27.9	8.20	24.8	6.21
31.9	5.92	24.4	5.92

Test the hypothesis of no correlation between body mass index and serum cholesterol at the 0.10 level. Explain your choice of a one- or two-sided alternative hypothesis. What is the *p*-value of the test?

8.15 Exercise 7.6 shows 15 hypothetical serum cholesterol values. For these data, test the hypothesis that the population variance equals 100 $(mg/dL)^2$ versus the alternative hypothesis that the population variance is greater than 100 $(mg/dL)^2$. Perform the test at the 0.025 level. Discuss the results of this test in relation to the confidence interval obtained in Exercise 7.6. Recall that this test requires that the cholesterol values follow a normal distribution. Examine the assumption of normality of the cholesterol values.

8.16 For the same data from Exercise 7.6, test the hypothesis that the measuring process works — that is, test the hypothesis that the population mean of the values measured by this process equals 190 versus the alternative hypothesis that the population mean is not equal to 190 mg/dL. Perform the test at the 0.02 significance level.

REFERENCES

Anderson, J. W., D. B. Spencer, C. C. Hamilton, et al. "Oat-Bran Cereal Lowers Serum Total and LDL Cholesterol in Hypercholesterolemic Men." *American Journal of Clinical Nutrition* 52:495–499, 1990.

Edwards, A. L. Note on the "Correction for Continuity" in Testing the Significance of the Difference between Correlated Proportions. *Psychometrika,* 13:185–187, 1948.

Frickhofen, N., J. P. Kaltwasser, H. Schrezenmeier, et al. "Treatment of Aplastic Anemia with Antilymphocyte Globulin and Methylprednisolone with or without Cyclosporine." *The New England Journal of Medicine* 324:1297–1304, 1991.

Hall, S. K. "Pulmonary Health Risk." *Journal of Environmental Health* 52:165–167, 1989.

Kirklin, J. K., E. H. Blackstone, J. W. Kirklin, et al. "Intracardiac Surgery under 3 Months of Age: Incremental Risk Factors for Hospitality Mortality." *The American Journal of Cardiology* 48:500–506, 1981.

Lee, E. S., and R. N. Forthofer. *Analyzing Complex Survey Data*, Second Edition. Thousand Oaks, CA: Sage Publications, 2006. Chapter 6, data processed from National Center for Health Statistics. The Third National Health and Nutrition Examination Survey, Adult sample of Phase I, 1999–2000.

National Center for Health Statistics. *Health, United States, 1990.* Hyattsville, MD: Public Health Service. DHHS Pub. No. 91-1232, 1991.

National Center for Health Statistics. *Health, United States, 1991 and Prevention Profile.* Hyattsville, MD: Public Health Service. DHHS Pub. No. 92-1232, 1992.

National Center for Health Statistics. *Health, United States, 2004 with Chartbook on Trends in the Health of Americans.* Hyattsville, MD: DHHS Pub. No. 2004-1232, 2004, Table 35.

Potter, C., and J. Porter. "American Perceptions of the British National Health Service: Five Myths." *Journal of Health Politics, Policy and Law* 14:341–365, 1989.

Public Citizen Health Research Group (PCHRG). "Work-Related Injuries Reached Record Level Last Year." *Public Citizen Health Research Group Health Letter* 8(12):1–3, 9, 1992.

Public Citizen Health Research Group (PCHRG). "Drug Recalls March 9–June 7, 1993." *Public Citizen Health Research Group Health Letter* 9(7):9–10, 1993.

Reisin, E., R. Abel, M. Modan, et al. "Effect of Weight Loss without Salt Restriction on the Reduction of Blood Pressure in Overweight Hypertensive Patients." *The New England Journal of Medicine* 298:1–6, 1978.

Nonparametric Tests 9

Chapter Outline

In this chapter we present several statistics for testing whether or not probability distributions have the same medians. The use of these statistics does not require that the sample data follow any particular probability distribution, and, thus, there are no distributional parameters to be estimated. Because of these features, these tests are called distribution-free or nonparametric tests. We still assume that the data come from continuous distributions. We begin with justification of using distribution-free methods.

9.1 Why Nonparametric Tests?

The methods studied in the previous chapter were mostly concerned with data from a normal distribution. In many situations the data may consist of a number of ordered categories such as a subjective rating of the amount of pain relief (none, a little, a lot, total) a patient perceives after receiving a treatment. In other cases the data may simply be the presence or absence of a condition. In such cases the investigator may be unwilling to use a numerical scale but still wants to test a hypothesis related to the effect of a treatment or to the effects of two different treatments. The sign test discussed in this chapter can be used for situations with two outcomes. Other methods in this chapter are used with ordered data or with numerical data that do not follow the normal distribution.

The methods for testing the mean and proportion in the previous chapter are based on normality assumptions. If there is obvious nonnormality in the data, distribution-free methods can be used. In some cases we may suspect that the data do not follow the normal distribution, but we cannot determine the lack of normality for sure because the sample size is too small. Distribution-free methods can be used then and are also often used for small samples when the central limit theorem may not apply.

9.2 The Sign Test

The sign test is one of the oldest tests used in statistics. For example, in 1710, John Arbuthnot, a British physician and collaborator of Jonathan Swift, performed what was in effect a sign test on the sex ratio of births over an 82-year period (Stigler 1986).

As we saw in the last chapter, the sign test can be used to compare different interventions for matched pairs. Individuals were assigned to a pair based on age, sex, weight, and exercise level, and then one member within the pair was randomly assigned to diet 1 and the other member assigned to diet 2. The sign test was then used to determine which of the two diets was more likely to be associated with the greater weight loss for each pair. Another way of stating this null hypothesis is that each difference of weight losses has a median of zero. The sign test can also be used with a single population — for example, in the comparison of multiple measurements made on the same individual or one set of measurements compared with some hypothesized value, as shown in the next examples.

As we saw in the last chapter, the p-value of the sign test can be exactly determined from the binomial distribution, or we can approximate the p-value by using the normal approximation to the binomial when the sample size is large.

Example 9.1

One problem often encountered in research designs involving pre- and posttest measurements is the reversion or regression toward the mean effect (Samuels 1991). Briefly, persons scoring high on one test tend not to score as high on a subsequent test and low scorers on the first test tend to score higher on the next test — that is, the test scores tend to revert toward the mean score. Reversion toward the mean is important because of its possible effect on test results (Davis 1976; Nesselroade, Stigler, and Baltes 1980).

We consider the caloric intake for 33 boys selected from a larger study (McPherson et al. 1990). Table 9.1 shows the caloric intake for the boys for the first two of three randomly selected days during a two-week period. The more extreme — the seven highest and seven lowest — day 1 values are marked. We can examine whether or not there is reversion toward the mean. Based on the descriptive statistics shown in Table 9.1, it appears that there could be a reversion toward the mean effect

Table 9.1 Two days of caloric intake for 33 boys enrolled in two middle schools outside of Houston.[a]

ID	Day 1	Day 2	ID	Day 1	Day 2	ID	Day 1	Day 2
10	1,823	1,623	39	2,330	2,339	118	1,781[L]	1,844
11	2,007	1,748	40	2,436	2,189	120	2,748	2,104
13	1,053[L]	2,484	41	3,076[H]	2,431	127	2,348	2,122
14	4,322[H]	2,926	44	1,843	2,907	130	2,773[H]	3,236
16	1,753[L]	1,054	46	2,301	4,120	137	2,310	1,569
17	2,685	2,304	47	2,546	1,732	139	2,594	2,867
26	2,340	3,182	50	1,292[L]	810	141	1,898	1,236
27	3,532[H]	3,289	51	3,049[H]	2,573	145	2,400	2,554
30	2,842[H]	2,849	101	3,277[H]	2,185	148	2,011	1,566
32	2,074	3,312	105	2,039	1,905	149	1,645[L]	2,269
33	1,505[L]	1,925	107	2,000	1,797	150	1,723[L]	3,163

		Mean	
	Number	**Day 1**	**Day 2**
[L]Lowest values	7	1,536	1,936
[H]Highest values	7	3,267	2,784

[a]Selected from a larger study by McPherson et al. (1990)

here. The seven lowest values had a mean of 1,536 calories on day 1 compared with a mean of 1,936 calories on day 2 — an increase. The seven highest values had a mean of 3,267 calories on day 1 compared with a mean of 2,784 calories on day 2 — a decrease. However, we wish to go beyond a descriptive presentation of the sample in our consideration of the question. We wish to test a hypothesis about the population values.

If there is no reversion toward the mean effect here, of the boys with extreme day 1 values, the proportion of those whose day 2 values move in the direction of the mean should be equal to 0.50 (ignoring the possibility that the day 1 and day 2 values are the same). If there is reversion toward the mean, the proportion should be greater than 0.50. The null and alternative hypotheses are therefore

$$H_0: \pi = 0.50 \text{ versus } H_a: \pi > 0.50.$$

If there are few ties (a subject has same values for day 1 and day 2) in the data, convention is that these observation pairs are dropped from the data. For example, if one out of the 14 boys had the same day 1 and day 2 values, the sample size for the binomial would then be 13 instead of 14, reflecting the deletion of the tied pair. When there are many ties, indicating no difference in the day 1 and day 2 values, there is little reason to perform the test for the remaining untied pairs.

The population from which this sample is drawn consists of middle schools in a northern suburb of Houston. Although the population is limited, perhaps the results from this population can be generalized to boys in suburban middle schools throughout the United States, not just to those in one suburb of Houston. As was mentioned in Chapter 6, this generalization does not flow from statistical properties because we did not sample this larger population, but it is based on substantive considerations. If there are differences in dietary practices between the one Houston suburb and others, this generalization to the larger population is then questionable.

We conduct the test of hypothesis at the 0.05 level. The test statistic is the number of boys with an extreme day 1 value whose day 2 value moves toward the mean, which is found to be 10 from the data. The critical region for the test can be found from the binomial distribution. For larger sample sizes, the normal approximation to the binomial can be used. We could use Table B2 to find the probabilities for a binomial distribution with $n = 14$ and $\pi = 0.50$. We are interested only in the upper tail of the binomial distribution; therefore, we consider only values above the expected value of 7. Because we wish to perform the test at the 0.05 level, the rejection region consists of the values of 11 to 14. If 10 were included in the rejection region, the probability of Type I error would exceed the significance level of 0.05. Ten of the 14 boys with an extreme day 1 value had day 2 values that moved in the direction of the mean. Since 10 is not included in the rejection region, we fail to reject the null hypothesis in favor of the alternative at the 0.05 significance level. Although we failed to reject the null hypothesis, the p-value of this result is 0.0898.

What is the power of the test — that is, what is the probability of rejecting the null hypothesis when it should be rejected? As we saw in Chapter 8, to find a value for power, we must provide a specific alternative. Let us work with the alternative that π is 0.70. Then the power is easily found from Table B2 ($n = 14$, $\pi = 0.3$, for

$x = 3, 2, 1, 0$, which is equivalent to x = 11, 12, 13, 14 under $\pi = 0.7$). The power is 0.3552 (= 0.1943 + 0.1134 + 0.0407 + 0.0068), not a large value.

It is even easier to perform the sign test using a computer program (see **Program Note 9.1** on the website).

Example 9.2

Two major questions in interlaboratory testing programs are (1) whether or not the measuring instruments are properly calibrated and (2) whether or not the technicians are properly trained. The first question concerns the validity or bias issue, and the second question deals with the reliability or precision issue. In Chapter 2 we looked at an interlaboratory testing program of the CDC. They distributed a blood sample to over 100 randomly selected laboratories throughout the country and asked to measure the lead concentration. The test samples were created to contain the lead concentration of exactly 41 μg/dL (Hunter 1980). The average reported by all participating laboratories was 44 μg/dL with a large variability, ranging from 30 to 60. It appears that both the validity and reliability problems are present. The sign test can be used with the 100 measurements compared with the true value to examine the bias issue between laboratories. We consider the precision issue within one laboratory in the following example.

Suppose that the same CDC sample was mixed in other samples in 13 consecutive days and the following measurements are recorded. The laboratory director wants to examine the bias issue (calibration of instruments) quickly — that is, whether or not these measurements differ significantly from the true value of 41:

$$\underline{45} \quad \underline{43} \quad 40 \quad \underline{44} \quad \underline{49} \quad 36 \quad \underline{51} \quad \underline{46} \quad 35 \quad \underline{50} \quad 41 \quad 38 \quad \underline{47}.$$

If the measuring instrument is properly calibrated, one would expect half the measurements to be above the value of 41 and half to be below the value. If there is a bias problem, the proportion should be greater or less than 0.50. The null and alternative hypotheses are therefore

$$H_0: \pi = 0.50 \text{ versus } H_a: \pi \neq 0.50.$$

In this case, 48 values are above 41 (underscored values) and four are below (8 positives and 4 negatives). One value is exactly 41, and we cannot assign a sign. We drop this observation from the data following the usual practice and analyze twelve observations for which we can determine a positive or negative sign. Thus, the test statistic, the number of positive signs, has a value of 8. We conduct the sign test at the 0.50 level. The critical region for the test can be found from the binomial distribution. For larger sample sizes, the normal approximation to the binomial can be used. We use Table B2 to find the probabilities for a binomial distribution with $n = 12$ and $\pi = 0.50$. Since it is a two-tailed test, we need to look at both tails of the distribution. The p-value of this test is then twice of the probability that X is 8 or more — that is, 2(0.1209 + 0.0537 + 0.0161 + 0.0029 + 0.0002), which gives 0.39. As the p-value is greater than 0.05, we fail to reject the null hypothesis of no difference in favor of the alternative hypothesis. There is no evidence that the early measurements are significantly different from the true value of 41.

We must realize how much information the sign test discards — a value of 42 is treated exactly the same as a value of 50 or 51. If the data are normal, the *t* test makes better use of the data and gets more out of the data. There are also other nonparametric tests that use more of the information in the data and those will be discussed in the following sections.

The sign test is easy to perform as the test statistic is simply a count of the occurrences of some event — for example, a move toward the mean or a positive difference. The test can also be used with nonnumerical data — for example, in situations in which the outcome is that the subject does or does not feel better. The simplicity of the test is attractive, but with numeric data, in ignoring the magnitude of the values, the sign test does not use all the information in the data. The other tests in this chapter use more of the available information in the data.

9.3   The Wilcoxon Signed Rank Test

Another much more recently developed test that can be used to examine whether or not there is reversion toward the mean in the data in Example 9.1 is the Wilcoxon Signed Rank (WSR) test. An American statistician, Frank Wilcoxon, who worked in the chemical industry, developed this test in 1945. Unlike the sign test which can be used with nonnumeric data, the WSR test requires that the differences in the paired data come from a continuous distribution.

To apply the WSR test to examine whether or not there is reversion toward the mean, we prepare the data as follows: The data for the 14 boys with an extreme day 1 value are shown in Table 9.2. In this table, the differences between day 1 and day 2 values are shown as either a change in the direction of the mean (+) or away from the mean (−). If the day 1 and day 2 values for a boy are the same, then we cannot assign a sign, and

Table 9.2   Days 1 and 2 caloric intakes for the 14 boys with the more extreme caloric intakes on day 1.

ID	Day 1	Day 2	Change (+) Toward the Mean	Change (−) Away from the Mean	Rank +	Rank −
13	1,053	2,484	1,431		13	
14	4,322	2,926	1,396		12	
16	1,753	1,054		699		10
27	3,532	3,289	243		3	
30	2,842	2,849		7		1
33	1,505	1,925	420		4	
41	3,076	2,431	645		9	
50	1,292	810		482		7
51	3,049	2,573	476		6	
101	1,277	2,185	1,092		11	
118	1,781	1,844	63		2	
130	2,773	3,236		463		5
149	1,645	2,269	624		8	
150	1,723	3,163	1,440		14	
				Sum of Ranks	82	23

such a pair would be excluded from the analysis. The absolute differences are ranked from smallest to largest, and the ranks are summed separately for those changes in the direction of mean and for those changes away from the mean. We use R_{WSR} to represent the signed rank sum statistic for the positive differences — in this case, those changes toward the mean.

We now consider the logic behind the testing of R_{WSR}. When there are n observations or pairs of data, the sum of the ranks is the sum of the integers from 1 to n and that sum is $n(n + 1)/2$. The average rank for an observation is therefore $(n + 1)/2$.

The null hypothesis is that the differences have a median of zero and the alternative hypothesis that the median is not equal to zero for a two-sided test or greater (or smaller) than zero for a one-sided test. If the null hypothesis is true, the distribution of the differences will be symmetric, and there should be $n/2$ positive differences and $n/2$ negative differences. Therefore, if the null hypothesis is true, the sum of the ranks for positive (or negative) differences, R_{WSR}, should be $(n/2)$ times the average rank: $(n/2)(n + 1)/2 = n(n + 1)/4$.

The test statistic is the sum of the ranks of positive (or negative) differences, R_{WSR}. For a small sample, Table B9 ($n < 30$) provides boundaries for the critical region for the sum of the ranks of the positive (or negative) differences. To give an idea how these boundaries were determined, let us consider five pairs of observations. The boundaries result from the enumeration of possible outcomes as shown in Table 9.3.

Table 9.3 Positive ranks for a sample of size 5 for 0, 1, and 2 positive ranks.

Number of Positive Ranks	Possible Ranks	Sum of Positive Ranks	Sum of Negative Ranks
0		0	15
1	1	1	14
	2	2	13
	3	3	12
	4	4	11
	5	5	10
2	1, 2	3	12
	1, 3	4	11
	1, 4	5	10
	1, 5	6	9
	2, 3	5	10
	2, 4	6	9
	2, 5	7	8
	3, 4	7	8
	3, 5	8	7
	4, 5	9	6

In Table 9.3, there is no need to show the sum of ranks for 3, 4, and 5 positive ranks because their values are already shown under the sum of the negative rank column. For example, when there are 0 positive ranks, there are 5 negative ranks with a sum of 15. But the sum of 5 positive ranks must also be 15. When there is 1 positive rank, there are 4 negative ranks with the indicated sums. But these are also the sum for the possibilities with 4 positive ranks. The same reasoning applies for 2 and 3 positive ranks.

Based on Table 9.3, we can form Table 9.4, which shows all the possible values of the sum and their relative frequency of occurrence. Using Table 9.4, we see that the smallest rejection region for a two-sided test is 0 or 15, and this gives the probability of

Table 9.4 All possible sums and their relative frequency.

Sum	Frequency	Relative Frequency
0 or 15	1	0.031
1 or 14	1	0.031
2 or 13	1	0.031
3 or 12	2	0.063
4 or 11	2	0.063
5 or 10	3	0.094
6 or 9	3	0.094
7 or 8	3	0.094

a Type I error of 0.062. Thus, in Table B9, there is no rejection region shown for a sample size of 5 and a significance level of 0.05. If the test of interest were a one-sided test, then it would be possible to have a Type I error probability less than 0.05.

Example 9.3

Let us return to the data prepared for the 14 pairs in Table 9.2. We shall perform the test at the 0.05 significance level, the same level used in the sign test. Since this is a one-sided test, we read the boundary above $\alpha \leq 0.05$ under one-sided comparisons shown at the bottom of the table, which is equivalent to $\alpha \leq 0.10$ under two-sided comparisons. Using the row $n = 14$, the critical values are (25, 80). Since our test statistic is 82, greater than 80, we reject the null hypothesis of no regression toward the mean in favor of the alternative that there is regression toward the mean.

This result is inconsistent with the result of the sign test in Example 9.1 and reflects the greater power of the WSR test. This greater power is due to the use of more of the information in the data by the WSR test compared to the sign test. The WSR test incorporates the fact that the average rank for the four changes away from the mean is 5.75 (= [1 + 5 + 7 + 10]/4), less than the average rank of 7.50. This lower average rank of these four changes, along with the fact that there were only four changes away from the mean, caused the WSR test to be significant. The sign test used only the number of changes toward the mean, not the ranks of these changes, and was not significant. Although the sign test failed to reject the null hypothesis, its p-value of 0.0898 was not that different from 0.05.

In applying the WSR test, two types of ties can occur in the data. One type is that some observed values are the same as the hypothesized value or some paired observations are the same — that is, the differences are zero. If this type of tie occurs in an observational unit or pair, that unit or pair is deleted from the data set, and the sample size is reduced by one for every unit or pair deleted. Again, this procedure is appropriate when there are only a few ties in the data. If there are many ties of this type, there is little reason to perform the test.

The other type of tie occurs when two or more differences have exactly the same nonzero value. This has an impact on the ranking of the differences. In this case, convention is that the differences are assigned the same rank. For example, if two differences were tied as the smallest value, each would receive the rank of 1.5, the average of ranks 1 and 2. If three differences were tied as the smallest value, each would receive the rank of 2, the average of ranks 1, 2, and 3. If there are few ties in the differences, the rank sum can still be used as the test statistic; however, the results of the test are now approximate. If there are many ties, an adjustment for the ties must be made (Hollander and Wolfe 1973), or one of the methods in the next chapter should be used.

Example 9.4

Let us apply the WSR test to the data in Example 9.2. The 13 measurements, the deviations from the true value of 41, and ranks of absolute differences are as follows:

Measures:	45	43	40	44	49	36	51	46	35	50	41	38	47
Differences:	+4	+2	−1	+3	+8	−5	+10	+5	−6	+9	0	−3	+6
Ranks:	5	2	1	3.5	10	6.5	12	6.5	8.5	11	—	3.5	8.5

Note that the average ranking procedure is used for the same values of absolute differences and the rank is not assigned to tenth observation.

Again the investigator wishes to test whether the repeated measurements are significantly different from the value of 41 at the 0.05 significance level. We delete one observation that has no rank. The test statistic (the sum of ranks for positive differences) is 58.5. Table B9 provides boundaries of the critical region. For $n = 12$ and $\alpha \leq 0.05$ under the two-sided comparison the boundaries are (13, 65). Since the test statistic is less than 65, we fail to reject the null hypothesis in favor of the alternative hypothesis at the 0.05 significance level. This conclusion is consistent with the result of the sign test in Example 9.2.

For a large sample, the normal approximation is used. If there are at least 16 pairs of observations used in the calculations, R_{WSR} will approximately follow a normal distribution. As we just saw, the expected value of R_{WSR}, under the assumption that the null hypothesis is true, is $n(n + 1)/4$, and its variance can be shown to be $n(n + 1)(2n + 1)/24$. Therefore, the statistic

$$\frac{|R_{WSR} - [n(n+1)/4]| - 0.5}{\sqrt{n(n+1)(2n+1)/24}}$$

approximately follows the standard normal distribution. The two vertical lines in the numerator indicate the absolute value of the difference — that is, regardless of the sign of the difference, it is now a positive value. The 0.5 term is the continuity correction term, required because the signed rank sum statistic is not a continuous variable.

Let us calculate the normal approximation to the pairs in Example 9.3. The expected value of R_{WSR} is 52.5 (= [14][15]/4), and the standard error is 15.93 (= $\sqrt{(14)(15)(29)/24}$). Therefore, the statistic's value is

$$\frac{|82 - 52.5| - 0.5}{15.93} = 1.82.$$

What is the probability that Z is greater than 1.82? This probability is found from Table B4 to be 0.0344. This agrees very closely with the exact p-value of 0.0338. The exact p-value is based on 554 of the 16,384 possible signed rank sums having a value of 82 or greater, applying the same logic illustrated in Tables 2 and 3 to the case $n = 14$. Thus, even though n is less than 16, the normal approximation worked quite well in this case. The WSR test can be performed by the computer (see **Program Note 9.1** on the website).

The sign and Wilcoxon Signed Rank tests are both used most frequently in the comparison of paired data, although they can be used with a single population to test that the median has a specified value. In the use of these tests with pre- and postintervention measurement designs, care must be taken to ensure that there are no extraneous factors that could have an impact during the study. Otherwise, the possibility of the confounding of the extraneous factor with the intervention variable is raised. In addition, the research designer must consider whether or not reversion to the mean is a possibility. If extraneous factors or reversion to the mean cannot be ruled out, the research design should be augmented to include a control group to help account for the effect of these possibilities.

9.4   The Wilcoxon Rank Sum Test

Another test developed by Wilcoxon is the Wilcoxon Rank Sum (WRS) test. This test is used to determine whether or not the probability that a randomly selected observation from one population being greater than a randomly selected observation from another population is equal to 0.5. This test is sometimes referred to as the Mann-Whitney test after Mann and Whitney, who later independently developed a similar test procedure for unequal sample sizes. The WRS test also requires that the data come from independent continuous distributions.

This test is appropriate for the following data situation. A nutritionist wishes to compare the proportion of calories from fat for boys in grades 5 and 6 and grades 7 and 8 that are shown in Table 9.5. Preparation of data involves (1) the ranking of all the observed values in the two groups from smallest to largest and (2) summing the ranks separately in each group. The ranks of these values are also shown in the table. We have rounded the proportions to three decimal places, and as a result there is one tie in the data. The tied values were the 16th and 17th smallest observations and hence were assigned the rank of 16.5, the average of 16 and 17. We could have used the fourth decimal place to break the tie, but we chose not to because we wanted to demonstrate how to calculate the ranks when there was a tie. The test statistic, R_{WRS}, is the sum of the ranks for the smaller sample ($n_1 = 14$) — in this case, for the 14 fifth- and sixth-grade boys. The total sample size n is 33 in this example.

If there were no differences in the magnitudes of the proportion of calories from fat variables in the two groups, the rank sum for the smaller sample would be the product of n_1 and the average rank of the n observations in the two groups — that is, $n_1(n + 1)/2$. For this example, the expected value of R_{WRS} under the null hypothesis of no difference would be 238. If the calculated R_{WRS} in Table 9.5 deviate greatly from 238 suggest that the null hypothesis of no difference in magnitudes should be rejected in favor of the alternative hypothesis that one group has larger values than the other. This test can be

Table 9.5 Proportion of calories from fat for boys in grades 5–6 and 7–8.

Grades 5–6		Grades 7–8	
Proportion from Fat	**Rank**	**Proportion from Fat**	**Rank**
0.365	21	0.311	13
0.437	30	0.278	6
0.248	4	0.282	8
0.424	26	0.421	25
0.403	23	0.426	28
0.337[a]	16.5	0.345	18
0.295	11	0.281	7
0.319	14	0.578	33
0.285	9	0.383	22
0.465	32	0.299	12
0.255	5	0.150	2
0.125	1	0.336	15
0.427	29	0.425	27
0.225	3	0.354	19
		0.337[b]	16.5
		0.289	10
		0.438	31
		0.411	24
		0.357	20
Sum of Ranks	224.5		336.5

[a]To four decimals, the value is 0.3373.
[b]To four decimals, the value is 0.3370.

done based on critical values shown in Table B10. In Table B10, the value 2α refers to the two-sided significance level and N_1 and N_2, respectively, refer to the number of observations in the smaller and larger groups. For a one-sided test at $\alpha = 0.05$, the page with $2\alpha = 0.10$ is used.

The critical regions, shown in Table B10, are determined in a similar manner to that for the Wilcoxon Signed Rank statistic. All possible arrangements of size n_1 of n ranks are listed, and the sum of n_1 ranks in each arrangement is found. The p-value of the R_{WRS} is then determined. For a two-sided test, if R_{WRS} is less than the expected sum, the p-value is twice the proportion of the rank sums that are less than or equal to the test statistic. If R_{WRS} is greater than the expected sum, the p-value is twice the proportion of the rank sums that are greater than or equal to R_{WRS}. For a lower tail, one-sided test, the p-value is the proportion of the rank sums that are less than or equal to R_{WRS}. For an upper tail, one-sided test, the p-value is the proportion of the rank sums that are greater than or equal to R_{WRS}.

As an example of determining the rejection region, consider a situation with four observations in each of two samples. The possible ranks are 1 through 8. Table 9.6 shows all possible arrangements of size 4 of these ranks, and Table 9.7 shows the relative frequency of the rank sums.

For a two-sided test that is to be performed at the 0.05 significance level, the rejection region consists of rank sums of 10 and 26. The probability of these two values is 0.0286, which is less than the 0.05 level. Including 11 and 25 in the rejection region increases the probability of the rejection region to 0.0571, which is greater than the 0.05 value. For a lower tail one-sided test to be performed at the 0.05 level, the rejection region is

Table 9.6   Listing of sets of size 4 from the ranks 1 to 8.

Set	Sum of Ranks	Set	Sum of Ranks
1,2,3,4	10	2,3,4,5	14
1,2,3,5	11	2,3,4,6	15
1,2,3,6	12	2,3,4,7	16
1,2,3,7	13	2,3,4,8	17
1,2,3,8	14	2,3,5,6	16
1,2,4,5	12	2,3,5,7	17
1,2,4,6	13	2,3,5,8	18
1,2,4,7	14	2,3,6,7	18
1,2,4,8	15	2,3,6,8	19
1,2,5,6	14	2,3,7,8	20
1,2,5,7	15	2,4,5,6	17
1,2,5,8	16	2,4,5,7	18
1,2,6,7	16	2,4,5,8	19
1,2,6,8	17	2,4,6,7	19
1,2,7,8	18	2,4,6,8	20
1,3,4,5	13	2,4,7,8	21
1,3,4,6	14	2,5,6,7	20
1,3,4,7	15	2,5,6,8	21
1,3,4,8	16	2,5,7,8	22
1,3,5,6	15	2,6,7,8	23
1,3,5,7	16	3,4,5,6	18
1,3,5,8	17	3,4,5,7	19
1,3,6,7	17	3,4,5,8	20
1,3,6,8	18	3,4,6,7	20
1,3,7,8	19	3,4,6,8	21
1,4,5,6	16	3,4,7,8	22
1,4,5,7	17	3,5,6,7	21
1,4,5,8	18	3,5,6,8	22
1,4,6,7	18	3,5,7,8	23
1,4,6,8	19	3,6,7,8	24
1,4,7,8	20	4,5,6,7	22
1,5,6,7	19	4,5,6,8	23
1,5,6,8	20	4,5,7,8	24
1,5,7,8	21	4,6,7,8	25
1,6,7,8	22	5,6,7,8	26

Table 9.7   Frequency and relative frequency of the rank sums for two samples of four observations each.

Rank Sum	Frequency	Relative Frequency
10 or 26	1	0.0143
11 or 25	1	0.0143
12 or 24	2	0.0286
13 or 23	3	0.0429
14 or 22	5	0.0714
15 or 21	5	0.0714
16 or 20	7	0.1000
17 or 19	7	0.1000
18	8	0.1143

10 and 11. It is not possible to perform the test at the 0.01 level because the probability of each rank sum in the Table 9.7 is greater than 0.01. The rejection region we have found here agrees with that shown in Table B10 ($2\alpha = 0.05$, $N_1 = 4$, $N_2 = 4$), the critical region for the WRS test at the 0.05 significance level.

Example 9.5

Now we return to the data regarding the proportion of calories coming from fat shown in Table 9.5. Let us perform the test of the null hypothesis of no difference in the magnitudes of the variable in the two independent populations at the 0.01 significance level. The alternative hypothesis is that there is a difference in the magnitudes. Since this is a two-sided test, extremely large or small values of the test statistic will cause us to reject the null hypothesis. The test statistic is the rank sum of the smaller sample, which is 224.5. Since the test is being performed at the 0.01 significance level, we use Table B10 ($2\alpha = 0.01$) with sample sizes of 14 and 19. The critical values are 168 and 308. If R_{WRS} is less than or equal to 168 or greater than or equal to 308, we reject the null hypothesis in favor of the alternative hypothesis. Since R_{WRS} is 224.5, a value not in the rejection region, we fail to reject the null hypothesis. Based on this test, there is no evidence that fifth- and sixth-grade boys differ from seventh- and eighth-grade boys in terms of the proportion of calories coming from fat.

Computer programs can be used to perform the Mann-Whitney test (see **Program Note 9.2** on the website).

Once we exceed the sample sizes shown in Table B10, or for both n_1 and n_2 greater than or equal to 8, we can use a normal distribution as an approximation for the distribution of the R_{WRS} statistic. As we just saw, the expected value of R_{WRS} is expressed in terms of the sample sizes. Let n_1 be the sample size of the smaller sample, n_2 be the sample size of the other sample, and n be their sum. The mean and variance of R_{WRS}, assuming that the null hypothesis is true, are $n_1(n + 1)/2$ and $n_1n_2(n + 1)/12$, respectively. Therefore, the statistic

$$\frac{|R_{WRS} - n_1(n+1)/2| - 0.5}{\sqrt{n_1 n_2 (n+1)/12}}$$

approximately follows the standard normal distribution. The 0.5 term is the continuity correction term, required since the rank sum statistic is not a continuous variable.

Let us calculate the normal approximation for the data in Example 9.5. The expected value of R_{WRS} is 238 ($= 14[34]/2$). The standard error is 27.453 ($= \sqrt{14 * 19 * 34/12}$). Therefore, the statistic's value is

$$\frac{|224.5 - 238| - 0.5}{27.453} = 0.4735.$$

Since this is a two-sided test, the p-value is twice the probability that a standard normal variable is greater than 0.4735. Using linear interpolation in Table B4, we find that

$$\Pr\{Z > 0.4735\} = 0.3179$$

and hence the p-value is twice that, or 0.6358.

If there are many ties between the data in the two groups, an adjustment for the ties should be made (Hollander and Wolfe 1973) or a procedure in the next chapter should be used in the analysis of the data.

9.5 The Kruskal-Wallis Test

The Wilcoxon Rank Sum test is limited to the consideration of two populations. In this section, a method for the comparison of the locations (medians) from two or more populations is presented. This method, the Kruskal-Wallis (KW) test, a generalization of the Wilcoxon test, is named after the two prominent American statisticians who developed it in 1952. The KW test also requires that the data come from continuous probability distributions. The hypothesis being tested by the KW statistic is that all the medians are equal to one another, and the alternative hypothesis is that the medians are not all equal.

We first introduce a data situation appropriate for this test. A study examined the effect of weight loss without salt restriction on blood pressure in overweight hypertensive patients (Reisin et al. 1978). Patients in the study all weighed at least 10 percent above their ideal weight, and all were hypertensive. The patients either were not taking any medication or were on medication that had not reduced their blood pressure below 140 mmHg systolic or 90 mmHg diastolic. Three groups of patients were formed. Group I consisted of patients who were not taking any antihypertensive medication and who were placed on a weight reduction program; Group II patients were also placed on a weight reduction program in addition to continuing their antihypertensive medication; and Group III patients simply continued with their antihypertensive medication. Patients already receiving medication were randomly assigned to Groups II or III. Patients were followed initially for two months, and the baseline value was the blood pressure reading at the end of the two-month period. Patients were then followed for four additional months. Changes in weight and blood pressure between Month 2 and Month 6 were measured.

Table 9.8 contains simulated values that are consistent with those reported in the study by Reisin et al. (1978). Besides using simulated values, the only data shown are from the female patients. We wish to determine whether or not there are differences in the median reductions in diastolic blood pressure in the populations of females from which these samples were drawn. To prepare the data for the test we rank all the simulated values in three groups from the smallest to the largest value (1 through 39 in this

Table 9.8 Simulated reductions (mmHg) in diastolic blood pressure for females from month two to month six of follow-up in each of the three treatment groups with ranks of simulated values and sums of ranks.

Only Weight Reduction ($n_1 = 8$)				Medication and Weight Reduction ($n_2 = 15$)				Only Medication ($n_3 = 16$)			
Simulated Values											
38	10	10	28	19	36	16	36	12	16	0	−12
6	8	33	8	38	28	36	22	14	16	−10	4
				42	24	40	34	−20	−6	18	16
				6	16	30		−14	6	−16	6
Ranks of Simulated Values											
36.5	15.5	15.5	28.5	25	34	21	34	17	21	7	4
10.5	13.5	31	13.5	36.5	28.5	34	26	18	21	5	8
				39	27	38	32	1	6	24	21
				10.5	21	30		3	10.5	2	10.5
Sums of Ranks											
$R_1 = 164.5$				$R_2 = 436.5$				$R_3 = 179$			

case) and sum the ranks separately in each group. Observations with the same value receive the same average rank as above. Table 9.8 also shows the ranks of the values and sums of ranks in each group.

It is possible, although not feasible for any reasonable sample sizes, to explore the rationale underlying this test by examining the sums of the ranks as we had done in the Wilcoxon tests. Since it is not feasible to determine the distribution of the rank sums, Kruskal and Wallis suggested that H, a statistic defined in terms of n_i and R_i, the sample size and rank sum for the ith group, be used as the test statistic. The definition of H is

$$H = \frac{12}{n(n+1)} \sum_{i=1}^{k} \frac{R_i^2}{n_i} - 3(n+1)$$

where n is the sum of the group sample sizes and k is the number of groups. This statistic follows the chi-square distribution with $(k-1)$ degrees of freedom when the null hypothesis is true. The statistic H follows the chi-square distribution because H can be shown to be proportional to the sample variance of the rank sums which follows a chi-square distribution. Thus, we reject the null hypothesis when the observed value of H is greater than $\chi^2_{k-1,1-\alpha}$, and we fail to reject the null hypothesis otherwise.

Example 9.6

Let us conduct the KW test for the weight loss data in Table 9.8. To calculate the test statistic H we need to use the rank sums for the three groups shown, and we must also choose the significance level for the test. Let us perform the test at the 0.10 significance level.

We already have the information required to calculate H. The observed value of H is

$$H = \frac{12}{39(39+1)} \left(\frac{164.5^2}{8} + \frac{436.5^2}{15} + \frac{179^2}{16} \right) - 3(39+1) = 19.133.$$

If the null hypothesis, equality of medians, is true, H follows the chi-square distribution with 2 degrees of freedom. Since 19.133 is greater than 4.61 ($=\chi^2_{2, 0.90}$), we reject the null hypothesis in favor of the alternative. From Table B7, we see that the p-value of H is less than 0.005. There appears to be a difference in the effects of the different interventions on diastolic blood pressure. Weight reduction can play an important role in blood pressure reduction for overweight patients.

Computer programs can be used to perform the Kruskal-Wallis test (see **Program Note 9.3** on the website).

9.6 The Friedman Test

While the Kruskal-Wallis test is designed to compare k independent groups, the Friedman test is for comparing k dependent groups. The groups are no longer independent when matched samples are assigned to k comparison groups. Referring to the experimental designs discussed in Chapter 6, the Kruskal-Wallis test is suitable for a completely randomized design, and the Friedman test is for a randomized block design. A distribution-free test for the randomized block design was given by Friedman (1937), and this test is a generalization of the sign test to more than two groups.

The Friedman test starts with ranking of observed values within blocks. The test statistic T suggested by Friedman is defined in terms of the sum of ranks for the ith comparison groups, R_i; the number of blocks, b; and the number of comparison groups, k, as follows:

$$T = \frac{12}{bk(k+1)} \sum_{i=1}^{k} R_i^2 - 3b(k+1).$$

It is similar to H statistic for the Kruskal-Wallis test, but the ranking procedure is different. The ranking in the Friedman test is done separately within blocks recognizing the randomized block design, whereas the Kruskal-Wallis test is based on a single overall ranking reflecting the completely randomized design. The T statistic follows the chi-square distribution with $(k-1)$ degrees of freedom when the null hypothesis is true. As in the case of H, we reject the null hypothesis when the observed value of T is greater than $\chi^2_{k-1,1-\alpha}$, and otherwise we fail to reject the null hypothesis.

Example 9.7

Effectiveness of insecticides is evaluated based on a randomized block design (Steel and Tome 1960). Four blocks of fields were used for this study. The numbers of living adult plum curculios emerging from separate caged areas of soil treated by five different insecticides and a control (check) were counted. The data in this example are count data ranging from 0 to 217, and the assumptions of normality and equality of variance may be in doubt. Therefore, we decided to use the Friedman to test the null hypothesis at the 0.05 significance level. Table 9.9 shows the data and the ranks within each block.

Based on the data in Table 9.9, the test statistic is calculated as follows:

$$T = \frac{12}{4(6)(6+1)} (12^2 + 6.5^2 + 5.5^2 + 20^2 + 16^2 + 24^2) - 3(4)(6+1) = 19.5.$$

Since the test statistic of 19.5 is greater than the critical value of 11.07 ($= \chi^2_{5,0.95}$), we reject the null hypothesis in favor of the alternative hypothesis that the treatment groups are different.

Table 9.9 The number of living adult plum curculios emerging from caged areas treated by different insecticides (the number in parentheses are ranks within each block).

Block	Insecticides					
	Lindane	Dieldrin	Aldrin	EPN	Chlordane	Check
1	14	7	6	95	37	212
	(3)	(2)	(1)	(5)	(4)	(6)
2	6	1	1	133	31	172
	(3)	(1.5)	(1.5)	(5)	(4)	(6)
3	8	0	1	86	13	202
	(3)	(1)	(2)	(5)	(4)	(6)
4	36	15	4	115	69	217
	(3)	(2)	(1)	(5)	(4)	(6)
Sums of ranks	12	6.5	5.5	20	16	24

Computer programs can be used to perform the Friedman test (see **Program Note 9.4** on the website).

Conclusion

In this chapter, we introduced several of the more frequently used nonparametric tests for continuous data. The nonparametric tests are attractive because they do not require an assumption of the normal distribution. Even when the data do come from normal distributions, these nonparametric tests do not sacrifice much power in comparison to tests based on the normality assumption. Although these tests were designed to be used with continuous data, they are often used with ordered data as well. Their use with ordered data can create problems as there are likely to be more ties for ordered data than for continuous data. In the next chapter, we introduce methods for testing hypotheses about ordered or nominal data, as well as about continuous data that are grouped into categories.

EXERCISES

9.1 The following table below shows the annual average fatality rate per 100,000 workers for each state, data originally introduced in Exercise 7.4. A state is placed into one of three groups according to the National Safe Workplace Institute (NSWI) score. Group 1 consists of states whose NSWI score was above 55, group 2 consists of states with scores of 31 to 55, and group 3 consists of states with scores less than or equal to 30. In Exercise 7.4, we examined the correlation between the fatality rates and the NSWI scores. Here we wish to determine whether or not we believe that the median fatality rates for the three groups of states are the same.

State Fatality Rates per 100,000 Workers by the National Safe Workplace Institute Scores

NSWI Groups								
Low (≤ 30)			Middle (31 to 55)			High (> 55)		
State	Rate	Rank	State	Rate	Rank	State	Rate	Rank
AR	12.5	41	LA	11.2	35	NH	4.5	8
WY	29.5	49	KY	11.9	39	WI	6.3	16
NM	12.0	40	GA	10.3	33	RI	3.3	4
KS	9.1	28	VT	6.8	19	AK	33.1	50
ND	13.8	43	AZ	4.1	6	VA	9.9	32
ID	17.1	47	DE	5.8	13	MI	5.3	10.5
TN	8.1	25	MO	5.3	10.5	OR	11.0	34
HI	6.0	14	MD	5.7	12	MN	4.3	7
AL	9.0	27	NC	7.2	21	CT	1.9	1
MS	14.6	44	IN	7.8	23.5	ME	7.8	23.5
SD	14.7	45	WV	16.2	46	TX	11.7	38
SC	6.7	18	FL	9.3	30.5	MA	2.4	2
UT	13.5	42	CO	9.3	30.5	NY	2.5	3
NE	11.3	36	OK	8.7	26	IL	6.9	20
MT	21.6	48	IA	9.2	29	NJ	3.4	5
NV	11.5	37	OH	4.8	9	CA	6.5	17
			PA	6.1	15			
			WA	7.7	22			
Sum of Ranks	584			420				271

Is there any need to use a statistical test of hypothesis to determine whether or not the median fatality rates of these three groups of states are the same? If there is, what test would you use?

9.2   A study was conducted to determine the effect of short-term, low-level exposure of demolition workers to asbestos fibers and silica-containing dusts (Kam 1989). Twenty-three demolition workers were exposed for 26 consecutive days during the destruction of a three-story building. The dependent variable is the percent reduction in the baseline value of the ratio of the forced expiratory volume in the first second to the forced vital capacity (FEV_1/FVC) compared to the same ratio at the end of the demolition project. None of the exposures to asbestos or silica were above the permissible values. The following table shows the data for the 23 workers, grouped according to the level of exposure to asbestos and silica.

Percent Reduction in Pre- and Postproject FEV_1/FVC Values by Level of Exposure to Asbestos and Silica-Containing Dusts

Higer Exposure ($n = 10$)					Lower Exposure ($n = 13$)				
0.73	0.72	0.70	0.33	0.54	0.42	0.70	0.65	0.62	0.81
0.75	0.67	0.73	0.69	0.59	0.64	0.63	0.60	0.66	0.61
					0.68	0.76	0.65		

Test the hypothesis that there is no difference in the median percent reduction for those with the higher level of exposure compared to those with the lower level of exposure. Use a 5 percent significance level.

9.3   A study was conducted to compare the effectiveness of the applied relaxation method and the applied relaxation method with biofeedback in patients with chronic low back pain (Strong, Cramond, and Mass 1989). Twenty female patients were randomly assigned to each treatment group, and the treatments were then provided. One of the dependent variables studied was the change in the pain rating index — based on the McGill Pain Questionnaire — between pre- and posttreatment. Patients were also followed for a longer period, but those results are not used in this exercise. The actual change data were not shown in the article, but the following table contains hypothetical changes for the two groups.

Hypothetical Data Showing the Changes in Pre- and Posttreatment Values of the McGill Pain Questionnaire for 40 Women Randomly Assigned to the Different Treatments

Relaxation Only										Relaxation with Biofeedback									
10	11	21	18	16	16	15	9	2	19	9	12	7	14	4	2	11	8	9	11
5	18	16	14	12	13	11	13	14	20	6	10	9	7	8	10	6	13	7	8

Use the appropriate one or two-sided test for the null hypothesis of no difference in the median changes in pain rating between the two groups at the 0.10 significance level. Provide the rationale for your choice of either the one-sided or two-sided test.

9.4   The following data are from the 1971 census for Hull, England (Goldstein 1982). The data show by ward, roughly equivalent to a census tract, the number of households per 1000 without a toilet and the corresponding incidence of infectious jaundice per 100,000 population reported between 1968 and 1973. Group the ward into three groups based on the rate of households without a toilet. Use the Kruskal-Wallis test to determine whether or not there is a difference in the median incidence of jaundice for the three groups at the 0.05 significance level.

Ward	Number of Toilets	Jaundice	Ward	Number of Toilets	Jaundice
1	222	139	12	1	128
2	258	479	13	276	263
3	39	88	14	466	469
4	389	589	15	443	339
5	46	498	16	186	189
6	385	400	17	54	198
7	241	80	18	749	401
8	629	286	19	133	317
9	24	108	20	25	201
10	5	389	21	36	419
11	61	252			

9.5 Exercise 9.4 provides an example of ecological data, data aggregated for a group of subjects. Care must be taken in the use of this type of data (Piantadosi, Byar, and Green 1988). For example, suppose in Exercise 9.4 there was a statistically significant difference in the median incidence of jaundice for the three groups of wards. Is it appropriate to conclude that there is an association between the presence or absence of a toilet in a household and the occurrence of jaundice? Provide the rationale for your answer.

9.6 In the study on Ramipril introduced in Chapter 7, there was a four-week baseline period during which patients took placebo tablets (Walter, Forthofer, and Witte 1987). Of the 160 patients involved in the study, 24 had previously taken medication for high blood pressure, but it had been greater than seven days since they had last taken their medication. These 24 patients had some expectation that medication works. We will examine hypothetical data based on the summary statistics reported to determine whether or not there is a placebo effect — a reduction in blood pressure values associated with taking the placebo — here. The hypothetical systolic blood pressure (SBP in mmHg) values are the following:

Patient Number	Week 0 SBP	Week 4 SBP	Patient Number	Week 0 SBP	Week 4 SBP
1	171	182	13	148	178
2	172	167	14	182	166
3	166	186	15	210	183
4	181	175	16	171	164
5	194	177	17	165	163
6	200	200	18	201	175
7	200	168	19	189	165
8	181	178	20	197	174
9	173	189	21	187	167
10	178	189	22	174	180
11	206	167	23	197	185
12	199	185	24	169	149

Use the sign test to test the hypothesis that the proportion of decreases in SBP between week 0 and week 4 is equal to 0.50 versus the alternative that the proportion of decreases in SBP is greater than 0.50. Use the 0.05 significance level. If there were reversion or regression to the mean here, would that affect our conclusion about the placebo effect? Test the hypothesis of no reversion to the mean at the 0.05 level.

9.7   Use the Wilcoxon Signed Rank test to test the hypothesis that the median change in SBP in Exercise 9.6 is zero versus the alternative hypothesis that the median change is greater than zero. Perform the test at the 0.05 level. Compare your results to those of the sign test. Do you think that there is a placebo effect here?

9.8   As an extension of Example 9.2, of the 13 measurements of lead concentration in the blood, 6 measurements were done in the morning, and the remaining 7 measurements were done in the afternoon. The measurements were as follows:

Mornings:	43	44	51	50	41	47	
Afternoons:	45	40	49	35	46	36	38

Is there any evidence that the morning measurements are different from the afternoon measurements in the differences from the true value of 41? Test the null hypothesis of no difference at the 0.05 significance level using the Wilcoxon Rank Sum test. Would you use a two-sample t test for this data? Why or why not?

9.9   A psychologist investigated the effect of three different patterns of reward (RR = full reinforcement, RU = reinforcement trial followed by unreinforcement trial, UR = unreinforcement trial followed by reinforcement trial) upon the extent of learning an opposing habit (Siegel 1956). Eighteen litters of rats, three in each litter, were trained under the three patterns of reward, and the three rats in each litter were randomly assigned to the three reinforcement patterns. The extent of learning was measured by counting the number of errors made in the trials to compare the three reward patterns. Because the count of errors is probably not an interval measure and the count data exhibits possible lack of homogeneity of variance, the error counts are ranked within each litter:

Letters:	1	2	3	4	5	6	7	8	9	10	11	12	13	14	15	16	17	18	Sum
RR:	1	2	1	1	3	2	3	1	3	3	2	2	3	2	2.5	3	3	2	39.5
RU:	3	3	3	2	1	3	2	3	1	1	3	3	2	3	2.5	2	2	3	42.5
UR:	2	1	2	3	2	1	1	2	2	2	1	1	1	1	1	1	1	1	26

What nonparametric test would you use for these data? Perform the test at the 0.05 significance level and interpret the test results

9.10   A group of researchers investigated the effectiveness of an educational intervention designed to improve physicians' knowledge of the costs of common medications and willingness to consider costs when prescribing (Korn et al. 2003). The researchers administered a written survey before and six months after the intervention. Physicians were asked to agree or disagree with the statements about the relevance of cost for prescribing, using a 5-point Likert scale (1 = strongly agree; 2 = somewhat agree; 3 = neutral; 4 = somewhat disagree;

5 = strongly disagree). They used "the Wilcoxon matched-pairs signed-rank test" which measures the effect of intervention considering the nature of the data. A total of 109 pairs of pre- and postsurvey responses were analyzed and *p*-values were reported for selected questions.

Discuss how the Wilcoxon signed-rank test could have been applied to the data. How many zeros do you think the investigators had in the differences in the 5-point scale? How many tied ranks do you think they had? Do you think the test was appropriate for the data?

REFERENCES

Davis, C. E. "The Effect of Regression to the Mean in Epidemiologic and Clinical Studies." *American Journal of Epidemiology* 104:493–498, 1976.

Friedman, M. "The Use of Ranks to Avoid the Assumption of Normality Implicit in the Analysis of Variance." *Journal of American Statistical Association* 32:675–701, 1937.

Goldstein, M. "Preliminary Inspection of Multivariate Data." *The American Statistician* 36:358–362, 1982.

Hollander, M., and D. Wolfe. *Nonparametric Statistical Methods.* New York: Wiley, 1973.

Hunter, J. S. "The National System of Scientific Measurement." *Science* 210:869–874, 1980.

Kam, J. K. "Demolition Worker Hazard: The Effect of Short-Term, Low-Level Combined Exposures." *Journal of Environmental Health* 52:162–163, 1989.

Korn, L. M., S. Reichert, T. Simon, and E. A. Halm. "Improving Physicians' Knowledge of the Costs of Common Medications and Willingness to Consider Costs When Prescribing." *Journal of General Internal Medicine* 18:31–37, 2003.

McPherson, R. S., M. Z. Nichaman, H. W. Kohl, D. B. Reed, and D. R. Labarthe. "Intake and Food Sources of Dietary Fat among Schoolchildren in The Woodlands, Texas." *Pediatrics* 86(4):520–526, 1990.

Nesselroade, J. R., S. M. Stigler, and P. B. Baltes. "Regression Toward the Mean and the Study of Change." *Psychological Bulletin* 88:622–637, 1980.

Piantadosi, S., D. P. Byar, and S. B. Green. "The Ecological Fallacy." *American Journal of Epidemiology* 127:893–904, 1988.

Samuels, M. L. "Statistical Reversion Toward the Mean: More Universal Than Regression Toward the Mean." *The American Statistician* 45:344–346, 1991.

Reisin, E., R. Abel, M. Modan, et al. "Effect of Weight Loss without Salt Restriction on the Reduction of Blood Pressure in Overweight Hypertensive Patients." *The New England Journal of Medicine* 298:1–6, 1978.

Siegel, S. *Nonparametric Statistics for the Behavioral Sciences.* New York: McGraw-Hill, 1956, p. 169–171.

Steel, R. G. D., and J. H. Torrie. *Principles and Procedures of Statistics.* New York: McGraw-Hill, 1960, p. 158–159.

Stigler, S. M. *The History of Statistics: The Measurement of Uncertainty before 1900.* Cambridge, MA: Belknap Press of Harvard University Press, 1986.

Strong, J., T. Cramond, and F. Mass. "The Effectiveness of Relaxation Techniques with Patients Who Have Chronic Low Back Pain." *The Occupational Therapy Journal of Research* 9:184–192, 1989.

Walter U., R. Forthofer, and R. U. Witte. "Dose-Response Relation of Amgiotensin Converting Enzyme Inhibitor Ramipril in Mild to Moderate Essential Hypertension." *American Journal of Cardiology* 59:125D–132D, 1987.

Analysis of Categorical Data

10

Chapter Outline

In this chapter, we present some additional nonparametric tests that are used with nominal, ordinal, and continuous data that have been grouped into categories. The data in this chapter are presented in the form of frequency or contingency tables. In Chapter 3, we demonstrated how one- and two-way frequency tables could be used in data description. In this chapter, we show how frequency or contingency tables can be used in the test of whether or not the distribution of the variable of interest agrees with some hypothesized distribution or whether or not there is an association among two or more variables. For example, in the material on the normal distribution in Chapter 5, we examined the distribution of blood pressure. In this chapter, we show how to test the null hypothesis that the data follow a particular distribution. In Chapter 4, we considered the association between birth weight and the timing of the initiation of prenatal care. In this chapter, we show how to test the null hypothesis that an association exists between two discrete variables versus the alternative hypothesis that there is no association. A goodness-of-fit statistic is used to test these hypotheses, and it follows a chi-square distribution if the null hypothesis is true.

10.1 The Goodness-of-Fit Test

The *goodness-of-fit* test can be used to examine the fit of a one-way frequency distribution for X, the variable of interest, to the distribution expected under the null hypothesis. This test, developed in 1900, is another contribution of Karl Pearson, also known for the Pearson correlation coefficient. The X variable is usually a discrete variable, but it could also be a continuous variable that has been grouped into categories.

To facilitate the presentation, we shall use the following symbols. Let O_i represent the number of sample observations at level i of X and E_i represent the expected number of observations at level i, assuming that the null hypothesis is true. The E_i are found by multiplying the population probability of level i, π_i, by n, the total number of observations. Since the sum of the π_i is one, the sum of the E_i is n.

A natural statistic for this comparison would seem to be the sum of the differences of O_i and E_i — that is, $\Sigma(O_i - E_i)$. However, since both the O_i and the E_i sum to n, the sum of their differences is always zero. Thus, this statistic is not very useful. However, the sum of the squares of the differences, $\Sigma(O_i - E_i)^2$, will be different from zero except when there is a perfect fit. Squaring the differences is the same strategy used in defining the variance in Chapter 3.

One problem remains with $\Sigma(O_i - E_i)^2$. If the sample size is large, even very small differences in the observed and expected proportions at each level of X become large in terms of the O_i and E_i. Therefore, we must take the magnitude of the O_i and E_i into account. Pearson suggested dividing each squared difference by the expected number for that category and using the result, $\Sigma(O_i - E_i)^2/E_i$ as the test statistic. It turns out that this statistic, for reasonably large values of E_i, follows the chi-square distribution. In the early 1920s, Sir Ronald A. Fisher showed that this statistic has $k - 1 - m$ degrees of freedom, where k is the number of levels of the X variable and m is the number of estimated parameters. For the chi-square distribution to apply, no cell should have an expected count that is less than five times the proportion of cells with E_i that are less than five (Yamold 1970). For example, if k is 10 and two cells have expected counts less than five, then no expected cell count should be less than one ($= 5[2/10]$). If some of the E_i are less than this minimum value, categories with small expected values may be combined with adjacent categories. The combinations of categories must make sense substantively; otherwise the categories should not be combined.

Note that the goodness-of-fit test is a one-sided test. Only large values of the chi-square test statistic will cause us to reject the null hypothesis of good agreement between the observed and expected counts in favor of the alternative hypothesis that the observed counts do not provide a good fit to the expected counts. Small values of the test statistic support the null hypothesis.

We consider the following two examples: In the first example, no parameter estimation is required, and two parameters are estimated in the second example.

Example 10.1

(No Parameter Estimation Required): The study of genetics has led to the discovery and understanding of the role of heredity in many diseases — for example, in hemophilia, color-blindness, Tay-Sachs disease, phenylketonuria, and diabetes insipidus (Snyder 1970). The father of genetics, Abbe Gregor Mendel, presented his research on garden peas in 1865, but the importance of his results was not appreciated until 1900. One of Mendel's discoveries was the 1 : 2 : 1 ratio for the number of dominant, heterozygous, and recessive offspring from hybrid parents — that is, from parents with one dominant and one recessive gene.

Although doubts have been raised about Mendel's data, we shall use data from one of his many experiments. Table 10.1 shows the number of offspring by type from

Table 10.1 Mendel's data on garden peas: number of smooth and wrinkled offspring from hybrid parents.

AA	Aa	aa	Total
138	256	126	529

the crossbreeding of smooth seeds, (A), the dominant type, with wrinkled seeds, (a), the recessive type (Bishop, Fienberg, and Holland 1975). Dominant means that when there are both a smooth and a wrinkled gene present, the pea will be smooth. The pea will be wrinkled only when both genes are wrinkled.

The question of interest is whether or not this experiment supports Mendel's theoretical ratio of 1 : 2 : 1. The null hypothesis is that the observed data are consistent with Mendel's theory. The alternative hypothesis is that the data are not consistent with his theory. Let us test this hypothesis at the 0.10 significance level.

We must first calculate the expected cell counts for this one-way contingency table. Since the expected counts are based on the theoretical 1 : 2 : 1 ratio, the ratio tells us that we expect 1/4 of the observations to be AA, 2/4 to be Aa or aA, and 1/4 to be aa. One-fourth of 529 is 132.25, and one-half of 529 is 264.5; therefore, the expected counts are 132.25, 264.5, and 132.25, respectively. The test statistic is

$$X^2 = \frac{(138-132.25)^2}{132.25} + \frac{(265-264.5)^2}{264.5} + \frac{(126-132.25)^2}{132.25} = 0.546.$$

This statistic follows the chi-square distribution if the null hypothesis is true. The number of degrees of freedom is $k - 1 - m$. In this example, the value of k is 3 for the three types of possible offspring. Since we did not estimate any parameters, m is 0. Therefore, there are 2 degrees of freedom. The critical value, $\chi^2_{2,0.90}$, is 4.61. Since 0.546 is less than 4.61, we fail to reject the null hypothesis. It appears that these data support Mendel's theoretical results.

The goodness-of-fit chi-square statistic can also be used to test the hypothesis that the data follow a particular probability distribution. Thus, it can be used to complement the graphical approaches — for example, the Poissonness and normal probability plots presented in Chapter 5. The test of hypothesis provides a number, the p-value, that can be used alone or together with the graphical approach, to help us decide whether or not we will reject or fail to reject the null hypothesis.

Example 10.2

Two Parameters Estimated: Let us test the hypothesis, at the 0.01 significance level, that the systolic blood pressure values for 150 typical 12-year-old boys in the United States, shown in Table 10.2, come from a normally distributed population. In testing the hypothesis that data are from a normally distributed population, we must specify the particular normal distribution. This specification means that the values of the population mean and standard deviation are required. However, since we do not know these values for the systolic blood pressure variable for U.S. 12-year-old boys, we will estimate their values. Table 10.2 shows the sample estimates of the mean and standard deviation.

The goodness-of-fit test is based on the variable of interest being discrete or being grouped into k categories. Therefore, we must group the systolic blood pressures into categories. We use ten categories, shown in Table 10.3.

Table 10.2 Systolic blood pressure values (mmHg) and their sample mean and standard deviation for 150 12-year-old boys.

Value	Freq.	Value	Freq.	Value	Freq.
80	3	100	19	118	2
82	1	102	7	120	7
84	1	104	9	122	2
86	1	105	6	124	2
88	2	106	4	125	3
90	7	108	10	126	2
92	2	110	17	128	2
94	2	112	7	130	5
95	6	114	3	134	2
96	2	115	2	136	1
98	3	116	6	140	2

Sample Mean   = 107.45
Sample Standard Deviation =   12.45

Table 10.3 Number of boys observed and expected[a] in the systolic blood pressure categories.

Systolic Blood Pressure (mmHg)	Number	
	Observed	Expected
≤80.5	3	2.28
80.51–87.5	3	5.90
87.51–94.5	13	14.18
94.51–101.5	30	25.10
101.51–108.5	36	32.57
108.51–115.5	29	31.13
115.51–122.5	17	21.83
122.51–129.5	9	11.26
129.51–136.5	7	4.28
≥136.5	3	1.47
Total	150	150.00

[a]Expected calculated assuming the data follow the N (107.45,12.45) distribution

The expected values are found by converting the category boundaries to standard normal values and then finding the probability associated with each category. For example, the probability associated with the first category, a systolic blood pressure less than 80.5 mmHg, is found in the following manner. First, 80.5 is converted to a standard normal value by subtracting the mean and dividing by the standard deviation. Thus, 80.5 is converted to $-2.165 (= [80.5 - 107.45]/12.45)$. The probability that a standard normal variable is less than -2.165 is 0.0152. The expected number of observations is found by taking the product of n, 150, and the probability of being in the category. Thus, the expected number of observations in the first category is 2.28 $(= 150[0.0152])$.

The expected number of observations in the second category is found in the following manner. The upper boundary of the second category, 87.5, is converted to the standard normal value of $-1.602 (= [87.5 - 107.45]/12.45)$. The probability that a standard normal variable is less than -1.602 is 0.0545. The probability of being in the second category is then 0.0393 $(= 0.0545 - 0.0152)$. This probability is multiplied by 150 to get the expected count of 5.90 for the second category. The other expected cell counts are calculated in this same way. If the sum of the expected counts does not equal the number of observations, with allowance for rounding, an error has been made. Note that three cells have expected counts less than 5. For the chi-square distribution to apply, no cell should have an expected count less than 1.5 (= 5[3/10]). The expected count in the last cell is 1.47, a value that is very close to 1.5. The difference between 1.47 and 1.50 is so slight that we may choose to apply the chi-square distribution, or we could combine the last two categories and use only nine categories in the calculation of the test statistic. Whichever choice we choose should not have much impact on the value of the test statistic.

The calculation of the chi-square goodness-of-fit statistic is

$$X^2 = \frac{(3-2.28)^2}{2.28} + \frac{(3-5.90)^2}{5.90} + \cdots + \frac{(3-1.47)^2}{1.47} = 8.058.$$

The value of k, the number of categories, is 10 and m, the number of parameters estimated, is 2. Therefore, there are 7 degrees of freedom (= 10 − 1 − 2). The value of this test statistic is compared to 18.48 (= $\chi^2_{7,0.99}$). Since 8.058 is less than 18.48, we fail to reject the null hypothesis. Based on this sample, there is no evidence to suggest that the systolic blood pressures of 12-year-old boys are not normally distributed.

In dealing with continuous variables — for example, the blood pressure variable — we have to decide how many intervals and what interval boundaries should be used. In the preceding example, we used 10 intervals. Some research has been conducted on the relation between power considerations and the number and size of intervals, and, as we might expect, the number of intervals depends on the sample size. Table 10.4, based on a review by Cochran (1952), shows the suggested number of intervals to be used with a continuous variable. The size of the intervals may also vary. The intervals can be chosen so that the expected number of observations in each interval is approximately equal. Thus, some intervals may be much narrower than other intervals. For ease of computation, it is reasonable to choose the intervals so that the observed number of observations in each interval is approximately equal. These suggestions for the choice of the number and size of intervals differ from those used in Example 10.2, but the goals of the analyses are also different. In Example 10.2, the equal size intervals were used regardless the number of observations in each interval to determine whether or not it appears that the data follow a particular distribution.

Table 10.4   Guideline for the number of intervals to be used with a continuous variable.

Sample size	200	400	600	800	1000	1500	2000
Number of intervals	15	20	24	27	30	35	39

Source: Cochran, 1952

10.2   The 2 by 2 Contingency Table

In this section, we extend the use of the chi-square goodness-of-fit test statistic to two-way contingency tables. This extension allows a determination of whether or not there is an association between two variables. We begin the study of the association of two discrete random variables with the simplest two-way table, the 2 by 2 contingency table. This statement by M. H. Doolittle in 1888 expresses the purpose of our analysis:

> The general problem may be stated as follows: Having given the number of instances respectively *in which things are both thus and so*, in which they are *thus but not so*, in which they are *so but not thus*, and in which they are *neither thus nor so*, it is required to eliminate the general quantitative relativity inhering in the mere thingness of the things, and to determine the special quantitative relativity subsisting between the *thusness and the soness* of the things (emphasis added; Goodman and Kruskal 1959).

A restatement of the purpose is that we wish to determine, at some significance level, whether or not there is an association between the variables.

For example, is there is an association between the occurrence of iron deficiency in women and their level of education? If we use two levels of education — for example, less than 12 years and greater than or equal to 12 years — the 2 by 2 table to use in this investigation would look like Table 10.5. The entries in the table, the n_{ij}, are the observed number of women in the ith row (level of education) and jth column (iron status) in the sample. The symbol $n_i.$ represents the sum of the frequencies in the ith row, $n_{\cdot j}$ is the sum of the frequencies in the jth column and n, the sample size, is the sum of the frequencies in the entire table.

Table 10.5 Iron status by level of education.

Education	Iron Status		Total
	Deficient	Acceptable	
<12 Years	n_{11}	n_{12}	$n_1.$
≥12 Years	n_{21}	n_{22}	$n_2.$
Total	$n_{\cdot 1}$	$n_{\cdot 2}$	n

There are several ways of answering the question about whether or not there is an association between these two variables. We begin with the approach from Chapter 7.

10.2.1 Comparing Two Independent Binomial Proportions

The 2 by 2 table is one way of presenting the data used in the calculation of two independent binomial proportions. If there is no association between iron status and education, then the probability of iron deficiency for women with less than 12 years of education, π_1, should equal the corresponding probability, π_2, for women with 12 or more years of education. We can construct a confidence interval for the difference of π_1 and π_2 using the method presented in Chapter 7. If the interval contains zero, there is no evidence of an association between iron status and education. The confidence interval is based on the sample estimates of π_1 and π_2 and these are $n_{11}/n_1.$ and $n_{21}/n_2.$, respectively.

10.2.2 Expected Cell Counts Assuming No Association: Chi-Square Test

Let us first consider first two common situations when a 2 by 2 table can be formed. The two choices for data collection that are used most often in practice are (1) an SRS of n observations and (2) stratified samples of $n_1.$ and $n_2.$ observations. In the SRS case, the test for no association is a test of the independence of the row and column variables. In the stratified sampling case, the test for no association is a test of the homogeneity of the proportions in the ith row with those in the jth row. Regardless of which of these two sample selection processes is used, the expected cell counts for the hypothesis of no association are calculated as shown below.

We use the symbol m_{ij} to represent the expected number of women in the ith row and jth column assuming that the null hypothesis is true. In the material on two-way tables, we are using n and m to represent the observed and expected cell counts instead of the O and E used in the previous section. For the null hypothesis of no association between iron status and education, the expected proportion of women with low iron status at each level of education, $m_{i1}/n_i.$, equals the overall proportion of iron deficient women, $n._1/n$. This is equivalent to saying that the proportion of women with low iron status is the same for those with less than 12 years of education as for those who have at least 12 years of education. Thus, when there is no association, the expected number of iron deficient women at the ith level of education can be found from the following relationship:

$$\frac{m_{i1}}{n_i.} = \frac{n._1}{n}$$

which yields

$$m_{i1} = \frac{n_i.n._1}{n}.$$

The same type of relation holds true for women with acceptable levels of iron. Therefore the general formula for the expected cell count, assuming no association, is

$$m_{ij} = \frac{n_i.n._j}{n}.$$

We can use these observed and expected values to calculate the chi-square goodness-of-fit statistic to test the hypothesis of no association between the two variables. We often use a modified version of the chi-square goodness-of-fit statistic. The modified form, called the Yates' corrected chi-square after the British statistician, Frank Yates (1984), who suggested it, is

$$X_{YC}^2 = \sum_i \sum_j \frac{(|n_{ij} - m_{ij}| - 0.5)^2}{m_{ij}}.$$

The modification consists of subtracting 0.5 from the absolute value of the difference of the observed and expected cell counts. The Yates' corrected chi-square statistic can be calculated directly from the frequencies without calculating the expected counts. The easier-to-use formula is

$$X_{YC}^2 = \frac{n(|n_{11}n_{22} - n_{12}n_{21}| - n/2)^2}{n_1.n_2.n._1n._2}.$$

The p-value associated with the Yates' corrected chi-square statistic agrees more closely with the p-value of the exact test statistic developed by Ronald Fisher (1935).

Some statisticians question the use of Fisher's exact test in 2 by 2 tables when the data arise from either of the two sampling methods just discussed. They question the application because Fisher's test was developed based on both the row and column margins being fixed in advance, a different sampling scheme than used in the two methods. Hence, they do not recommend the use of Yates' (1984) correction, but we believe that Yates' correction is appropriate.

Example 10.3

Suppose that we select an SRS of 100 women 20 to 44 years old and we obtain information on their educational level and iron status. The hypothetical data, based on a report (Life Sciences Research Office 1989), are shown in Table 10.6.

The estimated conditional probability of a woman being iron deficient given that she has less than 12 years of education is 0.133 (= 4/30). This is contrasted to the estimated probability of 0.057 (= 4/70) for a woman with 12 or more years of education. Using the procedure in Chapter 7, the 95 percent confidence interval for the difference of π_1 and π_2 is found by

$$(0.133 - 0.057) \pm z_{0.975} \sqrt{\frac{0.133(1 - 0.133)}{30} + \frac{0.057(1 - 0.057)}{70}}$$

which yields an interval from -0.057 to 0.209. Since zero is contained in the interval for the difference, there is no evidence of an association between iron status and education based on this sample.

Based on these data, the expected values, assuming the independence of the row and column variables, are

$$m_{11} = 30\ (8)\ /\ 100 =\ 2.4,$$
$$m_{12} = 30\ (92)\ /\ 100 = 27.6,$$
$$m_{21} = 70\ (8)\ /\ 100 =\ 5.6,$$
$$m_{22} = 70\ (92)\ /\ 100 = 64.4,$$
$$\text{Total} = 100.0.$$

The sum of the expected values in the first row is 30, the first row total. The sum of the expected values in the first column is 8, the first column total. Hence, once we calculate m_{11}, we know m_{12}'s value by subtracting m_{11} from 30. In the same way, we know m_{21}'s value by subtracting m_{11} from 8. Since we now know m_{12}'s value, we can also find m_{22}'s value by subtracting m_{12} from 92. Hence, once we calculate any cell's expected value, the expected values of the other three cells are determined.

Table 10.6   Hypothetical frequency data for iron status by education.

Education	Iron Status		Total
	Deficient	Acceptable	
<12 Years	4	26	30
≥12 Years	4	66	70
Total	8	92	100

This means that there is only one degree of freedom associated with the test of no association for a 2 by 2 contingency table.

The expected cell frequency for the cell in the intersection of the first row and first column is 2.4. This is the only expected frequency less than 5, and, according to the guideline just given, the minimum acceptable value for an expected cell frequency is 1.25 (= 5[1/4]). Since none of the expected frequencies are less than 1.25, we can use the chi-square test statistic.

Now that we have both the observed and expected cell counts, we can test the hypothesis of no association (independence) of iron status and education. We shall perform the test at the 0.05 significance level using the Yates' modified chi-square procedure.

The calculated X_{YC}^2 is compared to 3.84 (= $\chi_{1,0.95}^2$). If X_{YC}^2 is greater than 3.84, we reject the hypothesis of independence in favor of the alternative that there is some association between iron status and education. If X_{YC}^2 is less than 3.84, we fail to reject the null hypothesis. The test statistic is

$$X_{YC}^2 = \frac{(|4-2.4|-0.5)^2}{2.4} + \frac{(|26-27.6|-0.5)^2}{27.6}$$
$$+ \frac{(|4-5.6|-0.5)^2}{5.6} + \frac{(|66-64.4|-0.5)^2}{64.4} = 0.783.$$

Since X_{YC}^2 is less than 3.84, we fail to reject the null hypothesis. Based on this sample, it does not appear that there is any association between iron status and education. Note that the uncorrected X^2 value is 1.656 and we would therefore draw the same conclusion.

The chi-square test can easily be performed using computer packages (see **Program Note 10.1** on the website).

10.2.3   The Odds Ratio — a Measure of Association

A useful statistic for measuring the level of association in contingency tables is the *odds ratio*, θ. For example, in Table 10.5, an estimator of the odds that a woman with less than a high school education is iron deficient is n_{11}/n_{12}. The corresponding estimator of the odds that a woman with at least a high school education is iron deficient is n_{21}/n_{22}. If there is no association between education and iron status, these two odds should be equal. If the odds are equal, their ratio equals one. A sample estimator of the odds ratio *OR* is

$$\hat{\theta} = OR = \frac{n_{11}/n_{12}}{n_{21}/n_{22}} = \frac{n_{11}n_{22}}{n_{21}n_{12}}.$$

Thus, if *OR* is far from one, it calls into question the assumption (hypothesis) of no association. If the estimated odds ratio is much less than one, this means that the denominator is much larger than the numerator — that is, the product of the off-diagonal cells in the 2 by 2 table is larger than the product of the diagonal cells. For Table 10.5, an odds ratio of less than one indicates that the proportion of women with 12 or more years of education who are iron deficient is greater than the corresponding proportion for women with fewer than 12 years of education. An odds ratio of greater than one indicates that women with fewer than 12 years of education have the greater proportion of iron deficiency.

A problem with the estimated odds ratio occurs if any of the cell frequencies are zero. The estimated odds ratio is zero if n_{11} or n_{22} are zero, and it is undefined if n_{12} or n_{21} are zero. To avoid this problem, some statisticians base the calculation of the estimated odds ratio on $n_{ij} + 0.5$ instead of the n_{ij}.

We have to realize that there is sampling variation associated with the sample estimate of the odds ratio and this variation must be taken into account in interpreting the estimated odds ratio. Since the distribution of the natural logarithm of θ, $\ln(\theta)$, converges to the normal distribution for smaller sample sizes than the distribution of θ itself, we shall focus on the confidence interval for $\ln(\theta)$. After finding the confidence interval for $\ln(\theta)$, we can transform it to a confidence interval for θ. The estimated standard error for the sample estimate of $\ln(\theta)$ (Agresti 1990) is

$$\hat{\sigma}_{ln(OR)} = \sqrt{\frac{1}{n_{11}} + \frac{1}{n_{12}} + \frac{1}{n_{21}} + \frac{1}{n_{22}}}.$$

The $(1 - \alpha) * 100$ percent confidence interval for the $\ln(\theta)$ is

$$\ln(OR) \pm z_{1-\alpha/2}\hat{\sigma}_{ln(OR)}.$$

Example 10.4

The sample odds ratio for the data in Example 10.3 is 2.538 (= 4[66]/4[26]). This value seems to be different from one, and therefore it suggests that there is an association. However, we need to consider its confidence interval.

The estimated standard error for the sample estimate of $\ln(\theta)$ is 0.7441, which is obtained from $\sqrt{1/4 + 1/26 + 1/4 + 1/66}$. The value of the natural logarithm of the sample odds ratio, $\ln(2.538)$, is 0.9314. Therefore, the 95 percent confidence interval for $\ln(\theta)$ is $0.9314 \pm 1.96\,(0.7441)$, which ranges from -0.5270 to 2.3897. Taking the exponential of these limits provides the 95 percent confidence interval for θ and its limits are 0.5904 and 10.9104. The confidence interval for the odds ratio is quite large and does include the value of one. Hence, there is no evidence that the null hypothesis should be rejected.

Program Note 10.1 on the website provides a demonstration of these calculations using computer programs.

All three approaches agree that there is no evidence of an association between iron status and education based on this hypothetical sample. These approaches will almost always agree about whether or not an association exists between two variables. The confidence interval for the difference of the probabilities and the uncorrected chi-square statistic will always agree in their conclusions.

10.2.4   Fisher's Exact Test

The chi-square test of association just described relies on the test statistic having an approximate chi-square distribution. This approach is warranted when the expected cell counts are large. However, when very small cell counts (less than 5 times the proportion of cells with expected values less than 5) are involved, the use of chi-square distribution is no longer warranted. Fortunately, an alternative procedure suggested by Fisher is appropriate for such cases.

The basis of Fisher's exact test is to consider all configurations of cell counts given both row and column totals are fixed and to compute the probabilities of the observed configuration and more extreme configurations occurring by chance. If the sum of these probabilities turns out to be very small, we conclude that it is unlikely that such a small value could have occurred by chance, and we then reject the hypothesis of association between the row and column variables. The probability of each configuration is found from the hypergeometric distribution. This probability is based on the number of ways of observing an outcome conditional on fixed margins. To calculate this probability, we must find, using the notation in Table 10.5, the probability of selecting n_{11} elements from $n_1.$ and n_{21} elements from $n_2.$ given that the row margins are fixed. This probability is found by determining the number of ways selecting n_{11} elements from $n_1.$ and n_{21} from $n_2.$ and dividing that number by the number of ways selecting n_1 elements from n elements. In symbols, that is

$$\Pr(configuration) = \frac{\binom{n_1.}{n_{11}}\binom{n_2.}{n_{21}}}{\binom{n}{n_1.}}$$

which simplifies to

$$\Pr(configuration) = \frac{n_1.!\,n_2.!\,n_{.1}!\,n_{.2}!}{n!\,n_{11}!\,n_{12}!\,n_{21}!\,n_{22}!}.$$

Example 10.5

In a small company, the records of promotion in the past year are to be examined for a possible association of gender and promotion. The records show the following:

Gender	Promotion		Total	Percent Promoted
	Yes	No		
Male	5	1	6	83.3
Female	1	4	5	20.0
Total	6	5	11	

We want to test the hypothesis of no association between gender and promotion. We first calculate the expected cell counts under the hypothesis of no association, and they are 3.27 for the (1,1) cell, 2.73 for the (1,2) cell, 2.73 for the (2,1) cell, and 2.27 for the (2,2) cell. To use the chi-square test statistic, we require that none of the expected values is less than five times the proportion of cells with expectations less than five. In this example, all the expected cell counts are less than five. Therefore, the criterion is also five (= 5[4/4]), and since all the expectations are less than this criterion, we cannot use the chi-square test here. However, we can use Fisher's exact test here. Concentrating on (1,1) cell, we can calculate the probability of observed configuration and a more extreme configuration. The two configurations to be considered are

$$
\begin{array}{cc|c}
5 & 1 & 6 \\
1 & 4 & 5 \\
\hline
6 & 5 & 11
\end{array}
\quad \text{and} \quad
\begin{array}{cc|c}
6 & 0 & 6 \\
0 & 5 & 5 \\
\hline
6 & 5 & 11
\end{array}
$$

The combined p-value for these two configurations is

$$
p-value = \frac{6!5!6!5!}{11!5!1!1!4!} + \frac{6!5!6!5!}{11!5!0!0!5!} = 0.0649 + 0.0022 = 0.0671.
$$

The calculated p-value suggests the observed frequencies are somewhat unexpected. However, if we are using the 0.05 level for the test of hypothesis, there is not sufficient evidence to suggest an association between gender and promotion. Even if there were strong evidence of an association, it would not necessarily imply discrimination. There are many other variables that would need to be considered. For example, one important variable would be the date of last promotion. If all the women had been promoted the year before but none of the men, then our interpretation might change.

The calculation of Fisher's exact test statistic for 2 by 2 tables, or for its extension to r by c tables (Mehta and Patel 1983), is quite involved, and the use of computer is recommended. **Program Note 10.1** on the website also includes comments for the Fisher's exact test.

10.2.5 The Analysis of Matched-Pairs Studies

Surprisingly, we can also use the 2 by 2 table in situations with more than two variables. For example, the 2×2 table can be used when we wish to determine whether or not there is a relationship between two variables while controlling for other variables. The matched-pair study is one example of this situation.

In the health field we often wish to determine whether or not there is a relationship between disease status and a risk factor while controlling for variables that may affect the relationship. We may have some number of people with some disease of interest (the cases), and we select an equal number of people without the disease (the controls). In an effort to remove the effect of the extraneous variable(s), for each person in the disease group, we pair them with a person from the control group who is the best match on the extraneous variable(s). We present the paired data in a 2 by 2 table as follows where the entries in the table are the observed cell frequencies:

	Control Exposed to Risk Factor	
Case Exposed to Risk Factor	Yes	No
Yes	c_1	d_1
No	d_2	c_2

Pairs with the same exposure status for both case and control — the diagonal cells — are called concordant pairs (c_1 and c_2), and pairs with different exposures — the off-diagonal cells — are called discordant pairs (d_1 and d_2).

Let π be the probability that a discordant pair has an exposed case. Then, from the preceding table, π can be estimated by the following proportion,

$$\hat{\pi} = d_1/(d_1 + d_2).$$

Under the null hypothesis of no association between the risk factor and the disease, each discordant pair is just as likely to have a case exposed as to have a control exposed. Thus, the null hypothesis can be written as

$$H_0 : \pi = 1/2.$$

For large samples, we can use the normal approximation as discussed in Chapter 8. In this case the test statistic is

$$z = \frac{d_1/(d_1+d_2)-(1/2)}{\sqrt{(1/2)(1-1/2)/(d_1+d_2)}} = \frac{(d_1-d_2)}{\sqrt{d_1+d_2}}.$$

Alternatively, we could use the chi-square test with 1 degree of freedom by squaring the z statistic and incorporating Yates's correction for continuity. This chi-square test is referred as McNemar test (1947) for testing no association in a matched study of proportions — that is

$$X_M^2 = \frac{(|d_1-d_2|-1)^2}{d_1+d_2}.$$

We can use this test when $(d_1 + d_2)$ is large. However, this test is not recommended when $(d_1 + d_2)$ is less than 10, since the normal approximation is invalid as discussed in Chapter 5. In that case a preferable testing procedure is the binomial test illustrated in Chapter 5, based on a binominal distribution with $\pi = 0.5$ and $n = (d_1 + d_2)$. The same procedure was used for the sign test in Chapter 9. We also need to pay attention to the size of $(d_1 + d_2)$ in relation to the size of $(c_1 + c_2)$. When concordant pairs are predominant, there is little reason to analyze discordant pairs alone. The McNemar test is an approximate test and should be used with caution for a small data set with a relatively small number of discordant pairs.

Example 10.6

In Chapter 6 we discussed the case-control study design in which people with the disease under investigation (the cases) are compared with people who are free of the disease (the controls). A case-control study of presenile dementia by Forster et al. (1995) identified 109 clinically diagnosed patients aged below 65 years from hospital records. Each case was individually paired with a community control of the same sex and age. Steps were taken to ascertain that the control did not suffer from dementia. One of the risk factors explored in the study was family history of dementia. We wish to determine whether or not there is an association between the occurrence of presenile dementia and a family history of dementia. The following table shows a crosstabulation of the 109 pairs by the presence or absence of family history of dementia:

Family History of Dementia in Case	Family History of Dementia in Control	
	Present	Absent
Present	6	25
Absent	12	66

Since the cases were paired with the controls, information on the relationship between family history of dementia and disease comes from the 37 discordant pairs. Of these, 25 pairs had an exposed case, twice more than the pairs with an exposed control. The McNemar test statistic is

$$X_M^2 = \frac{(|25-12|-1)^2}{25+12} = 3.89.$$

Compared with 3.84 ($= \chi^2_{1,0.95}$, using Table B7), this is just significant at 0.05 level. Hence, there is evidence for an association between dementia and family history of the disease.

10.3   The *r* by *c* Contingency Table

We now consider the more general situation where two classification variables have more than two categories. First, we consider the situation where both variables are nominal followed by the situation when one of the variables is ordinal.

10.3.1   Testing Hypothesis of No Association

The same ideas used in the 2 by 2 table still apply to the *r* by *c* contingency table. If there is no association between a row variable and a column variable, the ratio of the expected cell frequency in the *i*th row and *j*th column, m_{ij}, to the *i*th row total, $n_{i\cdot}$, should equal the ratio of the *j*th column total, $n_{\cdot j}$, to the overall total. Thus, m_{ij} is still found from

$$\frac{m_{ij}}{n_{i\cdot}} = \frac{n_{\cdot j}}{n}$$

which yields

$$m_{ij} = \frac{n_{i\cdot} n_{\cdot j}}{n}.$$

There are $(r-1)(c-1)$ degrees of freedom for the *r* by *c* table because once we know the frequencies of any $(r-1)(c-1)$ cells, we can find the values of the other frequencies by subtraction from the row and column totals. The hypothesis of no association between the row and column variables is tested using the chi-square goodness-of-fit statistic. Most statisticians perform no adjustment to the test statistic when used with tables other than the 2 by 2 table. If the test statistic is greater than the value of $\chi^2_{(r-1)(c-1),1-\alpha}$, we reject the hypothesis of no association in favor of the alternative that the row and column variables are related. If the test statistic is less than $\chi^2_{(r-1)(c-1),1-\alpha}$, we fail to reject the null hypothesis.

Example 10.7

The data in Table 10.7 are from a study in Los Angeles conducted to determine the knowledge and opinion of women about mammography. The study was a response to concern raised in the media about the potential radiation hazards of the long-term use of mammography (Berkanovic and Reeder 1979). Two issues the study addressed were (1) whether or not these articles had caused women to refuse the use of mammography screening for breast cancer and (2) variables related to women's opinions about mammography. A telephone interview was conducted with a sample of women and approximately 60 percent of the women had heard or read something about mammography. Table 10.7 shows the opinion about mammography for those women who had heard or read about it. This is a 2 by 3 table. The question of interest for this table is whether or not there is an association between the woman's opinion about mammography screening and the variable knowing someone with breast cancer. We test this hypothesis at the 0.01 significance level.

There are two ($= [2 - 1][3 - 1]$) degrees of freedom for this table. Knowing the frequencies for the (1,1) and (1,2) cells allows us to find the value of the (1,3) cell by subtraction of the sum of the (1,1) and (1,2) frequencies from the total of the first row. Knowledge of the frequencies in the first row then allows us to find the cell frequencies in the second row by subtraction from the column totals. For example, the frequency of the (2,1) cell is found by subtracting the frequency of the (1,1) cell from the total of the first column. Similar logic applies to the calculation of expected counts. If we calculate the expected counts for any two cells, then the expected counts for the rest of the cells can be found by subtraction from the row and column totals. The expected counts are also shown in Table 10.7.

The value of the test statistic is found from

$$X^2 = \frac{(120-129.76)^2}{129.76} + \frac{(45-39.52)^2}{39.52} + \ldots + \frac{(8-12.29)^2}{12.29} = 6.65.$$

Table 10.7 Frequency of women by opinion about mammography and whether or not they know someone with breast cancer (expected counts are in parentheses).

	Opinion			
Know Someone with Breast Cancer	Positive	Neutral	Negative	Total
Yes	120	45	28	193
	(129.76)	(39.52)	(23.72)	
No	77	15	8	100
	(67.24)	(20.48)	(12.28)	
Total	197	60	36	293

Since 6.65 is less than 9.21 (= $\chi^2_{2,0.99}$), we fail to reject the null hypothesis. There does not appear to be a statistically significant association, at the 0.01 level, between opinion about mammography and whether or not someone with breast cancer was known.

We can use the computer to perform the test as shown in **Program Note 10.2** on the website. The p-value is $(1 - 0.9640)$ or 0.036. Although a p-value of 0.036 is significant at the 0.05 level, the association is not statistically significant at the 0.01 level.

10.3.2   Testing Hypothesis of No Trend

The hypothesis of no association is very general, and it is a reasonable hypothesis to test with nominal variables. However, when a variable conveys more information than the category name, it is possible to test a more specific hypothesis that uses more of the information contained in the variable. For example, in Example 10.5, opinion is an ordinal variable that ranges from positive to neutral to negative, and the test for no association ignores this ordering. In 2 by c contingency tables, there is a test, a test for trends that takes the ordering of the column variable into account. There is also a method that can be used for r by c contingency tables (Semenya et al. 1983).

In the test for no association in Example 10.5, we examined the unconditional cell probabilities. We also could have focused on the conditional probabilities — for example, the probability of women who knew someone with breast cancer conditional on their opinion of mammography. In calculating the conditional probabilities in this fashion, we are not implying that the probability of women who knew someone with breast cancer depends on their opinion of mammography. We are calculating the conditional probabilities in this fashion simply to see if there is a trend in the probabilities of women who knew someone with breast cancer by opinion category. The sample estimates of these conditional probabilities are easily found. For the women who are positive about mammography, the estimated probability of a woman knowing someone with breast cancer is 0.609 (= 120/197). The corresponding values for the women with neutral and negative opinions are 0.750 and 0.778, respectively. If the estimates of these probabilities are related to the opinion category, this suggests that an association exists between the row and column variables.

We now consider the *hypothesis of no linear trend*. By no linear trend, we mean that the proportion of women who knew someone with breast cancer does not increase (decrease) consistently with the changes in opinion from positive to neutral to negative. To perform a test of this hypothesis, we assign a numerical score to the categories of the opinion variable. For example, it seems reasonable to assign a score of +1 to the positive category, 0 to the neutral level and −1 to the negative category. This assignment of scores assumes that the distance from positive to neutral is the same as the distance from neutral to negative. The assignment of scores is subjective and, in unusual situations, the scoring system used can have an impact on the test of hypothesis. However, in most cases, different reasonable scoring systems will lead to the same conclusion about the test of hypothesis.

The hypothesis of no linear trend is basically a test of no correlation between the assigned scores and the conditional probabilities. Thus, the test statistic should look something like a correlation coefficient. The following notation is used in the representation of the test statistic. Let p_j be the sample estimate of the conditional probabilities of women who knew someone with breast cancer and S_j be the score assigned to the jth opinion category, where j equals 1, 2, and 3 for positive, neutral, and negative. The unconditional sample estimate of the women who knew someone with breast cancer is $\bar{p}$, and $\bar{q}$ is $(1 - \bar{p})$. Let $\bar{S}$ be the sample mean score.

The test statistic is

$$X^2 = \frac{\left(\sum_{j=1}^{c} n_{\cdot j}(p_j - \bar{p})(S_j - \bar{S}) \right)}{\bar{p}\bar{q} \sum_{j=1}^{c} n_{\cdot j}(S_j - \bar{S})^2}.$$

The numerator of this statistic is the square of the numerator of the correlation coefficient between the conditional proportion and the assigned score. Hence, we can see that this statistic is a measure of the linear trend between these two variables. For sufficiently large sample sizes, this statistic can be shown to follow the chi-square distribution with one degree of freedom if there is no linear trend. The sample size is sufficiently large if, given the value of $\bar{p}$, it is larger than that shown in Table 5.7. Large values of X^2 cause us to reject the null hypothesis of no linear trend in favor of the alternative hypothesis of a linear trend.

Example 10.8

Let us test the null hypothesis of no linear trend in the opinion about mammography data in Example 10.5 at the 0.01 significance level. The overall proportion, $\bar{p}$, of women who knew someone with breast cancer is 0.659 (= 193/293). Hence, $\bar{q}$ is 0.341. Since n is 293, much larger than the values in Table 6.7 for a proportion of 0.30 and 0.35, we can use the test statistic just shown. The p_j are 0.609, 0.750, and 0.778 for j values of 1, 2, and 3. S_1 is +1, S_2 is 0, and S_3 is −1, and the values of the column totals, $n_{\cdot j}$, are 197, 60, and 36, respectively. The mean of the scores, $\bar{S}$, is found by

$$\frac{197(1) + 60(0) + 36(-1)}{293} = 0.5495.$$

The test statistic is

$$X^2 = \frac{[197(-0.050)(0.4505) + 60(0.091)(-0.5495) + 36(0.119)(1.5495)]^2}{0.659(0.341)\left[197(0.4505)^2 + 60(-0.5495)^2 + 36(-1.5495)^2\right]}$$

$$= \frac{(-14.076)^2}{32.479} = 6.100.$$

This statistic is compared to 6.63 (= $\chi^2_{1,0.99}$). Since 6.100 is less than 6.63, we fail to reject the null hypothesis of no linear trend. The p-value of this test statistic is found to be 0.0135. Although there is not a statistically significant linear trend in these data at the 0.01 significance level, there is a strong inverse relationship between the conditional proportion of women who knew someone with breast cancer and their opinion about mammography. We know the relationship is inverse because the sign of the numerator, before squaring, is negative. The opinion about mammography is more likely to be neutral or negative as the proportion of women who knew someone with breast cancer increases.

For the use of computer for the preceding analysis, see **Program Note 10.3** on the website.

This test for trends is equivalent to creating a confidence interval for the difference in means from two independent populations. In this example, the two independent populations are the women who did not know someone with breast cancer and those who did know someone with breast cancer.

The test for trends is particularly appropriate for 2 by c contingency tables when there is an ordering among the column categories. If a linear trend exists, it may be missed by the general test for association, whereas the trend test has a greater chance of detecting it. The general test for association could cause us to say that there is no relationship between the rows and columns when there actually was a linear trend.

10.4   Multiple 2 by 2 Contingency Tables

Most studies involve the analysis of more than two variables at one time. Often we are interested in the relation between an independent variable and the dependent variable, but there is an extraneous variable that must also be considered. For example, consider a study to determine if there is any association between the occurrence of upper respiratory infections (URI) of young children and outdoor air pollution. There are several variables that could affect the relationship between the occurrence of infections and outdoor air pollution. One variable is the quality of the indoor air. One easily obtained variable that partially addresses the indoor air quality is whether or not someone smokes in the home. This variable is likely to be related to the dependent variable, the occurrence of URI, and it may also be related to the independent variable. Hypothetical data for this situation are based on an article by Jaakkola et al. (1991) and are shown in Table 10.8.

Table 10.8   Number of 6-year-old Finnish children by respiratory status and pollution level with a control for passive smoke in the home.[a]

| Passive Smoke in the Home | City Polluted | Upper Respiratory Infection during Previous 12 Months | | Total |
		Some	None	
Yes	High	100	20	120
	Low	124	40	164
	Total	224	60	284
No	High	128	62	190
	Low	166	119	285
	Total	294	181	475

[a]The entries in the table are based on an article by Jaakkola, but the data are hypothetical.

10.4.1   Analyzing the Tables Separately

If we ignore the passive smoke variable, the X_{YC}^2 for the combined table is 6.387, its p-value is 0.0115, and the estimate of the odds ratio is 1.524. There is a statistically significant relationship at the 0.05 level between the outdoor pollution variable and the occurrence of URI. The estimated odds ratio of 1.524 means that the odds of URI during the previous 12 months is about $1\frac{1}{2}$ times greater in the city with high pollution than in the city with low pollution. However, this analysis has excluded the passive smoke variable, a variable that we wish to take into account.

One way of taking the passive smoke variable into account is to analyze each 2 by 2 table separately. In this example, the X_{YC}^2 statistic is 2.039 and its p-value is 0.1533 for homes in which someone smoked. The X_{YC}^2 value is 3.645, and its p-value is 0.0562 for those without passive smoke in the home.

The corresponding estimates of the odds ratios for these two tables are 1.613 and 1.480. The 95 percent confidence intervals for the two odds ratios are from 0.887 to 2.933 and from 1.007 to 2.171, respectively. The first confidence interval, a much wider interval than the second interval, includes the value of one that suggests that there is no relation between the two variables. The second interval barely misses including one. The second interval's smaller size reflects the larger sample size associated with the home in which there was no passive smoke. Neither of these tables has a statistically significant association between the outdoor air pollution and the occurrence of URI at the 0.05 level based on the test statistics. The conclusion from the analyses of the separate tables is different from that of the combined table.

A problem with the use of the separate tables is that the analyses are based on the smaller sample sizes associated with each subtable, not on the sample size of the combined table. This makes it difficult to find the presence of small but consistent trends across tables. A method for eliminating this problem is discussed in the next section. However, before presenting the method, we should also consider a problem that can occur when subtables are combined.

Besides ignoring the passive smoke variable, a potential problem in using the combined table is that it can be misleading. For example, if the data are selected from a population that does not represent the target population, strange things can occur. Suppose that we want our results to apply to all children in Finland but that the children used in this study were sampled from those who had been hospitalized during the previous 12 months. If this were done, the population used in the study would not match the target population. Is that a problem? As we have said before, the decision on the generalizability of the results to the target population depends on substantive considerations, not on statistical ideas. Let us assume that the sample data are those in Table 10.9.

In both of the subtables, the city with the lesser pollution had the greater proportion of children with no URI during the past 12 months. If we ignore the passive smoke variable, the combined table is Table 10.10.

In the combined table, the city with the greater outdoor pollution now has the greater proportion of children with no URI during the past 12 months — 0.624 compared to 0.595 for the city with lesser pollution. This example points out that care must be exercised in combining tables when the population from which the sample was drawn is not

Table 10.9 Number of 6-year-old Finnish children by respiratory status and pollution level with a control for passive smoke in the home based on taking samples from a list of hospitalized children.[a]

Passive Smoke in the Home	City Polluted	Upper Respiratory Infection during Previous 12 Months		Total
		Some	None	
Yes	High	35	40	75
	Low	60	80	140
	Total	95	120	215
No	High	170	300	470
	Low	15	30	45
	Total	185	330	515

[a]Hypothetical data

Table 10.10 Number of children with occurrence of upper respiratory infection by pollution status of city ignoring the passive smoke variable.

City Polluted	Upper Respiratory Infection during Previous 12 Months		Total
	Some	None	
High	205	340	545
Low	75	110	185
Total	280	450	730

representative of the target population. This was clearly pointed out in an article by Berkson in 1946.

10.4.2 The Cochran-Mantel-Haenszel Test

Two biostatisticians, Nathan Mantel and William Haenszel, developed a method in 1959 for examining the relation between two categorical variables while controlling for another categorical variable (Mantel and Haenszel 1959). This method, similar to a method published by William Cochran in 1954, uses all the data in the combined table and produces one overall test statistic. The test is designed to detect the consistent effect of the independent variable on the dependent variable across the levels of the extraneous variable. Thus, this method should only be used when the estimated odds ratios in the subtables are similar to one another. One very attractive feature of this test is that it can be used with extremely small sample sizes. This test has also been generalized for application to three-way tables of size other than 2 by 2 by k (Landis, Heyman, and Koch 1978).

To facilitate the presentation of the test statistic, we shall use the following notation for the ith 2 by 2 contingency table, where i ranges from one to k, the number of levels of the extraneous variable. In our example, k is 2, since there are only two levels, presence or absence, of the passive smoke variable. The ith 2 by 2 table is shown here.

Polluted City	Upper Respiratory Infection		
	Some	None	Total
High	a_i	b_i	$a_i + b_i$
Low	c_i	d_i	$c_i + d_i$
Total	$a_i + c_i$	$b_i + d_i$	n_i

The test statistic is based on an overall comparison of the observed and expected in the (1,1) cell in each of the k subtables. As we saw earlier in this chapter, under the hypothesis of no association between the row and column variables, there is only one degree of freedom associated with the table. Hence, we key on only one cell in the table and the choice of which cell is arbitrary. A statistic that could be used to examine whether or not there is an association is

$$Z* = \frac{\sum_{i=1}^{k}(O_i - E_i)}{s.e.\left[\sum_{i=1}^{k}(O_i - E_i)\right]}$$

where O_i and E_i are the observed and expected values in the (1,1) cell in the ith subtable. This statistic is very similar to a standard normal variable where E_i is analogous to the hypothesized mean in the standard normal variable.

In terms of the entries in the ith table, E_i is defined as

$$E_i = \frac{(a_i + b_i)(a_i + c_i)}{n_i}$$

the product of the row total and the column total divided by the table's sample size. The observed (1,1) cell frequency, O_i, is a_i. V_i, with a variance of O_i minus E_i, can be shown to be

$$V_i = \frac{(a_i + b_i)(c_i + d_i)(a_i + c_i)(b_i + d_i)}{n_i^2(n_i - 1)}.$$

Because we are dealing with discrete variables, we should use the continuity correction with Z*. However, instead of using the continuity-corrected Z* statistic, we would prefer to use a chi-square statistic, since all the other tests associated with contingency tables use a chi-square statistic. This poses no problem, since the square of a standard normal variable follows a chi-square distribution with one degree of freedom. Thus, the statistic to be used to test the hypothesis of no association between air pollution and the occurrence of upper respiratory problems is the Cochran-Mantel-Haenszel chi-square statistic. Also called the Mantel-Haenszel statistic, it is defined by

$$X_{CMH}^2 = \frac{(|O - E| - 0.5)^2}{V}$$

where O, E, and V are defined as the sums of the O_i, the E_i and the V_i over the k subtables. If X_{CMH}^2 is greater than $\chi_{1,1-\alpha}^2$, we reject the hypothesis of no association between air pollution and the occurrence of upper respiratory infections. Otherwise we fail to reject the null hypothesis.

Example 10.9

Let us apply this method to the data in Table 10.8. Since the odds ratios in the two separate subtables were similar — 1.613 in homes with passive smoke and 1.480 in the other homes — we can use the X^2_{CMH} statistic. If the odds ratios had not been similar, the effect of the independent variable on the dependent variable is not consistent across the levels of the extraneous variable. Hence, it would not make sense to use the CMH statistic to test for a consistent effect of the independent variable when we already know that such an effect does not exist. Since the values of our odds ratios are similar, we can test the hypothesis of no association (no consistent effect) of air pollution with the occurrence of URI while controlling for passive smoke status, and we shall perform the test at the 0.05 significance level. From Table 10.8 we see that O_1 is 100 and O_2 is 128 and their sum is 228. The expected values are calculated to be

$$E_1 = \frac{120(224)}{284} = 94.65 \quad \text{and} \quad E_2 = \frac{190(294)}{475} = 117.60$$

and their sum is 212.25. The variances are calculated to be

$$V_1 = \frac{120(164)(60)(224)}{284^2(283)} = 11.59 \quad \text{and} \quad V_2 = \frac{190(285)(181)(294)}{475^2(474)} = 26.94$$

and their sum is 38.53. Thus we have the pieces needed to calculate X^2_{CMH} and its value is

$$X^2_{CMH} = \frac{(|228 - 212.26| - 0.5)^2}{38.53} = 6.036.$$

Since 6.036 is greater than 3.84 ($= \chi^2_{1,0.95}$), we reject the hypothesis of no association. At the 0.05 level, we conclude that there is an association between air pollution and URI even after controlling for passive smoke in the home.

10.4.3   The Mantel-Haenszel Common Odds Ratio

Mantel and Haenszel also showed how to combine the data from the separate subtables to form a common odds ratio for the data. Again this should only be done when the estimated odds ratios in the subtables are similar. If the estimated odds ratios for the subtables are not similar — for example, some are less than one and some are greater than one — the common odds ratio would not be very useful. The relation between the independent and dependent variable would depend on the level of the extraneous variable, and the use of a common odds ratio would mask this. The Mantel-Haenszel estimator of the common odds ratio, θ, is

$$OR_{MH} = \frac{\sum_{i=1}^{k}[a_i(d_i/n_i)]}{\sum_{i=1}^{k}[b_i(c_i/n_i)]}.$$

There are several approaches to finding an estimate of the variance of the Mantel-Haenszel estimator of the common odds ratio (Letters 1993; Mehta and Walsh 1992), but they are quite involved and will not be presented here.

Example 10.10

For the air pollution data in Table 10.8, the Mantel-Haenszel estimate of common odds ratio is found from

$$OR_{MH} = \frac{[100(40/284)]+[128(119/475)]}{[20(124/284)]+[62(166/475)]} = 1.517.$$

This value is similar to the individual odds ratios of 1.613 and 1.480 and also close to the value, 1.524, found from the overall table. The similarity of the values supports the finding that the passive smoke variable had little effect on the relation between air pollution and URI.

Program Note 10.4 on the website shows the commands used to perform the calculation shown in Examples 10.7 and 10.8. The standard error and confidence intervals for the common odds ratio are also provided.

Conclusion

In this chapter, we introduced another nonparametric test — the chi-square goodness-of-fit test — and showed its use with one- and two-way contingency tables. We also showed two related methods — comparison of two binomial proportions and the calculation of the odds ratio — for determining, at some significance level, whether or not there is a relation between two discrete variables with two levels each. The odds ratio is of particular interest as it is used extensively in epidemiologic research. We also presented the extension of the goodness-of-fit test for no interaction to r by c contingency tables. Another test shown was the trend test, and it is of interest because it has a greater chance of detecting a linear relationship between a nominal and an ordinal variable than does the general chi-square test for no interaction. We also showed different ways for testing the hypothesis of no relationship between two discrete variables with two levels each in the matched-pairs situation. The Cochran-Mantel-Haenszel test and estimate of the common odds ratio were introduced for multiple 2 by 2 contingency tables. These procedures are also used extensively by epidemiologists. In the next chapter, we conclude the material on nonparametric procedures with the presentation of several nonparametric methods for the analysis of survival data.

EXERCISES

10.1 The following data are from one of the hospitals that participated in a study performed by the Veterans Administration Cooperative Duodenal Ulcer Study Group (Grizzle, Starmer, and Koch 1969). The data from 148 men show the severity of an undesirable side effect, the dumping syndrome, of surgery for duodenal ulcer for four surgical procedures. The procedures are the following:

A is drainage and vagotomy; B is 25 percent resection (antrectomy) and vagotomy; C is 50 percent resection (hemigastrectomy) and vagotomy; and D is 75 percent resection.

	Severity of Dumping Syndrome			
Surgery	None	Slight	Moderate	Total
A	23	7	2	32
B	23	10	5	38
C	20	13	5	38
D	24	10	6	40
Total	90	40	18	148

Was the design used in this hospital a completely randomized design or a randomized block design? Explain your answer. Test the hypothesis of no association between the type of surgery and the severity of the side effect at the 0.05 significance level. Assuming that the procedures are equally effective, would you recommend any of the procedures over the others?

10.2 Test the hypothesis that the data from Gosset, shown in Table 5.4 and repeated here, come from a Poisson distribution at the 0.01 significance level (Poisson probabilities are shown in Table 5.4).

Observed Frequency of Yeast Cells in 400 Squares								
X	0	1	2	3	4	5	6	Total
Frequency	103	143	98	42	8	4	2	400

10.3 The following data, from an article by Cochran (1954), show the clinical change by degree of infiltration — a measure of a type of skin damage — present at the beginning of the study for 196 leprosy patients who received 48 weeks of treatment.

Degree of Infiltration	Improvement					
	Worse	Same	Slight	Moderate	Marked	Total
0–7	11	27	42	53	11	144
8–15	7	15	16	13	1	52
Total	18	42	58	66	12	196

Test the hypothesis of no association between the degree of infiltration and the clinical change at the 0.05 significance level. Is this a test of independence or homogeneity? Explain your answer. Now assign scores from −1 to +3 for the clinical change categories worse to marked improvement and test the hypothesis of no linear trend at the 0.05 significance level. Is there any difference in the results of the tests? Select another reasonable set of scores and perform the trend test again using the second set of scores. Is the result consistent with the result from the first set of scores?

10.4 Mantel (1963) provided data from a study to determine whether or not there is any difference in the effectiveness of immediately injecting or waiting 90 minutes before injecting penicillin in rabbits who have been given a lethal injection. An extraneous variable is the level of penicillin. The data are shown in the following table.

Penicillin Level	Delay	Response		Total
		Cured	Died	
1/8	None	0	6	6
	90 minutes	0	5	5
1/4	None	3	3	6
	90 minutes	0	6	6
1/2	None	6	0	6
	90 minutes	2	4	6
1	None	5	1	6
	90 minutes	6	0	6
4	None	2	0	2
	90 minutes	5	0	5

Is it appropriate to use the CMH statistic here to test the hypothesis of no association between the delay and response variables while controlling for the penicillin level? Explain your answer. If you feel that it is appropriate to use the CMH statistic here, test, at the 0.01 significance level, the hypothesis of no association between the delay and response variables while controlling for the penicillin level.

10.5 Your local health department conducts a course on food handling. To evaluate this course, you select an SRS of restaurants from the list of licensed restaurants. For these restaurants in your sample, you then search the list of violations found by the health department during the last two years as well as the list of restaurants with employees who have attended the course during the last two years. For the 86 sampled restaurants, the data can be presented as follows:

Attended Course	Violation		Total
	Yes	No	
Yes	9	28	37
No	36	13	49
Total	45	41	86

Use an appropriate procedure to test the hypothesis of no association between course attendance and the occurrence of a violation at the 0.10 significance level.

Based on these data, discuss whether or not course attendance had any effect on the finding of a restaurant's violation of the health code.

10.6 Cochran (1954) presented data on erythroblastosis foetalis, a sometimes fatal disease in newborn infants. The disease is caused by the presence of an anti-*Rh* antibody in the blood of an *Rh*+ baby. One treatment used for this disease is the transfusion of blood that is free of the anti-*Rh* antibody. In 179 cases in which this treatment was used in a Boston hospital, there were no infant deaths out of 42 cases when the donor was female compared to 27 deaths when the donor was male. One possible explanation for this surprising finding was that the male donors were used in the more severe cases. Therefore, the disease severity was taken into account and the data are shown in the following table:

Disease Severity	Donor's Sex	Survival Status		Total
		Dead	Alive	
None	M	2	21	23
	F	0	10	10
Mild	M	2	40	42
	F	0	18	18
Moderate	M	6	33	39
	F	0	10	10
Severe	M	17	16	33
	F	0	4	4
Total		27	152	179

Use the CMH statistic to test the hypothesis of no association between donor's sex and the survival status of the infant at the 0.05 significance level.

10.7 Group the blood pressure values shown in Table 10.2 into categories of <80, 80–89, 90–99, 100–109, 110–119, 120–129, ≥130 mmHg. Based on this grouping, test the hypothesis that the systolic blood pressure of 12-year-old boys follows a normal distribution using the 0.05 significance level. Compare your results to those based on the grouping shown in Table 10.3.

10.8 The following data show the relation between two types of media exposure and a person's knowledge of cancer (Forthofer and Lehnen 1981).

Media Exposure		Knowledge of Cancer	
Newspapers	Radio	Good	Poor
Read	Listen	168	138
	Do not listen	310	357
Do not read	Listen	34	72
	Do not listen	156	494
Total		668	1061

Based on these data, test the hypothesis of no association between newspapers and knowledge of cancer, ignoring the radio variable. Next, test the hypothesis of no association between radio and knowledge of cancer, ignoring the newspaper variable. Which variable has the stronger association with the knowledge of cancer variable? Based on these data, would you feel comfortable recommending one of these media over the other for the purpose of increasing the public's knowledge of cancer? If your answer is yes, what assumptions are you making about the data? If your answer is no, provide your rationale for your answer.

10.9 Two pathologists each examined coded material from the same 100 tumors and classified the material as malignant or benign. Pathologist A found that 18 are malignant, and pathologist B found 10 malignant cases. Both pathologists agreed on 8 cases as malignant. The investigator conducting the study is interested in determining the extent to which the pathologists differ in their assessments of the tumor material. Form an appropriate 2 by 2 table and test the null hypothesis of no difference at the 0.05 significance level.

REFERENCES

Agresti, A. *Categorical Data Analysis.* John Wiley & Sons, Inc., 1990, p. 54–55.

Berkanovic, E., and S. J. Reeder "Awareness, Opinion and Behavioral Intention of Urban Women Regarding Mammography." *American Journal of Public Health* 69:1172–1174, 1979.

Berkson, J. "Limitations of the Application of Fourfold Table Analysis to Hospital Data." *Biometrics Bulletin* (now *Biometrics*) 2:47–53, 1946.

Bishop, Y. M. M., S. E. Fienberg, and P. W. Holland. *Discrete Multivariate Analysis: Theory and Practice.* Boston, MA: The MIT Press, 1975, p. 328.

Cochran, W. G. "The χ^2 Test of Goodness of Fit." *Annals of Mathematical Statistics* 23:315–345, 1952.

——— "Some Methods for Strengthening the Common χ^2 Tests." *Biometrics* 10:417–451, 1954.

Fisher, R. A. "The Logic of Inductive Inference (with Discussion)." *Journal of the Royal Statistical Society* 98:39–82, 1935.

Forster, D. P., A. J. Newens, D. W. K. Kay, and J. A. Edwardson. "Risk Factors in Clinically Diagnosed Presenile Dementral of the Alzheimer Type: A Case-Control Study in Northern England." *Journal of Epidemiology and Community Health* 49: 253–258, 1995.

Forthofer, R. N., and R. G. Lehnen. *Public Program Analysis: A New Categorical Data Approach.* Belmont, CA. Lifetime Learning Publications, 1981, p. 36.

Goodman, L. A., and W. H. Kruskal. "Measures of Association for Cross-Classifications, II: Further Discussion and References." *Journal of the American Statistical Association* 54:123–163, 1959, p. 131.

Grizzle, J. E., C. F. Starmer, and G. G. Koch. "Analysis of Categorical Data for Linear Models." *Biometrics* 25:489–504, 1969.

Jaakkola, J. J. K., M. Paunio, M. Virtanen, and O. P. Heinonen. "Low-Level Air Pollution and Upper Respiratory Infections in Children." *American Journal of Public Health* 81:1060–1063, 1991.

Landis, J. R., E. R. Heyman, and G. G. Koch. "Average Partial Association in Three-Way Contingency Tables: A Review and Discussion of Alternative Tests." *International Statistical Review* 46:237–254, 1978.

Letters to the Editor. *The American Statistician* 47:86–87, 1993.

Life Sciences Research Office, Federation of American Societies for Experimental Biology: *Nutrition Monitoring in the United States — An Update Report on Nutrition Monitoring.* Prepared for the U.S. Department of Agriculture and the U.S. Department of Health and Human Services. DHHS Pub. No. (PHS) 89–1255, 1989, Figure 6–13.

Mantel, N. "Chi-Square Tests with One Degree of Freedom; Extensions of the Mantel-Haenszel Procedure." *Journal of the American Statistical Association* 58:690–700, 1963.

Mantel, N., and W. Haenszel. "Statistical Aspects of the Analysis of Data from Retrospective Studies of Disease." *Journal of the National Cancer Institute* 22:719–748, 1959.

McNemar, Q. "Note on the Sampling Error of the Difference Between Correlated Proportions or Percentages." *Psychometrika* 12:153–157, 1947.

Mehta, C. R., and N. R. Patel. "A Network Algorithm for Performing Fisher's Exact Test in r × c Contingency Tables." *Journal of the American Statistical Association* 78:427–434, 1983.

Mehta, C. R., and S. J. Walsh. "Comparison of Exact, Mid-p, and Mantel-Haenszel Confidence Intervals for the Common Odds Ratio Across Several 2 × 2 Contingency Tables." *The American Statistician* 46:146–150, 1992.

Semenya, K. A., G. G. Koch, M. E. Stokes, and R. N. Forthofer. "Linear Models Methods for Some Rank Function Analyses of Ordinal Categorical Data." *Communications in Statistics* 12:1277–1298, 1983.

Snyder, L. H. "Heredity." In *Collier's Encyclopedia* 12:68–76, 1970.

Yarnold, J. K. "The Minimum Expectation in χ^2 Goodness of Fit Tests and the Accuracy of Approximations for the Null Distribution." *Journal of the American Statistical Association* 65:864–886, 1970.

Yates, F. "Tests of Significance for 2×2 Contingency Tables (with Discussion)." *Journal of the Royal Statistical Society A* 147:426–463, 1984.

Analysis of Survival Data

Chapter Outline

This chapter introduces methods for analyzing data collected from a longitudinal study in which a group of subjects are followed for a defined time period or until some specified event occurs. We frequently encounter such data in the health field — for example, newly diagnosed cancer patients in a registry were followed annually until they died. Another example consists of smokers who completed a smoking cessation program and were then contacted every three months to find out whether or not they had relapsed. The focus in these studies is the length of time from a meaningful starting point until the time at which either some well-defined event happens, such as death or relapse to a certain condition, or the study ends. The data from such studies are called *survival data*. We have previously encountered survival data in our consideration of the life table in Chapter 4. In this chapter, we will consider a special type of life table: the follow-up life table.

We first discuss the collection and organization of the data. This discussion is followed by the presentation of two related methods for analyzing survival data. The life-table method is used for larger data sets, and the product-limit method is generally used for smaller data sets. We also show how the CMH test statistic from Chapter 10 can be used for comparing two survival distributions.

11.1 Data Collection in Follow-up Studies

Perhaps an example best illustrates the nature of the data required for a survival analysis.

Example 11.1

The California Tumor Registry (1963) identified a total of 2711 females with ovarian cancer initially diagnosed between 1942 and 1956 in 37 hospitals in California. The follow-up system of the Central Registry was designed to identify deaths through the statewide vital registration system and to facilitate the follow-up activities of the participating hospital registries. The Central Registry received yearly follow-up information on each case. The registry program served not only to furnish the information essential for statistical study of cancer cases, but also to stimulate periodic medical checkups of the cancer patients. Based on the data accumulated in the Central Registry up to 1957, the researchers were able to analyze ovarian cancer patients who had been followed for up to 17 years.

In this data set, patients were observed for different lengths of time and not all of the patients had died by 1957. In addition, others could not be contacted — that is, they were lost to follow-up. Despite the different lengths of observation and the incomplete observations, it is possible to analyze the survival experience of these patients. An appropriate survival analysis is not restricted to those who had died, but it incorporates all the patients who entered the study. It is essential to include all those who entered the study because the exclusion of any patient from the analysis could introduce a selection bias as well as reducing the sample size.

The survival time cannot be calculated for those patients who were still alive at the closing date of the study or for those patients whose survival status was unknown. For these incomplete observations, the survival time is said to be *censored*. Those patients who were still alive at the closing date are known as *withdrawn alive*, and those patients whose status could not be assessed (because, for example, they moved away or refused to participate) are known as *lost-to-follow-up*.

To include the censored observations in the analysis, we calculate a censored survival time from the date of diagnosis to (1) the closing date of the study for those withdrawn alive and (2) the last known date of observation for the lost-to-follow-up. This allows the number of years from the date of diagnosis to the date of death or to the termination date to be calculated for each patient in the study.

By tabulating the uncensored and censored survival times of all 2711 female ovarian cancer patients by one-year intervals, we obtain the data shown in Table 11.1. Within the first year after diagnosis, 1421 of 2711 patients had died and 68 were lost-to-follow-up. There were no patients in the category withdrawn alive since every patient was followed for at least one year. The last column of the table can be created by adding the total column entries from the bottom. This reverse cumulative total indicates the number of patients alive at the beginning of each interval. The entry in the first row of this column is the total number of patients in the study. The other entries in this last column can also be found by subtracting the sum of the number of deaths, lost-to-follow-up, and withdrawn alive from the number of persons who started the previous interval. For example, the second entry in this column is 1222, which is determined by subtracting the sum of 1421, 68, and 0 from 2711, the number of subjects who began the previous interval.

The essential data items required for a survival analysis include d_i, the number of deaths; l_i, the number of patients lost-to-follow-up; w_i, the number of patients withdrawn alive; and n_i, the number of patients alive at the beginning of the ith interval. These data, presented in Table 11.1, are analyzed by the life-table method presented in the next section.

Table 11.1   Survival times for ovarian cancer patients initially diagnosed 1942–1956, followed to 1957.

Years after Diagnosis	Death d_i	Censored Lost l_i	Censored Withdrawn w_i	Total	Number Entering Interval n_i
0–1	1,421	68	0	1,489	2,711
1–2	335	19	37	391	1,222
2–3	132	17	84	233	831
3–4	64	10	47	121	598
4–5	44	12	48	104	477
5–6	20	12	39	71	373
6–7	19	10	35	64	302
7–8	14	14	19	47	238
8–9	7	10	25	42	191
9–10	7	9	19	35	149
10–11	5	4	14	23	114
11–12	5	4	17	26	91
12–13	1	4	11	16	65
13–14	3	1	15	19	49
14–15	1	0	13	14	30
15–16	0	0	7	7	16
16–17	0	0	9	9	9

Source: California Tumor Registry, 1963

11.2   The Life-Table Method

In Chapter 4, the population life table was introduced to illustrate the idea of probability and its connection to life expectancy. The estimated life expectancy is generally used as a descriptive statistic. To use the life-table technique as an analytical tool, we shall combine ideas from Chapter 5 on probability distributions with the life-table analysis framework.

In survival analysis, our focus is on the length of survival. Let X be a continuous random variable representing survival time. Consider a new function, the *survival function*, defined in symbols as

$$S(x) = \Pr(X > x).$$

This function is the probability that a subject survives beyond time x. Since $F(x)$, the cdf, is defined as

$$F(x) = \Pr(X \leq x)$$

the survival function is one minus the cdf — that is,

$$S(x) = 1 - F(x).$$

It is more convenient to work with $S(x)$ rather than $F(x)$, since we usually talk about survival being greater than some value rather than being less than a value.

The idea of a survival function is contained in the population life table presented in Chapter 4. It is represented by the l_x column, the number of survivors at the beginning of each age interval. Specifically, $S(x)$ in the population life table is l_x/l_0. Recall that the l_x column starts with l_0, usually set at 100,000, and all subsequent l_x values are derived by multiplying the conditional probability of surviving in an age interval by the number of those who have survived all previous age intervals.

To analyze the data in Table 11.1 by the life-table method, we shall estimate the survival distribution in the same manner. The results of these calculations are shown in Table 11.2. The first two columns (the time interval and the number of deaths) are transferred from Table 11.1. The other columns show the results of the life-table analysis.

Table 11.2 Estimates of probabilities and standard errors for ovarian cancer patients.

(1) Years after Diagnosis	(2) Deaths d_i	(3) Exposed to Risk n_i'	Conditional Probability		(6) Cumulative Probability Surviving P_i	(7) Standard Error $SE(P_i)$
			(4) Dying q_i	(5) Surviving $(1 - q_i)$		
0–1	1,421	2,677.0	0.531	0.469	1.000	0.0000
1–2	335	1,194.0	0.281	0.719	0.469	0.0096
2–3	132	780.5	0.169	0.831	0.338	0.0092
3–4	64	569.5	0.112	0.888	0.280	0.0089
4–5	44	447.0	0.098	0.902	0.249	0.0087
5–6	20	347.5	0.058	0.942	0.224	0.0086
6–7	19	279.5	0.068	0.932	0.212	0.0086
7–8	14	221.5	0.063	0.937	0.197	0.0086
8–9	7	173.5	0.040	0.960	0.185	0.0087
9–10	7	135.0	0.052	0.948	0.177	0.0088
10–11	5	105.0	0.048	0.952	0.168	0.0090
11–12	5	80.5	0.062	0.938	0.160	0.0093
12–13	1	57.5	0.017	0.983	0.150	0.0097
13–14	3	41.0	0.073	0.927	0.147	0.0099
14–15	1	23.5	0.043	0.957	0.137	0.0109
15–16	0	12.5	0.000	1.000	0.131	0.0119
16–17	0	4.5	0.000	1.000	0.131	0.0119

The first task is to estimate the conditional probability of dying for each interval of observation. When there is no censoring in an interval, the estimate of the probability of dying in the interval is simply the ratio of the number who died during the interval to the number alive at the beginning of the interval. However, it is not appropriate to use this ratio as the estimator of the probability of dying if censoring occurred in the interval. The use of this denominator, the number alive at the beginning of the interval, means that those who were lost-to-follow-up or withdrawn alive during the interval are treated as if they survived the entire interval. Thus, using this ratio when there is censoring likely results in an underestimate of the probability of dying in the interval.

The problem with the censored individuals is that we do not know their actual length of survival during the interval. We do know that it is extremely unlikely that they all survived the entire interval. The assumption used most often in practice (although there are other more reasonable assumptions) is that the censored individuals survived to the midpoint of the interval. Under this assumption, we can calculate q_i, an estimator of the conditional probability of dying during the ith interval, as follows:

$$q_i = \frac{d_i}{n_i - \dfrac{l_i + w_i}{2}} = \frac{d_i}{n_i'}.$$

The denominator in the above equation is the effective number of subjects exposed to the risk of dying during the interval, denoted by n'_i. Table 11.2 shows the estimated effective number of patients exposed to the risk of dying in column 3 and the estimate of the conditional probability of dying in column 4. The use of n'_i implies that those patients who were lost or withdrawn were subjected to one half the risk of dying during the interval.

The estimator of the conditional probability of survival in the ith interval is one minus the estimator of the probability of dying, that is, $1 - q_i$. The result of this subtraction is shown in column 5.

Next, we calculate P_i, the sample estimator of the probability of surviving until the beginning of the ith interval. The set of the P_i are used to estimate the survival distribution $S(x)$. By definition, $P_1 = 1$, and the estimators of the other survival probabilities are calculated in the following manner:

$$P_2 = (1 - q_1), \ P_3 = (1 - q_2) \ (1 - q_1), \ \ldots$$

and in general

$$P_i = (1 - q_{i-1})(1 - q_{i-2}) \ldots (1 - q_1) = (1 - q_{i-1}) \ P_{i-1}.$$

The results of these products are shown in column 6 of Table 11.2. From column 6, we see that the estimate of the one-year survival probability for ovarian cancer patients in California who were diagnosed during the 1942–1956 period was 0.47 and the estimate of the five-year survival probability was 0.22. More recent statistics estimate the five-year survival probability for ovarian cancer to be 0.39 for white females and 0.38 for black females in the 1981–1986 period (National Cancer Institute 1990), suggesting some improvement in cancer treatment. However, this improvement may be due more to the early detection of ovarian cancer in recent years. Cancer-related statistics, including estimates of survival rates, are routinely provided by the National Cancer Institute's SEER (Surveillance, Epidemiology, and End Results) program, which includes many population-based cancer registries throughout the United States.

Besides knowing the point estimate of a population survival probability, we also wish to have a confidence interval for the survival probability. We shall assume that, in large samples, an estimated cumulative survival probability approximately follows a normal distribution. The variance of the estimated cumulative survival probability is estimated by

$$\hat{V}ar(P_i) = P_i^2 \sum_{j=1}^{i-1} \frac{q_i}{n'_j(1 - q_j)}.$$

The estimated standard errors (the square root of the estimated variance) of the P_i are shown in column 7 of Table 11.2.

Given these estimated standard errors plus the assumption of the approximate normality of the estimated survival probabilities, we can calculate confidence intervals for the survival probabilities. The approximate $(1 - \alpha)*100$ percent confidence interval for a survival probability is given by

$$P_i \pm (z_{1-\alpha/2}) \ s.e.(P_i)].$$

For example, an approximate 95 percent confidence interval for the five-year survival probability is

$$0.224 - 1.96\,(0.0086) \text{ to } 0.224 + 1.96\,(0.0086)$$

or from 0.207 to 0.241.

Although this procedure is adequate in most cases, there are other more complicated approaches to constructing a confidence interval for P_i that cause the actual confidence level to agree more closely with the nominal confidence level, especially for small samples (Thomas and Grunkemeier 1975).

It is also possible to calculate the confidence interval for the difference between two survival probabilities from different study groups — for example, the five-year survival probability of ovarian cancer for white females and black females — by using the procedure discussed in Chapter 7.

Let us further explore the estimated survival distribution by creating Figure 11.1, the plot of the cumulative survival probabilities against the years after diagnosis. Although we have values of P_i for only the integer values of t, we have connected the points to show the shape of the survival distribution. It starts with survival probability of 1 at time 0 and drops quickly as time progresses, indicating a very high early mortality for ovarian cancer patients. Note that the survival curve does not descend all the way to zero, since some women survive more than 17 years.

The rapid decrease in the estimated survival curve suggests that the mean and the median survival times will be short. To verify this, let us estimate the mean and the median survival times from the estimated survival distribution. Since some of the women survive longer than the 17 years of the study, this complicates the estimation of the population mean survival time. Instead of estimating the population mean, we shall therefore estimate the mean restricted to the time frame of 17 years, the length of the study. This restricted value will thus underestimate the true unrestricted mean. If no patient survived longer than the time frame of the study, the following procedure provides an estimate of the unrestricted mean.

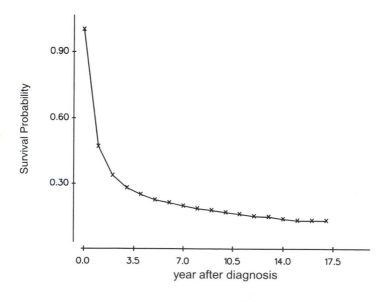

Figure 11.1 Estimated survival distribution of ovarian cancer patients.

The sample mean, restricted to the 17-year time frame, is found by summing the number of years (or other unit of time) survived during each time interval and dividing this sum by the sample size. However, the process of determining the number of years survived in an interval is complicated by the deaths, losses, and withdrawals that occurred during the interval. Instead of directly attempting to calculate the years survived, we shall use the following method to deal with this complication.

We calculate the sample mean by forming a weighted average of the years provided by each interval. The weight used with each interval is the cumulative survival probability associated with the interval. This approach deals with the complications mentioned above, since the probability takes the deaths, losses, and withdrawals into account. Since there are two cumulative survival probabilities associated with each interval — the probability at the beginning, P_i, and the probability at the end, P_{i+1} — we use their average. Thus, the formula for the restricted sample mean is

$$\bar{x}_r = \sum_{i=1}^{k} a_i \left(\frac{P_i + P_{i+1}}{2} \right)$$

where k is the number of intervals and a_i is the width of the ith interval.

This formula has an interesting geometrical interpretation: It provides an approximation to the area under the estimated survival curve. For example, consider a curve with three intervals (Figure 11.2). We are using rectangles to estimate the area under the curve. As we can see, some of the area under the curve is not included in the rectangles. However, this area is approximately offset by the areas included in the rectangles that are not under the curve. The formula for the area of a rectangle is the height multiplied by the width. In this case, the width is one unit or, in general, a_i units, and the height is taken to be the average of the points at the beginning and end of the interval, that is, $(P_i + P_{i+1})/2$. Hence, another way of interpreting the mean is that it is the area under the survival curve. We approximate this area by calculating the area of the rectangles that can be superimposed on the survival curve.

When the intervals are all of the same width — for example, a — then the formula can be simplified to

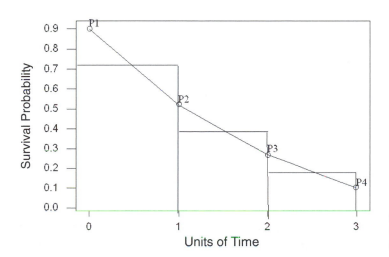

Figure 11.2 Survival curve with rectangles superimposed.

$$\bar{x}_r = a_i\left(\sum_{i=1}^{k+1} P_i - \frac{P_1 + P_{k+1}}{2}\right).$$

Because the intervals are all of width one in this example, the sample mean is simply the sum of the entries in column 6 of Table 11.2 minus one half of the first and last entries in the column. This is

$$(1.000 + 0.469 + 0.338 + \ldots + 0.131) - 0.5(1.000 + 0.131)$$

which equals 3.92 years. This restricted mean survival time appears to be larger than what the first-year survival probability might suggest. As we saw in Chapter 3, the mean can be affected by a few large observations, and that is the case here. The sample mean reflects the presence of a few long-term survivors. Let us now calculate the median length of survival.

The median survival time is estimated in the following manner. First, we read down the list of estimated cumulative survival probabilities, column 6 in Table 11.2, until we find the interval for which P_i is greater than or equal to 0.5 and P_{i+1} is less than 0.5. In Table 11.2, this is the first interval, since P_1 is greater than 0.5 and P_2 is less than 0.5. Thus, we know that the estimated median survival time is between 0 and 1 year. Since 47 percent of the patients survived the first year, we suspect that the estimated median survival time is much closer to one year than to zero years. To find a more precise value, we shall use linear interpolation.

In using linear interpolation, we are assuming that the deaths occurred at a constant rate throughout the interval. This is the same assumption we made when we connected the survival probabilities in Figure 11.1. In using linear interpolation, we know that to reach the median, we only require a portion of the interval, not the entire interval. The portion that we need is simply the ratio of the difference of P_i and 0.5 to the length of the interval. In symbols, this is

$$(P_i - 0.5) / (P_i - P_{i+1}).$$

We multiply this ratio by the width of the interval and add that to the survival time at the beginning of the interval. Replacing these words by symbols, the formula is

$$Sample\ median = x_i + a_i\left(\frac{P_i - 0.5}{P_i - P_{i+1}}\right)$$

where x_i is the survival time at beginning of the interval and a_i is the width of the interval. In this example, the sample median survival time is

$$Sample\ median = 0 + 1\left(\frac{1 - 0.5}{1 - 0.469}\right) = 0.94.$$

The sample median survival time of about one year is much shorter than the estimated restricted mean survival time. As we just mentioned, the mean survival time is affected by a small number of long-term survivors. This is why the median is more often used with survival data.

The median can also be obtained from the plot of the estimated survival curve shown in Figure 11.1. We move up the vertical axis until we reach the survival probability value of 0.5. We then draw a line parallel to the time axis and mark where it intersects the

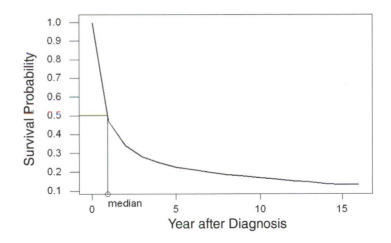

Figure 11.3 Using the estimated survival curve to find the median.

survival curve. We next draw a line, parallel to the vertical axis, from the intersection point to the time line. The sample median survival time is the value where the line intersects the time axis. Figure 11.3 shows the estimated survival curve plot with these lines used to find the sample median drawn in the plot as well. The accuracy of the estimate of the median is limited by the scales used in plotting the survival curve. In Figure 11.3, the precision of the estimate is likely not to be high because of the scales used. It appears that the sample estimate of the median is approximately one year.

Another statistic often used in survival analysis is the *hazard rate*, which is also known as the life-table mortality rate, force of mortality, and instantaneous failure rate. It is used to measure the proneness to failure during a very short time interval. It is analogous to an age-specific death rate or interval-specific failure rate. It is the proportion of subjects dying or failing in an interval per unit of time. The hazard rate is usually estimated by the following formula:

$$h_i = \frac{d_i}{a_i(n_i' - d_i/2)} = \frac{2q_i}{a_i(2 - q_i)}.$$

The denominator of this formula uses the number of survivors — again assuming that death is occurring at a constant rate throughout the interval — at the midpoint of the interval. When the interval is very short, it makes little difference whether the number of survivors at the beginning or at the midpoint of the interval is used in the denominator. The sample hazard rates are calculated and shown in Table 11.3 for the first 10 years of follow-up. The estimate of the first year hazard or mortality rate is quite high with 723 deaths per 1000 patients. The hazard is concentrated in the first five years after diagnosis and stabilizes at a low level after five years of survival. The variance of the sample hazard rate is estimated by

Table 11.3 Estimates of hazard rates and their standard errors.

Year	Hazard Rate	Standard Error	Year	Hazard Rate	Standard Error
0–1	0.723	0.0179	5–6	0.059	0.0132
1–2	0.326	0.0176	6–7	0.070	0.0161
2–3	0.185	0.0160	7–8	0.065	0.0174
3–4	0.119	0.0149	8–9	0.041	0.0156
4–5	0.104	0.0156	9–10	0.053	0.0201

$$\hat{Var}(h_i) = h_i^2 \left(\frac{1-(h_i a_i/2)^2}{n_i' q_i} \right).$$

The estimated standard errors (the square root of the estimated variance) of the sample hazard rates are calculated and shown in Table 11.3. If we assume that the sample hazard rates are asymptotically normally distributed, these sample standard errors can be used to calculate confidence intervals for the population hazard rates. For example, the 95 percent confidence interval for the first year hazard or mortality rate ranges from

$$0.723 - 1.96(0.0179) \text{ to } 0.723 + 1.96(0.0179)$$

or from 0.688 to 0.758.

These life table calculations can be performed by the computer (see **Program Note 11.1** on the website).

11.3 The Product-Limit Method

When we analyze a smaller data set — for example, a sample size less than 100 — the life-table method may not work very well because the grouping of survival times becomes problematic. Instead we use a method that is based on the actual survival time for each subject rather than grouping the subjects into intervals. The product-limit method, also known as the Kaplan-Meier method (Kaplan and Meier 1958), is used to estimate the cumulative survival probability from a small data set, without relying on groupings of survival times. As we can see following, the basic principles and computational procedures involved in the product-limit method are similar to the life-table method.

We start with an example.

Example 11.2

Suppose that 14 alcohol-dependent patients went through an intensive detoxification treatment for four years from 1990 to 1993 at a small clinic. There was a follow-up contact every month to check on their drinking status. The data shown in Table 11.4

Table 11.4 Status of 14 alcohol-dependent patients discharged from a clinic.

Patient Number	Date of Discharge	Date of Termination	Follow-up Status	Gender
1	9001	9312	2 Still sober (withdrawn)	1 Female
2	9003	9009	1 Relapsed	1 Female
3	9005	9209	1 Relapsed	2 Male
4	9009	9111	2 Lost-to-follow-up	2 Male
5	9011	9306	1 Relapsed	1 Female
6	9102	9312	2 Still sober (withdrawn)	1 Female
7	9104	9211	1 Relapsed	1 Female
8	9108	9304	1 Relapsed	1 Female
9	9110	9202	1 Relapsed	2 Male
10	9203	9308	2 Lost-to-follow-up	2 Male
11	9207	9311	1 Relapsed	2 Male
12	9212	9310	1 Relapsed	1 Female
13	9303	9312	2 Still sober (withdrawn)	2 Male
14	9304	9310	1 Relapsed	2 Male

were abstracted from the clinic patient records. The date of discharge and the date of termination are shown in year and month (9001 indicates 1990, January). The follow-up status is coded 2 if censored (withdrawn or lost-to-follow-up) and 1 if relapsed to drinking. Gender is coded 1 for females and 2 for males. The purpose of our study is to analyze the length of alcohol-free time among these 14 patients.

The first step of analysis is to calculate the survival time, x, in months for all subjects, censored and uncensored, and arrange them in order from the smallest to the largest with the censoring status indicated. If an uncensored subject and a censored subject have survival times of the same length, the uncensored one precedes the corresponding censored observation. For the data shown in Table 11.4, the ordered list of alcohol-free times in months, with the censored observations marked by asterisks, is as follows:

$$4, 6, 6, 9*, 10, 14*, 16, 17*, 19, 20, 28, 31, 34*, 47*.$$

The second step is to create a worksheet like that shown in Table 11.5. In Table 11.5, the column headings refer to death and survival. For this problem, death is equated with relapse and survival is remaining alcohol free. The first three columns in the worksheet are created according to the following procedures.

1. List the uncensored alcohol-free times in order. These are 4, 6, 10, 16, 19, 20, 28, and 31. We shall refer to these times as $x_1, x_2, \ldots, x_8$, respectively.
2. Count the number of relapses at each x_i. There is one relapse at each time unless there are ties. The numbers are 1, 2, 1, 1, 1, 1, 1, and 1.
3. Count the number of subjects who are at risk of relapse at x_i. For example, when the survival time is 10 months, three people have already relapsed and one person was withdrawn. Thus, there are only 10 persons at risk of relapse at 10 months. The list of these numbers is 14, 13, 10, 8, 6, 5, 4, and 3.

The fourth and fifth columns, estimates of the conditional probability of survival (1 − q_x) and the cumulative probability of survival (P_x) are calculated next, followed by the calculation of estimated standard error of P_x, shown in column 6. The estimator of the conditional probability of relapse is the number of relapses divided by the number at risk, that is, $q_x = d_x/n_x$. The estimator of the conditional probability of survival is

Table 11.5 Kaplan-Meier estimates of survival probabilities with standard errors.

(1) Survival Time x_i	(2) Number of Deaths d_x	(3) Number at Risk n_x	(4) Conditional Probability of Survival $(1 - q_x)$	(5) Cumulative Probability of Survival P_x	(6) Standard Error $SE(P_x)$	(7) Approx. Standard Error $SE(P_x)$
0	0	14	1.000	1.000	—	—
4	1	14	0.929	0.929	0.069	0.066
6	2	13	0.846	0.786	0.110	0.101
10	1	10	0.900	0.707	0.124	0.121
16	1	8	0.875	0.619	0.136	0.135
19	1	6	0.833	0.516	0.148	0.146
20	1	5	0.800	0.412	0.150	0.141
28	1	4	0.750	0.309	0.144	0.128
31	1	3	0.667	0.206	0.127	0.106

$$1 - q_x = 1 - \frac{d_x}{n_x} = \frac{n_x - d_x}{n_x}.$$

For example, $1 - q_4 = (1 - 1/14) = 0.929$ and $1 - q_6 = (1 - 2/13) = 0.846$. The estimator of the cumulative probability of survival is found from the estimators of the conditional probabilities of survival in the same way as in the life-table method — that is,

$$P_x = \prod_{t \le x}(1 - q_t) = \prod_{t \le x}\frac{n_t - d_t}{n_t}.$$

The product symbol, Π, means that we multiply each term in the expression by one another for the indicated values of t. For example,

$$P_6 = \prod_{t \le 6}(1 - q_t) = (1 - q_4)(1 - q_6) = (0.929)(0.846) = 0.786.$$

We could have included $1 - q_0$ in the product, but since q_0 is defined to be zero, its inclusion would not have changed the product.

As we have just seen, the censored observations have not been excluded from the analysis. They played a role in the determination of the number at risk at each time of relapse. If the censored observations were totally excluded from the analysis, the estimate of the conditional survival probabilities for the uncensored observations would be different.

The variance of P_x is estimated by

$$\hat{Var}(P_x) = P_x^2\left(\sum_{t \le x}\frac{d_t}{(n_t - d_t)n_t}\right) \cong P_x^2\left(\frac{1 - P_x}{n_x}\right).$$

The approximation shown in the preceding equation is much simpler to calculate, and it works reasonably well in most situations (Peto et al. 1977). Taking the square root of the variance, we obtain the estimated standard errors of the P_x that are shown in column 6. The approximate standard errors are shown in column 7. The approximate estimate of the standard error of P_4 is 0.066, compared to the value of 0.069 obtained from the use of the first expression for the sample variance.

Figure 11.4 graphically displays the estimated survival distribution shown in the fifth column of Table 11.5. The plot includes a survival probability of 1 at time 0. The plot of the survival probabilities is referred to as a step function, since it looks like a stair step. It has this appearance because the probability of survival stays the same over a time period — this causes the horizontal lines — and then drops whenever there is another relapse — the vertical lines. However, long horizontal lines, showing no change in survival probability for a long period of time, should not be interpreted as a period with no risk, for these may occur because of a small number of subjects under observation during those time periods.

We can estimate the mean survival time from the survival distribution. Again, just as in the life-table method, if the largest survival time is a censored time, we are really estimating a restricted mean. If the largest survival time is uncensored, then the survival probability will decrease to zero, and we will be estimating the unrestricted mean. As in the life table, the mean survival time is the area under the curve. We shall again use rect-

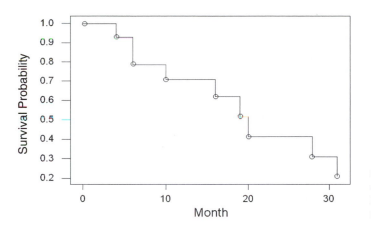

Figure 11.4 Survival distribution estimated by the product-limit method.

angles to approximate this area. Because of the step nature of the survival curve here, the rectangles are already formed for us. Unlike the life-table method, the widths of the intervals here are usually different. The following formula shows the area of each rectangle being calculated as the product of the height of the rectangle, the estimated cumulative survival probability associated with x_i, by the width, x_{i+1} minus x_i. In symbols, this is

$$\bar{x}_r = \sum_{i=0}^{k-1} P_{x_i}(x_{i+1} - x_i)$$

where k is the number of distinct time points when someone relapsed, x_0 is defined to be zero, and P_0 is defined to be one.

For these data, the estimate of the restricted mean alcohol-free time, restricted to a 31-month window, is given by

$$\bar{x}_r = 1(4 - 0) + 0.929(6 - 4) + \cdots + 0.309(31 - 28) = 18.4.$$

This is an underestimate of the true mean alcohol-free time because we are restricted to the study timeframe and there were still people free of alcohol at the end of the study.

From Table 11.5, we can see that the sample median survival time, the point at which the cumulative survival probability is 0.5, occurs between the 19th and 20th months and is closer to month 19. We shall interpolate to find the sample median in the same way as in the life-table method. From our data, the sample median survival time is found as follows:

$$Sample\ median = 19 + (20 - 19)\left(\frac{0.516 - 0.5}{0.516 - 0.412}\right) = 19.2\ months.$$

We should not use interpolation to find the median if there is a large gap in time between the two survival times in which we will be using the interpolation.

The computer can be used to calculate the entries in Table 11.5 as well as the sample mean, the median, and the graph in Figure 11.4 (see **Program Note 11.2** on the website).

Because the product-limit method is based on the ranking of individual survival times, it is cumbersome to apply with a large data set. We would not consider using it with the ovarian cancer data from the California Tumor Registry that had over 2000 observations. For a large data set, the life-table method simplifies the calculation and gives results similar to the product-limit method.

So far we have focused on describing the survival experience of a single population. However, we are often interested in comparing the survival experiences of two or more groups of subjects who differ on some account — for example, patients who have received different therapies for cancer or patients who belong to different age or sex groups. The comparison of two survival distributions is the topic of the following section.

11.4   Comparison of Two Survival Distributions

When comparing the survival experience of two or more groups, the description of the differences in the estimated survival distributions and the plot of the survival curves are only the beginning of the analysis. In addition to these descriptive techniques, researchers require a statistical test to determine whether the observed differences are statistically significant or due to chance variation.

In the analysis of survival data, we generally do not assume that the data follow any particular probability distribution. In the analysis, we also use the median survival time, rather than the mean, to summarize the survival experience. Because of these features, it seems as if a nonparametric test should be used when comparing survival distributions. If we know that the survival data follow a particular distribution, we should take advantage of that knowledge. There are parametric tests available that can be used when we know the probability distribution of the survival data (Lee 1992).

For small data sets with no censored observations, the Wilcoxon rank sum test (the Mann-Whitney test) can be used to test the null hypothesis of no difference in survival distributions for two independent samples. However, since survival time data usually contain censored observations, the Wilcoxon test cannot be directly applied. In this section, we show how the Cochran-Mantel-Haenszel (CMH) test statistic, described in Chapter 10, can be used in testing the hypothesis of no difference between two survival distributions (Mantel 1966). There are a number of other tests, extensions of the Wilcoxon and other rank tests, that could be used as well, but the CMH test seems to perform as well, if not better, than these other tests.

11.4.1   The CMH Test

The key to the use of the CMH method with survival data is to realize that the data in each time interval can be formulated as a 2 by 2 table. The number of deaths and the number of survivors (the number exposed minus the number of deaths) for the two groups can be put in a 2 by 2 table for each time interval as shown next.

	Number of Deaths	Number of Survivors	Total
Group 1	d_{1i}	$(n'_{1i} - d_{1i})$	n'_{1i}
Group 2	d_{2i}	$(n'_{2i} - d_{2i})$	n'_{2i}
Total	$d_{.i}$	$(n'_{.i} - d_{.i})$	$n'_{.i}$

It can be shown that the time intervals are uncorrelated with one another, which allows us to use the CMH statistic here.

Let us consider an example.

Example 11.3

The Hypertension Detection and Follow-up Program examined the effect of serum creatinine on eight-year mortality among hypertensive persons under care (Shulman et al. 1989). We are interested in testing whether or not the survival experience of persons with a serum creatinine concentration less than 1.7 mg/dL at the time of screening is more favorable than those with a serum creatinine concentration greater than or equal to 1.7 mg/dL. The data for testing this hypothesis are shown in Table 11.6.

Table 11.6 Sample sizes and numbers of deaths by year and level of serum creatinine concentration in the HDFP study.

| | Serum Creatinine (mg/dL) | | | | | |
| | <1.7 | | ≥1.7 | | Total | |
Year under Care	d_{1i}	n'_{1i}	d_{2i}	n'_{2i}	d_i	n'_i
0–1	93	10,469.5	21	297.0	114	10,766.5
1–2	115	10,374.5	16	276.0	131	10,650.5
2–3	125	10,254.0	13	260.0	138	10,514.0
3–4	181	10,121.5	14	246.5	195	10,368.0
4–5	160	9,930.5	17	232.0	177	10,162.5
5–6	212	9,763.0	10	215.0	222	9,978.0
6–7	191	9,551.0	14	205.0	205	9,756.0
7–8	203	9,147.5	8	186.5	211	9,334.0
Total	1,270		113		1,393	

First, we use the data in Example 11.3 to estimate the cumulative survival probabilities for the two groups, applying the methods discussed earlier. The estimated cumulative survival probabilities and their standard errors are shown in Table 11.7.

The estimated cumulative survival probabilities are also shown graphically in Figure 11.5. The survival distribution appears to be more favorable for the hypertensive persons

Table 11.7 Cumulative survival probabilities and standard errors by year and level of serum creatinine concentration in the HDFP study.

| | Creatinine Level <1.7 | | Creatinine Level ≥1.7 | |
Year under Care	Survival Probability	Standard Error	Survival Probability	Standard Error
0–1	1.0000	0	1.0000	0
1–2	0.9911	0.0009	0.9295	0.0148
2–3	0.9801	0.0014	0.8758	0.0191
3–4	0.9682	0.0017	0.8322	0.0216
4–5	0.9509	0.0021	0.7851	0.0238
5–6	0.9355	0.0024	0.7279	0.0258
6–7	0.9151	0.0027	0.6942	0.0267
7–8	0.8969	0.0028	0.6470	0.0277
8	0.8770	0.0032	0.6194	0.0282

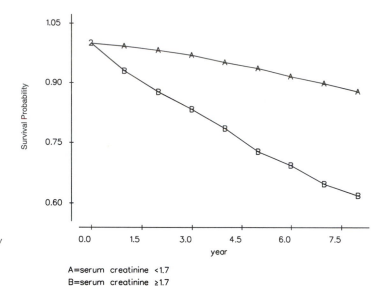

Figure 11.5 Estimated survival distributions by level of serum creatinine concentration.

A=serum creatinine <1.7
B=serum creatinine ≥1.7

with a serum creatinine concentration less than 1.7 mg/dL than those with a serum creatinine concentration greater than or equal to 1.7 mg/dL. The two survival curves are consistently diverging, suggesting that the odds ratios in each time interval are similar to one another. Therefore, we do not have any problem using the CMH test to compare the two survival distributions.

To apply this test to the data in Table 11.6, we need to find the expected number of deaths and the variance for the (1, 1) cell in each of the eight 2 by 2 tables. For example, the 2 by 2 table for the year 0–1 is shown next.

Creatinine Level	Number of Deaths	Number of Survivors	Total
<1.7 mg/dL	93	10,376.5	10,469.5
≥1.7 mg/dL	21	276.0	297.0
Total	114	10,652.5	10,766.5

The expected number of deaths in the (1, 1) cell is the product of the total of the first row and the first column divided by the table total. Thus, the expected value is

$$10469.5 \ (114) \ / \ 10766.5 = 110.86.$$

The estimated sample variance of the (1, 1) cell is the product of the four marginal totals divided by the square of the table total times the table total minus one. Thus, the sample variance is

$$\frac{10469.5(297)(114)(10652.5)}{10766.5^2(10766.5-1)} = 3.03.$$

Table 11.8 shows the expected number of deaths and the estimated variances for the eight (1, 1) cells based on the data in Table 11.6. The observed number of deaths in Group 1 (creatinine less than 1.7 mg/dL) is 1280 and the expected number of deaths is 1361, suggesting that Group 1 has a favorable survival experience. We shall test the hypothesis of no difference in the survival distributions of the two groups at the 0.01

Table 11.8 Expected values and variances of the (1, 1) cells.

Year under Care	Expected Value	Variance
0–1	110.86	3.03
1–2	127.61	3.27
2–3	134.59	3.28
3–4	190.36	4.44
4–5	172.96	3.88
5–6	217.22	4.58
6–7	200.69	4.13
7–8	206.78	4.04
Total	1,361.07	30.65

significance level. The test statistic, X^2_{CMH}, is calculated based on the data in Tables 11.6 and 11.8 as follows:

$$X^2_{CMH} = \frac{(|O - E| - 0.5)}{V} = \frac{(|1289 - 1361.07| - 0.5)^2}{30.65} = 211.80.$$

Since the test statistic is greater than 6.63 ($= \chi^2_{1,0.99}$), we reject the null hypothesis and conclude that persons with a serum creatinine concentration less than 1.7 mg/dL had a more favorable survival distribution than those with a higher creatinine value at the time of screening.

11.4.2   The Normal Distribution Approach

The individual survival probabilities of the two groups can be compared using the method discussed in Chapter 8. But this approach has the disadvantage that it focuses on a particular point in time and does not use all the information in the data set. For example, the two-year survival probability of the group with serum creatinine level less than 1.7 is 98 percent compared with 88 percent for the group with serum creatinine level greater than or equal to 1.7. Let us test whether these probabilities are significantly different at the 0.01 level. The test statistic for this comparison can be calculated from the data in Table 11.7 as follows:

$$z = \frac{p_1 - p_2}{\sqrt{[s.e.(p_1)]^2 + [s.e.(p_2)]^2}} = \frac{0.9801 - 0.8758}{\sqrt{0.0014^2 + 0.0191^2}} = 5.45.$$

The p-value for the calculated z statistic is 0.0001, which is statistically significant. If after two years the survival experience changed, this test would not provide any information about that change. One could use multiple tests but doing that has the disadvantage of not yielding a single overall test.

11.4.3   The Log-Rank Test

The CMH test for the comparison of survival curves is often called the log-rank test because of the similarity of these two test statistics. Peto and Peto's log-rank statistic is based on a set of scores derived from the logarithm of the survival function (Lee 1992; Peto and Peto 1972). Because of its complexity in calculation, researchers often use an approximate log-rank chi-square statistic that is easier to compute (Matthews and Farewell 1985, Chapter 7; Peto et al. 1977). Just as in the CMH approach, the approximate log rank test is based on the individual 2 by 2 tables, but it looks at the number of deaths and expected number of deaths for each group. The approximate log-rank statistic is calculated by

$$X_{LR}^2 = \frac{(O_1 - E_1)^2}{E_1} + \frac{(O_2 - E_2)^2}{E_2}$$

where O_1 is the sum of the observed numbers of deaths across the time points in the (1, 1) cell and O_2 is the corresponding sum for the (2, 1) cell. E_1 and E_2 are the corresponding sums of the expected number of deaths. The approximate log-rank test statistic looks like the goodness-of-fit chi-square statistic. Applying the approximate log-rank test chi-square procedure to the preceding data, we get

$$X_{LR}^2 = \frac{(1280 - 1361.07)^2}{1361.07} + \frac{(113 - 31.93)^2}{31.93} = 210.66.$$

It gives practically an identical result to the CMH chi-square value just shown. One advantage of the approximate log-rank test is that it can be extended to more than two group comparisons. The exact calculation of the statistic is more involved than we wish to present in this text, but different software packages often report the exact value.

11.4.4   Use of the CMH Approach with Small Data Sets

The CMH test statistic can also be used with a smaller data set along with the product-limit method. Let us consider an example.

Example 11.4

We reexamine the data used in Table 11.4 in comparing the survival distributions of male and female patients at the 0.05 significance level. The male and female survival distributions are shown in Figure 11.6. The median survival time for males is about 20 months, and it is 16 months for females.

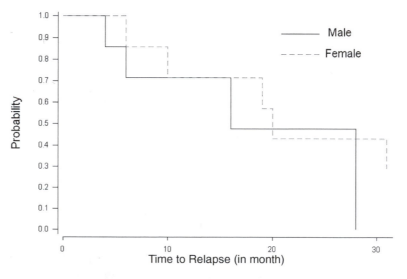

Figure 11.6   Estimated survival distributions by gender for the data in Table 11.4.

Table 11.9   Comparison of alcohol-free time distributions for females and males.

Survival Time (1) x_i		Number of Subjects			Observed No. of Relapses			Expected Relapses (8)	Variance (9)
		Total (2) n'_i	Female (3) n'_{1i}	Male (4) n'_{2i}	Total (5) d_i	Female (6) d_{1i}	Male (7) d_{2i}		
4	M	14	7	7	1	0	1	0.50	0.25
6	MF	13	7	6	2	1	1	1.08	0.46
9*	M	11	6	5	0	0	0	0	0
10	F	10	6	4	1	1	0	0.60	0.24
14*	M	9	5	4	0	0	0	0	0
16	M	8	5	3	1	0	1	0.63	0.23
17*	M	7	5	2	0	0	0	0	0
19	F	6	5	1	1	1	0	0.83	0.14
20	F	5	4	1	1	1	0	0.80	0.16
28	M	4	3	1	1	0	1	0.75	0.19
31	F	3	3	0	1	1	0	1.00	0.00
34*	F	2	2	0	0	0	0	0	0
47*	F	1	1	0	0	0	0	0	0
Total					9	5	4	6.19	1.67

*Censored observations

We wish to determine whether or not there is a significant difference between these two distributions. The data and the calculation of the test statistic for making this comparison are shown in Table 11.9.

The first column of the table shows the observed alcohol-free times (x_i) with the censoring status and gender indicated. The second column is the total number of subjects under observation at time x. The third and fourth columns show, respectively, the number of females (Group 1) and the number of males (Group 2) under observation at time x. The fifth column shows the observed number of relapses at time x. The numbers of relapses at time x in Group 1 and in Group 2 are shown in columns 6 and 7, respectively.

The eighth column shows the expected number of relapses at time x_i for females. It is calculated in the same manner as before. For example, at 6 months, two relapses are recorded. The proportion of females under observation at 6 months is 7/13. Therefore, the expected number of relapses for females is $2*(7/13)$, or 1.08. The variances of the observed numbers of relapses for females at time x_i are shown in column 9. These calculations are performed only for the uncensored survival times. The values are next summed and the CMH chi-square statistic is calculated as follows:

$$X^2_{CMH} = \frac{(|O - E| - 0.5)^2}{V} = \frac{(|5 - 6.19| - 0.5)^2}{1.67} = 0.29.$$

Since the test statistic is smaller than 3.84 ($= \chi^2_{1,0.95}$), we fail to reject the null hypothesis.

The approximate log-rank chi-square statistic gives

$$X_{LR}^2 = \frac{(5-6.19)^2}{6.19} + \frac{(4-2.81)^2}{2.81} = 0.73.$$

Computer programs provide the exact log-rank chi-square value of 0.84. Although the CMH chi-square value is smaller than the log-rank chi-square values (due mainly to the correction for continuity), we draw the same conclusion. Note that the CMH chi-square without the correction for continuity is 0.85.

In Chapter 10, we indicated that the CMH test statistic should be used only when the odds ratios are similar across the subtables. The same idea applies here, and the plot of the two survival functions gives a rough way of assessing the validity of this assumption. If the assumption is true, the plot of the two survival functions should be roughly parallel. If the lines representing the two survival functions cross one another, this definitely means that the assumption does not hold and the CMH test statistic should not be used. The reason for this is that one group has a better survival experience during part of the study period, and the other group has a better experience during another part of the period. Thus, it is difficult to say that one group has a better overall experience. The log-rank test also has the same requirement.

Comparison of two consistently different survival curves can be done by the computer (see **Program Note 11.3** on the website). Most statistics packages provide the log-rank chi-square and options for creating graphs of the survival functions.

Conclusion

In this chapter, we presented two methods for analyzing survival data: the life-table and product-limit methods. The life-table method is generally used for large data sets and the product-limit method for smaller data sets. In addition, we demonstrated the calculation of the sample median and restricted mean survival times. We also discussed why the median is preferred to the mean as a single summary statistic for use with survival data. We highly recommended the plotting of the survival distribution for a more complete description of survival data. Finally, we showed the use of the Cochran-Mantel-Haenszel test for comparing the equality of two survival distributions.

EXERCISES

11.1 In an effort to understand employment experience of nurses, personnel records of two large hospitals were reviewed (Benedict, Glasser, and Lee 1989). A total of 3221 nurses were hired during a 10-year period from 1970 to 1979 and employment records were reviewed 18 months beyond the end of 1979. In this cohort, only 780 nurses worked more than 33 months. The length of employment was presented by 3-month intervals as follows:

Month after Employment	Number Terminated	Number Censored	Number at Beginning of Interval
0–3	582	0	3,221
3–6	369	0	
6–9	247	0	
9–12	212	0	
12–15	182	0	
15–18	144	0	
18–21	129	75	
21–24	99	74	
24–27	85	59	
27–30	51	53	
30–33	45	35	
33+	0	780	

a. Prepare a worksheet for a life-table analysis and estimate the cumulative survival probabilities, the restricted mean, and the median length of employment. Also estimate the probability of termination for each of the intervals.

b. Estimate the standard errors of (1) the estimated cumulative survival probabilities and (2) the probability of termination for each interval.

c. Calculate 95% confidence intervals for (1) the 24-month cumulative survival probability and (2) the probability of termination during the first three months of employment.

d. What additional data, if any, do you need and what further analyses would you perform to assess the nursing employment situation?

11.2 The Hypertension Detection and Follow-up Program collected mortality data for eight years (Shulman et al. 1989). The following data show the survival experience of two subgroups formed by the level of serum creatinine concentration:

Year Care	Serum Creatinine Concentration (mg/dL)					
	2.00–2.49			≥2.5		
	Alive	Died	Censored	Alive	Died	Censored
0–1	78	3	0	72	8	0
1–2	75	4	0	64	8	0
2–3	71	6	0	56	3	0
3–4	65	3	0	53	3	0
4–5	62	5	0	50	8	0
5–6	57	4	0	42	3	0
6–7	53	2	0	39	5	0
7–8	51	3	3	34	1	1
8+	45	0	45	32	0	32

a. Analyze the survival pattern of each group using the life-table method: Estimate the cumulative survival probabilities and their standard errors, and compare the survival curves of these two groups graphically.

b. If it is appropriate, determine whether or not the two survival distributions are equal at the 0.01 significance level.

c. Comment on what factors may have confounded the preceding comparison and what further analyses you think are necessary before you can draw more defensible conclusions.

11.3 The SHEP (Systolic Hypertension in the Elderly Program) Cooperative Research Group (1991) assessed the ability of antihypertensive drug treatment to reduce the risk of stroke (nonfatal and fatal) in a randomized, double-blind, placebo-controlled experiment. A total of 4736 persons with systolic hypertension (systolic blood pressure 160 mmHg and above and diastolic blood pressure less than 90 mmHg) were screened from 447,921 elderly persons aged 60 years and above. During the study period, 213 deaths occurred in the treatment group and 242 deaths in the placebo group. The average follow-up period was 4.5 years. Total stroke was the primary end point and the following data were reported:

Year	Treatment Group			Placebo Group		
	Number Started	Strokes	Lost	Number Started	Strokes	Lost
0–1	2,365	28	0	2,371	34	0
1–2	2,316	22	0	2,308	42	0
2–3	2,264	21	0	2,229	22	2
3–4	2,153	18	0	2,193	34	2
4–5	1,438	13	5	1,393	24	1
5–6*	613	1	0	584	3	0

*The last stroke occurred during the 67th month of follow-up.

a. To analyze the above data by the life-table method, how would you set up the worksheet? It is obvious that there were censored observations other than the lost-to-follow-up, such as deaths and withdrawn alive. This can be seen since the difference in the number of persons starting one interval and the number starting the following interval decreased by more than the number of strokes in the interval. Would you include or exclude the data in the last reported interval?

b. If it is appropriate, test the hypothesis of the equality of the two survival distributions at the 0.05 significance level.

11.4 A group of 31 patients diagnosed with lymphoma and presenting with clinical symptoms ("B" symptoms) was compared with another group of 33 lymphoma patients diagnosed without symptoms ("A" symptoms) (Mattews and Farewell 1985, page 89). The recorded survival times (in months) for the 64 patients are as follows:

A symptoms:	3.2*	4.4*	6.2	9.0	9.9	14.4	15.8	18.5	27.6*	28.5	30.1*
	31.5*	32.2*	41.0	41.8*	44.5*	47.8*	50.6*	54.3*	55.0	60.0*	60.4*
	63.6*	63.7*	63.8*	66.1*	68.0*	68.7*	68.8*	70.9*	71.5*	75.3*	75.7*
B symptoms:	2.5	4.1	4.6	6.4	6.7	7.4	7.6	7.7	7.8	8.8	13.3
	13.4	18.3	19.7	21.9	24.7	27.5	29.7	30.1*	32.9	33.5	35.4*
	37.7*	40.9*	42.6*	45.4*	48.5	48.9*	60.4*	64.4*	66.4*		

Asterisks indicate censored observations.

a. Estimate the survival probabilities, plot the survival curves, and determine whether the use of the CMH or log-rank test is appropriate in comparing the two survival curves.

b. Carry out the test at the 0.01 significance level and interpret the results. How would you interpret the prolonged horizontal survival curve at the end of survival curves in both groups?

11.5 The following data were abstracted from the records of the neonatal intensive care unit (NICU) in a hospital during the month of February 1993 (day and 24-hour clock time are used to describe the timing of events — e.g., 0102 indicates the first day of February, 2/AM):

No.	Sex	Born	Last Observed	Status
1	Boy	0102	2210	Discharged
2	Girl	0306	1722	Died
3	Boy	0309	1517	Died
4	Boy	0523	2609	Discharged
5	Boy	0918	1001	Died
6	Girl	1004	2411	Died
7	Boy	1107	2512	Discharged
8	Girl	1110	1815	Discharged
9	Boy	1206	1408	Died
10	Girl	1307	2320	Died
11	Girl	1412	2823	Still in NICU
12	Boy	1500	1510	Died
13	Boy	1607	2220	Died
14	Girl	1819	2823	Still in NICU
15	Boy	1903	2009	Died
16	Boy	2009	2711	Discharged
17	Boy	2110	2823	Still in NICU
18	Girl	2208	2320	Died
19	Girl	2321	2823	Still in NICU
20	Girl	2323	2810	Discharged
21	Boy	2402	2823	Still in NICU
22	Girl	2509	2823	Still in NICU
23	Boy	2620	2823	Still in NICU
24	Girl	2701	2822	Died

a. Estimate the neonatal survival function for these NICU infants, estimate the median survival time, and form the 90 percent confidence interval for the 50-hour survival probability.

b. Plot the estimated neonatal survival functions separately for boys and girls and test the equality of the two survival distributions at the 0.10 significance level.

11.6 Quality of care for colorectal cancer was evaluated by comparing the survival experience of patients in two types of health plans (fee-for-service and health maintenance organization) offered by the same health care provider (Vemon et al. 1992). The following data were generated from the reported survival curves:

Practice	Survival Times in Months											
Fee-for-	2	5	10	12*	14	14	16	18	23	26*	27	31
Service	34	37*	39	42*	46	47*	50	53*				
HMO	4	10*	12	15	19	25	30*	35	38	43*	49	54*

Asterisks indicate censored observations.

a. Estimate the survival distributions by the product-limit method and graphically compare the survival curves.

b. Compare the equality of the survival distributions of the two medical services at the 0.01 significance level.

11.7 From April 1, 1999, family physicians are required to refer all patients who have suspected breast cancer in the United Kingdom to a hospital to be seen within 14 days of referral. Data from a cancer registry were used to examine whether the survival distributions of different length of delay groups (from referral to treatment) are different (Sainsbury, Johnston, and Haward 1999). Patients diagnosed with breast cancer during the 1986–1990 period were used for this analysis. Of the 9488 patients registered, 5708 had information on dates of referral and treatment. It was stated that "survival curves were estimated by the Kaplan-Meier method." Based on a survival analysis of the following data, the authors concluded that "delays of more than 90 days are unlikely to have an impact on survival and that, if delays can be kept to within this time, efforts to shorten delays further should not have priority."

| | Number of Survivors at the Beginning of the Interval | | | |
| | Delay Groups | | | |
Years of Survival	<30 Days	30–59 Days	60–89 Days	≥90 Days
0–1	3,534	1,578	345	251
1–2	3,113	1,490	328	235
2–3	2,743	1,370	301	217
3–4	2,470	1,274	275	198
4–5	2,235	1,182	254	186
5–6	2,062	1,101	239	176
6–7	1,897	1,050	225	168
7–8	1,769	982	212	157
8+	1,647	913	199	154

Assume that there were no censored observations.

a. More than one-half of the data were in the less than 30 days delay group. What are merits and demerits of splitting this group to <15 days and 15–29 days groups?

b. Do you think that the Kaplan-Meier method was appropriate for this analysis?

c. Estimate the survival distributions for <30 days delay group and the ≥90 days delay group, and test whether two survival distributions are significantly different at the 0.05 level.

d. Do you think the author's conclusions are supported by your analysis? Why or why not? What are possible confounders for the difference in survival distributions?

REFERENCES

Benedict, M. B., J. H. Glasser, and E. S. Lee. "Assessing Hospital Nursing Staff Retention and Turnover: A Life Table Approach." *Evaluation and the Health Professions* 12:73–96, 1989.

California Tumor Registry. *Cancer Registration and Survival in California.* Berkeley, CA: State of California Department of Public Health, 1963, p. 258–259.

Kaplan, E. L., and P. Meier. "Nonparametric Estimation from Incomplete Observations." *Journal of the American Statistical Association* 53:457–481, 1958.

Lee, E. T. *Statistical Methods For Survival Data Analysis*, 2nd ed. New York: Wiley, 1992.

Mantel, N. "Evaluation of Survival Data and Two New Rank Order Statistics Arising in Its Considerations." *Cancer Chemotherapy Reports* 50:163–170, 1966.

Matthews, D. E., and V. Farewell. *Using and Understanding Medical Statistics, Basel*. New York: Karger, 1985.

National Cancer Institute. *Cancer Statistics Review: 1973–1987*. Bethesda, MD: National Institutes of Health, U.S. Public Health Service, NIH Publication No. 90–2789, 1990.

Peto, R., and J. Peto. "Asymptotically Efficient Rank Invariant Test Procedures (with Discussion)." *Journal of Royal Statistical Society* A 135:185–206, 1972.

Peto, R., M. C. Pike, P. Armitage, N. E. Breslow, D.R. Cox, S.V. Howard, N. Mantel, K. McPherson, J. Peto, and P. G. Smith. "Design and Analysis of Randomized Clinical Trials Requiring Prolonged Observation of Each Patent, II: Analysis and Examples." *British Journal of Cancer* 35:1–39, 1977.

Sainsbury, R., C. Johnston, and B. Haward. "Effect on Survival of Delays in Referral of Patients with Breast-cancer Symptoms: A Retrospective Analysis." *Lancet* 353:1132–1135, 1999.

Shulman, N. B., C. E. Ford, W. D. Hall, M. D. Blaufox, D. Simon, H. G. Langford, and K. A. Schneider. On behalf of the Hypertension Detection and Follow-up Program Cooperative Group. "Prognostic Value of Serum Creatinine and Effect of Treatment of Hypertension on Renal Function." *Hypertension* 13 (Supplement I):80–93, 1989.

Systolic Hypertension in the Elderly Program Cooperative Research Group. "Prevention of Stroke by Antihypertenstive Drug Treatment in Older Persons with Isolated Systolic Hypertension." *Journal of American Medical Association* 265:3255–3264, 1991.

Thomas, D. R., and G. L. Grunkemeier. "Confidence Interval Estimation of Survival Probabilities for Censored Data." *Journal of the American Statistical Association* 70:865–871, 1975.

Vernon, S. W., J. I. Hughes, V. M. Heckel, and G. L. Jackson. "Quality of Care for Colorectal Cancer in a Fee-for-Service and Health Maintenance Organization Practice." *Cancer* 69:2418–2425, 1992.

Analysis of Variance

12

Chapter Outline

In Chapter 8, we used the t test for testing the equality of two population means based on data from two independent samples. In this chapter, we introduce a procedure for testing the equality of two or more means. The two experimental designs discussed in Chapter 6 — the completely randomized and the randomized block designs — will be considered.

The comparison of two or more means is based on partitioning the variation in the dependent variable into its components — hence, the method is called the analysis of variance (ANOVA). It was introduced by Sir Ronald A. Fisher and has been used in many fields of research. We begin this chapter with a presentation of the assumptions made when the ANOVA is used. This section is followed by an introduction to the one-way ANOVA. In conjunction with this analysis, we present three methods used in multiple comparison analysis. These topics are followed by the analysis of the randomized block design, an example of a two-way ANOVA, and a two-way ANOVA with interaction. We next provide a linear model representation of the ANOVA, followed by the use of the linear model with unequal group sizes.

12.1   Assumptions for Use of the ANOVA

The ANOVA is used to determine whether or not there is a statistically significant difference among the population means of two or more groups. The theoretical basis of the ANOVA requires that the data being analyzed are independent and normally distributed. We must also assume that the population variances in each of the groups have the same value, σ^2. The ANOVA procedure works reasonably well if there are small departures from the normality assumption. However, if the variances are very different, there is concern about the significance levels reported in the analysis (Scheffé 1959). This concern is consistent with the material presented in Chapters 7 and 8, where we saw different methods for comparing two means, depending on whether or not we assumed that the population variances were equal. One method for protecting against the effects of different values for the variances is to have approximately equal numbers of observations in each of the groups being analyzed. Another approach involves transformations of the dependent variable (Kleinbaum, Kupper, and Muller 1988; Lin and Vonesh 1989). One further assumption is that the groups being compared are the only groups of interest. This assumption means that the factors — the independent variables — are fixed factors. Fixed and random factors and their implications are discussed later in this chapter, and further information is available elsewhere (Steel and Torrie 1980).

There are no firm rules for the number of observations required by the ANOVA. It is possible to perform power calculations or to use the size of confidence intervals to estimate the required sample size. In general, we recommend that there be a minimum of 5 to 10 observations for each of the combinations of levels of the independent variables used in the analysis. For example, with two independent variables, if one has 3 levels and the other independent variable has 4 levels, there are 12 combinations of levels.

12.2   One-Way ANOVA

In a one-way ANOVA, there is only one independent variable. The data to be analyzed are obtained from either (1) a random sample of subjects who belong to different groups — for example, different racial groups — or (2) an experiment in which the subjects are randomly assigned to one of several groups. The latter situation arises when we use the completely randomized design discussed in Chapter 6. In the completely randomized design, subjects are randomly allocated to groups and the groups represent the levels of the independent variable. Observations of the continuous variable of interest, the dependent variable, are taken on the subjects and the subject's group membership is also recorded. In the following example, we consider data from a completely randomized design, and we wish to determine whether or not there is a difference in mean age among three groups.

Example 12.1

Data shown in Table 12.1 are based on an article by Kimball et al. (1986) and can be analyzed using a one-way ANOVA. In the article, the authors wished to evaluate ventricular performance after surgical correction of congenital coarctation of the aorta. The ventricular performance was compared to that found in two control groups. Because of the possible roles that age and gender play on ventricular performance, the authors wanted the age and sex distributions of the subjects who had undergone the surgery to be similar to those of the members of the two control groups. We wish to examine whether or not the authors were successful in obtaining groups that were similar on the age variable. The ages shown in Table 12.1 are hypothetical, based on the summary values reported by Kimball et al. In this example, the dependent variable is age, and the independent variable is the group to which the subjects belong.

The entries in Table 12.1 can be represented symbolically as y_{ij}, where the first subscript indicates the subject's group membership and the second subscript identifies the subject in the ith group. For example, y_{11} is 32 years old, y_{12} is 28 years old, y_{25} is 34 years old, y_{26} is 33 years old, and so on. The first subscript ranges from 1 to 3. When the first subscript has the value of 1, the range of the second subscript is

Table 12.1 Hypothetical ages for control and surgery subjects.

Group	Ages																	
Surgery	32	28	22	25	20	20	28	28	20	29	22	37	18	29	22	32	21	34
	19	23	23	26	41	20	33											
Control I	32	26	31	39	34	33	29	41	35	33	33	43	25	39	36	37	28	34
	27	45	22	29	51	28	35											
Control II	31	35	26	28	22	29	27	21	22	27	24	44	21	25	27	18	27	36

from 1 to 25, and this is also the case when the first subscript is 2. When the first subscript has the value of 3, the second subscript ranges from 1 to 18. In general, there are r groups and n_i observations in the ith group. We also use the · notation introduced in Chapter 10. For example, y_i. is a shorthand notation for $\sum_j y_{ij}$ and y.. is shorthand for $\sum_i \sum_j y_{ij}$. Thus, y_1. represents the sum of all the ages for the subjects in the surgery group, and y.. is the sum of all the 68 ages in the sample. It follows that $\bar{y}_i$. is the sample mean of the i-th group, and $\bar{y}$.. is the overall sample mean.

In the following section, we show how the data in Example 12.1 can be analyzed.

12.2.1 Sums of Squares and Mean Squares

As was just mentioned, this method, the analysis of variance, is based on a partitioning of the variation in the dependent variable. In the one-way ANOVA, there are two possible sources of variation in the dependent variable. One source is variation among (or between) the groups — that is, the groups may have different means that vary about the overall mean. The other possible source is variation within the groups. Not all the subjects in the same group will have exactly the same values, and the within-group variation reflects this.

The null hypothesis being tested here is that the population group means are equal to one another. If this hypothesis is true, all the observations come, in effect, from the same population. Thus, any variation that remains among the group means really reflects the random variation among the observations — that is, the within-groups variation. Thus, the adjusted among and within variations should be similar if the null hypothesis is true. If the null hypothesis is false, the adjusted among-groups variation should be larger than the adjusted within-group variation because it includes variation between the populations as well as the within-group variation. Thus, we can use the adjusted among- and within-group variations as the basis of a test of the equality of the group means.

We can represent the above idea in symbols as

$$\sum_{i=1}^{r}\sum_{j=1}^{n_i}(y_{ij}-\bar{y}..)^2=\sum_{i=1}^{r}\sum_{j=1}^{n_i}(\bar{y}_i.-\bar{y}..)^2+\sum_{i=1}^{r}\sum_{j=1}^{n_i}(y_{ij}-\bar{y}_i.)^2.$$

This equation shows the partitioning of the total variation in Y, the dependent variable, about its mean into an among (or between)-group component and a within-group component. These sum of squares are called the total sum of squares corrected for the mean (SST), the among (or between)-group sum of squares (SSB) and the within-group sum of squares (SSW).

If we adjust these two components for the number of independent observations used in their calculations — that is, divide each component sum of squares by its degrees of freedom — we have the mean square among (or between) and the mean square within. The mean square between is

$$MSB = \frac{\sum_{i=1}^{r} \sum_{j=1}^{n_i} (\bar{y}_{i\cdot} - \bar{y}_{\cdot\cdot})^2}{r-1} = \frac{\sum_{i=1}^{r} n_i (\bar{y}_{i\cdot} - \bar{y}_{\cdot\cdot})^2}{r-1}$$

where the second expression reflects the fact that the terms in the parentheses do not vary with j. The mean square within is

$$MSW = \frac{\sum_{i=1}^{r} \sum_{j=1}^{n_i} (\bar{y}_{i\cdot} - \bar{y}_{i\cdot})^2}{n-r} \qquad S/B \frac{(y_{ij} - y_i)}{n-r}$$

where n is the total number of observations — that is, the sum of the n_i. The degrees of freedom for the mean square between, $r - 1$, comes from the calculation of the variation in r means. The degrees of freedom for the mean square within, $n - r$, is the result of summing the $n_i - 1$ degrees of freedom associated with the ith group over the r groups.

The mean square within is particularly useful as it also provides an adjusted estimate of the variation within groups — that is, of σ^2, the variance of the dependent variable. It is based on the assumption that the variance of the dependent variable is the same within each group. If there is no difference between the group means, then the mean square between also estimates σ^2.

Example 12.2

For the data in Table 12.1, we have the following values of means and sums of squares. First, $\bar{y}_{1\cdot}$, the sample mean of the first group, is 26.08, $\bar{y}_{2\cdot}$ is 33.80, and $\bar{y}_{3\cdot}$ is 27.22. The overall sample mean, $\bar{y}_{\cdot\cdot}$, is 29.22 years. The sum of squares between is

$$SSB = 25(26.08 - 29.22)^2 + 25(33.80 - 29.22)^2 + 18(27.22 - 29.22)^2 = 842.9.$$

The sum of squares within involves too many terms to show, but its sum of squares is 2660.8 and the total sum of squares (corrected) is 3503.7.

12.2.2 The *F* Statistic

The comparison of these two mean squares provides information about whether or not the null hypothesis is true. One way of comparing the mean squares is to take their difference. If the null hypothesis were true, then the difference would be zero. However, the probability distribution of the difference is not widely available. Another way of comparing the mean squares is to take the ratio of the mean square between to the mean square within. If the null hypothesis were true, the ratio would equal one. If the null hypothesis were false, the ratio would be larger than one. Fortunately, the probability distribution of the ratio has been worked out, and it is an F distribution with $r - 1$ and $n - r$ degrees of freedom. Tables of the F distribution, named in honor of Sir Ronald Fisher, are shown in Appendix B11 for the 0.01, 0.05, and 0.10 significance levels for values of the numerator (f_1) and denominator (f_2) degrees of freedom parameters.

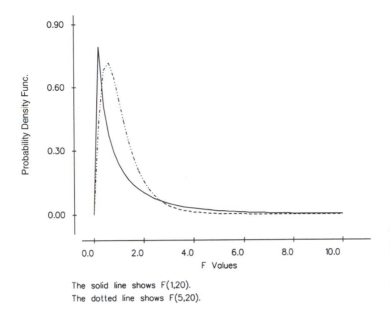

The solid line shows F(1,20).
The dotted line shows F(5,20).

Figure 12.1 Plot of the probability density functions of the F distribution for $F_{1,20}$ (solid line) and $F_{5,20}$ (broken line).

The F distribution has many different shapes, depending on the values of the degrees of freedom parameters. Figure 12.1 shows the shape of the F distribution for degrees of freedom pairs (1 and 20) and (5 and 20). We can see that the shapes are different, but most of the probability (area) is associated with values of F close to one.

There is also a relation between the t and F distributions that can be seen from the t and F tables. The relation is $t^2_{k,1-\alpha/2}$ is equal to $F_{1,k,1-\alpha}$. For example, when k is 10, $t_{10,0.95}$ is 1.8125, and its square is 3.2852. Examination of the F tables in Appendix B11 shows that $F_{1,10,0.90}$ is 3.29. This equivalence when there are two groups leads us to think that there may be a relation between the ANOVA and t test approaches in the two-group situation.

12.2.3   The ANOVA Table

The preceding sums of squares and mean squares are usually presented in tabular format, as shown in Table 12.2.

Table 12.2   Typical ANOVA table for a one-way analysis.

Source of Variation	Degrees of Freedom	Sum of Squares	Mean Square	F
Between Groups	$r-1$	SSB	$SSB/(r-1) = MSB$	MSB/MSW
Within Groups	$n-r$	SSW	$SSW/(n-r) = MSW$	
Total (Corrected)	$n-1$	SST		

The degrees of freedom and sums of squares associated with the between and within groups sum to the corresponding total values. If these values do not sum to the total, a mistake has been made in the calculations.

The F statistic is then used to test the null hypothesis that the group means are equal against the alternative hypothesis that the group means are not all equal. When the null hypothesis is true, the F statistic follows an F distribution with $r - 1$ and $n - r$ degrees of freedom. If the calculated F statistic is greater than $F_{r-1,n-r,1-\alpha}$, found in Appendix Table B11, we reject the null hypothesis in favor of the alternative hypothesis at the α significance level. If the calculated F statistic is less than this critical value, we do not have sufficient evidence to reject the null hypothesis.

Example 12.3

Based on the sums of squares presented in Example 12.2, we can complete the ANOVA table for the ages shown in Table 12.1. Let us test the hypothesis of the equality of the mean ages at the 0.01 significance level. Table 12.3 is the ANOVA table for the age data.

There are 68 observations in the three groups. Hence, there are 2 degrees of freedom for the factor (between groups), variable, 65 degrees of freedom for error (within groups), and 67 degrees of freedom for the total sum of squares. The table shows the sums of squares and mean squares as well as the F ratio. The exact critical value of this test is not shown in Table B11, but the closest value shown for F is 4.98 for $F_{2,60,0.99}$. From the table, we see that the exact F value for $F_{2,65,0.99}$ is slightly less than 4.98.

The calculated F statistic (10.29) is greater than the approximate critical value of 4.98. Therefore, we reject the equality of the mean ages in favor of the alternative hypothesis. It appears that the three groups differ on age. This finding means that it may be necessary to take age into account in the analysis of ventricular performance. The square root of the mean square for error is 6.395 ($= \sqrt{40.9}$), and it is an estimate of σ.

Computer packages can be used to perform the analysis of variance (see **Program Note 12.1** on the website). The computer output shows the ANOVA table with the p-value associated with the F ratio along with group means and standard deviations.

Figure 12.2 shows box plots for the data in Table 12.1. The group means are represented by dots in the box. It appears that the difference is due mainly to the first control group having a mean age that is much greater than the other two groups. When there is a statistically significant difference among the group means, we can perform additional tests to see if we can determine the source of the differences in the means. The next section describes three approaches to this additional testing.

Table 12.3   ANOVA table for the ages shown in Table 12.1.

Source of Variation	Degrees of Freedom	Sum of Squares	Mean Square	F
Between Groups	2	842.9	421.4	10.29
Within Groups	65	2660.8	40.9	
Total (Corrected)	67	3503.7		

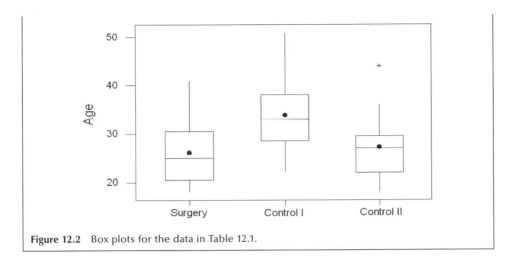

Figure 12.2   Box plots for the data in Table 12.1.

12.3   Multiple Comparisons

If the overall F statistic from the ANOVA is statistically significant, *multiple comparisons* procedures can be used in an attempt to discover the source of the significant differences among the group means. Most of these procedures are designed to examine the pairwise differences among group means, although there are more general procedures. The comparison of the group means is accomplished through the presentation of confidence intervals for pairwise differences of group means. The use of the multiple comparison procedures is generally not recommended when we fail to reject the null hypothesis. However, exceptions may occur when certain comparisons have been planned in the course of the experiment.

There are many different multiple comparison procedures, and we shall present three: the Tukey-Kramer method, Fisher's least significant difference (LSD) method, and Dunnett's method. The Tukey-Kramer method is the recommended procedure when one wishes to estimate simultaneously all pairwise differences among the means in a one-way ANOVA assuming that the variances are equal (Stoline 1981). We present the LSD method because it is frequently used in the literature. Dunnett's procedure is used when we wish to compare several groups with a specific group selected before the data were obtained (or the control group designated in the design). For example, if there were several new treatments and a standard treatment, we would use Dunnett's procedure to compare each of the new treatments with the standard. The multiple comparison procedures presented here use the mean square within as the estimator of σ^2. Before presenting these methods, we shall discuss error rates associated with the methods.

12.3.1   Error Rates: Individual and Family

In the pairwise comparison of the group means, many confidence intervals are formed. For example, when there are three groups, we form confidence intervals for the differences of groups 1 and 2, groups 1 and 3, and groups 2 and 3. When there are r groups, there are $_rC_2$ confidence intervals for the pairwise comparisons. Thus, we see that there are two probabilities of errors in multiple comparison procedures. One probability of error is associated with each individual confidence interval — the individual error rate. The other is associated with the $_rC_2$ intervals — the family of confidence intervals — the family error rate. This is the rate that is usually of primary interest — the rate that we want to be less than or equal to α.

It is clear that if we use the $t_{1-\alpha/2}$ value in the creation of the confidence intervals, the family error rate will be larger than α. If we wish to control the family error rate to be less than or equal to α, then we must use some value other (greater) than $t_{1-\alpha/2}$ in the calculation of the confidence intervals.

12.3.2 The Tukey-Kramer Method

The Tukey-Kramer method focuses on the family error rate. It replaces $t_{n-r,1-\alpha/2}$ in the confidence interval for the difference of two group means by $q_{r,\,n-r,\,1-\alpha}/\sqrt{2}$, where q is the upper α value from the studentized range distribution (r is equivalent to p in Table B12). Table B12 shows the upper α value from the studentized range distribution (at $\alpha = 0.01$ and 0.05). Note that the q value takes the number of possible comparisons into account, since its value depends on r, the number of groups.

The confidence interval for the difference of μ_i and μ_j is

$$(\bar{y}_{i\cdot} - \bar{y}_{j\cdot}) \pm \frac{q_{r,\,n-r,\,1-\alpha}}{\sqrt{2}} \sqrt{MSW} \sqrt{\frac{1}{n_i} + \frac{1}{n_j}}.$$

Example 12.4

Let us calculate the confidence intervals for the three pairwise comparisons for the hypothetical age data shown in Table 12.1. We shall set the family error rate to be 0.05. The value of $q_{3,65,0.95}$ is not found in Table B12. Since there is little variation in the value of q as $n - r$ changes from 40 to 60 to 120 in the table, we shall use 3.40 ($= q_{3,60,0.95}$) as an approximation to the desired value. The confidence interval for the difference of groups 1 and 2 is

$$(26.08 - 33.80) \pm \left(\frac{3.40}{\sqrt{2}} \sqrt{40.9} \sqrt{\frac{1}{25} + \frac{1}{25}} \right)$$

which yields

$$-7.72 \pm 4.35$$

and the interval ranges from -12.07 to -3.37. The corresponding interval for $\mu_1 - \mu_3$ is -5.89 to 3.61, and the interval for $\mu_2 - \mu_3$ is from 1.83 to 11.33. Both of the intervals involving μ_2 fail to contain zero, suggesting that the first control group differs significantly from both the study group and the second control group.

12.3.3 Fisher's Least Significant Difference Method

Fisher's LSD method focuses on the individual error rate. When the n_i are all equal, there is a value — the least significant difference — such that if any of the differences in sample means are greater than that value, the difference is statistically significant. If a difference is greater than that value, the corresponding confidence interval for the difference does not contain zero. If the number of sample observations differ across the groups, there is not a single least significant difference.

The LSD confidence interval looks like the ordinary confidence interval for the difference of two means with one exception. The mean square within is used as the estimator for the population variance instead of an estimator based on only data from the two groups being compared. The LSD confidence interval for $\mu_i - \mu_j$ is

$$(\bar{y}_{i\cdot} - \bar{y}_{j\cdot}) \pm t_{n-r,\, 1-\alpha/2} \sqrt{MSW} \sqrt{\frac{1}{n_i} + \frac{1}{n_j}}.$$

Example 12.5

Let us calculate the 0.05 individual error rate LSD confidence interval for $\mu_1 - \mu_2$. We have

$$(26.08 - 33.80) \pm \left(2.00\sqrt{40.9} \sqrt{\frac{1}{25} + \frac{1}{25}} \right)$$

which yields

$$-7.72 \pm 3.62$$

and the interval ranges from -11.34 to -4.10. This interval is narrower than the corresponding Tukey-Kramer interval as it must be, since it is based on the individual error rate, not the family error rate used by the Tukey-Kramer procedure. The corresponding LSD interval for $\mu_1 - \mu_3$ ranges from -5.10 to 2.82, and the interval for $\mu_2 - \mu_3$ ranges from 2.62 to 10.54.

12.3.4   Dunnett's Method

Dunnett's method is used in situations when we wish to compare the means of several groups with the mean of another group that was selected in advance. For example, we may wish to compare the means of different dosage levels of a new medication with the mean of a placebo group. In our example, there are two control groups and one treatment group. We wish to see if there is a difference between the two control groups and the treatment group (group 1). Thus, the comparisons of interest are $\mu_2 - \mu_1$ and $\mu_3 - \mu_1$.

The confidence interval for $\mu_i - \mu_j$ using Dunnett's procedure is given by

$$(\bar{y}_{i\cdot} - \bar{y}_{j\cdot}) \pm d_{r-1,\, n-r,\, 1-\alpha/2} \sqrt{MSW} \sqrt{\frac{1}{n_i} + \frac{1}{n_j}}$$

where the upper 0.005 and 0.025 levels of d are given in Table B13.

Example 12.6

Let us now calculate the confidence intervals using a family error rate of 0.05 with Dunnett's method for the data used in previous examples. For the comparison of the first control group with the treatment group, we have

$$(33.80 - 26.08) \pm \left(2.27\sqrt{40.9}\sqrt{\frac{1}{25} + \frac{1}{25}} \right)$$

where 2.27 is the value of $d_{2,60,0.975}$, and this is used as an approximation to $d_{2,65,0.975}$. This calculation yields

$$7.72 \pm 4.11$$

and the interval ranges from 3.61 to 11.83. The corresponding interval for $\mu_3 - \mu_1$ ranges from −3.35 to 5.63. The confidence intervals using Dunnett's procedure are narrower than those provided by the Tukey-Kramer method. This is reasonable, since we are doing fewer comparisons with Dunnett's procedure. Based on these intervals, there is a statistically significant difference between the first control group and the treatment group but no significant difference between the second control group and the treatment group.

These calculations can be performed by the computer packages in conjunction with analysis of variance (see **Program Note 12.1** on the website).

12.4   Two-Way ANOVA for the Randomized Block Design with *m* Replicates

As discussed in Chapter 6, in many situations the same experiment is conducted in several sites or under different conditions. In these situations, the random allocation of subjects takes place separately at each site or for each condition. These experiments are using what is called a *randomized block design*. The random allocation of the subjects to the treatments is performed separately for each block (site or condition) because it is thought that there may be an effect of the blocks on the outcome variable. If the subjects were randomly assigned ignoring the blocks, as in a completely randomized design, there is a chance that the block effects might be confounded with the treatment effects. Hence the random assignment is done separately.

Example 12.7

The data in Table 12.4 are from a randomized block design with five replicates per cell. The data are the changes in weight for moderately overweight female employees who participated in weight reduction programs. The women worked at one of two company sites: either the headquarters or a manufacturing plant. At each site, after a semiannual health examination, the women were randomly given memberships to a diet clinic or to a health club or to both. There was a control for company site because it was thought that there may be a difference in the effects of the weight

Table 12.4 Difference of pre- and postintervention weights (pounds) after 6 months of participation by intervention program at two sites.

Program	Office Site						Factory Site			
Diet Clinic	6	2	10	−1	8	3	15	4	8	6
	3	4	−2	6	−2	−4	6	8	−2	3
Both Programs	8	12	7	10	5	15	8	10	16	3

reduction programs for those who were less physically active — the headquarters group — compared to the women in the plant. After the next health examination, weight reduction was measured.

In this table, data are classified by program, the row variable, and site, the column variable. The type of intervention program is the treatment variable with three levels, and the site is considered to be the blocking variable with two levels. These two independent variables form six cells, and the cells all have the same number of observations. When there are the same numbers of observations in each cell, the design is said to be balanced. The analysis of unbalanced data is more complicated and will be discussed in the last section of this chapter.

The entries in Table 12.4 can be represented symbolically as y_{ijk}, where i is an indicator of the program (the row variable), j represents the site (the column variable), and k indicates the subject number within the ith program and jth site. The first subscript ranges from 1 to 3, the second subscript has the value 1 or 2, and the third subscript ranges from 1 to 5.

We continue to use the · notation. For example, $y_{\cdot 1 \cdot}$ represents $\Sigma_i \Sigma_k y_{i1k}$, the sum of weight losses for the female employees at the office site. Using this notation, the sample mean of the ith level of the program variable is $\bar{y}_{i\cdot\cdot}$, the sample mean of the jth level of the site variable is $\bar{y}_{\cdot j\cdot}$, and the overall sample mean is $\bar{y}_{\cdot\cdot\cdot}$. These are the values of these sample means:

Program Means		Site Means		Overall Mean
Diet	6.10	Office	5.07	5.83
Exercise	2.00	Factory	6.60	
Both	9.40			

To analyze this data set, we will use a two-way ANOVA. The method of analysis is called two-way because there are now two independent variables: the blocking variable with c levels and the treatment variable with r levels. The total sum of squares of the dependent variable about its mean is now partitioned into a sum of squares between treatment groups, a sum of squares between blocks and the within-cells (error or residual) sum of squares. This partitioning, based on m observations per cell, is

$$\sum_{i=1}^{r}\sum_{j=1}^{c}\sum_{k=1}^{m}(y_{ijk}-\bar{y}_{\cdots})^2 = cm\sum_{i=1}^{r}(\bar{y}_{i\cdot\cdot}-\bar{y}_{\cdots})^2 + rm\sum_{j=1}^{c}(\bar{y}_{\cdot j\cdot}-\bar{y}_{\cdots})^2 + SSW.$$

The total variation of Y about its mean (SST) is partitioned into the sum of squares for the row or treatment variable (SSR), the sum of squares for the column or block

Table 12.5 ANOVA table for a randomized block design.

Source of Variation	Degrees of Freedom	Sum of Squares	Mean Square	F
Treatments	$r - 1$	SSR	$SSR/(r - 1) = MSR$	MSR/MSW
Blocks	$c - 1$	SSC	$SSC/(c - 1) = MSC$	MSC/MSW
Residual	$n - r - c + 1$	SSW	$SSW/(n - r - c + 1) = MSW$	
Total (Corrected)	$n - 1$	SST		

variable (SSC), and the within or residual sum of squares (SSW). SSW is found by subtracting the sum of SSR and SSC from SST. The value of SSR is

$$SSR = 2(5)[(6.10 - 5.83)^2 + (2.00 - 5.83)^2 + (9.40 - 5.83)^2] = 274.9.$$

The value of SSC is similarly found and is

$$SSC = 3(5)[(5.07 - 5.83)^2 + (6.60 - 5.83)^2] = 17.56.$$

Too many terms are involved to show the calculation of SST, but its value is 768.2 and SSW, found by subtraction, is 475.7.

We use the same approach to the analysis in the two-way ANOVA as was used in the one-way ANOVA. To test the hypothesis of no difference in the treatments, we use the F statistic calculated as the ratio of the mean square for treatment to the residual mean square. If the null hypothesis of no difference in the treatment means, adjusted for the blocking variable, is true, this F statistic follows the F distribution. The mean square for treatments has $r - 1$ degrees of freedom, and the residual mean square has $n - r - c + 1 \ [= n - (r - 1) - (c - 1) - 1]$ degrees of freedom. Thus, the F statistic for the treatment variable will follow an F distribution with $r - 1$ and $n - r - c + 1$ degrees of freedom if there is no difference in the treatment group means. In the same way, we could also test the null hypothesis of no difference in the block means. The F statistic associated with this hypothesis follows the F distribution with $c - 1$ and $n - r - c + 1$ degrees of freedom if this null hypothesis is true. Usually, we are not as interested in the hypothesis about the block means as we are in the treatment group means.

The ANOVA table for a randomized block design with m replicates per cell is shown in Table 12.5. Let us perform the test of no treatment effect — that is, of no difference in the population means associated with the three interventions at the 0.05 significance level. The analysis for the change in weight data is shown in Table 12.6. As the calculated F-value of 7.51 is greater than the critical value of 3.37 ($= F_{2,26,0.95}$), we reject the null hypothesis and conclude that the intervention programs are significantly different. Alternatively, we can make the decision based on the p-value associated with 7.51. Since the p-value of 0.003 is smaller than 0.05, we draw the same conclusions. We are not interested in the site difference.

Since there is a difference in the treatment group means at the 0.05 significance level, we are interested in finding the source of the significant differences among the group means. Applying the Tukey-Kramer method of multiple comparisons, we find that the 95 confidence intervals for $(\mu_2 - \mu_1)$ is $(-8.85, 0.65)$, $(\mu_3 - \mu_1)$ is $(-1.45, 8.05)$, and $(\mu_3 - \mu_2)$ is $(2.65, 12.15)$. It appears that using both types of intervention is more effective than the intervention using exercise only.

What would have happened had we ignored the site variable in the preceding analysis? If we assume that we would have had the same assignment of the subjects to the

Table 12.6   ANOVA table for weight change data from Table 12.4: Three intervention programs at two sites.

Source of Variation	Degrees of Freedom	Sum of Squares	Mean Square	F	p-value
Between Programs	2	274.9	137.4	7.51	0.003
Between Sites	1	17.6	17.6		
Residual	26	475.7	18.3		
TOTAL	29	768.2			

different treatments, we can examine the effect of the use of the blocking variable. The residual sum of squares in the two-way ANOVA is less than or equal to the residual sum of squares in the corresponding one-way ANOVA, reflecting the removal of the between blocks sum of squares. If the sum of squares between the blocks is large and its degrees of freedom are small, then the residual mean square is much smaller in the two-way ANOVA. This means that if the blocking variable is important, there is a greater chance of detecting a difference in the treatment group means using the two-way ANOVA than using the corresponding one-way ANOVA. The computer packages can be used to perform the preceding analysis including the multiple comparisons (see **Program Note 12.2** on the website).

In the next section, we show a more general two-way analysis of variance that includes the interaction of the two independent variables.

12.5   Two-Way ANOVA with Interaction

In some instances, a researcher is interested in studying the effects of two factors. In these instances, the experimental subjects are randomly allocated to all combinations of levels of both factors. For example, if both the row and column factors have two levels each, then the subjects are randomly allocated to four groups. This type of experimental design is especially useful when we want to study the effects of each factor as well as the interaction effect of the factors with one another. *Interaction* exists when the differences in responses to the levels of one factor depend on the level of another factor. For example, in a study of byssinosis (brown-lung disease) in textile workers in North Carolina (Higgins and Koch 1977), two variables of interest were whether or not the worker smoked and whether or not the worker was exposed to dust in the workplace. Both of these variables were important — that is, both smoking and being exposed were associated with a higher occurrence of byssinosis. In addition, if a worker smoked and also was exposed to the dust, the occurrence of byssinosis was much higher than would have been expected by simply adding the effects of the smoking and exposure variables. In this case, there is a synergistic effect — that is, an interaction of these two independent variables.

We have previously been concerned about interaction, although we did not use the term *interaction* when we considered the Cochran-Mantel-Haenszel procedure. We said that the procedure should not be used when the odds ratios were not consistent across the subtables. If the odds ratios are not consistent, this means that the relation between the dependent and independent variables depends on the levels of an extraneous or confounding variable — that is, there is interaction between the independent and extraneous variable. If the interaction exists, it does not make sense to talk about an overall effect of the independent variable because its effect varies with the level of the extraneous or confounding variable.

Example 12.8

The data in Table 12.7 are from a two-factor experiment in a health education teacher-training program. Three new textbooks (factor A) were tested with two methods of instruction (factor B), and 36 trainees were randomly allocated to the six groups with six subjects per group. The trainees were tested before and after four weeks of instruction, and the increases in test scores were recorded as shown in the table. As in the randomized block design, data are classified by textbook, the row factor, and method of instruction, the column factor. In this experiment, the random allocation of subjects was done simultaneously to all combinations of the two sets of levels, whereas the randomization took place separately in each block in the randomized block design.

Table 12.7 Increase in test scores after four weeks of instruction using three textbooks and two teaching methods.

Textbook	Method of Instruction											
	Lecture						Discussion					
1	30	43	12	18	22	16	36	34	15	18	40	45
2	21	26	10	14	17	16	33	31	28	15	29	26
3	42	30	18	10	21	18	41	46	19	23	38	48

The entries in this table are also represented symbolically by y_{ijk} as in the randomized block design with replicates. Several means again will be used in the analysis. The means here include the cell means ($\bar{y}_{ij.}$), two sets of marginal means — row ($\bar{y}_{i..}$) and column ($\bar{y}_{.j.}$) — and the overall mean ($\bar{y}_{...}$). The values of these means are as follows:

Textbook	Methods of Instruction		Marginal Book Means
	Lecture	Discussion	
1	23.50	31.33	27.42
2	17.33	27.00	22.17
3	23.17	35.83	29.50
Marginal Method Means	21.33	31.39	26.36 (Overall Mean)

We analyze this data set using a two-way ANOVA with interaction. For the randomized block design, we used a two-way ANOVA, ignoring interaction. The researcher for this experiment could have used two separate completely randomized experiments (one-way ANOVAs) — one to compare the three textbooks and the other to compare the two types of instructional methods. However, based on these two separate experiments, the researcher would not know whether any textbook works better with one instructional method than the other. The effects of the textbooks may differ across the instructional methods. *Interaction* measures the difference in the textbook effects across the two instructional methods. If the distribution of the mean increase in test scores for the three textbook types for those taught by lecture differs from the corresponding distribution for those taught by discussion, there is interaction. The average effects of textbooks across both types of instruction and the average instructional effects across all textbooks are measures of the *main effects* of the two independent variables.

If there is an interaction of the two independent variables, then usually the interaction terms are of more interest than the main effects of the two independent variables. This is because, if there is an interaction, the effect of one independent variable depends — it changes — as the level of the other independent variable changes. Hence, in our analysis, we must first examine the test of hypothesis that there is no interaction before considering the test of no main effects of the independent variables.

If there is interaction, we can examine the cell means in an attempt to discover the nature of the interaction. If there is no evidence of an interaction, then we consider the hypotheses about the main effects. In this case, some statisticians would remove the interaction term from the analysis — that is, incorporate its sum of squares and degrees of freedom into the error term before calculating the F statistics for the main effects. The decision to incorporate or not to incorporate the nonsignificant interaction term into the error term usually has little effect on the results.

In order to include interaction in the analysis, the total sum of squares (SST) of the dependent variable about its mean is now partitioned into a sum of squares for the row factor R (SSR), a sum of squares for the column factor C (SSC), a sum of squares for interaction between factor R and factor C (SSRC), and the error sum of squares (SSE). As before, we shall use the symbols r and c for the numbers of levels for factors R and C, respectively, and use m to represent the number of replicates in each of the cells formed by the crosstabulation of factors R and C. This partitioning of the total sum of squares is expressed symbolically as

$$\sum_{i=1}^{r}\sum_{j=1}^{c}\sum_{k=1}^{k}(y_{ijk}-\bar{y}_{...})^2 = cm\sum_{i=1}^{r}(\bar{y}_{i..}-\bar{y}_{...})^2 + rm\sum_{j=1}^{c}(\bar{y}_{\cdot j\cdot}-\bar{y}_{...})^2$$

$$+ m\sum_{i=1}^{r}\sum_{j=1}^{c}(\bar{y}_{ij\cdot}-\bar{y}_{i..}-\bar{y}_{\cdot j\cdot}+\bar{y}_{...})^2 + \sum_{i=1}^{r}\sum_{j=1}^{c}\sum_{k=1}^{k}(y_{ijk}-\bar{y}_{ij\cdot})^2.$$

The rest of the analytic approach is the same as before. The mean squares for the main effects and the interaction are calculated by dividing the sums of squares by appropriate degrees of freedom. The mean squares for factors R and C have $r-1$ and $c-1$ degrees of freedom, respectively. The mean square for interaction has $(r-1)(c-1)$ degrees of freedom, and the error mean square has $n-rc$ [$= rc(m-1)$] degrees of freedom. The error mean square is then used as the denominator in the calculation of the F statistics for the two main effects and interaction. The ANOVA table for a two-factor experimental design with interaction is shown in Table 12.8.

The calculations of the sums of squares, similar to those shown previously in the randomized block analysis, are not shown here but are summarized in Table 12.9.

Let us perform the tests of hypotheses at the 0.05 significance level. The F statistic and its associated p-value for interaction indicate that there is no statistically significant interaction of the two independent variables. Since this is the case, we can now examine

Table 12.8 ANOVA table for a two-factor design with interaction.

Source of Variation	Degrees of Freedom	Sum of Squares	Mean Square	F
Factor R	$r - 1$	SSR	$SSR/(r - 1) = MSR$	MSR/MSE
Factor C	$c - 1$	SSC	$SSC/(c - 1) = MSC$	MSC/MSE
Interaction	$(r - 1)(c - 1)$	SSRC	$SSRC/(r - 1)(c - 1) = MSRC$	MSRC/MSE
Error	$n - rc$	SSE	$SSE/(n - rc) = MSE$	
Total (Corrected)	$n - 1$	SST		

Table 12.9 ANOVA table for test score increase data in Table 12.6 by combinations of three textbooks and two methods of instruction.

Source	DF	SS	MS	F	p-value
Textbooks	2	342.7	171.4	1.66	0.207
Methods of Instruction	1	910.0	910.0	8.81	0.006
Interaction	2	35.7	17.9	0.17	0.842
Error	30	3099.8	103.3		
Total	35	4388.3			

the F statistics associated with the hypotheses of no difference in the test score improvement between the two methods of instruction and among the three textbooks. There is a statistically significant effect for the methods of instruction — a p-value less than 0.05 — but no significant effect associated with the textbooks.

If we had removed the interaction term from the analysis after finding that it was not important, the error sum of squares would have been 3135.5 (= 35.7 + 3099.8), and there would have been 32 degrees of freedom associated with this error sum of squares. The error mean square would have been 97.98 instead of 103.3, and the F ratios for textbooks and methods of instruction would have been 1.75 and 9.29, respectively.

Let us explore further the preceding analytical results in relation to the cell means that were just calculated and are repeated here for our convenience. The lack of a significant main effect for textbooks is reflected in the marginal means for textbooks. The first and third textbooks appear to be a little more effective than the second book, but the ANOVA results indicated that these differences are not statistically significant. On the other hand, the discussion method was associated with a much greater increase — about 10 points — in test scores than the lecture method and this difference was statistically significant. The lack of an interaction effect is reflected in the cell means that are plotted in Figure 12.3.

	Methods of Instruction		
Textbook	Lecture	Discussion	Marginal Book Means
1	23.50	31.33	27.42
2	17.33	27.00	22.17
3	23.17	35.83	29.50
Marginal Method Means	21.33	31.39	26.36 (Overall Mean)

Interaction measures the degree of similarity between the responses to factor A at different levels of factor B. The lines connecting the three cell means for the discussion method are roughly parallel with the lines connecting cell means for the lecture method,

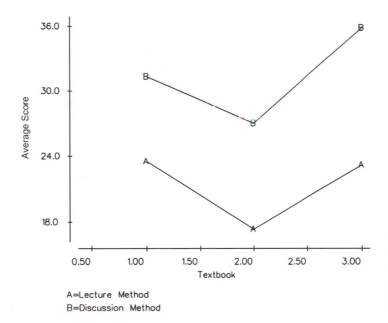

Figure 12.3 Plot of mean scores by methods of instruction on three textbooks (A = lecture method; B = discussion method).

A=Lecture Method
B=Discussion Method

reflecting the absence of interaction. If these two lines were not parallel or crossed each other, then the interaction effect would have been statistically significant. If a significant interaction is present, we need to examine the cell means carefully to draw appropriate conclusions. Computer packages can be used to conduct the preceding analysis (see **Program Note 12.2** on the website).

12.6 Linear Model Representation of the ANOVA

As shown in the last two sections, a two-way ANOVA can be used with or without interaction, which suggests that we need to specify the model to be used in the analysis. The choice of a model is dependent on how the data are collected and how we consider each effect to be specified. We consider this modeling aspect of ANOVA in this section.

In the ANOVA, we have partitioned the sum of squares of Y about its mean into within and between components in the completely randomized design or into treatment, blocks, and within components in the randomized block design. Underlying these partitions are linear models that show the relation between the dependent variable and the independent — treatment and/or blocking — variables. In the following sections, we show these models as well as the model with interaction. From these models, we can also see that it is possible to extend the ANOVA method of analysis to include combinations of the independent variables as well as including more than two independent variables.

12.6.1 The Completely Randomized Design

One representation of the linear model underlying the completely randomized design shows the dependent variable being equal to a constant plus a treatment effect plus individual variation — that is,

$$y_{ij} = \mu + \alpha_i + \varepsilon_{ij}$$

for i ranging from 1 to r and j going from 1 to n_i. The value of the jth observation of the dependent variable at the ith treatment level is y_{ij}. There are r levels of the treatment variable and n_i observations of Y at the ith treatment level. The constant is represented by μ, and the effect of the ith treatment level is represented by α_i. Since not everyone who has received the ith level of treatment will have the same value of the dependent variable, this individual variation, the departure from the sum of μ plus α_i, is represented by ε_{ij}.

Note that this model can be rewritten as follows:

$$y_{ij} = \mu + x_{ij}\,\alpha_i + \varepsilon_{ij}$$

where x_{ij} is an indicator variable which has the value of 1 if the ijth subject has received the ith level of the treatment and 0 otherwise. The X variable here simply indicates which level of treatment the person has received. We do not use this representation of the model here, but we shall refer to it in the next chapter.

In this linear model, there are $r + 1$ population parameters — the constant μ and the $r\alpha$'s; however, there are only r different treatment levels or groups. Since we can only estimate the same number of parameters as there are groups, to obtain estimators for r of the parameters, we must make some assumption about them. The appendix on the linear model in Forthofer and Lehnen (1981) provides a presentation of a number of assumptions that we could make. In this book, we shall measure the effect of the treatment levels from the effect of the rth treatment level. This means that α_r is assumed to be zero.

Now let us rewrite the linear model in terms of the population means. The equation for the ith level becomes

$$\mu_i = \mu + \alpha_i$$

and the representation of the model for all r levels is

$$\mu_1 = \mu + \alpha_1$$

$$\mu_2 = \mu + \alpha_2$$

$$\cdots$$

$$\mu_{r-1} = \mu + \alpha_{r-1}$$

$$\mu_r = \mu.$$

From these equations, we can see that the constant term is the mean of the rth level, and the effect of the other levels — $\alpha_1, \alpha_2, \ldots, \alpha_{r-1}$ — are measured from μ_r (or μ). For example, using the first of these equations to solve for α_1, we have

$$\alpha_1 = \mu_1 - \mu = \mu_1 - \mu_r.$$

This equation makes it clear that we are measuring the effects of the first level relative to the effect of the rth level, and the same is true for levels 2 through $r - 1$.

The sample estimator of the ith effect, $\hat{\alpha}_i$, is obtained by substituting the sample means for the population means — that is,

$$\hat{\alpha}_i = \bar{y}_{i\cdot} - \bar{y}_{r\cdot\cdot}$$

and the estimator of μ is simply $\bar{y}_{r\cdot\cdot}$.

The t test for comparing the means of two populations, assuming equal variances, also fits into the ANOVA framework. In this case, r is 2, and the preceding model still applies.

12.6.2   The Randomized Block Design with m Replicates

A linear model underlying the randomized block design has the dependent variable being equal to a constant plus the effect of the ith level of the treatment variable plus the jth block effect plus the individual variation term. In symbols, this is

$$y_{ijk} = \mu + \alpha_i + \beta_j + \varepsilon_{ijk}$$

where i goes from 1 to r, j ranges from 1 to c, and k ranges from 1 to m.

Just as in the completely randomized situation, the effects of the levels of the treatment variable are measured relative to the rth level of the treatment variable. In the same way, the effects of the levels of the blocking variable are measured relative to the cth level of the blocking variable. The definition of the parameters in terms of the μ_{ij} is complicated and will not be shown for this model, but it will be shown for the model in the next section.

12.6.3   Two-Way ANOVA with Interaction

The model for this situation is similar to the preceding two-way ANOVA model, except that it includes the interaction term, denoted by β_{ij}, in the model. The model is

$$y_{ijk} = \mu + \alpha_i + \beta_j + \alpha\beta_{ij} + \varepsilon_{ijk}$$

where i goes from 1 to r, j ranges from 1 to c, and k ranges from 1 to m.

The main effect terms in the model, the α_i and the β_j, again are all measured relative to their last level. The representation of this model in terms of the cell means, the μ_{ij}, for the first row is

$$\mu_{11} = \mu + \alpha_1 + \beta_1 + \alpha\beta_{11}$$
$$\mu_{12} = \mu + \alpha_1 + \beta_2 + \alpha\beta_{12}$$

$$\cdots$$

$$\mu_{1c} = \mu + \alpha_1.$$

Note that there is no β_c term or any $\alpha\beta_{1c}$ terms in the final equation. Since the cth level is the reference level for the column variable, β_c is taken to be zero. In addition, interaction terms having either an r or a c as a subscript are reference levels, and these interaction terms are also assumed to be zero. This pattern is repeated for the other rows except the last one.

$$\mu_{21} = \mu + \alpha_2 + \beta_1 + \alpha\beta_{21}$$
$$\mu_{22} = \mu + \alpha_2 + \beta_2 + \alpha\beta_{22}$$

$$\cdots$$

$$\mu_{2c} = \mu + \alpha_2$$

$$\cdots$$

$$\mu_{r1} = \mu + \beta_1$$

$$\mu_{r2} = \mu + \beta_2$$

$$\cdots$$

$$\mu_{rc} = \mu.$$

For the cells in the rth row, there is no α_r effect shown, since the rth level is the reference level for the row variable and α_r is taken to be zero. There are also no $\alpha\beta_{rj}$ terms in the last row, since the rth level is also a reference level for the interaction terms.

Using these equations, we obtain the following definitions of the parameters (μ, α, β, and $\alpha\beta$) in terms of the cell means. For example, from the last equation, we see that the constant term in the model is simply the mean of the cell formed by the rth row and cth column — that is, $\mu = \mu_{rc}$.

Once we have expressed μ in terms of the cell means, we can find the estimate of α_i from the equation $\mu_{ic} = \mu + \alpha_i$. This gives the solution that $\alpha_i = \mu_{ic} - \mu_{rc}$, where we have replaced μ_{rc} for μ. This definition for α_i is reasonable, as it compares the mean of the cell in the ith row and cth column with the mean of the cell in the rth row and cth column. It is comparing a cell in the ith row with its reference cell in the rth row. The column effect, β_j, is similarly defined as $\beta_j = \mu_{rj} - \mu_{rc}$.

The definition of the interaction term is $\alpha\beta_{ij} = (\mu_{ij} - \mu_{ic}) - (\mu_{rj} - \mu_{rc})$. The rcth cell is the reference cell and the other parameters are defined in terms of it. The ijth interaction parameter focuses on the difference of the jth and cth columns, and compares that difference for the ith and rth rows. If there is no interaction, the difference of the jth and cth columns is the same over all the rows.

12.7 ANOVA with Unequal Numbers of Observations in Subgroups

In the preceding discussion, we allowed the number of observations in each treatment to vary for the one-way ANOVA model, but we assumed an equal number of observations in each cell for the two-way ANOVA models. However, it is not always possible to have an equal number of observations on all treatment combinations. Even balanced designs often become unbalanced because people may drop out of the study or some of the data are missing. This imbalance in the size of subclasses introduces complications in the analysis. The main difficulty is that there is no unique way of finding the sums-of-squares corresponding to each main effect and each interaction.

One method of calculating the sums of squares is to consider the factors and interaction(s) sequentially. The effect of the first factor entered into the model is calculated unconditionally, but the second factor is evaluated conditional upon whatever factor was entered first. Using the notation used for conditional probability, the effect of the second factor (s) conditional on the first factor (f) can be written as $s|f$. The partitioned sums of squares obtained by sequential fitting are labeled as Type I SS in the computer output. The sequential sums of squares will add up to the total sum of squares.

One problem with the sequential approach is that the factors are treated differently depending on the order of entry, and effects of factors are difficult to interpret. An alternative approach is to consider the effect of each factor adjusted for all the other factors in the model. This approach produces adjusted sums of squares (called Type III SS) that do not add up to the total SS unless the data are balanced. The adjusted sums of squares are generally used when testing the effect of each factor (Maxwell and Delaney 1990). We will use this approach in examining an example following.

Computer packages provide Type III and Type I sums of squares by default. There are two other types of sums of squares, but they are generally of lesser interest. In cases where the interaction terms are unimportant and a main effect needs to be examined at each level of the other factors, Type I sums of squares are recommended (Nelder 1977).

Example 12.9

Let us consider the data shown in Table 12.10. This table represents a cross-classification of creatinine measurements by sex and age (categorized into two groups: under 56 and 56 & over) for the 40 patients in DIG40 that was introduced in Chapter 3. The numbers of observations in the four cells are not equal, and, therefore, we will use the general linear models procedure for these unbalanced data.

Table 12.10   Creatinine measurements (mg/dL) by sex and age group, DIG40.

Cell	Sex	Age	Creatinine Measurements (mg/dL)									
1,1	Male	<56	1.600	1.300	1.159	1.307	1.886	1.034	0.900	1.398	1.307	
1,2	Male	≥56	2.682	1.091	1.250	1.705	2.000	1.227	1.100	2.239	1.300	1.614
			1.200	1.455	1.489	1.700	1.307	1.200	1.273	1.300	1.659	1.261
			0.900									
2,1	Female	<56	1.386	0.900	1.000	1.148	1.170					
2,2	Female	≥56	1.534	0.900	0.900	1.352	0.909					

Before analyzing the data by ANOVA, let us look at the cell means. These are shown here in a 2 by 2 table:

	Ages		
Sex	**<56**	**≥56**	**Difference**
Male	1.3212	1.4739	−0.1527
Female	1.1236	1.1113	0.0123
Difference	0.1976	0.3626	

The effect of the sex variable (difference between male and female) can be calculated at each level of the age variable. Likewise, the effect of the age variable can be seen in the difference between the two age groups at each level of the sex variable. The

Table 12.11 ANOVA by the general linear models procedure for Table 12.10 (with interaction).

Source	DF	Adjusted (Type III)				Sequential SS (Type I)
		SS	MS	F	P	
Sex	1	0.5519	0.5519	4.21	0.048	0.7124
Age	1	0.0407	0.0407	0.31	0.581	0.1042
Sex*Age	1	0.0427	0.0427	0.33	0.572	0.0427
Error	36	4.7247	0.1312			4.7247
Total	39	5.5839				5.5839

Table 12.12 ANOVA by general linear models procedure for Table 12.10 (without interaction).

Source	DF	Adjusted (Type III)				Sequential SS (Type I)
		SS	MS	F	P	
Sex	1	0.5951	0.5951	4.62	0.038	0.7124
Age	1	0.1042	0.1042	0.81	0.374	0.1042
Error	37	4.7674	0.1288			4.7674
Total	39	5.5839				5.5839

effect of the sex variable is considerably larger than the age effect. The effects of the sex variable at two levels of age are in the same direction, although the effect of the sex variable is larger for the older age group, suggesting that the interaction of the sex and age variables on creatinine is likely to be nonsignificant.

Table 12.11 shows the results of ANOVA by the general linear models procedure. The model includes two main effects and the interaction of sex and age. Adjusted sums of squares are shown, and the analysis of variance is carried out based on Type III sums of squares to assess the effect of each term. Sequential sums of squares are shown in the last column. Note that sequential sum of squares of the last term listed in the model (the interaction term in this case) is the same as the Type III sum of squares for the factor because it is adjusted to all other factors in the model. The effect of the sex variable is significant, while the effect of the age variable is not.

Since the interaction effect is unimportant, we dropped the interaction term from the model and repeated the analysis. The results are shown in Table 12.12. Note that the sequential sum of squares for age is the same as the adjusted sum of squares for the factor because it is the last term in the model. Note also that the sum of squares due to the interaction is now included in the error term. Again, the sex variable effect is significant, while the age effect is not.

It is usually necessary to use a computer package for performing a general linear models analysis (see **Program Note 12.3** on the website).

Choosing an appropriate model for ANOVA is not straightforward, especially for an unbalanced data. As we saw in Chapter 6, it is important to strive for a balanced design to alleviate the complications in the analysis.

Conclusion

In this chapter we presented several basic models of analysis of variance. The one-way ANOVA is used to analyze data from a completely randomized experimental design. The two-way ANOVA can be used for a randomized block design as well as for a two-factor design with interaction. To use these analytical methods properly, we must be aware of how the data were collected and make sure that the data meet the ANOVA assumptions. Finally, we discussed the problems and methods for analyzing unbalanced data. In the next chapter, we will expand the linear model to regression models.

EXERCISES

12.1 The data shown here, taken from Brogan and Kutner (1980), are the change in the maximal rate of urea synthesis (MRUS) level for cirrhotic patients who underwent either a standard operation (a nonselective shunt) or a new procedure (a selective shunt). The purpose of the operations was to improve liver function, measured by MRUS. A low value of MRUS is associated with poor liver function. Patients in the nonselective shunt group are divided into two groups based on the preoperative MRUS values (≤ 40 and >40).

Change in Maximal Rate of Urea Synthesis (MRUS) Level (mg Urea N/hr/kg Body Weight) by Group							
Group	Change in MRUS Values						
Selective Shunt	−3	20	−6	−5	−3	−3	−6	12
Nonselective Shunt I	−18	−4	−18	−18	−6	−18		
Nonselective Shunt II	−24	−7	−15	4	−14	−8	−11	

Perform an analysis of variance of these data at the 0.05 significance level to determine if there is a difference in the three groups. If there is a significant difference, use an appropriate multiple comparison procedure to find the source of the difference.

12.2 In Chapter 8, we used the t test to compare the proportion of caloric intake from fat for fifth- and sixth-grade boys compared to seventh- and eighth-grade boys. The calculated t test statistic was -0.727 (Example 8.9). Perform a one-way ANOVA on these data in Table 7.7 and compare your results with the t test approach. How does the t statistic compare with the F statistic?

12.3 For the weight change data shown in Table 12.4, we were concerned about the level of physical activity of the women. Instead of using the site — headquarters or plant — as a way of controlling for physical activity, how else might we have controlled for the physical activity? Do you think that a control group — no intervention — should have been used? Explain your reasoning. Would you do anything to determine whether or not the women used the memberships? What, if any, other variables should be included in the analysis?

12.4 To investigate publication bias, 75 referees for one journal were randomly assigned to receive one of five versions of a manuscript (Dickersin 1990). All versions were identical in the Introduction and Methods sections but varied in either the Results or Discussion sections. The first and second groups received versions with either positive or negative results, respectively. The third and fourth groups received versions with mixed results and either positive or negative discussion. The fifth group was asked evaluate the manuscript on the basis of the Methods section, and no data were provided. The referees used a scale of 0 to 6 (low to high) to rate different aspects of the manuscript. The average scores for three aspects are shown here:

Manuscript Version	No. of Referees	Mean Ratings		
		Methods	Scientific Contribution	Publication Merit
Positive Results	12	4.2	4.3	3.2
Negative Results	14	2.4	2.4	1.8
Mixed Results with Positive Discussion	13	2.5	1.6	0.5
Mixed Results with Negative Discussion	14	2.7	1.7	1.4
Methods Only	14	3.4	4.5	3.4

State an appropriate linear model for this experiment using scientific contribution as the dependent variable. What are the null and alternative hypotheses of interest for this model? Assuming that the standard deviations for the scientific contribution score for the five groups are 1.1, 0.9, 0.7, 0.8, and 1.1, respectively, perform an analysis of variance of these data at the 0.05 significance level to determine if there is a bias in refereeing scientific papers for this journal. If there is a significant difference, use an appropriate multiple comparison procedure to find the source of the bias. State your conclusions clearly.

12.5 In an investigation of the effect of smoking on work performance under different lighting conditions in a large company, a random sample of nine male workers was selected from each of the three smoking status groups: nonsmokers, moderate smokers, and heavy smokers. Each sample was randomly assigned to three working environments with different levels of lighting. The time to complete a standard assembling task was recorded in minutes. The sums of squares were as follows:

Source	df	SS	MS	F	p-value
Smoking Status		84.90			
Lighting Conditions		298.07			
Interaction		2.81			
Error		59.25			
Total		445.03			

Perform an analysis of variance for these data to examine the interaction of the variables at the 0.05 significance level. If there is no significant interaction, test whether or not the smoking and lighting conditions variables have significant effects on the workers' performance and state your conclusions.

12.6 The midterm and final test scores of 12 students in a class are recorded as follows:

	Students											
Test	1	2	3	4	5	6	7	8	9	10	11	12
Midterm	80	85	65	77	58	98	91	72	62	82	45	42
Final	78	90	72	80	71	92	93	70	73	85	60	61

These are paired data. First, perform the paired t test (discussed in Chapter 8) to compare the two sets of test scores. Now perform a two-way ANOVA for the randomized block design with two replicates (an additive model), considering students as blocks, and compare your results with the paired t test approach. How does the t statistic compare with the F statistic for the test variable? Can you draw the same conclusion?

12.7 A randomized study was conducted to compare the effects of two intervention procedures (tracking with outreach and provider prompting) to raise immunization in primary care clinics serving impoverished children (Rodewald et al. 1999). The study used a 2 by 2 factorial design, and each intervention had two levels (1 = no intervention; 2 = intervention). After 18 months of intervention the immunization status was assessed. Two major outcome measures were the percentage of immunization and the number of days of delay in immunization. The authors claim that the two-way ANOVA was used to test for effects of each of the interventions on the outcome measures. They stated that the interaction was insignificant. The number of children allocated to each group was slightly different and the number of children who completed the assessment of the outcomes also varied as shown here:

Treatment Groups		Number Allocated	Number Completed	Percent Immunized	Mean Days of Delay[a]
Prompting	Outreach (Group)				
1	1 (no intervention)	769	719	74.0	140.0
1	2 (outreach only)	715	630	95.1	76.5
2	1 (prompting only)	801	744	75.9	133.2
2	2 (both interventions)	732	648	95.1	69.1

[a]Mean days of delay in immunization

Discuss whether the two-way ANOVA was appropriate for the two major outcome measures shown. If you think the ANOVA is inappropriate for any of the outcome measures, what statistical method would you recommend? If you think the ANOVA is appropriate for any of outcome measure, would you accept the claim of no interaction based on these data? A considerable number of subjects were lost during the course of the study, and the number of dropouts varies between the four groups. Discuss how the differential loss of the subjects might impact the study outcome.

REFERENCES

Brogan, D. R., and M. H. Kutner. "Comparative Analyses of Pretest-Posttest Research Designs." *The American Statistician* 34:229–232, 1980.

Dickersin, K. "The Existence of Publication Bias and Risk Factors for Its Occurrence." *Journal of American Medical Association* 263:1385–1389, 1990.

Forthofer, R. N., and R. G. Lehnen. *Public Program Analysis: A New Categorical Data Approach.* Belmont, CA: Lifetime Learning Publications, 1981.

Higgins, J. E., and G. G. Koch. "Variable Selection and Generalized Chi-Square Analysis of Categorical Data Applied to a Large Cross-Sectional Occupational Health Survey." *International Statistical Review* 45:51–62, 1977.

Kimball, B. P., B. L. Shurvell, S. Houle, J. C. Fulop, H. Rakowski, and P. R. McLaughlin. "Persistent Ventricular Adaptations in Postoperative Coarctation of the Aorta." *Journal of the American College of Cardiology* 8:172–178, 1986.

Kleinbaum, D. G., L. L. Kupper, and K. E. Muller. *Applied Regression Analysis and Other Multivariable Methods*, 2nd ed. Boston: PWS-Kent, 1988, Chapter 12.

Lin, L. I., and E. F. Vonesh. "An Empirical Nonlinear Data-Fitting Approach for Transforming Data to Normality." *The American Statistician* 43:237–243, 1989.

Maxwell, S. E., and H. D. Delaney. *Designing Experiments and Analyzing Data.* Belmont, CA: Wadsworth, 1990.

Nelder, J. A. "A Reformulation of Linear Models." *Journal of the Royal Statistical Society* 140:48–63, 1977.

Rodewald, L. E., P. G. Szilagyi, S. G. Humiston, R. Barth, R. Kraus, and R. F. Raubertas. "A Randomized Study of Tracking with Outreach and Provider Prompting to Improve Immunization Coverage and Primary Care." *Pediatrics* 103 (1):31–38, January 1999.

Scheffé, H. *The Analysis of Variance.* New York: John Wiley & Sons, 1959, Chapter 10.

Steel, R. G. D., and J. H. Torrie. *Principles and Procedures of Statistics: A Biometrical Approach*, New York: McGraw-Hill, 1980, Chapter 9.

Stoline, M. R. "The Status of Multiple Comparisons: Simultaneous Estimation of All Pairwise Comparisons in One-Way ANOVA Designs." *The American Statistician* 35:134–141, 1981.

Linear Regression

13

In this chapter we present methods for examining the relation between a response or dependent variable and one or more predictor or independent variables. The methods are based on the linear model introduced in Chapter 12. In linear regression, we examine the relation between a normally distributed response or dependent variable and one or more continuous predictor or independent variables. In a sense, linear regression is an extension of the correlation coefficient. Although linear regression was created for the examination of the relation between continuous variables, in practice, people often use the term linear regression even when continuous and discrete independent variables are used in the analysis.

Linear regression is one of the more frequently used techniques in statistics today. These methods are often used because problems, particularly those concerning humans, usually involve several independent variables. For example, in the creation of norms for lung functioning, age, race, and sex are taken into account. Linear regression is one approach that allows multiple independent variables to be used in the analysis. In the linear regression model, the dependent variable is the observed pulmonary function test value and age, race, and sex are the independent variables. When the dependent variable is a discrete variable as in the disease status (presence or absence), logistic regression (the topic of the next chapter) is used to consider many possible risk factors related to the disease.

13.1 Simple Linear Regression

Simple linear regression is used to examine the relation between a normally distributed dependent variable and a continuous independent variable. An example of a situation where simple linear regression is useful is the following.

Some physicians believe that there should be a standard — a value that only a small percentage of the population exceeds — for blood pressure in children (NHLBI Task Force 1987). When a standard is used, it is desirable that it be easy for the physician to quickly and accurately determine how the patient relates to the standard. Therefore, the standards should be based on a small number of variables that are easy to measure. Since it is known that blood pressure is related to maturation, the variables used in the development of the standard should, therefore, reflect maturation. Two variables that are related to maturation and are easy to measure are age and height. Of these two variables, height appears to be the more appropriate variable for the development of standards (Forthofer 1991; Gillum, Prineas, and Horibe 1982; Voors et al. 1977). Because of physiological differences, the standards are developed separately for females and males. In the following, we shall focus on systolic blood pressure (SBP).

In developing the standards, we are going to assume that the mean SBP for girls increases by a constant amount for each one unit increase in height. The use of the mean instead of the individual SBP values reflects the fact that there is variation in the SBP of girls of the same height. Not all the girls who are 50 inches tall have the same SBP value; their SBPs vary about the mean SBP of girls who are 50 inches tall. The assumption of a constant increase in the mean SBP for each one unit increase in height is characteristic of a linear relation. Thus, in symbols, the relation between Y, the SBP variable, and X, the height variable, can be expressed as

$$\mu_{Y|X} = \beta_0 + \beta_1 X$$

where $\mu_{Y|X}$ is the mean SBP for girls who are X units tall, β_0 is a constant term, and β_1 is the coefficient of the height variable — that is, β_1 is the increase in the mean SBP for each one unit change in height. The β_0 coefficient is the Y intercept and β_1 is the slope of the straight line.

In general, the X variable shown in the preceding expression may represent the square, the reciprocal, the logarithm, or some other nonlinear transformation of a variable. This is acceptable in linear regression because the expression is really a linear combination of the β_i's, not of the independent variables.

The preceding equation is similar to the linear growth model in Chapter 3 and the linear model representation of ANOVA. In the ANOVA model, values of the X variables, 1 or 0, indicate which effect should be added in the model. In the regression model, the values of the X variable are the individual observations of the continuous independent variable. The parameters in the ANOVA model are the effects of the different levels of the independent variable. In the regression model, the parameters are the Y-intercept and the slope of the line.

Figure 13.1 shows the graph of this simple linear regression equation. The $\otimes$ symbols show the values of the mean SBP for the different values of height that we are considering. As we can see, a straight line does indeed have a rate of increase in the mean SBP that is constant for each one unit increase in height. The ■ symbols show the projected values of the mean SBP, assuming that the relationship holds for very small height values as well. It is usually inappropriate to estimate the values of $\mu_{Y|X}$ for values of X outside the range of observation. The point at which the projected line intersects the $\mu_{Y|X}$ axis is β_0. Since β_1 is the amount of increase in $\mu_{Y|X}$ for each one unit increase in X, the bracketed change in $\mu_{Y|X}$ is $8\,\beta_1$, since X has increased 8 units from x_1 to x_2. Note that if the regression line is flat — that is, parallel to the X axis — this means that there is no change in $\mu_{Y|X}$ regardless of how much X changes. Thus, if the regression line is flat, then β_1 is zero and there is no linear relation between $\mu_{Y|X}$ and X.

If we wish to express this relationship in terms of individual observations, we must take the variation in SBP for each height into account. The model that does this is

$$y_i = \beta_0 + \beta_1 x_i + \varepsilon_i$$

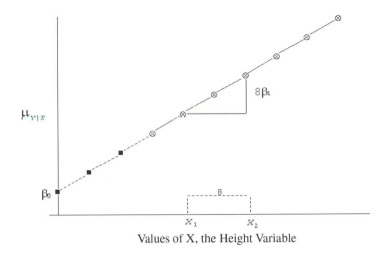

Figure 13.1 Line showing the regression of $\mu_{Y|X}$ on X.

where ε_i represents the difference between the mean SBP value at height x_i and the SBP of the ith girl who is also x_i units tall. The ε term is also referred to as the residual or error term. Knowledge of β_0 and β_1 is necessary in developing the standards for SBP. However, we do not know them and we have to collect data to estimate these values.

13.1.1   Estimation of the Coefficients

There are a variety of ways of estimating β_0 and β_1. We must decide on what criterion we will use to find the "best" estimators of these two coefficients. Possible criteria include minimization of the following:

1. The sum of the differences of y_i and $\hat{y}_i$, where y_i is the observed value of the SBP and $\hat{y}_i$ is the estimated value of the SBP for the ith girl. The value of $\hat{y}_i$ is found by substituting the estimates of β_0 and β_1 in the simple linear regression equation — that is, $\hat{y}_i = \hat{\beta}_0 + x_i \hat{\beta}_1$, where x_i is the observed value of height for the ith girl.
2. The sum of the absolute differences of y_i and $\hat{y}_i$.
3. The sum of the squared differences of y_i and $\hat{y}_i$.

The first criterion can be made to equal zero by setting $\hat{\beta}_1$ to zero and letting $\hat{\beta}_0$ equal to the sample mean. The use of the absolute value yields interesting estimators, but the testing of hypotheses is more difficult with these estimators. Based on considerations similar to those discussed in Chapter 3 in the presentation of the variance, we are going to use the third criterion to determine our "best" estimators.

Thus our estimators of the coefficients will be derived based on the minimization of the sum of squares of the differences of the observed and estimated values of SBP. In symbols, this is the minimization of

$$\sum_i (y_i - \hat{y}_i)^2.$$

The use of this criterion provides estimators that are called *least squares estimators* because they minimize the sum of squares of the differences.

The least squares estimators of the coefficients are given by

$$\hat{\beta}_1 = \frac{\sum_{i=1}^{n}(x_i - \bar{x})(y_i - \bar{y})}{\sum_{i=1}^{n}(x_i - \bar{x})^2} = \frac{\sum_{i=1}^{n} x_i y_i - n\bar{x}\bar{y}}{\sum_{i=1}^{n} x_i^2 - n\bar{x}^2}$$

and

$$\hat{\beta}_0 = \bar{y} - \hat{\beta}_1\bar{x}.$$

The second formula for $\hat{\beta}_1$ is provided because it is easier to calculate. Let's use these formulas to calculate the least squares estimates for the data in Table 13.1. The hypothetical values of the SBP and height variables for the 50 girls are based on data from the NHANES II (Forthofer 1991).

The value of $\hat{\beta}_1$ is found from

$$\hat{\beta}_1 = \frac{\sum_{i=1}^{n} x_i y_i - n\bar{x}\bar{y}}{\sum_{i=1}^{n} x_i^2 - n\bar{x}^2} = \frac{269902 - 50(52.5)(101.5)}{142319 - 50(52.5)^2} = 0.7688.$$

The calculation of $\hat{\beta}_0$ is easier to perform, and its value is found from

$$\hat{\beta}_0 = \bar{y} - \hat{\beta}_1\bar{x} = 101.5 - 0.7688(52.5) = 61.138.$$

Table 13.1   Hypothetical data — SBP and predicted SBP[a] (mmHg) and height (inches) for 50 girls.

SBP	Predicted SBP	Height	SBP	Predicted SBP	Height	SBP	Predicted SBP	Height
105	88.8	36	120	98.0	48	94	106.5	59
90	89.6	37	114	98.8	49	88	107.3	60
82	90.4	38	78	98.8	49	110	107.3	60
96	90.4	38	116	99.6	50	124	107.3	60
82	91.1	39	74	99.6	50	86	108.0	61
74	91.1	39	80	100.3	51	120	108.0	61
104	91.9	40	98	101.1	52	112	108.8	62
100	91.9	40	90	101.9	53	100	109.6	63
80	92.7	41	92	102.7	54	122	110.3	64
98	93.4	42	80	102.7	54	122	110.3	64
96	94.2	43	88	102.7	54	110	111.1	65
86	95.0	44	104	103.4	55	124	111.1	65
88	95.0	44	100	104.2	56	122	111.9	66
128	95.0	44	126	105.0	57	94	112.6	67
118	95.7	45	108	105.7	58	110	112.6	67
90	96.5	46	106	106.5	59	140	114.2	69
108	98.0	48	98	106.5	59			

[a]Predicted using the least squares estimates of the regression coefficients.

The estimated coefficient of the height variable is about 0.8, which means that there is an increase of 0.8 mmHg in SBP for an increase of 1 inch in height for girls between the heights of 36 and 69 inches. The estimate of the β_0 coefficient is about 60 mmHg and that is the Y intercept. Based on projecting the regression line beyond the data values observed, the Y intercept gives the value of SBP for a girl 0 inches tall. However, it does not make sense to talk about the SBP for a girl 0 inches tall, and this shows one of the dangers of extrapolating the regression line beyond the observed data.

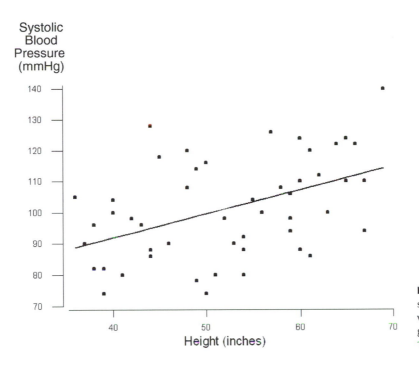

Systolic Blood Pressure (mmHg)

Height (inches)

Figure 13.2 Plot of systolic blood pressure versus height for 50 girls shown in Table 13.1.

Figure 13.2 is a plot of SBP versus height for the data shown in Table 13.1. From this plot, we can see that there is a slight tendency for the larger values of SBP to be associated with the larger values of height, but the relationship is not particularly strong. The path of the regression line is shown within the range of observations.

We can use the preceding estimates of the population coefficients in predicting SBP values for the hypothetical data shown in Table 13.1. For example, the predicted value of SBP for the first observation in Table 13.1, a girl 36 inches tall, is

$$61.138 + 0.7688(36) = 88.82 \text{ mmHg.}$$

The other predicted SBP values are found in the same way, and they are also shown in Table 13.1.

13.1.2   The Variance of $Y|X$

Before going forward with the use of the regression line in the development of the standards, we should examine whether or not the estimated regression line is an improvement over simply using the sample mean as an estimate of the observed values. One way of obtaining a feel for this is to examine the sum of squares of deviations of Y from $\hat{Y}$ — that is,

$$\sum_{i=1}^{n}(y_i - \hat{y}_i)^2.$$

If we subtract and add $\bar{y}$ in this expression, we can rewrite this sum of squares as

$$\sum_{i=1}^{n}[(y_i - \bar{y}) - (\hat{y}_i - \bar{y})]^2$$

and we have not changed the value of the sum of squares. However, this sum of squares can be rewritten as

$$\sum_{i=1}^{n}(y_i - \hat{y}_i)^2 = \sum_{i=1}^{n}(y_i - \bar{y})^2 - \sum_{i=1}^{n}(\hat{y}_i - \bar{y})^2$$

because the crossproduct terms, $(y_i - \bar{y})(\hat{y}_i - \bar{y})$, sum to zero. In regression terminology, the first sum of squares is called the *sum of squares about regression* or the *residual or error sum of squares*. The second sum of squares, about the sample mean, is called the *total sum of squares* (corrected for the mean) and the third sum of squares is called the *sum of squares due to regression*. If we rewrite this equation, putting the total sum of squares (corrected for the mean) on the left side of the equal sign, we have

$$\sum_{i=1}^{n}(y_i - \bar{y})^2 = \sum_{i=1}^{n}(y_i - \hat{y}_i)^2 + \sum_{i=1}^{n}(\hat{y}_i - \bar{y})^2.$$

This equation shows the partition of the total sum of squares into two components, the sum of squares about regression, and the sum of squares due to regression.

Figure 13.3 is a graph which shows the differences, $(y_i - \bar{y})$, $(y_i - \hat{y}_i)$ and $(\hat{y}_i - \bar{y})$, for one y_i. In Figure 13.3, the regression line is shown as well as a horizontal line that shows the value of the sample mean. We have focused on the last point, the girl who is 69 inches tall and who has an SBP of 140 mmHg. For this point, the deviation of the observed SBP of 140 from the sample mean of 101.5 can be partitioned into two components. The first component is the difference between the observed value and 114.2, the value predicted from the regression line. The second component is the difference between this predicted value and the sample mean. This partitioning cannot be done for many of the points, since, for example, the sample mean may be closer to the observed point than the regression line is.

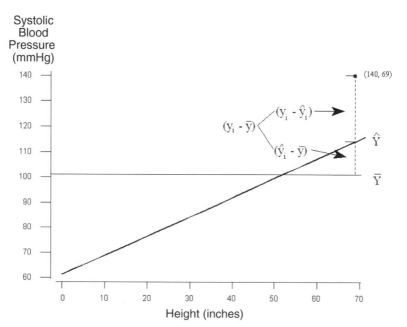

Figure 13.3   An observed value in relation to the regression line and the sample mean.

Ideally, we would like the sum of squares about the regression line to be close to zero. From the last preceding equation, we see that the sum of the square deviations from the regression line must be less than or equal to the sum of the square deviations from the sample mean. However, the direct comparison of the sum of squares is not fair, since they are based on different degrees of freedom. The sum of squares about the sample mean has $n - 1$ degrees of freedom, as we discussed in the material about the variance. Since we estimated two coefficients in obtaining the least squares estimator of Y, there are thus $n - 2$ degrees of freedom associated with sum of squares about $\hat{Y}$. Thus, let us compare s_Y^2 with $s_{Y|X}^2$ — that is,

$$\frac{\sum_{i=1}^{n}(y_i - \bar{y})^2}{n-1} \quad \text{versus} \quad \frac{\sum_{i=1}^{n}(y_i - \hat{y}_i)^2}{n-2}.$$

If $s_{Y|X}^2$ is much less than s_Y^2, then the regression was worthwhile; if not, then we should use the sample mean as there appears to be little linear relation between Y and X.

Let us calculate the sample variance of Y and the sample variance of Y, taking X into account. The sample variance of Y is

$$\frac{\sum_{i=1}^{n}(y_i - \bar{y})^2}{n-1} = \frac{12780}{50-1} = 260.827$$

and the sample variance of Y given X is

$$\frac{\sum_{i=1}^{n}(y_i - \hat{y}_i)^2}{n-2} = \frac{10117}{50-2} = 210.772.$$

Thus, $s_{Y|X}^2$ is less than s_Y^2. The use of the height variable has reduced the sample variance from 260.827 to 210.772, about a 20 percent reduction. It appears that the inclusion of the height variable has allowed for somewhat better estimation of the SBP values.

13.1.3 The Coefficient of Determination (R^2)

An additional way of examining whether or not the regression was helpful is to divide the sum of squares due to regression by the sum of squares about the mean — that is,

$$\frac{\sum_{i=1}^{n}(\hat{y}_i - \bar{y})^2}{\sum_{i=1}^{n}(y_i - \bar{y}_i)^2} = \frac{\sum_{i=1}^{n}(y_i - \bar{y}_i)^2 - \sum_{i=1}^{n}(y_i - \hat{y}_i)^2}{\sum_{i=1}^{n}(y_i - \bar{y}_i)^2}.$$

If the regression line provides estimates of the SBP values that closely match the observed SBP values, this ratio will be close to one. If the regression line is close to the mean line, then this ratio will be close to zero. Hence, the ratio provides a measure that varies from 0 to 1, with 0 indicating no linear relation between Y and X, and 1 indicating a perfect linear relation between Y and X. This ratio is denoted by R^2, and is called the *coefficient of determination*. It is a measure of how much of the variation in Y is accounted for by X. R^2 is also the square of the sample Pearson correlation coefficient between Y and X.

For the SBP example, the value of R^2 is

$$\frac{12780 - 10117}{12780} = 0.2084.$$

Approximately 21 percent of the variation in SBP is accounted for by height for girls between 36 to 69 inches tall. This is not an impressive amount. Almost 80 percent of the variation in SBP remains to be explained. Even though this measure of the relation between SBP and height is only 21 percent, it is larger than its corresponding value for the relation between SBP and age.

The derivation of the R^2 term is based on a linear model that has both a β_0 and a β_1 term. If the model does not include β_0, then a different expression must be used to calculate R^2.

The sample Pearson correlation coefficient, r, is defined as

$$\frac{\sum_{i=1}^{n}(x_i - \bar{x})(y_i - \bar{y})}{\sqrt{\sum_{i=1}^{n}(x_i - \bar{x})^2 \sum_{i=1}^{n}(y_i - \bar{y})^2}}$$

and its numerical value is

$$r = \frac{3464.5}{\sqrt{4506.6(12780)}} = 0.4565.$$

If we square r, r^2 is 0.2084, which agrees with R^2, as it must.

Although, symbolically, R^2 is the square of the sample Pearson correlation coefficient, R^2 does not necessarily measure the strength of the linear association between Y and X. In correlation analysis, the observed pairs of values of Y and X are obtained by simple random sampling from a population. In correlation analysis, we don't necessarily consider one of the variables to be the dependent variable and the other the independent variable. The sample r simply measures the strength of the linear association between the two variables. In contrast, linear regression provides a formula that describes the linear relation between a dependent variable and an independent variable(s). To discover that relationship, we often use stratified random sampling — that is, we select simple random samples of Y for specified values of X; however, as Ranney and Thigpen (1981) show, the value of R^2 depends on the range of the values of X used in the analysis, the number of repeated observations at given values of X, and the location of the X values. Hence, although symbolically R^2 is the square of the correlation coefficient between two variables, it does not necessarily measure the strength of the linear association between the variables. It does reflect how much of the variation in Y is accounted for by knowledge of X. Korn and Simon provide more on the interpretation of R^2 (Korn and Simon 1991).

There is also a relation between the sample correlation coefficient and the estimator of β_1. From Chapter 3, we had another form for r than the defining formula given above and it was

$$r = \frac{\sum (x_i - \bar{x})(y_i - \bar{y})/(n-1)}{s_x s_y}.$$

The estimator of β_1 is

$$\hat{\beta}_1 = \frac{\sum (x_i - \bar{x})(y_i - \bar{y})}{\sum (x_i - \bar{x})^2} = \frac{\sum (x_i - \bar{x})(y_i - \bar{y})/(n-1)}{s_x^2}.$$

If we multiply r by s_y and divide r by s_x, we have

$$\left(\frac{s_y}{s_x}\right) r = \left(\frac{s_y}{s_x}\right) \frac{\sum (x_i - \bar{x})(y_i - \bar{y})/(n-1)}{s_x s_y}$$

or

$$\left(\frac{s_y}{s_x}\right) r = \frac{\sum (x_i - \bar{x})(y_i - \bar{y})/(n-1)}{s_x^2} = \hat{\beta}_1.$$

As the preceding relation shows, if the correlation coefficient is zero, the slope coefficient is also zero and vice versa.

13.2   Inference about the Coefficients

The parametric approach to testing hypotheses about a parameter requires that we know the probability distribution of the sample estimator of the parameter. The standard approach to finding the probability distributions of the sample estimators of β_0 and β_1 is based on the following assumptions.

13.2.1   Assumptions for Inference in Linear Regression

We assume that the y_i's are independent, normally distributed for each value of X, and that the normal distributions at the different values of X all have the same variance, σ^2. Figure 13.4 graphically shows these assumptions. The regression line, showing the relation between $\mu_{Y|X}$ and X, is graphed as well as the distributions of Y at the selected values of

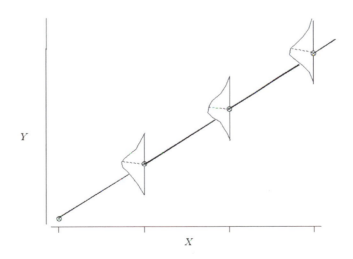

Figure 13.4
Distribution of Y at selected values of X.

X. Note that Y is normally distributed at each of the selected X values and that the normal distributions have the same shapes — that is, the same variance, σ^2. The mean of the normal distribution, $\mu_{Y|X}$, is obtained from the regression equation and is $\beta_0 + \beta_1 X$.

In the following, we shall consider the values of the X variable to be fixed. There are two ways that X can be viewed as being fixed. First, we may have used a stratified sample, stratified on height, to select girls with the heights shown in Table 13.1. Since we have chosen the values of the height variable, they are viewed as being fixed. In a second way, we consider our results to be conditional on the observed values of X. The conditional approach is usually used with simple random samples in which both Y and X otherwise would be considered to be random variables. This is the conventional approach, and it means that the error or residual term, ε, also follows a normal distribution with mean 0 and variance σ^2. Note that the least squares estimation of the regression coefficients did not require this specification of the probability distribution of Y.

Before testing hypotheses about the regression coefficients, we should attempt to determine whether or not the assumptions just stated are true. We should also examine whether or not any single data point is exercising a large influence on the estimates of the regression coefficients. These two issues are discussed in the next section.

13.2.2 Regression Diagnostics

In our brief introduction to regression diagnostics — methods for examining the regression equation — we consider only two of the many methods that exist. More detail on other methods is given in Kleinbaum et al. (1998). The first method we shall present involves plotting of the residuals. Plots are used in an attempt to determine whether or not the residuals or errors are normally distributed or to see if there are any patterns in the residuals. The second method tries to discover the existence of data points that play a major role in the estimation of the regression coefficients.

Residuals and Standardized Residuals: The sample estimator of ε_i is the residual e_i, defined as the difference between y_i and $\hat{y}_i$, and the e_i can be used to examine the regression assumptions. Since we are used to dealing with standardized variables, people often consider a standardized residual, $e_i/s_{Y|X}$, instead of e_i itself. The standardized residuals should approximately follow a standard normal distribution if the regression assumptions are met. Thus, values of the standardized residuals larger than 2.5 or less than -2.5 are unusual. Table 13.2 shows these residuals and a quantity called leverage (described in the next section) for the data in Table 13.1.

We use the standardized residuals in our examination of the normality assumption. Other residuals could also be used for this examination (Kleinbaum 1998). The normal scores of the standardized residuals are plotted in Figure 13.5. The normal scores plot looks reasonably straight; thus the assumption that the error term is normally distributed does not appear to be violated.

If this plot deviates sufficiently from a straight line to cause us to question the assumption of normality, then it may be necessary to consider a transformation of the dependent variable. There are a number of mathematical functions which can be used to transform nonnormally distributed data to normality (Kleinbaum 1998; Lin and Vonesh 1989; Miller 1984).

Table 13.2   Residuals and leverage for the data in Table 13.1.

Y	Residual	Standardized Residual	Leverage h_i	Y	Residual	Standardized Residual	Leverage h_i
105	16.1848	1.16253	0.08041	92	−10.6532	−0.74143	0.02049
90	0.4161	0.02977	0.07331	80	−22.6532	−1.57659	0.02049
82	−8.3527	−0.59552	0.06665	88	−14.6532	−1.01982	0.02049
96	5.6473	0.40264	0.06665	104	0.5781	0.04025	0.02138
82	−9.1215	−0.64818	0.06044	100	−4.1907	−0.29199	0.02271
74	−17.1215	−1.21667	0.06044	126	21.0405	1.46735	0.02449
104	12.1097	0.85790	0.05467	108	2.2717	0.15861	0.02671
100	8.1097	0.57452	0.05467	106	−0.4971	−0.03475	0.02937
80	−12.6590	−0.89430	0.04934	98	−8.4971	−0.59407	0.02937
98	4.5722	0.32218	0.04446	94	−12.4971	−0.87373	0.02937
96	1.8034	0.12678	0.04002	88	−19.2658	−1.34912	0.03248
86	−8.9654	−0.62897	0.03603	110	2.7342	0.19146	0.03248
88	−6.9654	−0.48866	0.03603	124	16.7342	1.17184	0.03248
128	33.0346	2.31756	0.03603	86	−22.0346	−1.54585	0.03603
118	22.2658	1.55920	0.03248	120	11.9654	0.83944	0.03603
90	−6.5029	−0.45465	0.02937	112	3.1966	0.22473	0.04002
108	9.9595	0.69457	0.02449	100	−9.5722	−0.67450	0.04446
120	21.9595	1.53144	0.02449	122	11.6590	0.82366	0.04934
114	15.1907	1.05843	0.02271	122	11.6590	0.82366	0.04934
78	−20.8093	−1.44991	0.02271	110	−1.1097	−0.07862	0.05467
116	16.4219	1.14344	0.02138	124	12.8903	0.91320	0.05467
74	−25.5781	−1.78096	0.02138	122	10.1215	0.71924	0.06044
80	−20.3468	−1.41608	0.02049	94	−18.6473	−1.32950	0.06665
98	−3.1156	−0.21679	0.02005	110	−2.6473	−0.18874	0.06665
90	−11.8844	−0.82693	0.02005	140	25.8152	1.85426	0.08041

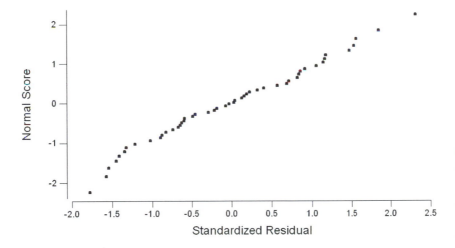

Figure 13.5 Normal scores plot of the standardized residuals from the linear regression of systolic blood pressure on height.

It is also of interest to plot the standardized residuals against the values of the X variable(s). If any pattern is observed in this plot, it suggests that another term involving the X variable — for example, X^2, might be needed in the model. Figure 13.6 shows the plot of the standardized residuals versus the height variable. No pattern is immediately obvious from an examination of this plot. Again, there is no evidence to cause us to reject this model. If the data have been collected in time sequence, it is also useful to examine a plot of the residuals against time.

Leverage: The predicted values of Y are found from

$$\hat{\beta}_0 + \hat{\beta}_1 X$$

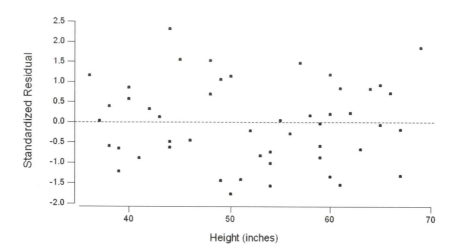

Figure 13.6 Plot of standardized residuals versus height.

where the estimators of β_0 and β_1 are linear combinations of the observed values of Y. Thus, the predicted values of Y are also linear combinations of the observed values of Y. An expression for the predicted value of y_i reflecting this relation is

$$\hat{y}_i = h_{i1}y_1 + h_{i2}y_2 + \ldots + h_{ii}y_i + \ldots + h_{in}y_n$$

where h_{ij} is the coefficient of y_j in the expression for $\hat{y}_i$. For simplicity, h_{ii} is denoted by h_i. The effect of y_i on its predicted value is denoted by h_i and this effect is called *leverage*. Leverage shows how much change there is in the predicted value of y_i per unit change in y_i. The possible values of the h_i are greater than or equal to zero and less than or equal to one. The average value of the leverages is the number of estimated coefficients in the regression equation divided by the sample size. In our problem, we estimated two coefficients and there were 50 observations. Thus the average value of the leverages is 0.04 (= 2/50). If any of the leverages are large — some statisticians consider large to be greater than twice the average leverage and others say greater than three times the average — the points with these large leverages should be examined. Perhaps there was a mistake in recording the values or there is something unique about the points that should be examined. If there is nothing wrong or unusual with the points, it is useful to perform the regression again excluding these points. A comparison of the two regression equations can be made, and the effect of the excluded points can be observed.

In our problem, we can see from Table 13.2 that there are two points, the first and the last, with the larger leverages. Both of these points had leverages slightly larger than twice the average leverage value. The first girl had a large SBP value relative to her height, and the last girl had the highest SBP value. At this stage, we will assume that there was no error in recording or entering the data. We could perform the regression again and see if there is much difference in the results. However, since the leverages are only slightly larger than twice the average leverage, we shall not perform any additional regressions.

Based on these looks at the data, we have no reason to doubt the appropriateness of the regression assumptions and there do not appear to be any really unusual data points that would cause us concern. Therefore, it is appropriate to move into the inferential part of the analysis, that is, to test hypotheses and to form confidence and prediction intervals. We begin the inferential stage with consideration of the slope coefficient.

13.2.3   The Slope Coefficient

Even though there is an indication of a linear relation between SBP and height — that is, it appears that β_1 is not zero — we do not know if β_1 is statistically significantly different from zero. To determine this, we must estimate the standard error of $\hat{\beta}_1$, which is used in both confidence intervals and tests of hypotheses about β_1. To form the confidence interval about β_1 or to test a hypothesis about it, we also must know the probability distribution of $\hat{\beta}_1$.

Since we are assuming that Y is normally distributed, this means that $\hat{\beta}_1$, a linear combination of the observed Y values, is also normally distributed. Therefore, to form a confidence interval or to test a hypothesis about β_1, we now need to know the standard error of its estimator. The standard error (s.e.) of $\hat{\beta}_1$ is

$$s.e.(\beta_1) = \frac{\sigma}{\sqrt{\sum_{i=1}^{n}(x_i - \overline{x})^2}}$$

and, because σ is usually unknown, the standard error is estimated by substituting $s_{Y|X}$ for σ. From the above equation, we can see that the magnitude of the standard error depends on the variability in the X variable. Larger variability decreases the standard error of $\hat{\beta}_1$. Thus, we should be sure to include some values of X at the extremes of X over the range of interest.

To test the hypothesis that β_1 is equal to β_{10} — that is,

$$H_0: \beta_1 = \beta_{10},$$

we use the statistic

$$t = \frac{\hat{\beta}_1 - \beta_{10}}{est.s.e.(\hat{\beta}_1)} = \frac{(\hat{\beta}_1 - \beta_{10})\sqrt{\sum(x_i - \overline{x})^2}}{s_{Y|X}}.$$

If σ were known, the test statistic, using σ instead of $s_{Y|X}$, would follow the standard normal distribution; however, σ is usually unknown, and the test statistic using $s_{Y|X}$ follows the t distribution with $n - 2$ degrees of freedom. The degrees of freedom parameter has the value of $n - 2$, since we have estimated two coefficients, β_0 and β_1.

If the alternative hypothesis is

$$H_a: \beta_1 \neq \beta_{10}$$

the rejection region consists of values of t less than or equal to $t_{n-2,\alpha/2}$ or greater than or equal to $t_{n-2,1-\alpha/2}$.

The hypothesis usually of interest is that β_{10} is zero — that is, there is no linear relation between Y and X. If, however, our study is one attempting to replicate previous findings, we may wish to determine if our slope coefficient is the same as that reported in the original work. Then β_{10} will be set equal to the previously reported value. Let us test the hypothesis that there is no linear relation between SBP and height versus the alternative hypothesis that there is some linear relation at the 0.05 significance level.

The test statistic, t, is

$$t = \frac{\left(\hat{\beta}_1 - \beta_{10}\right)\sqrt{\sum (x_i - \bar{x})^2}}{s_{X|Y}}$$

which is

$$t = \frac{(0.7688 - 0)\sqrt{4506.5}}{14.518} = 3.555.$$

This value is compared with -2.01 ($= t_{48,0.025}$) and 2.01 ($= t_{48,0.975}$). Since 3.555 is greater than 2.01, we reject the hypothesis of no linear relation between SBP and height. The p-value of this test is approximately 0.001.

The $(1 - \alpha)*100$ percent confidence interval for β_1 is formed by

$$\hat{\beta}_1 \pm t_{n-2,1-\alpha/2} * est.\ s.e.\ (\hat{\beta}_1)$$

which is

$$\hat{\beta}_1 \pm t_{n-2,1-\alpha/2} * \frac{s_{Y|X}}{\sqrt{\sum (x_i - \bar{x})^2}}.$$

The 95 percent confidence interval for β_1 is found using

$$0.7688 \pm 2.01 \frac{14.518}{\sqrt{4506.5}} = 0.7688 \pm 0.4347$$

and this gives a confidence interval from 0.3341 to 1.2035. The confidence interval is consistent with the test given above. Since zero is not contained in the confidence interval for β_1, there appears to be a linear relation between SBP and height. Since there is evidence to suggest that β_1 is not zero, this also means that the correlation coefficient between Y and X is not zero.

13.2.4   The *Y*-intercept

It is also possible to form confidence intervals and to test hypotheses about β_0, although these are usually of less interest than those for β_1. The location of the Y intercept is relatively unimportant compared to determining whether or not there is a relation between the dependent and independent variables. However, sometimes we wish to compare whether or not both our coefficients — slope and Y intercept — agree with those presented in the literature. In this case, we are interested in examining β_0 as well as β_1.

Since the estimator of β_0 is also a linear combination of the observed values of the normally distributed dependent variable, $\hat{\beta}_0$ also follows a normal distribution. The standard error of $\hat{\beta}_0$ is estimated by

$$est.s.e.\left(\beta_0\right) = s_{Y|X}\sqrt{\frac{\sum x_i^2}{n\sum (x_i - \bar{x})^2}}.$$

The hypothesis of interest is

$$H_0: \beta_0 = \beta_{00}$$

versus either a one- or two-sided alternative hypothesis. The test statistic for this hypothesis is

$$t = \frac{\hat{\beta}_0 - \beta_{00}}{s_{Y|X}\sqrt{\sum x_i^2 \big/ \left[n\sum(x_i - \bar{x})^2\right]}}$$

and this is compared to $\pm t_{n-2,1-\alpha/2}$ for the two-sided alternative hypothesis. If the alternative hypothesis is that β_0 is greater than β_{00}, we reject the null hypothesis in favor of the alternative when t is greater than $t_{n-2,1-\alpha}$. If the alternative hypothesis is that β_0 is less than β_{00}, we reject the null hypothesis in favor of the alternative when t is less than $-t_{n-2,1-\alpha}$.

The $(1 - \alpha/2)*100$ percent confidence interval for β_0 is given by

$$\hat{\beta}_0 \pm t_{n-2,1-\alpha/2}s_{Y|X}\sqrt{\frac{\sum x_i^2}{n\sum(x_i - \bar{x})^2}}.$$

Let us form the 99 percent confidence interval for β_0 for these SBP data. The 0.995 value of the t distribution with 48 degrees of freedom is approximately 2.68. Therefore, the confidence interval is found from the following calculations

$$61.14 \pm 2.68(14.52)\sqrt{\frac{142319}{50(4506.5)}} = 61.14 \pm 30.93$$

which gives an interval from 30.21 to 92.07, a wide interval.

13.2.5   An ANOVA Table Summary

Table 13.3 shows the information required to test the hypothesis of no relation between the dependent and independent variables in an ANOVA table similar to that used in Chapter 12. The test statistic for the hypothesis of no linear relation between the dependent and independent variables is the F ratio, which is distributed as an F variable with 1 and $n - 2$ degrees of freedom. Large values of the F ratio cause us to reject the null hypothesis of no linear relation in favor of the alternative hypothesis of a linear relation. The F statistic is the ratio of the mean square due to regression to the mean square about regression (mean square error or residual mean square). The degrees of freedom parameters for the F ratio come from the two mean squares involved in the ratio. The degrees of freedom due to regression is the number of parameters estimated minus one. The degrees of freedom associated with the about regression source of variation is the sample size minus the number of coefficients estimated in the regression model.

Table 13.3   An ANOVA table for the simple linear regression model.

Source of Variation	Degrees of Freedom	Sum of Squares	Mean Square	F Ratio[a]
Due to Regression	1	$\sum(\hat{y}_i - \bar{y})^2$	$\sum(\hat{y}_i - \bar{y})^2/1$	MSR/MSE
About Regression or Error	$n - 2$	$\sum(y_i - \hat{y}_i)^2$	$\sum(y_i - \hat{y}_i)^2/(n - 2)$	
Corrected Total	$n - 1$	$\sum(y_i - \bar{y})^2$		

[a]MSR is the mean square due to regression, and MSE is the mean square error term.

Table 13.4 ANOVA table for the regression of SBP on height.

Source of Variation	Degrees of Freedom	Sum of Squares	Mean Square	F Ratio
Due to Regression	1	2,663	2,663	12.63
About Regression or Error	48	10,117	210.77	
Corrected Total	49	12,780		

The ANOVA table for the SBP and height data is shown in Table 13.4. If we perform this test at the 0.05 significance level, we will compare the calculated F ratio to $F_{1,48,0.95}$, which is approximately 4.04. Since the calculated value, 12.63, is greater than the tabulated value, 4.04, we reject the null hypothesis in favor of the alternative hypothesis. There appears to be a linear relation between SBP and height at the 0.05 significance level.

Note that if we take the square root of 12.63, we obtain 3.554. With allowance for rounding, we have obtained the value of the t statistic calculated in the section for testing the hypothesis that β_1 is zero. This equality is additional verification of the relation, pointed out in Chapter 12, between the t and F statistics. An F statistic with 1 and $n - p$ degrees of freedom is the square of the t statistic with $n - p$ degrees of freedom. Examination of the t and F tables shows that $t^2_{n-p,1-\alpha/2}$ equals $F_{1,n-p,1-\alpha}$. Hence, we have two equivalent ways of testing whether or not the dependent and independent variables are linearly related at a given significance level. As we shall see in the multiple regression material, the F statistic directly extends to simultaneously testing several variables, whereas the t can be used with only one variable at a time.

These calculations associated with regression analysis require much time, care, and effort. However, they can be quickly and accurately performed with computer packages (see **Program Note 13.1** on the website).

13.3 Interval Estimation for $\mu_{Y|X}$ and $Y|X$

Even though the relation between SBP and height is not impressive, we will continue with the idea of developing a height-based standard for SBP for children. We would be much more comfortable doing this if the relation between height and SBP were stronger. The height-based standards that we shall create are the SBP levels such that 95 percent of the girls of a given height have lower SBP and 5 percent have a higher SBP. This standard is not based on the occurrence of any disease or other undesirable property. When using a standard created in this manner, approximately 5 percent of the girls will be said to have undesirably high SBP, regardless of whether or not that is really a problem.

The standard will be based on a one-sided prediction interval for the SBP variable. Also of interest is the confidence interval for the SBP variable and we shall consider the confidence interval first.

13.3.1 Confidence Interval for $\mu_{Y|X}$

The regression line provides estimates of the mean of the dependent variable for different values of the independent variable. How confident are we about these estimates or predicted values? The confidence interval provides one way of answering this question. To create the confidence interval, we require knowledge of the distribution of $\hat{Y}$ and also an estimate of its standard error.

Since the predicted value of $\mu_{Y|X}$ at a given value of x, say x_k, is also a linear combination of normal values, it is normally distributed. Its standard error is estimated by

$$est.s.e.(\mu_{Y|X_k}) = s_{Y|X}\sqrt{\frac{1}{n} + \frac{(x_k - \overline{x})^2}{\sum(x_i - \overline{x})^2}}.$$

The estimated standard error increases with increases in the distance between x_k and $\overline{x}$, and there is a unique estimate of the standard error for each x_k.

Because we are using $s_{Y|X}$ to estimate σ, we must use the t distribution in place of the normal in the formation of the confidence interval. The confidence interval for $\mu_{Y|X}$ has the form

$$\hat{\mu}_{Y|X} \pm t_{n-2,1-\alpha/2} est. \ s.e. \ (\hat{\mu}_{Y|X}).$$

Figure 13.7 shows the 95 percent confidence interval for SBP as a function of height. As we can see from the graph, the confidence interval widens as the values of height move away from the mean of the height variable. This is in accord with the expression for the confidence interval, which has the term $(x_k - \overline{x})^2$ in the numerator. We are thus less sure of our prediction for the extreme values of the independent variable. The confidence interval is about 17 mmHg wide for girls 35 or 70 inches tall and narrows to about 8 mmHg for girls about 50 to 55 inches tall.

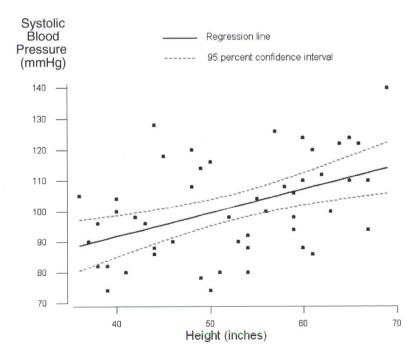

Figure 13.7 Ninety-five percent confidence interval for $\mu_{Y|X}$.

13.3.2   Prediction Interval for Y|X

In the preceding section, we saw how to form the confidence interval for the mean of SBP for a height value. In this section, we shall form the prediction interval — the interval for a single observation. The prediction interval is of interest to a physician because the physician is examining a single person, not an entire community. How does the person's SBP value relate to the standard?

As we saw in Chapter 7 in the material on intervals based on the normal distribution, the prediction interval is wider than the corresponding confidence interval because we must add the individual variation about the mean to the mean's variation. Similarly, the formula for the prediction interval based on the regression equation adds the individual variation to the mean's variation. Thus, the estimated standard error for a single observation is

$$est.\,s.e.(\hat{y}_k) = s_{Y|X}\sqrt{1 + \frac{1}{n} + \frac{(x_k - \overline{x})^2}{\sum(x_i - \overline{x})^2}}.$$

The corresponding two-sided $(1 - \alpha)*100$ percent prediction interval is

$$\hat{y}_k \pm t_{n-2,1-\alpha/2}\; est.\; s.e.\; (\hat{y}_k).$$

Figure 13.8 shows the 95 percent prediction interval for the data in Table 13.1. The prediction interval is much wider than the corresponding confidence interval because of the addition of the individual variation in the standard error term. The prediction interval here is about 60 mmHg wide. Note that most of the data points are within the prediction interval band. Inclusion of the individual variation term has greatly reduced the effect of the $(x_k - \overline{x})^2$ term in the estimated standard error in this example. The upper

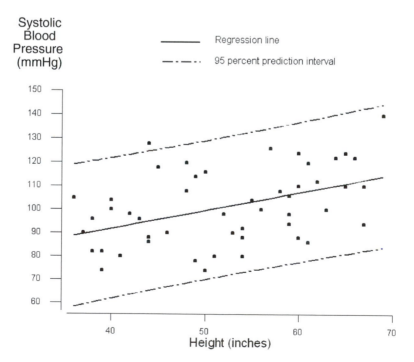

Figure 13.8 Ninety-five percent prediction interval for y_k.

and lower limits are essentially straight lines, in contrast to the shape of the upper and lower limits of the confidence interval.

Software packages can be used to perform the calculations necessary to create the 95 percent confidence and prediction intervals (see **Program Note 13.2** on the website).

Example 13.1

We apply the prediction interval to develop the standard for systolic blood pressure. Since we are only concerned about systolic blood pressures that may be too high, we shall use a one-sided prediction interval in the creation of the height-based standard for SBP for girls. The upper $(1 - \alpha) * 100$ percent prediction interval for SBP is found from

$$\hat{y}_k \pm t_{n-2,1-\alpha} \; est. \; s.e.(\hat{y}_k).$$

Because the standard is the value such that 95 percent of the SBP values fall below it and 5 percent of the values are greater than it, we shall use the upper 95 percent prediction interval to obtain the standard.

The data shown in Figure 13.8 can be used to help create the height-based standards for SBP. The difference between the one- and two-sided interval is the use of $t_{n-2,1-\alpha}$ in place of $t_{n-2,1-\alpha/2}$. Thus, the amount to be added to $\hat{y}_k$ for the upper one-sided interval is simply 0.834 ($= t_{48,0.95}/t_{48,0.975}$) times the amount added for the two-sided interval. To find the amount added for the two-sided interval, we subtract the predicted SBP value shown from the upper limit of the 95 percent prediction interval. For example, for a girl 35 inches tall, the amount added, using the two-sided interval, is found by subtracting 88.05 (predicted value) from 118.50 (upper limit of the two-sided prediction interval). This yields a difference of 30.45 mmHg. If we multiply this difference by 0.834, we have the amount to add to the 88.05 value. Thus, the standard for a girl 35 inches tall is

$$0.834 \, (118.50 - 88.05) + 88.05 = 113.45 \text{ mmHg}.$$

Table 13.5 shows these calculations and the height-based standards for SBP for girls. As just shown, the calculations in Table 13.5 consist of taking column 2 minus

Table 13.5   Creation of height-based standards for SBP (mmHg) for girls.

x_k (Inches) (1)	Upper Limit of Prediction Interval (2)	$\hat{y}_k$ (3)	Difference (4)	Difference Times 0.834 (5)	Standard (6)
35	118.50	88.05	30.45	25.40	113.45
40	121.87	91.89	29.98	25.00	116.89
45	125.40	95.93	29.67	24.74	120.67
50	129.09	99.58	29.51	24.61	124.19
55	132.93	103.42	29.51	24.61	128.03
60	136.93	107.27	29.66	24.74	132.01
65	141.09	111.11	29.98	25.00	136.11
70	145.41	114.95	30.46	25.40	140.35

column 3. This is stored in column 4. Column 5 contains 0.834 times column 4. The standard, column 6, is the sum of column 3 with column 5.

The upper one-sided prediction interval is one way of creating height-based standards for SBP. It has the advantage over simply using the observed 95th percentiles of the SBP at the different heights in that it does not require such a large sample size to achieve the same precision. If SBP is really linearly related to height, standards based on the prediction interval also smooth out random fluctuations that may be found in considering each height separately.

The standards developed here are illustrative of the procedure. If one were going to develop standards, a larger sample size would be required. We would also prefer to use additional variables or another variable to increase the amount of variation in the SBP that is accounted for by the independent variable(s). In addition, as we just stated, the rationale for having standards for blood pressure in children is much weaker than that for having standards in adults. In adults, there is a direct linkage between high blood pressure and disease, whereas in children no such linkage exists. Additionally, the evidence that relatively high blood pressure in children carries over into adulthood is inconclusive. Use of the 95th percentile or other percentiles as the basis of a standard implies that some children will be identified as having a problem when none may exist.

So far we have focused on a single independent variable. In the next section, we consider multiple independent variables.

13.4   Multiple Linear Regression

For many chronic diseases, there is no one single cause associated with the occurrence of the disease. There are many factors, called risk factors, that play a role in the development of the disease. In the study of the occurrence of air pollution, there are many factors — for example, wind, temperature, and time of day — that must be considered. In comparing mortality rates for hospitals, factors such as the mean age of the patients, severity of the diseases seen, and the percentage of patients admitted from the emergency room must be taken into account in the analysis. As these examples suggest, it is uncommon for an analysis to include only one independent variable. Therefore, in this section we introduce multiple linear regression, a method for examining the relation between one normally distributed dependent variable and more than one continuous independent variable. We also extend the mode to include categorical independent variables.

13.4.1   The Multiple Linear Regression Model

The equation showing the hypothesized relation between the dependent and $(p - 1)$ independent variables is

$$y_i = \beta_0 + \beta_1 x_{1i} + \beta_2 x_{2i} + \ldots + \beta_{p-1} x_{p-1,i} + \varepsilon_i.$$

The coefficient β_i describes how much change there is in the dependent variable when the ith independent variable changes by one unit and the other independent variables are held constant. Again, the key hypothesis is whether or not β_i is equal to zero. If β_i is equal to zero, we probably would drop the corresponding X_i from the equation because there is no linear relation between X_i and the dependent variable once the other independent variables are taken into account.

The regression coefficients of $(p - 1)$ independent variables and the intercept can be estimated by the least squares method, the same approach we used in the simple model presented above. We are also making the same assumptions — independence, normality, and constant variance — about the dependent variable and the error term in this model as we did in the simple linear regression model. We can also partition the sums of squares for the multiple regression model similarly to the partition used in the simple linear regression situation. The corresponding ANOVA table is

Source	DF	Sum of Squares	Mean Square	F-ratio
Regression	$p - 1$	$\sum(\hat{y}_i - \bar{y})^2 = SSR$	$SSR/(p - 1) = MSR$	MSR/MSE
Residual	$n - p$	$\sum(y_i - \hat{y}_i)^2 = SSE$	$SSE/(n - p) = MSE$	
Total	$n - 1$	$\sum(y_i - \bar{y})^2$		

and the overall F ratio now tests the hypothesis that the $p - 1$ regression coefficients (excluding the intercept) are equal to zero.

A goal of multiple regression is to obtain a small set of independent variables that makes sense substantively and that does a reasonable job in accounting for the variation in the dependent variable. Often we have a large number of variables as candidates for the independent variables, and our job is to reduce that larger set to a parsimonious set of variables. As we just saw, we do not want to retain a variable in the equation if it is not making a contribution. Inclusion of redundant or noncontributing variables increases the standard errors of the other variables and may also make it more difficult to discern the true relationship among the variables. A number of approaches have been developed to aid in the selection of the independent variables, and we show a few of these approaches.

The calculations and the details of multiple linear regression are much more than we can cover in this text. For more information on this topic, see books by Kleinbaum, Kupper, and Muller and by Draper and Smith, both excellent texts that focus on linear regression methods. We consider examples for the use of multiple linear regression based on NHANES III sample data that are shown in Table 13.6.

13.4.2   Specification of a Multiple Regression Model

There are no firm sample size requirements for performing a multiple regression analysis. However, a reasonable guideline is that the sample size should be at least 10 times as large as the number of independent variables to be used in the final multiple linear regression equation. In our example, there are 50 observations, and we will probably use no more than three independent variables in the final regression equation. Hence, our sample size meets the guideline, assuming that we do not add interaction terms or higher-order terms of the three independent variables.

Before beginning any formal analysis, it is highly recommend that we look at our data to see if we detect any possible problems or questionable data points. The

Table 13.6   Adult (≥18 years of age) sample data from NHANES III, Phase II (1991–1994).

Row	Race[a]	Sex[b]	Age[c]	Education[d]	Height[e]	Weight[f]	Smoke[g]	SBP[h]	BMI[i]
1	1	1	28	16	68	160	7	111	24.33
2	1	1	26	12	68	165	1	101	25.09
3	2	2	31	15	68	175	1	120	26.61
4	2	1	18	12	76	265	7	158	32.26
5	1	1	50	17	67	145	1	125	22.71
6	2	1	42	12	69	247	1	166	36.48
7	1	2	20	12	66	156	7	114	25.18
8	1	1	29	12	76	180	1	143	21.91
9	1	2	35	12	63	166	2	111	29.41
10	1	1	47	16	66	169	1	133	27.28
11	1	2	20	14	69	120	7	95	17.72
12	1	2	33	16	68	133	7	113	20.22
13	4	1	24	13	71	185	7	128	25.80
14	1	1	28	14	72	150	1	110	20.34
15	1	2	32	8	61	126	1	117	23.81
16	2	1	21	10	68	190	1	112	28.89
17	1	1	28	17	71	150	7	110	20.92
18	1	2	60	12	61	130	7	117	24.56
19	1	1	55	12	66	215	2	142	34.70
20	1	2	74	12	65	130	7	105	21.63
21	1	2	38	16	68	126	7	94	19.16
22	1	1	26	14	66	160	2	131	25.82
23	1	1	52	9	74	328	2	128	42.11
24	1	2	25	16	69	125	7	93	18.46
25	1	2	24	12	67	133	1	103	20.83
26	1	2	26	16	59	105	1	114	21.21
27	1	2	51	13	64	119	7	130	20.43
28	2	2	29	16	62	98	7	105	17.92
29	4	1	26	0	64	150	7	117	25.75
30	1	2	60	12	64	175	1	124	30.04
31	1	1	22	9	70	190	1	122	27.26
32	1	2	19	12	65	125	7	112	20.80
33	3	1	39	12	73	210	1	135	27.71
34	3	2	77	4	62	138	7	150	25.24
35	1	1	39	12	73	230	2	125	30.34
36	1	1	40	11	69	170	1	126	25.10
37	1	2	44	13	62	115	7	99	21.03
38	3	2	27	9	61	140	7	114	26.45
39	1	1	29	14	73	220	7	139	29.03
40	1	2	78	11	63	110	7	150	19.49
41	1	1	62	13	65	208	7	112	34.61
42	1	1	22	10	71	125	1	127	17.43
43	1	2	37	11	64	176	7	125	30.21
44	1	1	38	17	72	195	7	136	26.45
45	3	1	22	12	65	140	7	108	23.30
46	3	1	79	0	61	125	2	156	23.62
47	1	2	24	12	62	146	7	108	26.70
48	1	2	32	13	67	141	2	105	22.08
49	1	1	42	16	70	192	7	121	27.55
50	1	1	42	14	68	185	7	126	28.13

[a](1 = white, 2 = black, 3 = Hispanic, 4 = other); [b](1 = male; 2 = female); [c]Age in years; [d]Number of years of education; [e]Height (inches); [f]Weight (pounds); [g](1 = current smoker, 2 = never, 7 = previous); [h]Systolic blood pressure (mmHg); [i]Body mass index

descriptive statistics, such as the minimum and maximum, along with different graphical procedures, such as the box plot, are certainly very useful. A simple examination of the data in Table 13.6 finds that there are two people with zero years of education. One of these people is 26 years old and the other is 79 years old. Is it possible that someone 26 years old didn't go to school at all? It is possible but highly unlikely. Before using the education variable in any analysis, we should try to determine more about these values.

We consider building a model for SBP based on weight, age, and height. Before starting with the multiple regression analysis, it may be helpful to examine the relationship among these variables using a *scatterplot matrix* shown in Figure 13.9. It is essentially a grid of scatterplots for each pair of variables. Such a display is often useful in assessing the general relationships between the variables and in identifying possible outliers. The individual relationships of SBP to each of the explanatory variables shown in the first column of the scatterplot matrix do not appear to be particularly impressive, apart perhaps from the weight variable.

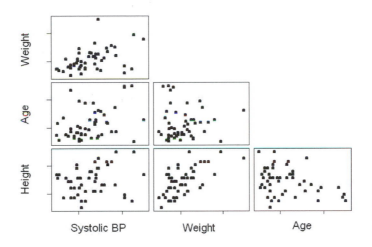

Figure 13.9   Scatterplot matrix for systolic blood pressure, weight, age, and height.

It may also be helpful to examine the correlation among the variables under consideration. The simple correlation coefficients among these variables can be represented in the format shown in Table 13.7. The correlation between SBP and weight is 0.465, the largest of the correlations between SBP and any of the variables. The correlation between height and weight is 0.636, the largest correlation in this table. It is clear from these estimates of the correlations among these three independent variables that they are not really independent of one another. We prefer the use of the term *predictor* variables, but the term *independent* variables is so widely accepted that it is unlikely to be changed.

Table 13.7   Correlations among systolic blood pressure, weight, age, and height for 50 adults in Table 13.6.

	Systolic Blood Pressure	Weight	Age
Weight	0.465		
Age	0.393	−0.004	
Height	0.214	0.636	−0.327

In this multiple regression situation, we have three variables that are candidates for inclusion in the multiple linear regression equation to help account for the variation in SBP. As just mentioned, we wish to obtain a parsimonious set of independent variables that account for much of the variation in SBP. We shall use a stepwise regression procedure and an all possible regressions procedure to demonstrate two approaches to selecting the independent variables to be included in the final regression model.

There are many varieties of stepwise regression, and we shall consider forward stepwise regression. In forward stepwise regression, independent variables are added to the equation in steps, one per each step. The first variable to be added to the equation is the independent variable with the highest correlation with the dependent variable, provided that the correlation is high enough. The analyst provides the level that is used to determine whether or not the correlation is high enough. Instead of actually using the value of the correlation coefficient, the criterion for inclusion into the model is expressed in terms of the significance levels of the F ratio for the test that the regression coefficient is zero.

After the first variable is entered, the next variable to enter the model is the one that has the highest correlation with the residuals from the earlier model. This variable must also satisfy the significance level of the F ratio requirement for inclusion. This process continues in this stepwise fashion, and an independent variable may be added or deleted at each step. An independent variable that had been added previously may be deleted from the model if, after the inclusion of other variables, it no longer meets the required F ratio.

Table 13.8 shows the results of applying the forward stepwise regression procedure to our example. In the stepwise output, we see that the weight variable is the independent variable that entered the model first. It is highly significant with a t-value of 3.64, and the R^2 for the model is 21.61 percent. In the second step the age variable is added to the model. The default significance level of the F ratio for adding or deleting a variable is 0.15. The age variable is also highly significant with a t-value of 3.42 and as a result the R^2 value increased to 37.23 percent. Thus, this is the model selected by the forward stepwise process.

Table 13.8 Forward stepwise regression: Systolic blood pressure regressed on weight, age, and height.

Predictor	Step 1	Step 2
Constant	92.50	77.18
Weight	0.177	0.177
(*t*-value)	(3.64)	(4.04)
(*p*-value)	(0.001)	(<0.001)
Age		0.41
(*t*-value)		(3.42)
(*p*-value)		(0.001)
$S_{Y|X}$	15.1	13.7
R^2	21.61	37.23
Adjusted R^2	19.98	34.55
C_p	11.8	2.3

In Table 13.8 there are four different statistics shown: R^2, adjusted R^2, C_p, and $s_{Y|X}$. Adjusted R^2 is similar to R^2, but it takes the number of variables in the equation into account. If a variable is added to the equation, but its associated F ratio is less than one, the adjusted R^2 will decrease. In this sense, the adjusted R^2 is a better measure than R^2. One minor problem with adjusted R^2 is that it can be slightly less than zero. The formula for calculating the adjusted R^2 is

$$Adjusted\ R_p^2 = 1 - (1 - R_p^2)\left(\frac{n}{n-p}\right)$$

where R_p^2 is the coefficient of determination for a model with p coefficients.

The statistic C_p was suggested by Mallows (1973) as a possible alternative criterion for selecting variables. It is defined as

$$C_p = \frac{SSE_p}{s^2} - (n - 2p)$$

where s^2 is the mean square error from the regression including all the independent variables under consideration and SSE_p is the residual sum of squares for a model that includes a given subset of $p - 1$ independent variables. It is generally recommended that we choose the model where C_p first approaches p.

The all possible regression procedure in effect considers all possible regressions with one independent variable, with two independent variables, with three independent variables, and so on, and it provides a summary report of the results for the "best" models. "Best" here is defined in statistical terms, but the actual determination of what is best must use substantive knowledge as well as statistical measures. Table 13.9 shows the results of applying the all possible regression procedure to our example.

Table 13.9   All possible (best subsets) regression: Systolic blood pressure regressed on weight, age, and height.

| Number of | Adjusted | | | | Variables Entered | | |
Variables Entered	R^2	R^2	Cp	$s_{Y|X}$	Weight	Age	Height
1	21.6	20.0	11.8	15.110	X		
1	15.5	13.7	16.4	15.692		X	
2	37.2	34.6	2.3	13.665	X	X	
2	28.6	25.6	8.7	14.573	X		X
3	37.7	33.6	4.0	13.764	X	X	X

From the all possible regressions output, we see that the model including weight was the best model with one independent variable. The second best model, with only one independent variable, used the age variable. The best two-independent-variable model used weight and age. The second best model, with two independent variables, used weight and height. The only three-independent-variable model has the highest R^2 value, but its adjusted R^2 is less than that for the best two independent variable model. Thus, on statistical grounds, we should select the model with weight and age as independent variables. It has the highest adjusted R^2 and the lowest value of $s_{Y|X}$. It also has C_p value closest to 2.

Again, these automatic selection procedures should be used with caution. We cannot treat the selected subset as containing the only variables that have an effect on the dependent variable. The excluded variables may still be important when different variables are in the model. Often it is necessary to force certain variables to be included in the model based on substantive considerations.

We also must realize that, since we are performing numerous tests, the p-values now only reflect the relative importance of the variables instead of the actual significance level associated with a variable.

13.4.3 Parameter Estimates, ANOVA, and Diagnostics

Let us now proceed to the multiple regression analysis with the full three-independent-variable model and compare it with the selected model that uses weight and age. Table 13.10 shows the regression with the three independent variables. The main features of interest are the tests of hypotheses and the parameter estimates. In the ANOVA table the F ratio of 9.27 is the value of the test statistic for the hypothesis that all the coefficients are simultaneously zero. Since its associated p-value is <0.001, we reject the hypothesis in favor of the alternative hypothesis that at least one of the coefficients is not zero. In general, however, this overall test is of little real interest because it is unlikely that none of the independent variables are related to the response variable. Of greater interest is the examination of the regression coefficients to see which independent variables are related to the response variable. In this model with the three independent variables, weight and age are statistically significant, but height is not, as shown by the t-values and the associated p-values. We should remove the statistically unimportant variables from the model unless there is a substantive reason to retain them. In fitting the model with the statistically unimportant variables eliminated, the estimated coefficients and standard errors will likely change in value due to the lack of independence of the predictor variables.

Table 13.10 also shows the sequential sum of squares. These sums of squares show the added contribution of the variables when they are entered in the order specified in the model. The contribution of height is very small after weight and age are already in the model. The table also shows *VIF* (variance inflation factor), and this term is discussed in the next section.

Table 13.10 Multiple regression analysis I: Systolic blood pressure regressed on weight, age, and height.

Predictor	Coef	SE Coef	T	p	VIF
Constant	53.96	41.54	1.30	0.200	
Weight	0.15435	0.05969	2.59	0.013	1.8
Age	0.4381	0.1319	3.32	0.002	1.2
Height	0.3845	0.6725	0.57	0.570	2.0

S = 13.76   R − Sq = 37.7%   R − Sq(adj) = 33.6%

Analysis of Variance:

Source	DF	SS	MS	F	p
Regression	3	5,266.5	1,755.5	9.27	<0.001
Residual Error	46	8,714.4	189.4		
Total	49	13,980.9			

Source	DF	Seq SS
Weight	1	3,021.9
Age	1	2,182.6
Height	1	61.9

Table 13.11 shows the multiple regression analysis of the model selected by the variable selection procedures. In this model, the coefficients for the weight and age variables are highly significant with the t-values of 4.04 and 3.42, respectively, and an F ratio for the overall test of the model is 13.94. The estimated coefficient for the weight variable (0.177) increased slightly from its value in the three-independent-variable model (0.154), and its standard error has decreased to 0.044 from 0.060 in the previous model. Inclu-

Table 13.11   Multiple regression analysis II: Systolic blood pressure regressed on weight and age.

Predictor	Coef	SE Coef	T	p	VIF
Constant	77.185	8.668	8.91	<0.001	
Weight	0.17727	0.04391	4.04	<0.001	1.0
Age	0.4064	0.1189	3.42	0.001	1.0

S = 13.66   R – Sq = 37.2%   R – Sq(adj) = 34.6%

ANOVA table:

Source	DF	SS	MS	F	p
Regression	2	5,204.5	2,602.3	13.94	<0.001
Residual Error	47	8,776.3	186.7		
Total	49	13,980.9			

Source	DF	Seq SS
Weight	1	3,021.9
Age	1	2,182.6

sion of an unnecessary term in the three-independent-variable model has caused the increase in the estimated standard errors and thus makes it harder to discern the significance of any of the independent variables.

The R^2 statistic indicates that the selected model is not able to account for the great majority of the variation in SBP. Much work needs to be done to discover these additional sources of variation before standards are created. It is likely that the effects of weight and age would be altered if we include other variables that have not been considered in the current models. A key message is that conclusions drawn about the importance of independent variables depend on the model that is being considered.

Having arrived at a final multiple regression model for the data set, it is important to go further and check the assumptions we made in selecting the important variables. Most useful at this stage is an examination of residuals from the fitted model. Among many regression diagnostics now available in computer packages, the following graphic plots are often used.

(1) A normal probability plot of the residuals: In creating the regression model, we assume that the errors (ε_i) are distributed normally. After the systematic variation associated with the independent variables in the model has been removed from the data, the residuals should therefore resemble a sample from a normal distribution. The normal probability plot of standardized residuals is shown in Figure 13.10. The points appear to lie along a line with the exception of the one large residual value, giving support to the normality assumption. If the normality assumption does not appear to be valid, then we may need to transform the response variable. However, transformations are not innocuous and must be done with care (Kleinbaum et al. 1998).

(2) A plot of the residuals against the fitted values: Figure 13.11 shows the standardized residuals plotted against the estimated values of the dependent variable with a useful reference line at zero. There is no strong pattern shown in the plot although the larger residuals in absolute value show a tendency to occur with estimated systolic blood pressure values over 130. If the trend were stronger, the equal variance assumption might be invalid and a transformation of the response variable might be required. If any clear patterns are shown in this plot, it raises concerns.

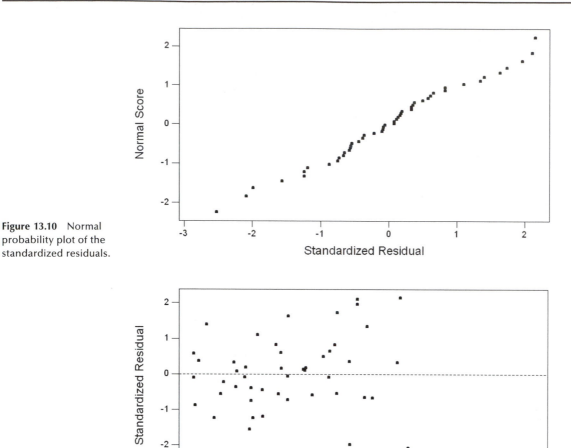

Figure 13.10 Normal probability plot of the standardized residuals.

Figure 13.11 Plot of standardized residual versus the fitted value.

(3) A plot of residuals against each independent variable in the model: This plot helps in determining whether or not there may be a nonlinear relationship between the response variable and the independent variable used in the plot. Figure 13.12 shows the plot of residuals with the weight variable, and Figure 13.13 is a plot of the standardized residuals with the age variable. Neither plot shows the existence of any pattern. The presence of a curvilinear relationship, for example, would suggest that a higher-order term such as a quadratic term in the independent variable may be needed.

13.4.4   Multicollinearity Problems

In a multiple regression situation, it is not uncommon to have independent variables that are interrelated to a certain extent especially when survey data are used. *Multicollinearity* occurs when an explanatory variable is strongly related to a linear combination of the other independent variables. Multicollinearity does not violate the assumptions of the model, but it does increase the variance of the regression coefficients. This increase means that the parameter estimates are less reliable. Severe multicollinearity also makes determining the importance of a given explanatory variable difficult because the effects of explanatory variables are confounded.

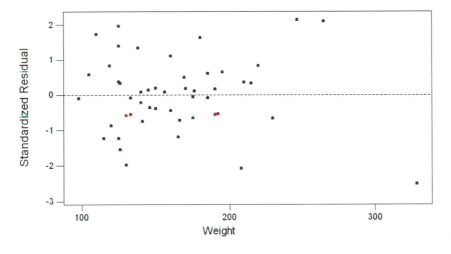

Figure 13.12 Plot of the standardized residual versus the weight variable.

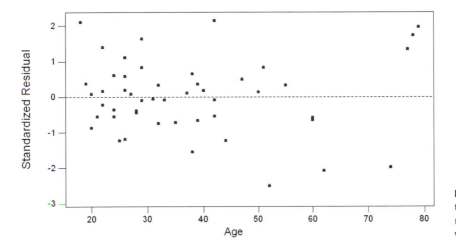

Figure 13.13 Plot of the standardized residual versus the age variable.

Recognizing multicollinearity among a set of explanatory variables is not necessarily easy. Obviously, we can simply examine the scatterplot matrix or the correlations between these variables, but we may miss more subtle forms of multicollinearity. An alternative and more useful approach is to examine what are known as the *variance inflation factors* (*VIF*) of the explanatory variables. The *VIF* for the *j*th independent variable is given by

$$VIF_j = \frac{1}{1 - R_j^2}$$

where R_j^2 is the R^2 from the regression of the *j*th explanatory variable on the remaining explanatory variables. The *VIF* of an explanatory variable indicates the strength of the linear relationship between the variable and the remaining explanatory variables. A rough rule of thumb is that the *VIF*s greater than 10 give some cause for concern.

Now let us review the multiple regression results shown in Tables 13.10 and 13.11. The *VIF*s shown in these tables are all less than 10, indicating that the multicollinearity does not pose a serious problem for those models. As a demonstration for a severe multicollinearity, we added to the model shown in Table 13.10 another independent variable that is closed associated with weight and height. Table 13.12 shows the multiple regression analysis of SBP on weight, age, height, and the body mass index (BMI) defined as your weight in kilograms divided by the square of your height in meters. The *VIF*s for weight, height, and BMI are all greater than 10 in Table 13.12. More important, the variances of the regression coefficients for weight and height increased, and these variables are no longer statistically significant. The effect of weight on SBP shown in the earlier model cannot be demonstrated if we add BMI. A solution to a severe multicollinearity is to delete one of correlated variables. If we drop the BMI variable, we would eliminate the extreme multicollinearity.

Table 13.12   Multiple regression analysis III: Systolic blood pressure versus weight, age, height, body mass index.

Predictor	Coef	SE Coef	T	p	VIF
Constant	105.2	154.4	0.68	0.499	
Weight	0.3052	0.4413	0.69	0.493	97.6
Age	0.4364	0.1333	3.27	0.002	1.2
Height	−0.354	2.246	−0.16	0.875	22.3
BMI	−1.040	3.016	−0.34	0.732	60.9

S = 13.90   R − Sq = 37.8%   R − Sq(adj) = 32.3%

Analysis of Variance:

Source	DF	SS	MS	F	p
Regression	4	5,289.4	1,322.4	6.85	<0.001
Residual Error	45	8,691.4	193.1		
Total	49	13,980.9			

Source	DF	Seq SS
Weight	1	3,021.9
Age	1	2,182.6
Height	1	61.9
BMI	1	23.0

13.4.5   Extending the Regression Model: Dummy Variables

So far we limited our analysis to continuous independent variables. As we discussed briefly in the previous chapter in conjunction with unbalanced ANOVA models, the independent variables can be categorical as well as continuous. It is easy to incorporate categorical explanatory variables into a multiple regression equation, provided we code the categorical variables with care. Let us consider the smoking status variable shown in Table 13.6. It has three levels: current smoker, never smoked, and previous smoker. Let us consider the never smoked category the baseline level and measure the effects of being a current smoker or a previous smoker from the never smoked level. We will then create two indicator variables to represent smoking status. The first indicator variable will have the value of 1 if the person is a current smoker and a value of 0 otherwise. The second indicator will have the value of 1 if the person is a former smoker and 0 otherwise. If the person has never smoked, both the indicator variables are 0.

Category	Indicator Variables		
	x_1	x_2	
Never Smoked	0	0	(reference)
Current Smoker	1	0	
Previous Smoker	0	1	

Table 13.13   Multiple regression analysis IV: Systolic blood pressure versus weight, age, sex (dummy variable).

Predictor	Coef	SE Coef	T	p	VIF
Constant	81.218	8.689	9.35	<0.001	
Weight	0.11749	0.05287	2.22	0.031	1.5
Age	0.4295	0.1162	3.69	0.001	1.0
Sex	8.990	4.685	1.92	0.061	1.5

S = 13.29   R – Sq = 41.9%   R – Sq(adj) = 38.1%

Analysis of Variance:

Source	DF	SS	MS	F	p
Regression	3	5,854.8	1,951.6	11.05	<0.001
Residual Error	46	8,126.1	176.7		
Total	49	13,980.9			

Source	DF	Seq SS
Weight	1	3,021.9
Age	1	2,182.6
Sex	1	650.3

The number of indicator variables we need to represent a categorical variable is one less than the number of categories, corresponding to the degrees of freedom for the variable.

To demonstrate the use of an indicator variable into a regression analysis, we added the gender variable (female = 0; male = 1) to the multiple regression model shown in Table 13.11. The regression analysis of systolic blood pressure on weight, age, and gender is shown in Table 13.13. The gender variable accounted for some variation in SBP, although it did not quite reach statistical significance at the 0.05 level. The estimated regression equation is

$$SBP = 81.218 + 0.117 \text{ (weight)} + 0.430 \text{ (age)} + 8.990 \text{ (sex)}.$$

The predicted SBP for females with weight of 100 lbs and age of 50 is

$$81.218 + 0.117(100) + 0.430(50) + 8.990(0) = 114.418.$$

The predicted SBP for males with the same weight and age is

$$81.218 + 0.117(100) + 0.430(50) + 8.990(1) = 123.408.$$

The predicted value for males is 8.990 mmHg higher than the predicted value for females. In other words, the regression coefficient for sex represents the difference in the mean SBP between the indicated category (coded as 1, males in this case) and the reference category (coded as 0, females in this case), holding the other independent variables constant.

Multiple regression analysis is a very useful technique. It becomes even more useful through its ability to incorporate categorical predictor variables along with continuous predictor variables. If only categorical explanatory variables are used, we have the analysis of variance situation. All of these situations — linear regression, ANOVA, and multiple linear regression with a mixture of continuous and discrete predictor variables — fit under the rubric of the *General Linear Model* (GLM).

See **Program Note 13.3** on the website for conducting multiple regression analysis including the use of variable selection procedures and residual plots.

Conclusion

In this chapter, we showed how to examine the relation between a normally distributed dependent variable and a continuous independent variable via linear regression analysis. We also demonstrated how this method could be extended to include many independent variables. We further expanded the linear regression model to include discrete predictor variables. These discrete predictor variables are incorporated through binary coding. Often we wish to use the linear regression or ANOVA idea, but the dependent variable is a binary variable — for example, the occurrence of a disease. In this case, the logistic regression method, discussed in the next chapter, can be used.

EXERCISES

13.1 Restenosis — narrowing of the blood vessels — frequently occurs after coronary angioplasty, but accurate prediction of which individuals will have this problem is problematic. In a study by Simons et al. (1993), the authors hypothesized that restenosis is more likely to occur if activated smooth-muscle cells in coronary lesions at the time of surgery are present. They used the number of reactive nuclei in the coronary lesions as an indicator of the presence of the activated smooth-muscle cells. The number of reactive nuclei in the lesions and the degree of stenosis at follow-up for 16 patients who underwent a second angiography are shown here.

Patient	Degree of Stenosis (%) at Follow-up	Number of Reactive Nuclei at Initial Surgery
1	28	5
2	15	3
3	22	2
4	93	10
5	60	12
6	90	25
7	42	8
8	53	3
9	72	15
10	0	13
11	79	17
12	28	0
13	82	13
14	28	14
15	100	17
16	21	1

Are you suspicious of any of these data points? If so, why? Does there appear to be a linear relation between the degree of stenosis and the number of reactive nuclei? If there is, describe the relation. Are there any points that have a large influence on the estimated regression line? If there are, eliminate the point with the greatest leverage and refit the equation. Is there much difference between the two regression equations? Are there any points that have a large standardized residual? Explain why the residuals are large for these points. Do you think that Simons et al. have a promising lead for predicting which patients will undergo restenosis?

13.2 Use the following data (NCHS 2005) to determine whether or not there is a linear relation between the U.S. national health expenditures as a percent of gross domestic product (GDP) and time.

Year	National Health Expenditures as Percentage of GDP	Year	National Health Expenditures as Percentage of GDP
1960	5.1	1999	13.2
1970	7.0	2000	13.3
1980	8.8	2001	14.1
1990	12.0	2002	14.9
1995	13.4		
1997	13.1		
1998	13.2		

What is your predicted value for national health expenditures as a percent of GDP for 2010? What is the 95 percent confidence interval for your estimate? What data have you used as the basis of your predictions? What assumptions have you made?

13.3 The estimated age-adjusted percent of persons 18 years of age and over who smoke cigarettes are shown below for females and males for selected years (NCHS 2005).

	Estimated Age-Adjusted Percent Smoking Cigarettes	
Year	Female	Male
1965	33.7	51.2
1974	32.2	42.8
1979	30.1	37.0
1985	27.9	32.2
1990	22.9	28.0
1995	22.7	26.5
1998	22.1	25.9
1999	21.6	25.2
2000	21.1	25.2
2001	20.7	24.6
2002	20.0	24.6
2003	19.4	23.7

Describe the linear relation between the estimated age-adjusted percent smoking and time for females and males separately. How much of the variation in the percents is accounted for by time for females and for males? Do females and males appear to have the same rate of decrease in the estimated age-adjusted percent smoking? Provide an estimate when the age-adjusted percent of males who smoke will equal the corresponding percent for females. What assumption(s) have you made in coming up with the estimate of this time point? Do you think this assumption is reasonable? Explain your answer.

13.4 Use the data in Table 13.1 to construct height-based standards for systolic blood pressure for girls. In constructing these standards, you should be concerned about values that may be too low as well as too high.

13.5 Anderson et al. (Anderson 1990) provide serum cholesterol and body mass index (BMI) values for subjects who participated in a study to examine the effects of oat-bran cereal on serum cholesterol. The values of serum cholesterol and BMI for the 12 subjects included in the analysis are shown next.

Subject	Serum cholesterol (mmol/L)	BMI
1	7.29	29.0
2	8.04	26.3
3	8.43	21.6
4	7.96	21.8
5	5.43	27.2
6	5.77	24.8
7	6.96	25.2
8	6.23	24.5
9	6.65	25.1
10	6.26	23.5
11	8.20	27.9
12	6.21	24.8

Plot serum cholesterol versus BMI. Calculate the correlation coefficient between serum cholesterol and BMI. Regress serum cholesterol on BMI. Does there appear to be any linear relation between these two variables? Form a new variable that is BMI minus its mean. Square this new variable. Include this new independent variable in the regression equation along with the BMI variable. Does there appear to be any linear relation between these two independent variables and serum cholesterol? Why do you think that we suggested that this new variable be added to the regression equation?

13.6 The following data are a sample of observations from the NHANES II. We wish to determine whether or not diastolic blood pressure (DBP) of adults can be predicted based on knowledge of the person's body mass index (BMI — weight in kilograms divided by the square of height in meters), age, sex (females coded as 0 and males coded as 1), smoking status (not currently a smoker is coded as 0 and currently a smoker is coded as 1), race (0 represents nonblack and 1 represents black), years of education, poverty status (household income expressed as a multiple of the poverty level for households of the same size), and vitamin status (0 indicates not taking supplements and 1 indicates taking supplements).

Select an appropriate multiple regression model that shows the relation between DBP and the set or a subset of the independent variables shown here. Note that the independent variables include both continuous and discrete variables. Provide an interpretation of the estimated regression coefficients for each discrete independent variable used in the model. From these independent variables, are we able to do a good job of predicting DBP? What other independent variables, if any, should be included to improve the prediction of DBP?

Vitamin Status	BMI	Sex	Race	Education	Age	Poverty Index	DBP	Smoking Status
1	18.46	0	0	13	24	1.93	50	0
0	32.98	1	0	14	24	3.97	98	0
1	29.48	1	0	12	39	1.71	80	1
1	19.20	0	0	12	29	1.62	62	1
0	24.76	0	0	12	45	5.49	90	0
1	20.60	0	0	14	24	4.78	70	0
0	24.80	1	0	8	65	3.63	80	0
1	24.24	0	0	12	25	4.55	56	1
0	29.95	1	0	16	24	2.77	90	0

(continued)

Vitamin Status	BMI	Sex	Race	Education	Age	Poverty Index	DBP	Smoking Status
0	21.80	1	0	17	29	2.15	78	0
0	23.19	1	0	13	29	1.09	56	0
0	28.34	0	0	12	18	1.71	78	0
0	22.00	1	0	12	28	5.49	70	1
0	24.60	1	0	8	65	3.35	70	1
1	21.83	0	0	16	26	0.77	74	0
0	30.50	0	0	3	73	1.10	70	0
1	19.63	0	0	13	33	5.48	62	1
0	27.92	0	0	12	65	3.83	78	0
1	26.77	1	0	12	59	3.57	90	0
1	21.02	1	0	15	21	1.25	64	0
1	19.40	0	0	16	26	3.25	70	0
0	31.12	0	0	12	58	1.91	100	0
0	20.68	0	0	7	57	4.63	74	0
0	22.48	0	0	12	28	1.75	75	0
0	24.89	0	0	14	23	3.25	74	0
1	21.08	0	0	12	56	5.04	68	0
1	23.67	1	0	14	23	4.47	86	1
1	28.19	1	0	12	24	3.38	82	1
0	22.09	0	1	7	58	1.73	80	0
0	23.46	1	0	14	66	5.12	70	0
1	21.11	1	0	13	18	0.64	70	1
1	21.35	0	1	12	20	0.26	60	1
0	20.36	0	1	14	23	2.85	78	0
1	25.00	0	1	4	36	0.72	80	0
1	20.47	0	0	17	37	3.97	88	1
0	24.73	0	1	8	44	1.36	82	0
0	27.87	0	0	12	50	3.31	70	1
0	28.22	1	0	15	50	3.41	112	0
0	26.05	1	0	13	33	5.85	80	0
0	24.51	0	0	12	42	3.17	92	0
1	28.09	0	1	16	46	2.39	92	0
1	18.85	0	1	11	36	1.62	56	1
0	25.99	0	1	12	74	1.40	80	0
1	23.47	1	0	16	35	1.97	96	1
0	26.57	0	0	12	55	6.11	86	0
0	25.09	1	0	12	33	2.15	104	1
0	30.78	0	0	12	38	1.37	74	0
0	28.89	1	0	14	49	1.82	90	1
1	23.82	1	0	17	35	2.85	70	0
0	28.29	1	0	12	62	6.89	60	0

13.7 Find an article from a health-related journal that used a multiple regression analysis and review it thoroughly. Is the multiple regression model an appropriate choice of analysis? Would you conduct the analysis or interpret the result differently? Did your article report all the necessary analytical results that would convince you to accept the author's conclusions?

13.8 The following data set consists of infant mortality rates (IMR) for 50 states in 1997–1998, along with the following eight potential explanatory variables (NCHS 2004).

1. Low birthweight: Percent of live births with weight less than 2500 grams, 1997–1999

2. Vaccination: Percent of children 19–35 months of age vaccinated against selected diseases, 1998

3. Medicaid expenditures as percent of total personal health care expenditures, 1998

4. Prenatal care: Percent of live births with prenatal care started in the first trimester, 1998
5. Uninsured: Percent of people under 65 years of age without health insurance, 1998
6. Hospital care: Per capita expenditure in dollars for hospital care, 1998
7. Personal care: Per capita expenditure for personal health care, 1998
8. Personal care: Per capita expenditure for personal health care, 1996

States are grouped in regions and the region can be another potential explanatory variable(s). Build an appropriate multiple regression model that would show the relationship between infant mortality rate and a subset of the potential explanatory variables (five or fewer considering the total number of observations). Apply different criteria for selecting a subset and see whether different criteria give different results. Check whether various assumptions are met in your final model. Interpret your analytical results, taking into account that these variables are measurements made at the state level. Do you think all relevant explanatory variables are represented in your model?

State	IMR	Low Birth Weight	Vaccination	Medicaid	Prenatal Care	Uninsured	Hospital Care	Personal Care 1998	Personal Care 1996
East (New England & Mideast)									
CT	6.8	7.56	90	17.5	88.8	14.3	1,478	4,656	4,250
ME	5.3	5.93	86	21.1	89.0	14.6	1,501	4,025	3,512
MA	5.1	6.99	87	19.3	89.3	11.6	1,807	4,810	4,347
NH	4.5	5.91	82	15.6	90.0	12.5	1,234	3,840	3,441
RI	6.5	7.43	86	21.6	90.1	7.6	1,626	4,497	3,978
VT	6.7	6.15	86	18.0	87.8	11.0	1,328	3,654	3,273
DE	6.5	8.01	82	12.5	81.4	17.1	1,581	5,258	3,847
MD	6.6	7.83	85	12.7	80.9	18.9	1,486	3,848	3,573
NJ	7.5	7.69	83	14.0	84.6	18.0	1,481	4,197	4,009
NY	8.5	7.96	78	31.5	82.6	19.7	1,769	4,706	4,346
PA	8.1	7.84	78	16.3	80.2	12.1	1,599	4,168	3,791
Midwest (Great Lakes & Plains)									
IL	8.2	7.84	78	14.8	84.1	16.6	1,558	3,801	3,535
ID	7.8	7.78	78	12.0	85.7	16.1	1,413	3,566	3,196
MI	7.0	6.53	78	14.9	84.3	14.9	1,489	3,676	3,457
OH	6.5	6.31	82	15.6	87.5	11.7	1,437	3,747	3,542
WI	7.6	7.01	82	13.4	85.7	13.2	1,377	3,845	3,476
IA	5.9	5.92	82	15.4	84.4	10.9	1,520	3,765	3,368
KS	7.6	7.75	85	10.8	86.4	12.2	1,428	3,707	3,412
MN	7.8	6.75	76	15.4	84.1	10.3	1,254	3,986	3,614
MO	6.8	6.31	79	14.4	85.6	12.1	1,566	3,754	3,390
NE	7.4	5.75	74	14.4	82.7	10.2	1,507	3,627	3,287
ND	8.4	8.57	79	13.8	83.2	16.5	1,741	3,881	3,540
SD	7.3	8.09	79	13.4	83.8	16.3	1,534	3,650	3,253
South (Southeast & Southwest)									
AL	8.7	8.68	80	13.0	86.5	19.5	1,432	3,630	3,422
AR	8.6	8.82	77	15.5	87.8	21.7	1,430	3,540	3,177
FL	9.2	8.84	83	10.4	84.5	21.1	1,371	4,046	3,774
GA	9.2	9.52	88	12.2	80.9	19.4	1,329	3,505	3,291
KY	7.7	7.80	80	16.9	85.2	16.0	1,479	3,711	3,300
LA	8.3	8.12	82	19.1	83.6	21.3	1,601	3,742	4,396
MS	10.0	9.28	82	15.8	82.6	22.9	1,551	3,474	3,145
NC	7.3	8.06	82	16.9	86.3	17.0	1,373	3,535	3,232
SC	10.5	10.18	84	16.6	80.7	17.4	1,480	3,529	3,131
TN	8.4	9.01	82	17.4	84.0	14.3	1,375	3,808	3,569
VA	9.0	8.62	73	9.9	77.5	15.8	1,286	3,284	3,009

(continued)

State	IMR	Low Birth Weight	Vaccination	Medicaid	Prenatal Care	Uninsured	Hospital Care	Personal Care 1998	Personal Care 1996
WV	9.2	10.09	78	17.3	82.1	20.8	1,693	4,044	3,649
AZ	8.1	7.28	75	12.0	79.2	26.9	1,085	3,100	2,862
NM	6.3	7.35	74	17.7	79.0	24.0	1,389	3,209	2,942
OK	7.4	6.86	76	11.8	75.5	21.2	1,307	3,397	3,188
TX	6.8	8.60	76	12.5	82.2	26.9	1,274	3,397	3,117
West (Rocky Mountains & Far West)									
CO	7.0	6.15	76	11.4	79.3	16.4	1,147	3,331	3,071
ID	7.1	6.71	82	12.1	82.9	19.7	1,163	3,035	2,765
MT	6.6	7.59	76	13.8	75.3	21.9	1,440	3,314	2,917
UT	6.6	7.68	71	11.8	68.2	15.1	1,016	2,731	2,506
WY	5.9	6.72	76	12.3	82.1	18.8	1,436	3,881	3,046
AK	6.6	8.75	80	16.9	82.3	17.9	1,496	3,442	3,227
CA	6.9	5.90	81	12.7	80.4	24.4	1,145	3,429	3,200
HI	5.9	6.17	76	14.2	82.6	11.3	1,391	3,770	3,656
NV	6.5	7.44	79	9.1	84.8	23.7	1,033	3,147	2,949
OR	5.5	5.41	76	15.3	80.7	16.0	1,112	3,334	3,019
WA	5.7	5.72	81	16.2	83.1	13.4	1,116	3,382	3,142

REFERENCES

Anderson, J. W., D. B. Spencer, C. C. Hamilton, S. F. Smith, J. Tietyen, C. A. Bryant, and P. Oeltgen. "Oat-Bran Cereal Lowers Serum Total and LDL Cholesterol in Hypercholesterolemic Men." *American Journal of Clinical Nutrition* 52:495–499, 1990.

Draper, N. R., and H. Smith. *Applied Regression Analysis*. New York: John Wiley & Sons, 1981.

Forthofer, R. N. "Blood Pressure Standards in Children." Paper presented at the American Statistical Association Meeting, August 1991.

Gillum, R., R. Prineas, and H. Horibe. "Maturation vs Age: Assessing Blood Pressure by Height." *Journal of the National Medical Association* 74:43–46, 1982.

Kleinbaum, D. G., L. L. Kupper, K. E. Muller, and A. Nizam. *Applied Regression Analysis and Other Multivariable Methods*, 3rd ed. Pacific Grove, CA: Brooks/Cole, 1998.

Korn, E. L., and R. Simon. "Explained Residual Variation, Explained Risk, and Goodness of Fit." *The American Statistician* 45:201–206, 1991.

Lin, L. I., and E. F. Vonesh. "An Empirical Nonlinear Data-Fitting Approach for Transforming Data to Normality." *The American Statistician* 43:237–243, 1989.

Mallows, C. L. "Some Comments on Cp." *Technometrics* 15:661–675, 1973.

Miller, D. M. "Reducing Transformation Bias in Curve Fitting." *The American Statistician* 38:124–126, 1984.

National Center for Health Statistics. *Health, United States, 2004 with Chartbook on Trends in the Health of Americans*. Hyattsville, MD: Public Health Service. DHHS Pub. No. 2004-1232, September 2004, Tables 7, 14, 23, 72, 119, 143, 148, and 153.

National Center for Health Statistics. *Health, United States, 2005 with Chartbook on Trends in the Health of Americans*. Hyattsville, MD: Public Health Service. DHHS Pub. No. 2005-1232, November 2005, Tables 63 and 118.

The NHLBI Task Force on Blood Pressure Control in Children. "The Report of the Second Task Force on Blood Pressure Control in Children, 1987." *Pediatrics* 79:1–25, 1987.

Ranney, G. B., and C. C. Thigpen. "The Sample Coefficient of Determination in Simple Linear Regression." *The American Statistician* 35:152–153, 1981.

Simons, M., G. Leclerc, R. D. Safian, J. M. Isner, L. Weir, and D. S. Baim. "Relation between Activated Smooth-Muscle Cells in Coronary-Artery Lesions and Restenosis after Atherectomy." *The New England Journal of Medicine* 328:608–613, 1993.

Voors, A., L. Webber, R. Frerichs, and G. S. Berrenson. "Body Height and Body Mass as Determinants of Basal Blood Pressure in Children — the Bogalusa Heart Study." *American Journal of Epidemiology* 106:101–108, 1977.

Logistic and Proportional Hazards Regression

14

Chapter Outline

In this chapter we present logistic regression, a method for examining the relationship between a dependent variable with two levels and one or more independent variables. Logistic regression represents another application of the linear model idea used in the two previous chapters. We also provide an introduction to proportional hazards regression (or Cox's regression). Proportional hazards regression is an extension of the survival analysis method presented in Chapter 11, and it also uses the linear model approach.

14.1 Simple Logistic Regression

Joseph Berkson did much to advance the use of logistics in the 1940s and 1950s (Berkson 1944; 1951). However, it was D. R. Cox (1969) who popularized the logit transformation for modeling binary data. Since the 1980s, logistic regression has become one of the more widely used analysis techniques in public health and the biomedical sciences because it allows for an examination of the relation between disease status (presence or absence) and a set of possible risk factors for the disease based on data from cross-sectional, case-control, or cohort studies.

Let's consider a simple example to introduce the topic because it allows us to show the logistic regression in terms of statistics that we already know.

Example 14.1

Suppose that we wish to determine whether or not there is a relationship between a male's pulmonary function test (PFT) results and air pollution level at his residence — lead in the air serving as a proxy for overall air pollution. The data for this situation are shown in Table 14.1 (Forthofer and Lehnen 1981).

Table 14.1 Pulmonary function test results by ambient air pollution.

Pulmonary Function Test Results	Pollution (Lead) Level	
	Low	High
Normal	368	82
Abnormal	19	10
Total	387	92
Proportions Normal	0.9509	0.8913
Odds (normal)	19.367	8.200
Logits (normal)	2.964	2.104

We cannot use ordinary linear regression for this situation because the dependent variable — PFT results categorized as normal or nonnormal — has only two levels, and, hence, the assumption of a continuous and normally distributed dependent variable does not hold. We can use categorical data analysis, since the independent variable, lead level categorized as low or high, is discrete. More generally, if there were several independent variables, some of which were continuous, then the categorical data approach would no longer be appropriate.

One categorical data approach is to compare the odds of having a normal PFT between those exposed to low and those exposed to high levels of air pollution — that is, to calculate the odds ratio and then test the hypothesis that the odds ratio is equal to one. In the following, we shall consider the relation between logistic regression and the odds ratio.

In logistic regression the underlying model is that the natural logarithm, written as ln, of the odds of a normal (or nonnormal) PFT is a linear function of a constant and the effect of lead pollution. The logarithm of the odds is also referred to as the *log odds* or *logit*. In this example, a larger logit value indicates a more favorable outcome because it indicates a greater proportion of males having a normal PFT. Hence those with low exposures to lead (logit = 2.964) have a more favorable outcome than those with higher exposure to lead (logit = 2.104) for this sample.

This model is

$$\ln\left(\frac{\pi_{i1}}{\pi_{i2}}\right) = constant + i^{th} \ lead \ pollution \ effect$$

where π_{i1} is the probability of a normal PFT and π_{i2} is the probability of a nonnormal PFT for the ith lead level. The ratio of π_{i1} to π_{i2} is the odds of a normal PFT for the ith level of lead.

Substituting symbols for all the terms in the preceding equation yields

$$\ln\left(\frac{\pi_{i1}}{\pi_{i2}}\right) = \mu + \alpha_i \tag{14.1}$$

where μ represents the constant and α_i is the effect of the ith level of lead. This model has the same structure that we used in the ANOVA where we are measuring the effect of the levels of a variable from a reference level. For the lead variable, we consider the high level of pollution to be the reference level. This means that α_2 is taken to be 0 and that μ is the logit for the high lead level as can be seen from the following two equations:

$$\ln\left(\frac{\pi_{11}}{\pi_{12}}\right) = \mu + \alpha_i$$

$$\ln\left(\frac{\pi_{21}}{\pi_{22}}\right) = \mu.$$

It is clear from the second of these two equations that μ is the logit of a normal PFT for those exposed to the high lead pollution level. If we subtract the second equation from the first, we see that α_1 is simply the difference of the two logits — that is,

$$\ln\left(\frac{\pi_{11}}{\pi_{12}}\right) - \ln\left(\frac{\pi_{21}}{\pi_{22}}\right) = \alpha_1.$$

Since the difference of two logarithms is the logarithm of the ratio, we have

$$\alpha_1 = \ln\left(\frac{\pi_{11} \cdot \pi_{22}}{\pi_{12} \cdot \pi_{21}}\right).$$

Thus, α_1 is the natural logarithm of the odds ratio, and this is one of the reasons that logistic regression is so useful. In this example, the estimate of α_1 is 0.860 and the estimate of μ is 2.104. If we take the exponential of the estimate of α_1, we obtain the value 2.362, the estimated odds ratio. This value is much greater than one, and it strongly supports the idea that those with the lower lead exposure have the greater proportion of a normal PFT. The estimate of the constant term is the logit for the high level of lead, and the exponential of the estimate of μ is 8.2, the odds of a normal PFT result for those with high lead exposures. Thus the logistic regression model leads to parameters that are readily interpretable.

14.1.1   Proportion, Odds, and Logit

Before proceeding with the extension of the logistic regression model to multiple independent variables, it is helpful to examine the relationship between probabilities (π_i), odds [$o_i = \pi_i/(1 - \pi_i)$] and logits [$\lambda_i = \ln(o_i)$] shown in Table 14.2.

Note that when the probability is 0.5, the odds equal 1 or are even. As the probabilities increase toward 1, the odds increase quite rapidly. As the probabilities decrease toward 0, the odds also approach 0. When the odds equal 1, the logit is 0. As the odds decrease below 1, the logit takes a negative value, approaching negative infinity. As the odds increase above 1, the logit takes a positive value, approaching positive infinity.

The relationship between probabilities and logits is graphically shown in Figure 14.1. The relationship is essentially linear for probabilities between 0.3 and 0.7 and nonlinear

Table 14.2   A comparison of probabilities, odds, and log odds (logits).

Probabilities (π_i)	0.01	0.10	0.20	0.30	0.40	0.50	0.60	0.70	0.80	0.90	0.99
Odds (o_i)	0.01	0.11	0.25	0.43	0.67	1.00	1.50	2.33	4.00	9.00	99.00
Logits (λ_i)	−4.59	−2.20	−1.39	−0.85	−0.41	0.00	0.41	0.85	1.39	2.20	4.59

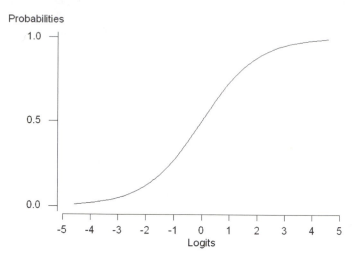

Figure 14.1 Plot of probabilities versus logits.

for lower and greater probabilities. A unit change in the logit results in greater differences in probabilities at levels in the middle than at high and low levels.

Manipulating the formula for the odds allows us to express probabilities in terms of odds

$$\pi_i = \frac{o_i}{1 + o_i}. \tag{14.2}$$

Since $o_i = \exp(\lambda_i)$, we can also express probabilities in terms of logits

$$\pi_i = \frac{\exp(\lambda_i)}{1 + \exp(\lambda_i)}. \tag{14.3}$$

This expression for the probability is one that is often seen in the literature when dealing with logistic regression.

14.1.2   Estimation of Parameters

Example 14.1 showed that the estimation of parameters for the case where both the outcome variable and the exposure variable have two levels is quite simple. However, the estimation of parameters in logistic regression becomes more complex when we incorporate continuous independent variables and discrete variables with multiple levels in the model.

It turns out that the least squares estimation procedure doesn't yield the best estimates for the parameters in logistic regression. Instead of least squares, logistic regression uses the *maximum likelihood* procedure to obtain the parameter estimates. The maximum likelihood approach finds estimates of the model parameters that have the greatest likelihood of producing the observed data. The estimation procedure usually begins with the least squares estimates of coefficients and then uses an iterative algorithm to successively find new sets of coefficients that have higher likelihood of producing the observed data. Computer programs typically show the number of iterations required to find the estimated coefficients with the greatest likelihood. However, it is beyond the scope of this book to provide the details of the estimation. For more information on logistic regression, see the excellent book by Hosmer and Lemeshow (1999a).

14.1.3   Computer Output

The following two examples are applications of logistic regression with a single independent variable. In the first example, the independent variable has only two levels, whereas in the second example, the independent variable is continuous.

Example 14.2

Table 14.3 presents a summary of the computer output for a logistic regression analysis of the data used in Example 14.1 (see **Program Note 14.1** on the website).

The estimates for the intercept and the effect of the low lead level are 2.104 and 0.860, respectively. These estimates are the same as in Example 14.1. Table 14.3 also shows the standard errors for the coefficients, test statistics, p-value, and confidence interval for the odds ratio. These will be explained in the next section.

Table 14.3   Estimates resulting from the fitted logit model for the PFT data in Table 14.1.

Variable	Coefficient	Standard Error	Wald Statistic	p-value	Odds Ratio	95% Confidence Interval
Intercept	2.104	0.335	6.28	<0.001	—	—
Lead (low)	0.860	0.409	2.10	0.036	2.362	(1.06, 5.27)

Likelihood ratio: chi-square = 4.025, df = 1, p-value = 0.045

In the following example, we consider the case with a continuous covariate in the model.

Example 14.3

We want to explore the relationship between diabetes (presence or absence) and body mass index (BMI) using individuals from the DIG200 dataset. In the DIG200 dataset, BMI is a continuous variable and will serve as the independent variable. The symbolic representation of this model is

$$\ln\left(\frac{\pi}{1-\pi}\right) = \beta_0 + \beta_1 x_1$$

where x_1 represents the value of the BMI. For simplicity we rounded the values of BMI to the nearest whole number. The results of fitting the logistic regression model are shown in Table 14.4. See **Program Note 14.1** on the website for fitting this model.

The fitted logit model is

$$\ln\left(\frac{\hat{\pi}}{1-\hat{\pi}}\right) = -3.034 + 0.075 x_1.$$

Table 14.4   Estimates resulting from the logistic regression analysis of diabetes on body mass index, DIG200.

Variable	Coefficient	Standard Error	Wald Statistic	p-value	Odds Ratio	95% Confidence Interval
Intercept	−3.034	0.893	−3.40	0.001	—	—
BMI	0.075	0.032	2.35	0.019	1.08	(1.01, 1.15)

Log-Likelihood: Intercept only   −116.652
BMI term added   −113.851
Likelihood ratio: chi-square = 5.602, df = 1, p-value = 0.018

This estimated equation means that for a 1 kg/m^2 increase in BMI, the log odds of having diabetes increases by 0.075 units. However, a 5 kg/m^2 increase in BMI may be more meaningful than a change of 1 kg/m^2. A 5 kg/m^2 increase in BMI increases the log odds by 0.375 (= 5 ∗ 0.075) units. The estimated change in the odds is easily calculated by $\exp(0.375) = 1.45$. This value means that the estimated odds of diabetes increases by 45 percent for every 5 kg/m^2 increase in BMI.

14.1.4   Statistical Inference

In this section we are interested in examining if a significant relationship exists between the dependent variable and independent variable(s) contained in the logistic model. The two tests commonly used in the tests of hypotheses in logistic regression are the *Wald test* and the *likelihood ratio test* (LRT). We are interested in testing the null hypothesis that the coefficient of the independent variable is equal to zero versus the alternative hypothesis that the coefficient is nonzero — that is,

$$H_0: \beta_1 = 0 \text{ versus } Ha: \beta_1 \neq 0.$$

We begin with the Wald test.

The test statistic for the Wald test is obtained by dividing the maximum likelihood estimate (MLE) of the slope parameter $\hat{\beta}_1$ by the estimate of its standard error, se $(\hat{\beta}_1)$. Under the null hypothesis, this ratio follows a standard normal distribution.

Example 14.4

Let us reexamine the material from Example 14.2. As shown in Table 14.3, the value of $\hat{\beta}_1$ is 0.860 and se $(\hat{\beta}_1)$ is 0.409. Therefore, the Wald test statistic is calculated as follows:

$$\frac{\hat{\beta}_1}{se(\hat{\beta}_1)} = \frac{0.860}{0.409} = 2.10.$$

If the null hypothesis is true, this statistic follows the standard normal distribution. The p-value for this test is 0.036 [= 2 ∗ Prob(Z > 2.10)], suggesting that β_1 is significantly different from zero at the 0.05 level. These values are shown in Table 14.3.

We can use the confidence interval for the odds ratio to determine whether or not the odds ratio equals one. If the confidence interval does not contain one, then we conclude that the odds ratio is statistically significant. The use of the confidence interval is equivalent to testing the hypothesis that $\beta_1 = 0$. The $100 * (1 - \alpha)$ percent confidence interval for the odds ratio [$\exp(\beta_1)$] is calculated by

$$[\exp\{\hat{\beta}_1 - z_{1-\alpha/2} \cdot se(\hat{\beta}_1)\}, \exp\{\hat{\beta}_1 + z_{1-\alpha/2} \cdot se(\hat{\beta}_1)\}].$$

Using the estimates in Table 14.3, the 95 percent confidence interval for the odds ratio is

$$[\exp\{0.860 - 1.96 * (0.409)\}, \exp\{0.860 + 1.96 * (0.409)\}]$$

or from 1.059 to 5.269. Since the interval does not contain one, the odds ratio is considered to be statistically significant at the 0.05 level. Note that the confidence interval for the odds ratio is not symmetric around the sample estimate. We also did not use the usual approach and base the confidence interval on the estimated odds ratio itself and its estimated standard error because the estimated odds ratio does not follow a normal distribution.

The LRT is used to test the hypothesis that an independent variable is zero. The LRT test statistic is

$$\chi^2_{LR} = -2\ln\left(\frac{\text{Likelihood of the reduced model}}{\text{Likelihood of the full model}}\right) \quad (14.4)$$

a quantity that follows the chi-square distribution under the null hypothesis. The degrees of freedom for the chi-square distribution is the difference between the number of parameters in the full model and the number of parameters in the reduced model. In the simple case of only one covariate in the model, the null hypothesis is that the covariate's coefficient is equal to zero. Although the Wald test's p-values are commonly reported, we recommend the use of the p-values from the likelihood ratio test (Hauck and Donner 1977; Jennings 1986). The following example demonstrates the use of the LRT.

Example 14.5

Let us revisit the results of the logistic model for the BMI data shown Table 14.4 in Example 14.3. We wish to determine whether or not there is a significant relationship between the independent variable BMI and the presence or absence of diabetes. We shall test the hypothesis of no relationship at the 0.05 level. In symbols, the hypothesis is

$$H_0 : \beta_1 = 0 \text{ versus } H_a : \beta_1 \neq 0.$$

We begin with the model containing only the constant term and compare it to a model containing both the constant and the BMI variable. The log of the likelihood for the constant only model is -116.652, and the log of the likelihood for the model with the BMI variable is -113.851. The test statistic is found by applying Equation (14.4) — that is,

$$X^2_{LR} = -2[\ln(\text{likelihood of reduced model}) - \ln(\text{likelihood of full model})]$$

$= -2 * [-116.652 - (-113.851)] = -2 * (-2.801)$, which is 5.602. There is one degree of freedom for this test of hypothesis because the full model contains only one covariate and the reduced model does not contain any covariates. In this case, the p-value for a chi-square value of 5.602 with one degree of freedom is 0.018. Therefore, we reject the null hypothesis and conclude that β_1 is significantly different from zero — that is, the occurrence of diabetes is related to the BMI variable at the 0.05 level.

14.2 Multiple Logistic Regression

Regression models are useful because they help us explore the relationships between a dependent or response variable and one or more independent or predictor variables of interest. In particular, logistic regression models allow medical researchers to help clinicians in the choice of an appropriate treatment strategy for individual patients.

14.2.1 Model and Assumptions

In the previous section we introduced the simple logistic regression model with only one independent variable. For multiple logistic regression with k independent variables, $x_1, x_2, \ldots, x_k$, the model, taking the form of Equation (14.3), is

$$\pi = \frac{\exp(\beta_0 + \beta_1 x_1 + \beta_2 x_2 + \cdots + \beta_k x_k)}{1 + \exp(\beta_0 + \beta_1 x_1 + \beta_2 x_2 + \cdots + \beta_k x_k)}. \tag{14.5}$$

By obtaining estimates for the betas in the linear combination, $\beta_0 + \beta_1 x_1 + \cdots + \beta_k x_k$, we can calculate the estimated or predicted probability of the outcome of interest.

We present two examples here. The first example includes discrete independent variables only, whereas the second example has both discrete and continuous independent variables.

Example 14.6

We reconsider Example 14.1 and now introduce a covariate. In the example, we found a significant lead effect, a finding that is somewhat surprising, since lead has not been shown to have a negative impact on the respiratory system in other studies. However, during the period 1974–1975 when this study was performed, automobile emissions were a major source of lead pollution. Thus, a possible explanation for this finding is that lead pollution is serving as a proxy for nitrogen dioxide or other pollutants that have adverse respiratory effects. Another possible explanation is that we have not controlled for possible confounding variables. Smoking status is a key variable that has been ignored in the analysis so far. Table 14.5 shows the smoking status by lead level and PFT result.

We begin by considering a model containing the main effects of lead and smoking. Because smoking status contains four levels, we must create three dummy variables in order to obtain a symbolic representation of this model. The dummy variables can be expressed as shown in Table 14.6.

Table 14.5   Pulmonary function test (PFT) results by smoking status and lead exposure.

Lead Level	Smoking Status	PFT Results		Total	Proportions Normal	Odds (normal)	Logits (normal)
		Normal	Abnormal				
Low	Heavy	84	3	87	0.9655	28.000	3.332
	Light	75	6	81	0.9260	12.500	2.526
	Former	49	6	55	0.8910	8.167	2.100
	Never	160	4	164	0.9756	40.000	3.689
High	Heavy	16	3	19	0.8421	5.333	1.674
	Light	21	2	23	0.9130	10.500	2.351
	Former	12	2	14	0.8571	6.000	1.792
	Never	33	3	36	0.9167	11.000	2.398

Table 14.6   Dummy variables for the smoking status variable.

	Smoking Status	D_1	D_2	D_3
Smoking status 0	Heavy	0	0	0
Smoking status 1	Light	1	0	0
Smoking status 2	Former	0	1	0
Smoking status 3	Never	0	0	1

We will use the heavy smoking status as the reference category and measure the effects of the other smoking categories from it. Thus, the dummy variable D_1 is 1 when the smoking status is light and 0 otherwise, D_2 is 1 when the smoking status is former and 0 otherwise, and D_3 1 when the smoking status is never and 0 otherwise. Statistical software packages can create these dummy variables for the user (see **Program Note 14.2** on the website for more details).

Therefore, the estimated logit can be expressed as

$$\ln\left(\frac{\hat{\pi}_i}{1-\hat{\pi}_i}\right) = \hat{\beta}_0 + \hat{\beta}_1 x_{1i} + \hat{\beta}_2 D_{1i} + \hat{\beta}_3 D_{2i} + \hat{\beta}_4 D_{3i}.$$

The estimated values of the logit model's parameters are the following:

$$\ln\left(\frac{\hat{\pi}_i}{1-\hat{\pi}_i}\right) = 2.18 + 0.84 x_{1i} - 0.29 D_{1i} - 0.77 D_{2i} + 0.50 D_{3i}.$$

The addition of the smoking variable has not changed the parameter estimates much. The estimate of the constant was previously 2.104 (versus 2.18 now), and the previous estimate of the low lead effect was 0.860 (versus 0.84 now).

In this situation, the estimate of β_1 is the natural logarithm of the odds ratio if the high and low lead levels had contained the same distributions of the smoking status variable. Examination of Table 14.5 shows that the distributions of the smoking status variable are similar for the high and low lead levels. Hence it is not surprising that the estimates of the odds ratio for high lead levels compared to low lead levels are approximately the same for the simple model shown in Table 14.3 and the model shown in Table 14.7. Individuals in residences with low lead levels are 2.3 times more likely to have normal PFT results compared to individuals in residences with high lead levels after adjusting for smoking status.

Table 14.7   Estimates for the logit model parameters and odds ratio for the data in Table 14.5.

Variable	Coefficient	Standard Error	Wald Statistic	p-value	Odds Ratio	95% Confidence Interval
Intercept	2.178	0.510	4.27	<0.0001	—	—
Lead (low)	0.837	0.414	2.03	0.043	2.31	(1.03, 5.20)
Smoke 1[a]	−0.289	0.562	−0.51	0.607	0.75	(0.25, 2.25)
Smoke 2[b]	−0.767	0.567	−1.35	0.176	0.46	(0.15, 1.41)
Smoke 3[c]	0.508	0.572	0.89	0.375	1.66	(0.54, 5.10)

Likelihood ratio: chi-square = 9.914, df = 4, p-value = 0.042
Goodness of fit tests: Pearson chi-square = 2.276, df = 3, p-value = 0.517
Deviance chi-square = 2.256, df = 3, p-value = 0.521

[a]light smoker; [b]former smoker; [c]never smoked

There is no suggestion that the effect of any of the three levels of the smoking variables differ from the effect of the heavy smoking level. The 95 percent confidence intervals for the odds ratios of the smoking effects all contain one, and none of the Wald statistics suggest statistical significance at the 0.05 level. The likelihood ratio test statistic shown in the output is used to test the hypothesis that all four model coefficients (the lead effect and the three smoking effects) are simultaneously equal to zero. We reject this hypothesis at the 0.05 level. The lead variable still appears to be related to the PFT variable. We will explain the other two test statistics later in this chapter.

Example 14.7

Suppose that we would like to develop a logistic regression model to predict diabetes using BMI, treatment, and race using the DIG200 dataset. The literature suggests that individuals with larger values of BMI, who are on a placebo, and who are non-white are more likely to have diabetes. As we did in Example 14.3, we rounded the values of BMI to the nearest whole number. Table 14.8 shows information about the three predictor variables and the presence or absence of diabetes.

We will consider the placebo level of the treatment variable to be the reference level and measure the effect of the digoxin treatment from it. We will also consider

Table 14.8   Patient characteristics by diabetes status, DIG200.

	Diabetes	
Characteristics	Yes	No
Mean BMI[a] ± SD[b]	28.0 ± 5.5	26.1 ± 4.6
Range	(18 – 43)	(15 – 45)
Treatment: Placebo	34	66
Digoxin	20	80
Race: White	42	131
Nonwhite	12	15

[a]BMI—Body Mass Index rounded to the nearest whole number
[b]SD—Standard Deviation

Table 14.9   Logistic regression analysis of diabetes on BMI, treatment, and race, DIG200.

Variable	Coefficient	Standard Error	Wald Statistic	p-value	Odds Ratio	95% Confidence Interval
Intercept	−2.948	0.914	−3.22	0.001		
BMI (kg/m²)	0.081	0.033	2.45	0.014	1.08	(1.02, 1.16)
Treatment (digoxin)	−0.796	0.339	−2.35	0.019	0.45	(0.23, 0.88)
Race (nonwhite)	0.904	0.440	2.05	0.040	2.47	(1.04, 5.85)

Likelihood ratio: chi-square = 15.471, df = 3, p-value = 0.001

Goodness of fit tests: Pearson chi-square = 44.485, df = 57, p-value = 0.886
Deviance chi-square = 57.816, df = 57, p-value = 0.445
H-L chi-square = 2.532, df = 8, p-value = 0.960

the white race to be the reference level and measure the nonwhite effect from it. The results of the logistic regression analysis are shown in Table 14.9 (see **Program Note 14.3** on the website).

The fitted logit model is

$$\ln\left(\frac{\hat{\pi}}{1-\hat{\pi}}\right) = -2.948 + 0.081x_1 - 0.796x_2 + 0.904x_3.$$

From Table 14.9, we see that the odds of having diabetes is higher for larger values of BMI even after adjusting for treatment and race. The estimated adjusted odds ratios are greater than one for the BMI and race variables, whereas the adjusted odds ratio is below one for the treatment variable. This indicates that patients receiving digoxin are less likely (specifically 45 percent less likely) to have diabetes compared to patients on the placebo after adjusting for BMI and race.

All three of the independent variables are statistically significant at the 0.05 level. The likelihood ratio chi-square statistic (= 15.47 with three degrees of freedom) suggests that the three coefficients associated with the independent variables are not simultaneously equal to zero at the 0.05 level.

The probability of diabetes given an individual's BMI, treatment group, and race can also be estimated based on the estimated model parameters using Equation (14.5)

$$\hat{\pi}_i = \frac{\exp(-2.948 + 0.081x_{1i} - 0.796x_{2i} + 0.904x_{3i})}{1 + \exp(-2.948 + 0.081x_{1i} - 0.796x_{2i} + 0.904x_{3i})}.$$

As an example let us consider calculating the probability of having diabetes given a patient with a BMI of 24 kg/m², on digoxin treatment, and being of a nonwhite race. The calculation is

$$\hat{\pi} = \frac{\exp[-2.948 + 0.081(24) - 0.796(1) + 0.0904(1)]}{1 + \exp[-2.948 + 0.081(24) - 0.796(1) + 0.904(1)]} = 0.290.$$

Therefore, the odds of diabetes given a patient with a BMI of 24 kg/m², on digoxin treatment, and being nonwhite is [0.290/(1 − 0.290)] = 0.408. We explain the other two test statistics in the following sections.

14.2.2   Residuals

In logistic regression, we can get a feel for how well the model agrees with the data by comparing the observed and predicted logits or probabilities for all possible covariate patterns. For example, in Example 14.6 the eight possible covariate patterns are listed again in Table 14.10 along with observed and predicted logits and probabilities. The observed logits and probabilities come from Table 14.5.

Table 14.10   List of covariate patterns for PFT data in Example 14.6.

Covariate Pattern	Lead Level	Smoking Status			Logit		Probability	
					Observed	Predicted	Observed	Predicted
(i)	x_i	D_{1i}	D_{2i}	D_{3i}	l_i	$\hat{\lambda}_i$	p_i	$\hat{\pi}_i$
1	1	0	0	0	3.332	3.015	0.9655	0.9532
2	1	1	0	0	2.526	2.726	0.9260	0.9385
3	1	0	1	0	2.100	2.248	0.8910	0.9045
4	1	0	0	1	3.689	3.523	0.9756	0.9713
5	0	0	0	0	1.674	2.178	0.8421	0.8982
6	0	1	0	0	2.351	1.889	0.9130	0.8686
7	0	0	1	0	1.789	1.411	0.8571	0.8038
8	0	0	0	1	2.398	2.686	0.9167	0.9362

In multiple linear regression, the residuals provided useful information about possible problems with the model. We can also use the residuals in logistic regression to examine the fit of the logistic model. Two common forms of residuals used in logistic regression are *Pearson residuals* and *deviance residuals*. These residuals are useful for identifying outlying and influential points (Pregibon 1981). The *Pearson residual* is defined as

$$r_i = \frac{y_i - n_i \hat{\pi}_i}{\sqrt{n_i \hat{\pi}_i (1 - \hat{\pi}_i)}}$$

where n_i is the number of observations with the ith covariate pattern, y_i is the number of observations with the outcome of interest among n_i observations, and $\hat{\pi}_i$ is the predicted probability of the outcome of interest for the ith covariate pattern. The form of the Pearson residual is familiar — dividing the difference in the observed and predicted cell counts by the standard error of the observed count. We did the same calculations in converting statistics to a standard normal variable. Note that we can also express the numerator of r_i as $y_i - \hat{y}_i$, where $\hat{y}_i$ is equal to $n_i \hat{\pi}_i$.

Some recommend a slightly different form of the Pearson residual. For example, according to Collett (2003), a better procedure is to divide the raw residual, $y_i - \hat{y}_i$, by its standard error, $se(y_i - \hat{y}_i)$. This standard error is complicated to derive, but it is used in many of the logistic regression programs. Residuals based on the $se(y_i - \hat{y}_i)$ are known as the *standardized Pearson residuals*.

The *deviance residual* is defined as

$$d_i = sgn(y_i - n_j \hat{\pi}_j) \left[2 y_i \ln\left(\frac{y_i}{n_i \hat{\pi}_i} \right) + 2(n_i - y_i) \ln\left(\frac{n_i - y_i}{n_i (1 - \hat{\pi}_i)} \right) \right]^{1/2}$$

where sgn is plus if the quantity in the parenthesis is positive and negative if the quantity is negative.

Table 14.11   Pearson and deviance residuals for the multiple logistic regression model from Example 14.6.

Covariate Pattern	Residual	
	Pearson	Deviance
1	0.54	0.57
2	−0.47	−0.46
3	−0.34	−0.34
4	0.33	0.34
5	−0.81	−0.75
6	0.63	0.67
7	0.50	0.52
8	−0.48	−0.46
Sum of Squares	2.28	2.25

Since there are only eight covariate patterns for the PFT data in Example 14.6, we can easily show the Pearson and deviance residuals in Table 14.11 (see **Program Note 14.2** on the website).

14.2.3   Goodness-of-Fit Statistics

We can also use the residuals in testing the goodness of fit of the model. A Pearson test statistic can be calculated by summing the squares of the residuals, that is, Σr_i^2. A similar test statistic based on the deviance residuals is then Σd_i^2. If the model fits, both of these statistics follow a chi-square distribution with degrees of freedom equal to number of covariate patterns minus the number of parameters in the model plus one.

Let's now test the goodness of fit of the model. The null and alternative hypotheses are

H_0: the model fits the data *versus* Hâ: the model does not fit the data.

Because we estimated four parameters in the model and there are eight covariate patterns, there are three degrees of freedom for the chi-square test. If we test the hypothesis that the model fits at the 0.05 level, a value of the test statistic greater than 7.81 is required to reject the null hypothesis. Since both test statistics (values of 2.28 and 2.25 for the Pearson statistic and the deviance statistic, respectively) are smaller than this critical value, we fail to reject the hypothesis that the model fits.

In logistic situations with continuous independent variables, it is likely that the number of distinct covariate patterns will be close to the number of observations. The next example considers this situation.

Example 14.8

We are going to plot the Pearson and deviance residuals by individual for the multiple logistic regression model considered for the diabetes data in Example 14.7 (see **Program Note 14.3** on the website).

Since these residuals have, in effect, been divided by their standard errors — it is hard to see that this statement applies to the deviance residuals, but it does —

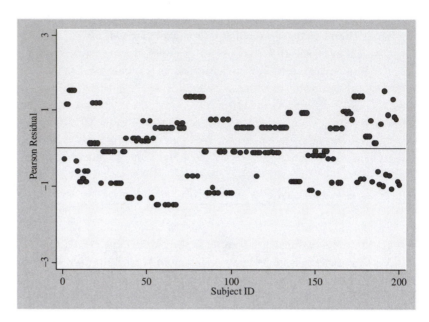

Figure 14.2 Pearson residual by subject for the data in Example 14.7.

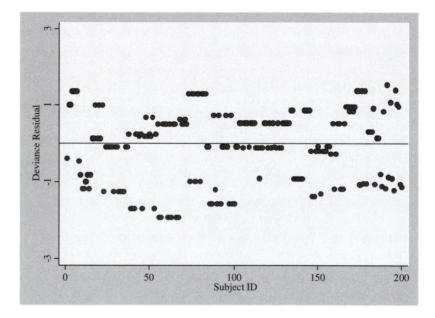

Figure 14.3 Deviance residual by subject for the data in Example 4.7.

residuals that have a value greater than two are of interest. Residuals with a value greater than two could result because of a coding error or simply represent a rare occurrence. We are looking for any patterns in the residuals, similar to the analysis of residuals in multiple linear regression. Since there don't appear to be any large residuals in Figure 14.2 or Figure 14.3, it does not appear that any of the observations require further inspection. If there were large residuals, we could try other plots such as the residuals versus the independent variables as well as doing univariate analysis on the original data looking for anomalies.

In some cases, particularly those with continuous independent variables, we prefer not to use the Pearson and deviance chi-square statistics to test the fit of the model. In these cases, we believe that other tests — for example, the *Hosmer-Lemeshow (H-L)* goodness-of-fit test — have better statistical properties (Hosmer and Lemeshow 1999a). The H-L procedure groups the data into g categories where g is usually 10. The grouping is based on the values of the predicted probabilities from the model. In one approach, the data are grouped into equal-sized ordered categories with the first category having the subjects with the smallest estimated probabilities and so forth to the last group containing the subjects with the largest estimated probabilities. In another approach suggested by Hosmer and Lemeshow, the categories are formed by specific cutpoints — for examples, $0.10, 0.20, \ldots, 0.90$. The first group contains all subjects with predicted probabilities less than or equal to 0.10, the second group contains all subjects with predicted probabilities greater than 0.10 and less than or equal to 0.20 and so on, to the last group that contains all the subjects with predicted probabilities greater than 0.90. The H-L test statistic is defined as

$$\hat{C} = \sum_{k=1}^{g} \frac{(o_k - n_k'\bar{\pi}_k)^2}{n_k'\bar{\pi}_k(1 - \bar{\pi}_k)}$$

where n_k' is the number of covariate patterns in the kth group, o_k is the number of subjects with the condition of interest in the n_k' covariate patterns, and $\bar{\pi}_k$ is the average predicted probability in the kth group. Based on extensive simulations, the H-L statistic follows the chi-square distribution with $g - 2$ degrees of freedom.

Let's now test the goodness-of-fit of the logistic model used in Example 14.7 at the 0.05 level. We will use the first method of grouping — that is, dividing the data into 10 equal-sized categories. As shown in Table 14.9, the H-L test statistic is 2.532 with 8 degrees of freedom. Since the H-L statistic is less than the critical value of 15.51, we fail to reject the goodness of fit of the model at the 0.05 level.

14.2.4 The ROC Curve

Another measure of how well a logistic regression model performs can be obtained by examining the area under the receiver operating characteristic (ROC) curve, originally presented in Chapter 4, for that model. Recall that the ROC curve is created by plotting 1-specificity against sensitivity at different cutoff points for determining a positive or negative test result. In the logistic model, the sensitivity and specificity can be evaluated at different levels of predicted probabilities by comparing the predicted classification with the observed classification of the dependent variable. The area under the ROC curve provides a measure of the discriminative ability of the logistic model. Hosmer and Lemeshow (1999a) suggest the following guidelines for assessing the discriminatory power of the model:

If the area under the ROC curve (AUROC) is 0.5, the model does not discriminate.
If $0.5 < \text{AUROC} < 0.7$, the model has poor to fair discrimination.
If $0.7 < \text{AUROC} < 0.8$, the model has acceptable discrimination.
If $0.8 \leq \text{AUROC} < 0.9$, the model has excellent discrimination.
If $\text{AUROC} \leq 0.9$ — a very rare outcome — the model has outstanding discrimination.

Example 14.9

Let's consider the logistic regression model including lead levels and smoking status as predictors of the PFT results shown in Example 14.6 to see how we create the ROC curve. As shown in Table 14.10, there are eight predicted probabilities in this example and we can evaluate sensitivity and specificity at eight different cutoff points. At the lowest predicted probability of 0.8038 (high lead level and former smoker), the predicted PFT status is determined to be "normal" if the predicted probabilities are greater than or equal to 0.8038. The 2 by 2 table shown here can be formed from the cross-tabulation of the data in Table 14.5 by the predicted and observed PFT status. Sensitivity and specificity are calculated from the table using the procedure explained in Chapter 4:

Predicted PFT Status	Observed PFT Status		
	Normal	Abnormal	
Normal	450	29	Sensitivity = 450/450 = 1.00
Abnormal	0	0	Specificity = 0/29 = 0.00
Total	450	29	

Similarly, we can evaluate sensitivity and specificity at the second lowest predicted probability of 0.8686 (high lead level and light smoker) as follows:

Predicted PFT Status	Observed PFT Status		
	Normal	Abnormal	
Normal	438	27	Sensitivity = 438/450 = 0.973
Abnormal	12	2	Specificity = 2/29 = 0.069
Total	450	29	

For the rest of the cutoff points the sensitivity and specificity are

Cutoff Point	Sensitivity	Specificity
0.8982	0.927	0.138
0.9045	0.891	0.241
0.9363	0.782	0.448
0.9385	0.709	0.552
0.9532	0.542	0.759
0.9713	0.356	0.862
1.0000	0.000	1.000

From these data the ROC curve can be plotted. We can use a computer program to create the ROC curve and calculate AUROC, as shown in Figure 14.4 (see **Program Note 14.2** on the website).

AUROC can be interpreted as the likelihood that an individual who has a non-normal PFT result will have a higher predicted probability of having a nonnormal PFT than an individual who does not have a nonnormal PFT result (Pregibon 1981). The AUROC value for this example is approximately 0.68, a value suggesting poor to fair discrimination.

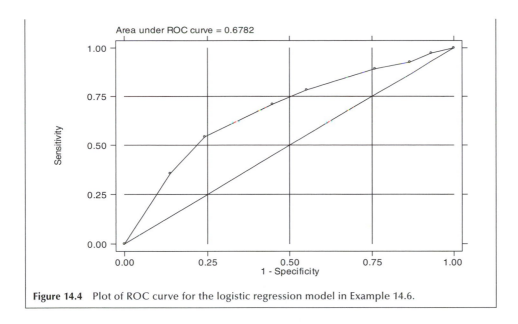

Figure 14.4   Plot of ROC curve for the logistic regression model in Example 14.6.

Many programs also report a pseudo-R^2. Statisticians tend to give less attention to this measure because it may suggest the model has poor explanatory power, whereas other measures such as the AUROC suggest good discriminatory power. The goodness-of-fit tests, the examination of residuals, and the AUROC are three tools with good acceptance by statisticians for examining multiple logistic regression models.

We have provided a few of the numerous diagnostic tools available to the researcher for examining the logistic regression model. The use of additional plots and many other statistics shown in Chapter 13 for examining the fit of the model carry over to logistic regression. To learn more about the application of these other tools, the reader is encouraged to check other sources on logistic regression (Hosmer and Lemeshow 1999; Pregibon 1981). However, we should not automatically delete those subjects identified using these diagnostic methods. Any elimination of subjects must be done very carefully and be based on substantive considerations as well as on diagnostic methods.

14.3   Ordered Logistic Regression

In previous sections, we introduced logistic regression models that have a dependent variable with a dichotomous outcome. However, more complicated forms of logistic regression are also available, and we begin this section by considering an ordinal-dependent variable with more than two levels.

Example 14.10

Let us examine the perceived health status reported in the National Health and Nutrition Examination Survey. The health status is reported as "excellent," "very good," "good," "fair," and "poor." Based on an NHANES III Phase II adult sample, 23.0 percent of U.S. adults reported that their health status is "excellent," 30.3 percent "very good," 31.3 percent "good," and 15.5 percent for the "fair or poor" categories combined. We want to determine whether or not there is a relationship between health status and the use of vitamin or mineral supplements (1 = use, 0 = nonuse) and education reflected by the number of years of schooling. Note that if a relationship exists, it does not necessarily imply any causal relationship between the variables we are labeling as independent and the variable we are labeling as dependent. It may be that supplement use is a function of health status, or it could be that health status is a function of supplement use or there could be a mixture of relationships. We can't tell from these data the direction of the relationship even if a relationship actually exists.

Given the relatively small number of people in the fair and poor categories we have combined them into one category. Hence, we are now working with four ordered health status categories. Let's start our investigation by looking at health status and supplement use. Before examining the relationship between these two variables, we must decide how to handle this ordinal health status variable. Since there are four levels, there are really only three pieces of independent information. This means that we could create three independent functions that would contain all the information in the health status variable. One such set of functions is the following:

Pr (excellent) versus Pr (all other levels)

Pr (excellent plus very good) versus (good plus fair or poor)

Pr (excellent plus very good plus good) versus (fair or poor).

Given the sample values just mentioned for the probabilities of the various health status states, we would expect the first function to be much less than one, the second function to be close to one, and the third function to be much greater than one. If we take the natural logarithm of the three functions, we would expect the first to be negative, the second close to zero, and the third to be positive.

We could then perform three separate binary logistic regressions to examine the relationships to supplement use. A logistic model that could be used to examine the relationship is

ln (*health-status function$_i$*) = *constant$_i$* + *effect of supplement use$_i$*.

However, if the effect of supplement use on health status is consistent for these three functions, we could estimate this "average" effect of supplement use by considering a single model that included the supplement use effect plus three separate constant terms. In effect, this model is

ln (*health-status function$_i$*) = *constant$_i$* + *effect of supplement use*.

This representation reflects the idea that the regression lines for the different outcome functions are parallel to each other but that they have different intercepts. Table 14.12 shows basic data for this analysis and for checking of the assumption of a consistent effect of supplement use — that is, the odds ratios for each of the health status functions with supplement use are similar. This assumption is called the proportional odds assumption.

Since the odds ratios of 1.19, 1.36, and 1.51 are reasonably similar, we can conclude that the proportional odds assumption seems to be acceptable. We see that

Table 14.12   Perceived health status by use and nonuse of vitamin or mineral supplements, NHANES, Phase II adult subsample (n = 988).

Vitamin Use	Perceived Health Status				
	Excellent	Very Good	Good	Fair or Poor	Total
User	105	139	127	53	424
Nonuser	122	160	182	100	564
Total	227	299	309	153	988[a]
	(28.0%)	(30.3%)	(31.3%)	(15.5%)	(100.1%)
Comparisons	I		II		III
	105	319	244	180	371	53
	122	442	282	282	464	100
Odds Ratio	1.19		1.36		1.51

[a]Excluding cases with missing values

those taking vitamin or mineral supplements are more likely to feel better about their health and vice versa. Given that the proportional odds assumption seems to hold, we can estimate the common odds ratio that summarizes the effect.

Note that it is not uncommon for an ordered logistic regression model not to satisfy the proportionality assumption (or parallel regression assumption). If this assumption is not satisfied, other alternative models should be considered, such as the multinomial logistic model (Hosmer and Lemeshow 1999a).

Table 14.13 shows the results of ordered logistic regression analysis (see **Program Note 14.4** on the website). The top panel shows the ordered logistic regression of health status on supplement use based on the reduced model that assumes the equality of the supplement coefficients for the three health-status variables. In this example, the equality of the supplement coefficients is another way of saying that the lines are parallel or that the odds are proportional for the three health-status variables.

In examining the results, we first look at the test for the goodness of fit of the model. In this case, the goodness-of-fit test examines whether or not the three coefficients for supplement use in the full model are all equal. Based on the goodness-of-fit values from the Pearson and deviance tests, we fail to reject the equality of the coefficients (or that the lines are parallel or that the odds are proportional), a result we expected, since the preceding three odds ratios were fairly similar.

The maximum likelihood estimates of coefficients include the three intercepts and the common supplement effect. The intercepts don't hold much interest for us, but their values are consistent with the expected pattern mentioned above (negative, close to zero, and positive). The estimated coefficient for vitamin use is 0.2835, and the corresponding estimated odds ratio is 1.33. This is the estimated common odds ratio for healthier status, comparing those taking supplements with those not taking supplements. The 95 percent confidence interval for the common odds ratio, the p-value for the test that the coefficient for supplement use is zero, and the g statistic (follows a chi-square distribution) all suggest that there is a significant relationship between supplement use and health status at the 0.05 level.

The preceding analysis could be done using the CMH method presented in Chapter 10. But the ordered logistic regression model allows us to include continuous

Table 14.13　Ordered logistic regression analysis of perceived health status on use of vitamin or mineral supplements and years of schooling, NHANES III, Phase II adult subsample ($n = 988$).

Model I (health status on vitamin use)

Predictor	Coef	SE Coef	Z	p	Odds Ratio	95% CI Lower Upper
Constant (1)	−1.3384	0.0923	−14.49	<0.001	—	—
Constant (2)	0.0063	0.0808	0.08	0.938	—	—
Constant (3)	1.5808	0.0993	15.92	<0.001	—	—
Vitamin use	0.2835	0.1160	2.44	0.015	1.33	(1.06, 1.67)

Log-likelihood = −1332.777
　Test that all slopes are zero: G = 6.004, DF = 1, p-Value = 0.014

Pseudo R-Square = 0.002

Goodness-of-Fit Tests:
　Pearson　　Chi-Square = 1.354, df = 2, p-Value = 0.508
　Deviance　Chi-Square = 1.357, df = 2, p-Value = 0.507

Model II (health status on vitamin use and years of schooling)

Predictor	Coef	SE Coef	Z	p	Odds Ratio	95% CI Lower Upper
Constant (1)	−4.4268	0.2768	−15.99	<0.001	—	—
Constant (2)	−2.9434	0.2586	−11.38	<0.001	—	—
Constant (3)	−1.1679	0.2452	−4.76	<0.001	—	—
Vitamin use	0.0425	0.1192	0.36	0.722	1.04	(0.83, 1.32)
Schooling	0.2476	0.0205	12.08	<0.001	1.28	(1.23, 1.33)

Score test for the proportional odds assumption:
　Chi-Square = 1.594, df = 4, p-Value = 0.810

Log-likelihood = −1254.178
　Test that all slopes are zero: G = 163.202, DF = 2, p-Value = <0.001

Pseudo R-Square = 0.061

Goodness-of-Fit Tests:
　Pearson　　Chi-Square = 130.426, df = 100, p-Value = 0.022
　Deviance　Chi-Square = 119.519, df = 100, p-Value = 0.089

explanatory variables. The results of the logistic regression of health status on supplement use and the number of years of schooling are shown in the bottom panel of Table 14.13. First, our attention is called to goodness-of-fit statistics. Since the Pearson and deviance residual statistics are larger than the degrees of freedom, the key finding here is that this model does not provide a good fit to the data. Given that the model does not fit, there is little reason to place much emphasis on the parameter estimates. However, note that the supplement variable's effect has been greatly reduced when the years of schooling variable is considered. As just stated, it is difficult using data from a point in time to examine relationships over time. In this situation, it is even not clear what variable should be used as the response or dependent variable.

In general, if the outcome variable is ordered and has g categories, we can form $(g - 1)$ independent functions from the outcome variable. The proportional odds model assumes that the odds ratio across all $(g - 1)$ cut-points is the same. Applying the same approach as previously, the proportional odds model for the $j = 1, 2, \ldots, g - 1$ functions and p explanatory variables is

$$\ln\left(\frac{\pi_{\leq j}}{\pi_{\leq j+1}}\right) = \beta_{0j} + \sum_{i=1}^{p} \beta_i x_i.$$

The functions used as the dependent variables are the logits of being in the g category or lower versus being in higher categories.

14.4   Conditional Logistic Regression

Data from matched studies can also be analyzed by a logistic regression approach. As discussed in Chapter 6, *matching* is a way of balancing certain characteristics between two groups. If matching is used in the design phase of a study, a treatment is given to one member of a matched pair and a placebo is given to the other. In case-control studies, a case with a particular outcome is matched to a control without the outcome of interest and an examination of a possible relationship to an exposure is assessed retrospectively. Matching can be one to one or one to several controls.

One way of analyzing matched studies is *conditional logistic regression*, a method illustrated in the following example.

Example 14.11

The DIG200 data set contains 27 subjects with cardiovascular disease (CVD) — cases who can be perfectly matched to 27 subjects without CVD — controls based on age, sex, and race. The matched data are shown in Table 14.14.

Table 14.14   Twenty-seven controls and matched cases of cardiovascular disease, DIG200.

	Control (without CVD)						Case (with CVD)				
Set	Age	Sex	Race	SBP	MI	Set	Age	Sex	Race	SBP	MI
1	43	1	1	120	1	1	43	1	1	90	0
2	45	1	1	122	0	2	45	1	1	160	1
3	46	1	1	96	1	3	46	1	1	110	1
4	47	2	1	120	0	4	47	2	1	116	0
5	49	1	1	140	0	5	49	1	1	122	1
6	50	1	1	148	1	6	50	1	1	140	0
7	51	2	1	124	1	7	51	2	1	95	0
8	54	1	1	120	1	8	54	1	1	106	0
9	57	1	1	136	0	9	57	1	1	140	1
10	58	2	2	100	0	10	58	2	2	100	1
11	59	1	1	100	1	11	59	1	1	100	0
12	60	1	1	102	1	12	60	1	1	140	1
13	63	1	1	105	0	13	63	1	1	114	0
14	64	1	1	150	1	14	64	1	1	130	0
15	65	1	1	132	0	15	65	1	1	130	1
16	66	1	1	130	1	16	66	1	1	160	0
17	67	1	1	130	1	17	67	1	1	130	1
18	68	2	1	144	1	18	68	2	1	152	1
19	69	1	1	130	0	19	69	1	1	116	0
20	70	1	1	150	1	20	70	1	1	110	0
21	71	1	1	90	0	21	71	1	1	90	0
22	72	2	1	140	0	22	72	2	1	155	0
23	73	1	1	140	1	23	73	1	1	150	0
24	74	1	1	100	1	24	74	1	1	140	1
25	76	1	1	140	0	25	76	1	1	130	0
26	79	1	1	130	0	26	79	1	1	150	1
27	80	2	1	118	1	27	80	2	1	165	0

Let us first look at the relationship between CVD and prior MI. As we discussed in Chapter 10 (Section 10.2.5), the relationship can be summarized in the following 2 by 2 table:

	Prior MI in Controls		
Prior MI in Cases	**Yes**	**No**	**Total**
Yes	5 (c_1)	6 (d_1)	11
No	10 (d_2)	6 (c_2)	16
Total	15	12	27

Relevant information for the analysis of this table is contained in discordant cells (d_1 and d_2), and we used the McNemar chi-square test to test the hypothesis of no relationship between CVD and MI. For the preceding table, the McNemar test statistic is

$$X_M^2 = \frac{(|d_1 - d_2| - 1)^2}{d_1 + d_2} = \frac{(|6 - 10| - 1)^2}{6 + 10} = 0.56.$$

The p-value for this test statistic is 0.454. If we ignore the correction for continuity, the test statistic is 1.00 with p-value of 0.317. There is no statistically significant relationship between CVD and prior MI — that is, a previous MI is not predictive of the occurrence of CVD. The odds ratio for the preceding table is $d_1/d_2 = 6/10 = 0.6$ and the corresponding 95 percent confidence interval is (0.179, 1.82). Since the confidence interval contains the value of 1, there does not appear to be a significant relationship.

Conditional logistic regression offers an alternative method of analysis for matched studies. For example, if we wish to examine whether or not there may be a relationship between the occurrence of CVD (1 = yes, 0 = no) and MI (1 = yes, 0 = no), we will focus on the difference of the variables within each of the 27 pairs because of the matching. The idea of focusing on the differences is similar to the use of differences in the paired t test. The CVD difference is always equal to +1 by definition. The difference in the MI variable can have the value of +1, 0, or −1 and this difference variable is now treated as a continuous variable by the computer software. We can use ordinary logistic regression using the differences as the variables. Since we are using differences, there is no need to include the constant term in the analysis.

The first panel of Table 14.15 shows the results of the logistic regression analyses of the presence and absence of cardiovascular disease on prior myocardial infarction (see **Program Note 14.5** on the website).

The estimated coefficient is −0.5108 (se = 0.5164), which gives the estimated odds ratio as exp(−0.5108) = 0.6. The 95 percent confidence interval is found from the exp(−0.5108 ± 1.96 ∗ 0.5164) or (0.22, 1.65). The odds ratio is exactly the same as found from the 2 by 2 table. The test results also turn out to be very similar to those obtained from the 2 by 2 table. The p-value for McNemar test was 0.317 compared to 0.3147 from the likelihood ratio test for the conditional logistic regression and to 0.323 based on the normal test. Note that we entered the data for 54 observations (27 pairs), but we could have entered just the 16 discordant pairs and obtained the same results, since data for concordant pairs do not contribute anything to the analysis.

Table 14.15   Conditional logistic regression analysis of matched cases of cardiovascular disease on prior myocardial infarction and systolic pressure, 57 pairs from DIG200.

Model I (CVD on prior MI)

Conditional (fixed-effects) logistic regression		Number of obs		=	54
		LR chi2 (1)		=	1.01
		Prob > chi2		=	0.3147
Log likelihood = −18.209631		Pseudo R2		=	0.0270

	Coef.	Std. Err.	z	$p > |z|$	Odds Ratio	[95% CI]
Prior MI	−0.5108	0.5164	−0.99	0.323	0.600	(0.218, 1.651)

Model II (CVD on prior MI and systolic blood pressure)

Conditional (fixed-effects) logistic regression		Number of obs		=	54
		LR chi2 (2)		=	2.00
		Prob > chi2		=	0.3683
Log likelihood = −17.716158		Pseudo R2		=	0.0534

	Coef.	Std. Err.	z	$p > |z|$	Odds Ratio	[95% CI]
Prior MI	−0.6496	0.5546	−1.17	0.242	0.522	(0.176, 1.549)
SBP	0.0187	0.0195	0.96	0.337	1.019	(0.981, 1.059)

For a simple situation like in the above 2 by 2 table, there is really no need to use the conditional logistic regression model. However, the conditional logistic model is very useful for more complicated situations where multiple predictor variables (including continuous variables) are used or for predictor variables with more than two levels. In the case of a discrete variable, such as the smoking variable in Table 14.5, we use three dummy variables like those shown in Table 14.6 to show the smoking status of a person. But now in our conditional logistic regression model, we are subtracting the smoking status of the control from that of the case. This means that we are now creating three new difference variables having either the value of +1, 0 or −1. Each of these three difference variables reflecting the smoking status would then be entered into the model and treated as if they were continuous variables.

In the model shown in the lower panel in Table 14.15, we entered two predictor variables (prior MI and systolic blood pressure). The results show that the estimated coefficient for MI changed slightly. The estimated odds ratio for prior MI adjusted for systolic blood pressure is 0.52, and its confidence interval still includes one. The normal test for prior MI has a *p*-value of 0.242, and the *p*-value for the two-degree-of-freedom test of hypothesis that both the prior MI coefficient and the SBP coefficient are simultaneously zero is 0.368. Hence we may conclude that prior MI appears to have no statistically significant effect on CVD, whether or not we adjust for SBP.

14.5   Introduction to Proportional Hazard Regression

The proportional hazards model introduced by D. R. Cox (1972) is an extension of the material in Chapter 11, and the Cox approach has become the most widely used regression model in survival analysis. In Chapter 11, we introduced the hazard function, defined as the probability of failure during an interval of time divided by the size of the interval. Cox's regression allows the examination of the possible relationship between the hazard function and a set of independent variables. We use the following example in the introduction of the Cox model.

Example 14.12

The DIG200 data set was introduced in Chapter 3 as part of the digoxin trial. Mortality was monitored and the number of days to death or to the end of the trial for those who were still living. Mortality and the number of days of survival for 200 subjects in the DIG200 dataset are shown in Table 14.16, along with age and BMI rounded to the whole number.

Table 14.16 Survival data for 200 subjects in the Digoxin trial, DIG200.

	Placebo Group				Digoxin Group			
ID	Death	Days to Death	Age	BMI	Death	Days to Death	Age	BMI
1	0	631	70	26	1	627	45	33
2	0	1,166	74	30	0	1,501	66	29
3	1	1,025	65	26	1	431	62	27
4	0	1,508	51	30	1	149	63	23
5	0	1,727	73	28	0	1,335	72	22
6	0	1,167	52	30	1	620	31	27
7	0	1,117	62	29	0	1,157	58	23
8	0	1,544	70	23	0	1,215	55	21
9	0	1,578	52	31	1	1,216	74	26
10	0	1,192	62	22	1	165	28	29
11	1	1,075	65	28	0	880	57	28
12	0	1,052	66	28	0	1,518	63	29
13	1	338	71	33	1	586	69	27
14	0	1,131	58	27	0	1,181	60	23
15	0	1,173	50	27	0	1,136	47	31
16	0	1,432	29	41	0	1,475	79	38
17	0	1,432	68	28	1	169	73	27
18	0	970	46	22	0	1,194	58	26
20	0	1,279	71	21	0	879	71	26
21	1	940	70	19	1	562	63	30
22	0	1,328	57	24	0	1,697	61	23
23	0	1,454	51	20	0	1,591	63	28
24	1	1,516	72	27	0	1,523	58	28
25	0	1,598	84	32	1	415	50	32
26	0	1,355	57	27	0	1,542	66	33
27	0	1,013	59	18	0	1,353	61	27
28	1	901	52	24	0	1,390	77	27
29	1	50	63	20	0	1,060	71	27
30	0	1,726	50	26	0	1,748	73	27
31	0	1,188	46	26	0	1,559	57	26
32	1	825	68	25	0	1,034	68	24
33	1	33	79	30	0	1,680	51	26
34	0	1,501	88	33	1	300	65	24
35	0	1,318	54	31	1	644	56	26
36	1	538	53	34	1	132	66	27
37	1	629	79	27	0	1,528	60	27
38	1	1,359	76	31	1	951	67	26
39	1	374	78	23	0	969	49	15
40	0	887	69	36	0	958	53	26
41	1	790	63	25	0	989	66	29
42	1	966	55	27	0	1,566	66	32
43	0	1,250	60	30	0	1,157	45	24

Table 14.16 *Continued*

	Placebo Group				Digoxin Group		
ID	Death	Days to Death	Age	BMI	Death	Days to Death	Age	BMI
44	0	1,192	55	20	1	949	68	21
45	0	1,108	51	22	0	537	73	22
46	1	1,176	55	22	0	1,279	49	27
47	0	1,160	71	25	0	1,629	57	24
48	1	8	72	23	0	1,277	60	22
49	1	609	69	22	0	1,342	77	22
50	0	1,649	79	19	1	943	72	23
51	1	609	64	21	0	1,626	66	27
52	0	1,374	74	36	0	1,147	42	24
53	0	1,168	78	43	0	867	52	24
54	1	1,268	68	26	0	1,144	54	27
55	0	871	71	22	0	1,152	65	29
56	0	1,516	65	27	1	295	46	36
57	0	1,090	44	26	1	447	67	30
58	1	1,007	62	25	1	511	75	26
59	0	1,391	65	29	0	899	54	27
60	1	547	61	32	0	1,622	58	28
61	1	531	52	25	1	1,567	66	24
62	1	848	64	27	0	1,328	57	24
63	1	305	57	19	0	1,203	61	26
64	1	392	69	25	1	229	65	28
65	0	1,500	76	30	1	1,003	80	25
66	1	1,464	50	34	1	335	77	27
67	0	982	68	27	1	543	46	29
68	0	1,259	54	22	1	1,004	70	19
69	0	1,125	42	24	1	10	35	26
70	0	1,508	43	29	0	895	69	20
71	0	1,559	55	19	0	984	63	21
72	1	299	67	24	0	872	71	23
73	0	1,405	56	35	0	881	53	25
74	0	1,489	47	23	0	1,598	58	29
75	0	1,012	57	23	0	947	80	27
76	1	270	56	20	0	1,588	70	29
77	0	1,298	64	18	0	1,116	38	31
78	0	1,567	81	23	0	1,587	68	25
79	0	873	75	23	1	636	50	24
80	1	1,553	69	22	0	344	54	23
81	0	1,340	43	21	0	1,097	67	33
82	1	340	69	24	1	970	65	45
83	1	188	81	38	0	1,341	76	40
84	0	1,522	59	27	0	1,339	75	38
85	0	1,504	77	24	0	898	59	22
86	1	59	74	29	0	975	47	32
87	1	1,254	67	22	0	1,131	70	37
88	0	949	53	27	0	1,486	49	23
89	0	1,553	76	25	0	1,570	79	27
90	1	895	46	21	1	477	45	27
91	0	1,270	68	28	0	1,287	67	20
92	0	1,228	55	23	0	1,678	37	27
93	1	1,298	83	26	0	1,585	67	34
94	1	1,144	59	25	0	1,350	70	24
95	0	1,669	69	32	0	1,166	69	22
96	0	1,262	61	28	1	1,032	68	24
97	1	253	55	26	1	681	34	20
98	1	495	46	27	0	42	48	30
99	0	1,180	54	32	0	538	60	24
100	1	346	45	23	0	1,612	77	28

Death: 1 = died, 0 = survived; BMI is rounded to the whole number

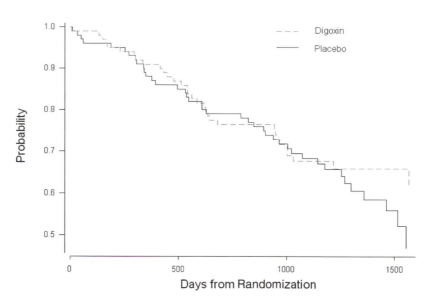

Figure 14.5 Kaplan-Meier curves for digoxon and placebo groups, DIG200.

In the DIG200 dataset, there are 72 deaths: 40 deaths in the placebo group and 32 deaths in the treatment group. To compare survival experience of the two groups, we can use the methods discussed in Chapter 11. As explained in Chapter 11, we need to treat those subjects who were still living at the end of the follow-up period as censored observations. Figure 14.5 shows the Kaplan-Meier survival curves by treatment group.

The Kaplan-Meier curves do not show a noticeable difference in the survival experience between the placebo and treatment group, although survival appears to favor the treatment group slightly after 1200 days. In addition, the hazard plots shown in Figure 14.6 do not show an appreciable difference between the two groups except for later time periods.

Descriptive statistics in Table 14.17 show slightly better survival probabilities for the treatment group. However, the p-value of the log-rank test for comparing the two survival distributions is 0.398, indicating that there is no statistically significant benefit to being treated with digoxin.

The Cox proportional hazards regression model offers an alternative method to compare the survival experience of the two groups. The model focuses on the hazards in the two groups. Let $h_0(t)$ be the hazard at time t for the placebo group and $h_1(t)$ be the hazard at time t for the digoxin group. Then the ratio of these two hazards, the *hazard ratio*, can be modeled under the assumption that it is constant at all survival times, t. It implies that

$$\frac{h_1(t)}{h_0(t)} = \phi.$$

The hazard function in the denominator is called the *baseline hazard*. We already encountered this proportional hazards assumption in applying the CMH and log rank

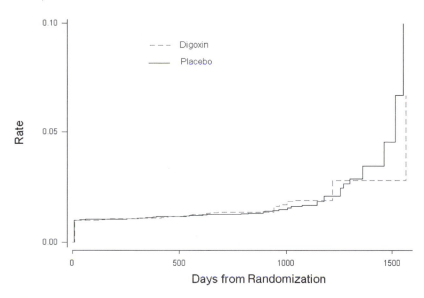

Figure 14.6 Hazard rate plot for digoxin and placebo groups, DIG200.

Table 14.17   Descriptive analysis of survival for digoxin and placebo groups, DIG200.

Descriptor	Digoxin Group ($n = 100$)	Placebo Group ($n = 100$)
Number of Deaths	32	40
Survival Probabilities at		
360 days	0.909	0.880
900 days	0.763	0.749
1440 days	0.660	0.583
Survival Percentiles		
25th	949 days	895 days
50th	—	1553 days
Log Rank Test	Chi-square	0.715
	p-value	0.398

tests in Chapter 11. Since the Cox procedure is based on this assumption, it behooves us to examine this assumption. Based on the plot of the hazard rates in Figure 14.6, it appears as if the ratio of the rates is a constant at least as far out as 1300 days. After that, the ratio changes slightly from around one to less than one. We can separate our investigation into two parts, before and after 1300 days, if we want to be safe. If we limit the analysis to the first 1300 days, the log-rank test chi-square value is 0.0062 with a p-value of 0.938. We conclude that there is no difference in survival between the placebo and digoxin groups. There is one death in the digoxin group and four deaths in the placebo group after 1300 days. For purposes of demonstration, we will simply consider the entire follow-up period in our analysis.

Since hazards are always positive, we can substitute e^{β} where β is a parameter with no restrictions (can be positive, zero, or negative) for the quantity ϕ. Using this notation, we can express Cox's regression model as

$$\ln\left(\frac{h_1(t)}{h_0(t)}\right) = \beta x$$

where x is an indicator variable (0 if an individual received a placebo or 1 if an individual received the digoxin treatment). Note that this linear model has no intercept term unlike the general regression model. No intercept is necessary here because we are only concerned with estimating the hazard ratio.

Just as when using the Kaplan-Meier procedure, in the Cox regression model we also must specify the censored observations — that is, those who were still living at the end of follow-up period, when entering the data for analysis. The results of fitting this model to the survival data for the two groups are shown in Table 14.18 (Model I). (see **Program Note 14.6** on the website.) The estimated coefficient is −0.2007 with standard error of 0.2377. The estimated hazard ratio is exp(−0.2007) = 0.82, suggesting that the hazard is 18 percent lower for the treatment group. This is consistent with the slightly favorable survival probabilities for the treatment group shown previously. The 95 percent confidence interval for the estimated hazard ratio is exp(0.2007 ± 1.96 ∗ 0.2377), or (0.51, 1.30). Since the confidence interval contains the value of one, there is not sufficient evidence to conclude that the use of digoxin lowers the risk of dying. Finally, notice that the p-values for the Wald test statistic (0.399) and likelihood ratio test statistic (0.397) are very close to the p-value for the log rank test (0.398), and they all cause us to fail to reject the null hypothesis of no treatment effect.

The Cox model allows us to incorporate additional predictor variables besides the treatment variable. To demonstrate the inclusion of additional variables, we next carry out Cox's regression analyses of survival experience in DIG200 considering treatment status and two continuous variables, age and body mass index. Model II in Table 14.18 considers treatment group status and age as predictor variables. Model III includes treatment group status, age, and body mass index in the model. Since the digoxin trial randomly allocated patients into the two groups, we do not expect that incorporation of age and BMI would change the difference in survival between

Table 14.18 The fit of the proportional hazards regression of survival on digoxin treatment, age, and body mass index: DIG200.

Model I (survival on digoxin treatment)

					LR chi^2 (1)	=	0.72
Log likelihood = −353.81122					Prob > chi^2	=	0.3972

	Coef.	Std. Err.	z	$p > |z|$	Odds Ratio	[95% CI]
Treatment	−0.2007	0.2377	−0.84	0.399	0.8182	(0.5134, 1.3037)

Model II (survival on digoxin treatment and age)

					LR chi^2 (2)	=	0.82
Log likelihood = −353.76215					Prob > chi^2	=	0.6653

	Coef.	Std. Err.	z	$p > |z|$	Odds Ratio	[95% CI]
Treatment	−0.2015	0.2378	−0.85	0.397	0.8175	(0.5130, 1.3029)
Age	−0.0033	0.0105	−0.31	0.754	0.9967	(0.9765, 1.0174)

Model III (survival on digoxin treatment, age, and body mass index)

					LR chi^2 (3)	=	0.93
Log likelihood = −353.70398					Prob > chi^2	=	0.8178

	Coef.	Std. Err.	z	$p > |z|$	Odds Ratio	[95% CI]
Treatment	−0.1980	0.2380	−0.83	0.405	0.8204	(0.5146, 1.3080)
Age	−0.0031	0.0106	−0.29	0.771	0.9969	(0.9765, 1.0178)
BMI	−0.0085	0.0249	−0.34	0.735	0.9916	(0.9443, 1.0413)

treatment and control groups. We are considering these two additional models to illustrate the usefulness of the Cox approach.

In Model II, the estimated hazard ratio for digoxin treatment, for a fixed age, is 0.82, which is the same as the unadjusted hazard ratio in Model I. The estimated hazard ratio for age, in the same group, is 1.00, suggesting that age variable does not make any difference at all. The comparison in the two likelihood values suggests there is no significant age effect. Similarly, in Model III, addition of body mass index to the model does not make a difference. The estimated hazard ratio for digoxin treatment, holding age and BMI constant, is still 0.82.

In general, the proportional hazards model considering k independent variables is expressed in terms of the hazard function

$$h(t) = h_0(t) \cdot \exp(\beta_1 x_1 + \beta_2 x_2 + \ldots + \beta_k x_k)$$

where $h_0(t)$ is referred to as the baseline hazard and is multiplied by the exponential of the k independent variables. This model can also be expressed as

$$\ln\left(\frac{h(t)}{h_0(t)}\right) = \beta_1 x_1 + \beta_2 x_2 + \ldots + \beta_k x_k.$$

The natural log of the hazard ratio is linearly related to the sum of the k independent variables. This equation is similar to the formula for the logit model we used in logistic regression. Independent variables may be discrete or continuous. Discrete independent variables with more than two levels require dummy coding as in the general regression model. For more detailed treatment of proportional hazards regression, we refer to more advanced books (Collett 1994; Cox and Oakes 1984; Hosmer and Lemeshow 1999a).

Conclusion

In this chapter, we showed that logistic regression is a part of the larger general linear model approach for analyzing data. Logistic regression is an important method, particularly in epidemiology, as it allows the investigator to examine the relation between a binary dependent variable and a set of continuous and discrete independent variables. The interpretation of the model parameters in terms of the odds and odds ratios is a key attraction of the logistic regression procedure. Many of the diagnostic procedures used to examine the appropriateness and fit of the multiple linear regression model have also been adapted to logistic regression, making it an even more attractive method. We also briefly introduced the Cox's proportional hazards model as a method that goes beyond the survival analysis methods in Chapter 11. This model allows us to examine multiple factors to determine whether or not there appears to be an association with the length of survival.

This chapter provides an introduction to both of these topics. It is not meant to be exhaustive, particularly regarding the presentation of the proportional hazards model. Interested readers are encouraged to avail themselves of several books that focus on these topics.

EXERCISES

14.1 Data from an article by Madsen (1976) are used here to examine the relation between survival status — less than 10 years or greater than or equal to 10 years — and the type of operation — extensive (total removal of the ovaries and the uterus) and not extensive — for 299 patients with cancer of the ovary. Other factors could be included — for example, stage of the tumor, whether or not radiation was used, and whether or not the tumor had spread — in a logistic regression analysis. However, we begin our consideration with only the one independent variable. The data are

	Survival Status	
Type of Operation	**<10 years**	**≥10 years**
Extensive	29	122
Not Extensive	20	28

In a logistic regression analysis — using the logit for >10 years of survival and the not extensive type of operation as the base level — the estimates of the constant term and the regression coefficient for the type of operation (extensive) are 0.3365 and 0.3920, respectively. Provide an interpretation for these estimates. Demonstrate that your interpretations are correct by relating these estimates to the preceding table.

14.2 Based on DIG200, investigate how previous myocardial infarction (MI) is related to age, race, sex, and BMI. Summarize the computer output in a table and interpret the results. Explain the odd ratios for each independent variable. What is the predicted proportion of having had an MI for a nonwhite female 60 years of age with a BMI of 30?

14.3 The story of the Donner party, stranded in the Sierra Nevada in the winter of 1846–1847, illustrates the hardship of the pioneers' journey to California. Of the 83 members of the Donner party, only 45 survived to reach California. The following data represent sex, age, and survival status of adult members (15 years of age and older).

Person	Sex	Age	Status	Person	Sex	Age	Status
1	M	62	died	23	M	32	survived
2	F	45	died	24	F	23	survived
3	M	56	died	25	M	30	died
4	F	45	died	26	F	19	survived
5	M	20	survived	27	M	30	died
6	M	25	died	28	M	30	survived
7	M	28	died	29	F	30	survived
8	F	32	survived	30	M	57	died
9	F	25	survived	31	F	47	died
10	M	24	died	32	F	20	survived
11	M	28	died	33	M	18	survived
12	M	25	died	34	F	15	survived
13	M	51	survived	35	F	22	survived
14	F	40	survived	36	M	23	died
15	M	35	died	37	M	25	died
16	M	28	survived	38	M	23	died
17	F	25	died	39	M	18	survived
18	F	50	died	40	M	46	survived
19	M	15	died	41	M	25	survived
20	F	23	survived	42	M	60	died
21	M	28	survived	43	M	25	died
22	F	75	died				

Source: http://members.aol.com/DanMRosen/donner/survivor.htm

Run a logistic regression analysis using sex and age as predictor variables for survival and interpret the results. How does a female's odds of survival compare with a male's odds while controlling for age? How does a 45-year-old person's odds of survival compare with a 15-year-old person's odds while controlling for sex?

14.4 Woodward et al. (1995) investigated prevalence of coronary heart diseases (CHD) in men. Prevalent CHD was defined on a four-point graded scale in decreasing order of severity: myocardial infarction (MI), angina grade II, angina grade I, no CHD. One of several risk factors examined was parental history of CHD before age 60. The data are

Parental History	CHD Categories				
of CHD	MI	Angina II	Angina I	No CHD	Total
Present	104	17	45	830	996
Absent	192	30	122	3,376	3,720
Total	296	47	167	4,206	4,716

Analyze the data using an ordered logistic regression model treating the CHD categories as levels of an ordinal dependent variable and parental history of CHD as the independent variable. If the proportional odds assumption is satisfied, summarize the results and interpret the findings. What is the predicted risk of CHD for a person with no CHD?

14.5 Kellermann et al. (1993) investigated the effect of gun ownership on homicide in the home using a retrospective matched-pairs design. They compared 388 cases of homicides with control subjects matched according to neighborhood, sex, race, and age range. They presented the following information in the article:

Number of Gun Owners		Odds Ratio (95% CI)
Case	Control	
174	139	1.6 (1.2 – 2.2)

Is it possible to verify that the reported crude odds ratio is correct? If yes, verify it. If not, what information is lacking? For a multivariate analysis, the following information is shown:

Variable	Odds Ratio (95% CI)	
	Crude	Adjusted
Gun ownership	1.6 (1.2 – 2.2)	2.7 (1.6 – 4.4)
Home rented	5.9 (3.8 – 9.2)	4.4 (2.3 – 8.2)
Lived alone	3.4 (2.2 – 5.1)	3.7 (2.1 – 6.6)
Domestic violence	7.9 (5.0 – 12.7)	4.4 (2.2 – 8.8)
Any household member arrested	4.4 (3.0 – 6.0)	2.5 (1.6 – 4.1)
Any member used illicit drugs	9.0 (5.4 – 15.0)	5.7 (2.6 – 12.6)

Explain what statistical method is used to calculate the adjusted odds ratios and their confidence intervals. How would you interpret the adjusted odds ratio of 2.7 for gun ownership?

14.6 A case-control study of presenile dementia was introduced in Chapter 10 (Example 10.6). Each dementia case was individually paired with a community control of the same sex and age, and family history of dementia was ascertained in both groups, retrospectively. The following cross-tabulation of the 109 pairs by the presence or absence of family history of dementia was analyzed. Based on the McNemar chi-square test statistic, we concluded that there is evidence for an association between dementia and family history of the disease:

Family History of Dementia in Case	Family History of Dementia in Control	
	Present	Absent
Present	6	25
Absent	12	66

The following table shows the data for the 37 discordant pairs. Analyze the data using the conditional logistic regression approach and see whether the same conclusion can be drawn.

Control (without Dementia)			Case (with Dementia)		
Set	Dementia[a]	History[b]	Set	Dementia[a]	History[b]
1	0	1	1	1	0
2	0	1	2	1	0
3	0	1	3	1	0
4	0	1	4	1	0
5	0	1	5	1	0
6	0	1	6	1	0
7	0	1	7	1	0
8	0	1	8	1	0
9	0	1	9	1	0
10	0	1	10	1	0
11	0	1	11	1	0
12	0	1	12	1	0
13	0	0	13	1	1
14	0	0	14	1	1
15	0	0	15	1	1
16	0	0	16	1	1
17	0	0	17	1	1
18	0	0	18	1	1
19	0	0	19	1	1
20	0	0	20	1	1
21	0	0	21	1	1
22	0	0	22	1	1
23	0	0	23	1	1
24	0	0	24	1	1
25	0	0	25	1	1
26	0	0	26	1	1
27	0	0	27	1	1
28	0	0	28	1	1
29	0	0	29	1	1
30	0	0	30	1	1
31	0	0	31	1	1
32	0	0	32	1	1
33	0	0	33	1	1
34	0	0	34	1	1
35	0	0	35	1	1
36	0	0	36	1	1
37	0	0	37	1	1

[a]Codes: 0 = without dementia; 1 = with dementia
[b]Codes: 0 = without history; 1 = with history

14.7 The survival of 64 lymphoma patients was analyzed for two different symptom groups (A and B) in Exercise 11.4. The survival times (in months) for the two symptom groups is shown here:

A symptoms:	3.2*	4.4*	6.2	9.0	9.9	14.4	15.8	18.5	27.6*	28.5	30.1*
	31.5*	32.2*	41.0	41.8*	44.5*	47.8*	50.6*	54.3*	55.0	60.0*	60.4*
	63.6*	63.7*	63.8*	66.1*	68.0*	68.7*	68.8*	70.9*	71.5*	75.3*	75.7*
B symptoms:	2.5	4.1	4.6	6.4	6.7	7.4	7.6	7.7	7.8	8.8	13.3
	13.4	18.3	19.7	21.9	24.7	27.5	29.7	30.1*	32.9	33.5	35.4*
	37.7*	40.9*	42.6*	45.4*	48.5	48.9*	60.4*	64.4*	66.4*		

Asterisks indicate censored observations.

Analyze the data using Cox's regression method and see whether the same conclusion can be drawn as in Exercise 11.4. Do you think that the proportional hazards assumption is acceptable in your analysis?

REFERENCES

Berkson, J. "Application of the Logistic Function to Bio-Assay." *Journal of the American Statistical Association* 39:357–365, 1944.

Berkson, J. "Why I Prefer Logits to Probits." *Biometrics* 7:327–339, 1951.

Collett, D. *Modelling Survival Data in Medical Research*. London: Chapman & Hall, 1994.

Collett, D. *Modelling Binary Data*, 2nd ed. London: Chapman & Hall, 2003.

Cox, D. R. "Regression Models and Life Table (with discussion)." *Journal of Royal Statistical Society* B 74:187–220, 1972.

Cox, D. R. *Analysis of Binary Data*. London: Chapman and Hall, 1969.

Cox, D. R., and D. Oakes. *Analysis of Survival Data*. London: Chapman & Hall, 1984.

Forthofer, R. N., and R. G. Lehnen. *Public Program Analysis: A New Categorical Data Approach*. Belmont, CA: Lifetime Learning Publications, 1981, Table 7.1.

Hauck, W. W., and A. Donner. "Wald's Test as Applied to Hypotheses in Logit Analysis." *Journal of the American Statistical Association* 72:851–853, 1977.

Hosmer, D. W., and S. Lemeshow. *Applied Logistic Regression*, 2nd ed. New York: John Wiley & Sons, 1999a.

Hosmer, D. W., and S. Lemeshow. *Applied Survival Analysis: Regression Modeling of Time to Event*. New York: John Wiley & Sons, 1999b.

Jennings, D. E. "Judging Inference Adequacy in Logistic Regression." *Journal of the American Statistical Association* 81:471–476, 1986.

Kellermann, A. L., F. P. Rivara, N. B. Rushforth, J. G. Banton, D. T. Reay, J. T. Francisco, A. B. Locci, J. Prodzinski, B. B. Hackman, and G. Somes. "Gun Ownership as Risk Factor for Homicide in the Home." *New England Journal of Medicine* 329(15):1084–1091, October 7, 1993.

Madsen, M. "Statistical Analysis of Multiple Contingency Tables: Two Examples." *Scandanavian Journal of Statistics* 3:97–106, 1976.

Pregibon, D. "Logistic Regression Diagnostics." *The Annals of Statistics* 9:705–724, 1981.

Woodward, M., K. Laurent, and H. Tunstall-Pedoe. "An Analysis of Risk Factors for Prevalent Coronary Heart Disease using Proportional Odds Model." *The Statistician* 44:69–80, 1995.

Analysis of Survey Data 15

Chapter Outline

All of the statistical methods we have discussed so far are based on the assumption that the data were obtained by simple random sampling with replacement. As we discussed in Chapter 6, simple random sampling can be very expensive, if not infeasible, to implement in community surveys. Consequently, survey statisticians often use alternative sample selection methods that use such design features as stratification, clustering, and unequal selection probabilities. Some of these features were briefly discussed in Chapter 6. Sample designs that use some of these more advanced design features are referred as *complex sample designs*. These complex designs require adjustments in the methods of analysis to account for the differences from simple random sampling. Once these adjustments are made, all the analytic methods discussed in this book can be used with complex survey data. We introduce several different ways of making these adjustments in this chapter, with a focus on two specific topics: the use of sample weights and the calculation of estimated variances of parameter estimates based on complex sample designs.

Our treatment of the material in this chapter differs from the treatment in the other chapters in that we provide few formulas here. Instead, we attempt to provide the reader with a feel for the different approaches. We also provide some examples pointing out how ignoring the sample design in the analysis can yield very misleading conclusions. We follow this nonformulaic path because of the mathematical complexity of the procedures. In addition, we do not go into detail about procedures for addressing two important problems in the analysis of survey data — nonresponse and missing data. There are several approaches for dealing with these problems (Levy and Lemeshow 1999; Little and Rubin 2002), but they all make assumptions about the data that are difficult to check. We cannot stress too highly the importance of reducing nonresponse in surveys. Even after reading this chapter, we think the reader will need to work with a survey statistician when carrying out the analysis of survey data.

15.1 Introduction to Design-Based Inference

There are two general approaches for dealing with the analytic complexities in survey data and these can be loosely grouped under the headings of "design-based" and "model-based." We are presenting only the design-based approach because it is the standard way of analyzing complex surveys, although the model-based approach also has many supporters. Several sources discuss the model-based approach (Korn and Graubard 1999; Lee and Forthofer 2006; Lohr 1999).

The design-based approach requires that the sample design be taken into account in the calculation of estimates of parameters and their variances. As we just mentioned, a key feature of the complex sample design is the sample weight, which is based on the probability of selection of the units in the sample. The calculation of the estimated variance for a parameter estimate from complex survey data usually cannot be done through applying a simple formula. The following special procedures are used.

15.2　Components of Design-Based Analysis

As just mentioned, most community surveys utilize complex sample designs to facilitate the conduct of the surveys. As a result of using stratification and clustering, the selection probabilities of units are unequal. In some surveys, unequal selection probabilities are used intentionally to achieve certain survey objectives. For example, the elderly, children, and women of childbearing ages are often oversampled to obtain a sufficient number of people in those categories for detailed analysis.

15.2.1　Sample Weights

The weight is used to account for differential representation of sample observations. The weight is defined as the inverse of selection probability for each observation. Let us explore the concept of the sample weight in the simple random sampling situation. Suppose that an SRS of $n = 100$ households was selected from a population of $N = 1000$ households to estimate the total medical expenditure for a year for the population. The selection probability of each sample observation is $n/N = 0.1$, and the sample weight is therefore 10 ($= N/n$). The sample weights add up to N. If the average annual medical expenditure for the sample ($\bar{y}$) was found to be $2000, then the estimated total medical expenditure for the population would be $N\bar{y} = \$2,000,000$. Another way of writing the estimate is

$$N\bar{y} = N\left(\frac{\sum y_i}{n}\right) = \sum \left(\frac{N}{n}\right) y_i$$

or the weighted total of sample observations. Since the weight is the same for all sample observations in simple random sampling, we don't need to weight each observation separately.

The situation is slightly different with a disproportionate stratified random sample design. Suppose the population of 1000 households consists of two strata: 200 (N_1) households with at least one senior citizen and 800 (N_2) households without any seniors. Suppose further that 50 households were randomly selected from each stratum. The selection probability in the first stratum is 50/200 and the weight is 4 ($= N_1/n_1$). In the second stratum the selection probability is 50/800 and the weight is 16. If the average medical expenditure for the first and second stratum were found to be $5000 ($\bar{y}_1$) and $1250 ($\bar{y}_2$), respectively, then the estimated total medical expenditure for the population would be $2,000,000 ($= N_1\bar{y}_1 + N_2\bar{y}_2 = 200\{1250\} + 800\{5000\}$). The following relationship shows the role of the weight in estimation:

$$N_1\bar{y}_1 + N_2\bar{y}_2 = N_1\left(\frac{\sum y_{1i}}{n_1}\right) + N_2\left(\frac{\sum y_{2i}}{n_2}\right) = \sum\left(\frac{N_1}{n_1}\right)y_{1i} + \sum\left(\frac{N_2}{n_2}\right)y_{2i}.$$

Although we have used SRS and stratified sample designs to introduce the sample weights, the same concept extends to more complex designs. When each observation in the sample has a different weight (w_i), the estimates for the population can be obtained using the following general estimator:

$$\text{population estimate} = \sum w_i y_i.$$

This procedure applies to all sample designs.

In survey analysis, the weight is often modified further by poststratification adjustments discussed in the following sections.

15.2.2   Poststratification

In the health field, many of the variables of interest vary by, for example, a person's age, sex, and race. If we knew these variables before we carried out the survey, we could use them to create a stratified design that would take these variables into account. Unfortunately, we often don't know the values of these variables before the survey, and this fact prevents us from using a stratified sample design.

However, we still wish to take these variables into account in the analysis. We do this by adjusting the sample distributions so that they match their population distributions for these variables. We accomplish this matching by using a technique called poststratification that adjusts the sample weights after (post) the sample data have been collected.

The following example shows how poststratification adjustment is created for the different categories.

Example 15.1

A telephone survey was conducted in a community to estimate the average amount spent on food per household in a week. Telephone surveys are popular because they are quick and easy to perform. Unfortunately, they exclude the small percentage of the households without a landline telephone, and this exclusion could introduce some small degree of bias in the results. With the increasing use of cell phones, the potential for bias in telephone surveys is increasing unless cell phone numbers are also included. Given the goal of this survey, one desirable stratification variable would be household size because larger households likely spend more than smaller households. Since information on household size was not readily available before the survey was conducted, we could not stratify on this variable in the survey design.

The survey failed to obtain responses from 12 percent of the households. It was thought that these nonrespondents were more likely to be living in smaller households, and this idea is supported by the data shown in Table 15.1. These data show the distribution of sample households by household size and the corresponding distribution from a previous study involving household size in the community. Smaller

Table 15.1 Poststratification adjustments by household size for a telephone survey.

Number of Persons in Household	Number of Households in Sample	Sample Distribution	Population Distribution	Adjustment Weight[a]	Average Food Expenditure
1	63	0.2072	0.2358	1.13803	$38
2	81	0.2664	0.3234	1.21396	52
3	63	0.2072	0.1700	0.82046	78
4	52	0.1711	0.1608	0.93980	98
5+	45	0.1480	0.1100	0.74324	111

[a]Population distribution divided by sample distribution

households are indeed underrepresented in the sample and this suggests the average food expenditure would be overestimated unless we make an adjustment for household size. Table 15.1 shows the procedure of poststratification adjustment.

The postratification adjustment for single-person households is to multiply the number of single-person households by 1.138, reflecting that this category of households is underrepresented by 14 percent. This adjustment is equivalent to multiplying the sample weights by the same factor. The adjusted number of households for this category is then 71.7 (= 63{1.138}); for the rest of household size categories the adjusted numbers are 98.3, 51.7, 48.9, and 33.4. The distribution of these adjusted numbers of household by household size now matches the distribution in the population. The average food expenditure based on these adjusted numbers of households is $67.00, compared with the unadjusted average of $71.08. As expected, the adjusted average is lower than the unadjusted estimate.

We have not addressed the nonresponders directly through the poststratification adjustment. Given that the nonresponse rate was only 12 percent, it is unlikely that the average food expenditure estimate would change much if the nonresponders were included. However, it would be good to do more follow-up with a sample of the nonresponders in an effort to determine if they differed drastically from the responders.

Multiplying the sample weight by the poststratification adjustment factors causes the weighted sample distribution to match the population distribution for the variables used in the poststratification.

15.2.3 The Design Effect

In Chapter 6 we demonstrated in one example that a stratified sample could provide a more precise — have smaller sampling variance — estimator for the sample mean than a simple random sample of the same sample size. In this section we provide a measure, the *design effect*, for comparing a sample design to a simple random sample design with replacement. To introduce this idea, we will begin by comparing simple random sampling without replacement to simple random sampling with replacement.

In Chapters 7 and 8, we used s^2/n as the estimator for the variance of the sample mean ($\bar{x}$) and $s/\sqrt{n}$ as the estimator for the standard error for data resulting from simple random sampling with replacement. When simple random sampling without replacement is used, the formula for the estimated variance is

$$\hat{\mathrm{Var}}(\overline{x}) = \frac{s^2}{n}\left(1 - \frac{n}{N}\right). \tag{15.1}$$

The term $(1 - n/N)$, called the *finite population correction* (FPC), adjusts the formula to take into account that we are no longer sampling from an infinite population. Use of this term decreases the magnitude of the variance estimate. For samples from large populations, the FPC is approximately one, and it can be ignored in these cases.

The ratio of the sampling variance of SRSWOR to that of SRSWR is the FPC, and it reflects the effect of using SRSWOR compared to using SRSWR. This ratio comparing the variance of some statistic from any particular sample design to that of SRSWR is called the *design effect* for that statistic. It is used to assess the loss or gain in precision of sample estimates from the sample design used. A design effect less than one indicates that fewer observations are needed to achieve the same precision as SRSWR whereas a design effect greater than one implies that more observations may be needed to yield the same precision. Extending this concept to sample size, the *effective sample size* of a design is the size of a simple random sample with replacement that would have produced the same estimated sample variance for the estimator under consideration. The effective sample size is the actual sample size of the design being used divided by the design effect.

The design effect can be examined theoretically for some simple sample designs. As was just mentioned, we pointed out in Chapter 6 that stratified random sampling often produces smaller sampling variance than SRS. Cluster sampling will lead to a greater sampling variability when the sampling units are similar within clusters. The *intraclass correlation coefficient* (ICC) is used to assess the variability within the clusters. The ICC is the Pearson correlation coefficient based on all possible pairs of observations within a cluster.

The design effect of single-stage cluster sample design with equal size clusters is

$$1 + (M - 1)ICC$$

where M is the size of each cluster. Given this design, the ICC ranges from $-1/(M - 1)$ to 1. When ICC is positive, the design effect will be greater than one. If the clusters were formed at random, then ICC = 0; when all the units within each cluster have the same value, then ICC = 1 and the design effect is the same as the size of the cluster. Most clusters used in community surveys consist of houses in the same area, and these generally yield positive ICCs for many survey variables such as socioeconomic and some demographic characteristics.

Since the determination of the design effect requires that we have an estimate of the sample variance for a given design, this calculation is usually not a simple task for a complex sample design. The complexity of the design often means that we cannot use the variance estimating formulas presented in previous chapters; rather, special techniques that utilize unfamiliar strategies are required. The next section presents several strategies for estimating sampling variance for statistics from complex sample designs.

15.3   Strategies for Variance Estimation

The estimation of the sampling variance of a survey statistic is complicated not only by the complexity of the sample design but also by the form of the statistic. Even with an SRS design, the variance for some sample statistics requires nonstandard estimating techniques. For example, the sampling variance of sample median was not covered in previous chapters. Moreover, the variance estimator for a weighted statistic is complicated because both the numerator and denominator are random variables. We will present several techniques for estimating sampling variances: (1) from complex samples and (2) for nonlinear statistics. These techniques include replicated sampling, balanced repeated replication, jackknife repeated replication, and the linearization method (Taylor series approximation).

15.3.1   Replicated Sampling: A General Method

The replicated sampling method requires the selection of a set of subsamples from the population with each subsample being drawn independently following the same sample selection design. Then an estimate is calculated for each subsample, and the sampling variance of the overall estimate based on all the subsamples can be estimated from the variability of these independent subsample estimates. The repeated systematic sampling discussed in Chapter 6 represents this strategy.

The standard error of the mean ($\bar{u}$) of t replicate estimates, $u_1, u_2, \ldots, u_1$ of the parameter U can be estimated by

$$\sqrt{\sum_{i=1}^{t}(u_i - \bar{u})^2 / [t(t-1)]}. \tag{15.2}$$

This estimator can be applied to any sample statistic obtained from independent replicates for any sample design.

In applying this estimator, ten replicates are recommended by Deming (1960) and a minimum of four by Sudman (1976) for descriptive statistics. An approximate estimate of the standard error can be calculated by dividing the range in the replicate estimates by the number of replicates when the number of replicates is between 3 and 13 (Kish 1965). However, because this estimator with t replicates is based on $t-1$ degrees of freedom, a larger number of replicates may be needed for analytic studies, perhaps 20 to 30 (Kalton 1983).

Example 15.2

In this artificial example, we demonstrate the use of replicated sampling for the estimation of the sample variance of a statistic. In this case, we are going to estimate the population proportion of male births and the sample variance of this statistic based on replicated samples. Instead of collecting actual data, we will use the random digits in Table B1 to create our replicated samples. In our simulation process, we are going to assume that the population proportion of male births is 0.5. We will take 10 replicated samples of size 40 using the first eight 5-digit-columns of lines 1

Table 15.2   Estimation of standard errors for the proportion of boys from 10 replicated samples of size 40.

Replicate	n	Number of Boys	Proportion of Boys	Standard Error
Full sample	400	205	0.512	0.025[a]
1	40	21	0.525	
2	40	16	*0.400*	
3	40	21	0.525	
4	40	20	0.500	
5	40	20	0.500	
6	40	17	0.425	
7	40	21	0.525	
8	40	26	*0.650*	
9	40	24	0.600	
10	40	19	0.475	0.022[b]
			$(0.650 - 0.400)/10 \quad = $	0.025[c]

[a]Based on SRS
[b]Based on Equation (15.2)
[c]Based on the range in replicate proportions divided by the number of replicates

through 10. Table 15.2 shows the number and proportion of boys with estimates of the standard error by three different methods.

For the full sample of 400 — combining the data from the 10 separate samples — the proportion of boys is 0.512 and its standard error is 0.025 based on simple random sampling. The standard error estimated from the 10 replicate estimates using Equation (15.2) is 0.022. An approximate estimate can also be obtained by taking the range in replicate estimates divided by the number of replicates. This value is 0.025 ([0.650 − 0.400]/10). Of course replicated sampling is not needed for estimating standard errors for a simple random sample design. But this strategy also works for complex sample designs.

The chief advantage of replicated sampling is the ease in estimation of the standard errors for complex sample designs. This strategy is especially useful in systematic sampling, since there is no way to estimate the standard error of an estimator from a systematic sample with only one replicate. Replicated systematic sampling can easily be implemented by randomly selecting multiple starting points. In applying Equation (15.2), the sample statistic for the full sample is generally used instead of the mean of replicate estimates when sample weights are present.

However, replicated sampling is difficult to implement in multistage cluster sampling designs and is seldom used in large-scale surveys. Instead, the replicated sampling idea can be applied in the data analysis stage where pseudo-replication methods for variance estimation are used. The next two sections present two such methods.

15.3.2   Balanced Repeated Replication

The balanced repeated replication (BRR) method represents an application of the replicated sample idea to a paired selection design in which two primary sampling units (PSU) are sampled from each stratum. The paired selection design is often used to simplify the calculation of variance within a large number of strata. The variance between two units within a stratum is one-half of the squared difference between the units. McCarthy (1966) originally proposed the BRR method for the National Center for Health Statistics for analyzing the National Health Examination Survey that used a paired selection design.

To apply the replicated sampling idea, half-sample replicates are created by taking one PSU from each stratum. From the paired selection design, we can create only two half-sample replicates. Since the estimate of standard error based on two replicates is unstable, we repeat the process of forming half-sample replicates in such a way that replicates are independent of each other (Plackett and Burman 1946).

Replicate estimates, $u_1, u_2, \ldots, u_t$, for a sample statistic are calculated by doubling the sample weights, since each replicate contains one-half of the total observations. Then the standard error of the statistic ($\bar{u}$) for the full sample can be calculated by

$$\sum_{i=1}^{t} (u_i - \bar{u})^2 / t.$$

Since this process involves so much manipulation of the data, it is usually necessary to use specialized computer software created to carry out the BRR approach.

15.3.3 Jackknife Repeated Replication

Another replication-based procedure is called the jackknife repeated replication method (JRR). This procedure creates pseudo-replicates by deleting one unit from the sample, then calculating the sample statistic of interest. That unit is put back into the sample, another unit is deleted and the statistic is calculated, and so on. The estimate of the variance is then based on the variation in these sample statistics. The term *jackknife* may be used because this procedure can be used for a variety of purposes. The idea of jackknifing was introduced by Quenouille in 1949 in the estimation of bias for a sample estimator. Frankel (1971) first applied the jackknife procedure to the computation of sampling variance in complex surveys, using it in a manner similar to the BRR method. The following example illustrates the principle of jackknifing.

Example 15.3

We consider a small data set — the ages of the first 10 patients in DIG40 shown in Table 3.1. Assuming that these data are from a simple random sample, the sample mean is 58.2 and the sample median is 59.5. If we ignore the FPC, the estimated standard error of the sample mean is 4.27. These statistics are shown in Table 15.3 along with the 10 observations. We next estimate the standard error of the sample mean by the jackknife procedure.

We create a jackknife replicate by deleting the first observation (age 55) and calculate the mean for the replicate, which gives 58.56, as shown in the table. By deleting the second observation (age 78) we get the second jackknife replicate estimate of 56. Repeating the same procedure we have 10 replicate estimates, $\bar{y}_{(1)}, \bar{y}_{(2)}, \ldots, \bar{y}_{(10)}$. Let the mean of the replicate estimates be $\bar{\bar{y}} (= \sum \bar{y}_{(i)} / n)$, and this value is 58.2,

Table 15.3   Estimation of standard error by the jackknife procedure for the mean and median age for 10 patients in DIG40.

Patient	Age	Jackknife Replicate Estimates	
		Mean	Median
1	55	58.56	60
2	78	56.00	59
3	50	59.11	60
4	60	58.00	59
5	31	61.22	60
6	70	56.89	59
7	46	59.56	65
8	59	58.11	59
9	68	57.11	59
10	65	57.44	59
Mean	58.2	58.2	59.9
Median	59.5		

Standard error estimates for the mean:
   From the sample   4.27
   From jackknife replicates   4.27
Standard error estimate for the median:
   From jackknife replicates   5.27

which is the same as the sample mean. The standard error can be estimated by $\sqrt{(n-1)\Sigma(\bar{y}_{(i)} - \bar{y})^2/n}$, which equals 4.27. The standard error estimated from replicate estimates is the same as the estimate obtained directly from the sample, suggesting the jackknife procedure works.

The jackknife procedure also allows us to estimate the standard error for the median. The first replicate estimate for the median is based on the nine observations remaining after deleting the first observation as before. Deleting each observation in turn allows us to determine ten replicate estimates of the median as shown in Table 15.3. The mean of the replicate medians is 59.9. Using the same formula shown above, we can estimate the standard error of the median and this value is 5.27.

For a complex sample design, the JRR method is generally applied at the PSU level. The JRR method is not restricted to a paired selection design but is applicable to any number of PSUs per stratum. Let us consider a situation with L strata. If u_{hi} is the estimate of the parameter U from the hth stratum and ith replicate, n_h is the number of sampled PSUs in the hth stratum, and r_h be the number of replicates formed in the hth stratum, then the standard error is estimated by

$$\sqrt{\sum_{h=1}^{L}\left(\frac{n_h-1}{r_h}\right)\sum_{i=1}^{r_h}(u_{hi} - \bar{u})^2}.$$

If each of the PSUs in the hth stratum is removed to form a replicate, r_h is the same as n_h in each stratum, but the formation of n_h replicates in the hth stratum is not required. When the number of strata is large and n_h is two or more, we can reduce the computational burden by using only one replicate in each stratum. However, a sufficient number of replicates must be used in analytic studies to ensure that there are adequate degrees of freedom.

15.3.4 Linearization Method

A completely different approach from the pseudo-replication methods for estimating variances from complex survey designs follows a more mathematical approach. This mathematical approach, called Taylor series linearization, is used in statistics to obtain a linear approximation to nonlinear functions. The beauty of the Taylor series is that many nonlinear functions are approximated quite well by only the first few terms of the series. This approach has gained wide acceptance in the analysis of weighted data from complex surveys because many of the statistics that we estimate, including regression coefficients, are nonlinear, and their estimated variances are also nonlinear. This approach to variance estimation has several other names in the literature, including the linearization method and the delta method. A brief presentation of the Taylor series approach and an example is presented in Appendix A.

The following example demonstrates how the linearization works for the calculation of sampling variance of a *ratio estimate*.

Example 15.4

We consider a small sample for this illustration. A simple random sample of eight health departments was selected from 60 (N) rural counties to estimate the total number of professional workers with a master of public health degree. It is known that the 60 health departments employ a total of 1150 (Y) professional workers. The sample data shown in Table 15.4 are the number of professional workers (y_i) and the number of professional workers with an MPH degree (x_i).

Based on the sample data on the number of workers with an MPH degree, we can estimate the total number of professional workers with an MPH degree, that is 630 ($= N\bar{x} = 60 * 10.5$). The variance of this estimate is $\hat{Var}(N\bar{x}) = N^2\hat{Var}(\bar{x})$ as shown in Chapter 4. Using Equation 15.1 for $\hat{Var}(\bar{x})$, the estimated standard error of this estimate is

$$\sqrt{\frac{N^2\sum(x_i - \bar{x})^2}{n(n-1)}\left(1 - \frac{n}{N}\right)} = 97.3.$$

Table 15.4 Numbers of professional workers and professionals with an MPH degree for 8 sample health departments.

Health Department	Number of Professional Workers (y_i)	Number of Workers with MPH (x_i)
1	21	14
2	18	8
3	9	3
4	13	6
5	15	8
6	22	13
7	30	17
8	27	15
Mean	19.375	10.5

Since the total number of professional workers for the population is known and x and y are highly correlated, we prefer to use a ratio estimate. The ratio estimate of the total number of professional workers with an MPH is

$$\hat{X} = \left(\frac{\overline{x}}{\overline{y}} \right) Y = \left(\frac{10.5}{19.375} \right) 1150 = 623.$$

The standard error calculation for the ratio estimate is complicated because both the numerator and denominator of the ratio are random variables. The linearization method can provide an approximation. Using Stata we obtained the following results (see **Program Note 15.1** on the website):

Ratio	Estimate	Std. Err.	[95% Conf. Interval]		Deff
totxmph/prof	623.2258	29.92157	552.4725	693.9791	1

The estimated standard error for the ratio estimate is 29.9, which is much smaller than that obtained by simple random sample estimator (97.3), suggesting that the ratio estimate is the preferred method of estimation for this case.

The Taylor series approximation is applied to PSU totals within strata — that is, the variance estimate is a weighted combination of the variation across PSUs within the same stratum. This calculation is complex but can require much less computing time than the replication methods just discussed. This method can be applied to any statistic that is expressed mathematically — for example, the mean and the regression coefficient. But it cannot be used with nonfunctional statistics such as the median and other percentiles.

15.4   Strategies for Analysis

We introduce the Third National Health and Nutrition Examination Survey (NHANES III) here to illustrate the methods of survey data analysis. NHANES III, sponsored by NCHS (1994), collected information on a variety of health-related subjects from a large number of individuals through personal interviews and medical examinations. Its sample design was complex to accommodate the practical constraints of cost and survey requirements, resulting in a stratified, multistage, probability cluster sample of eligible persons in households. The PSUs were counties or small groups of contiguous counties and a total of 2812 PSUs were divided into 47 strata based on demographic characteristics. Thirteen of the 47 strata contained one large urban county, and these urban PSUs were automatically included in the sample. Two PSUs were sampled from each of the remaining 34 strata. The subsequent hierarchical sampling units included census enumeration districts, clusters of households, households, and eligible persons. Preschool children, the aged, and the poor were oversampled to provide sufficient numbers of persons in these subgroups. The NHANES III was conducted in two phases. The 13 large urban counties were rearranged into 21 survey sites, subdividing some large counties. Combining with nonurban PSUs, 89 survey sites were randomly divided into two sets: 44 sites were surveyed in 1988–1991 (Phase I) and the remaining 45 sites in 1991–1994 (Phase

II). Each phase sample can be considered an independent sample, and the combined sample can be used for a large-scale analysis.

We chose to use the Phase II adult sample (17 years of age and over) of NHANES III. It included 9920 observations that are arranged in 23 pseudo-strata with 2 pseudo-PSUs in each stratum. The sample weight contained in the public-use micro data files is the expansion weight (inverse of selection probability adjusted for nonresponse and poststratification). We created a working data file by selecting variables and calculating new variables such as body mass index. The expansion weight was converted to the relative weight by dividing the expansion weight by the average weight.

15.4.1 Preliminary Analysis

Survey data analysis begins with a preliminary exploration to see whether the data are suitable for a meaningful analysis. One important consideration in the preliminary examination of sample survey data is to examine whether there is a sufficient number of observations available in the various subgroups to support the proposed analysis. Based on the unweighted tabulations, the analyst determines whether sample sizes are large enough and whether categories of the variables need to be collapsed. The unweighted tabulations also give the number of the observations with missing values and those with extreme values, which could indicate either measurement errors or errors of transcription.

It is also necessary to examine if all the PSUs have a sufficient number of observations to support the planned analysis. Some PSUs may contain only a few or no observations because of nonresponse and exclusion of missing values. If necessary, the PSUs with none or only a few observations may be combined with an adjacent PSU within the same stratum. If a stratum contains only a single PSU as a result of combining PSUs, it may be combined with an adjacent stratum. However, collapsing too many PSUs and strata is not recommended because the resultant design may now differ substantially from the original design.

The number of observations that is needed in each PSU is dependent on the type of analysis planned. The required number is larger for analytic studies than for estimation of descriptive statistics. A general guideline is that the number should be large enough to estimate the intra-PSU variance for a given estimate.

The first step in a preliminary analysis is to explore the distributions of key variables. The tabulations may point out the need for refining operational definitions of variables and for combining categories of certain variables. Based on summary statistics, one may discern interesting patterns in the distributions of certain variables in the sample. After analyzing the variables one at a time, we can use standard graphs and SRS-based statistical methods to examine relations among variables. However given the importance of sampling weights in survey data, any preliminary analysis ignoring the weights may fail to uncover important aspects of the data.

One way to conduct a preliminary analysis taking weights into account is to select a subsample of manageable size with the probability of selection proportional to the magnitude of the weights (PPS). The PPS subsample can be explored with the regular descriptive and graphic methods, since the weights are now reflected in the selection of the subsample.

Example 15.5

For a preliminary analysis, we generated a PPS sample of 1000 from the 9920 persons in the adult file of Phase II of NHANES III. We first sorted the total sample by stratum and PSU and then selected a PPS subsample systematically using a skipping interval of 9.92 on the scale of cumulative relative weights. The sorting by stratum and PSU preserved in essence the integrity of the original sample design.

Table 15.5 demonstrates the use of our PPS subsample analyzed by conventional statistical methods. In this demonstration, we selected several variables that are likely to be most affected by the weights. Because of oversampling of the elderly and ethnic minorities, the weighted estimates are different from the unweighted estimates for mean age and percent Hispanic. The weights also make a difference for vitamin use and systolic blood pressure as well as for the correlation between body mass index and systolic blood pressure. The subsample estimates, although not weighted, are very close to the weighted estimates in the total sample, supporting the use of a PPS subsample for preliminary analysis.

Table 15.5   Comparison of sample statistics based on the PPS subsample and the total sample, NHANES III, Phase II (adults 17 years of age and older).

Sample	Sample Statistics				
	Mean Age	Percent Hispanic	Mean SBP[a]	Percent Vitamin Use	Correlation BMI[b] & SBP
PPS subsample (n = 1000)					
Unweighted	42.9	5.9	122.2	43.0	0.235
Total sample (n = 9920)					
Weighted	43.6	5.4	122.3	42.9	0.243
Unweighted	46.9	26.1	125.9	38.4	0.153

[a]Systolic blood pressure
[b]Body mass index

15.4.2   Subpopulation Analysis

When we analyze the data from a simple random sampling design, it is customary to perform some specific subdomain analysis — that is, to analyze separately, for example, different age groups or different sexes. However we have to be careful how we carry out this practice with complex survey data. Elimination of observations outside the specific group of interest — say, Hispanics, for example — does not alter the sample weights for Hispanics, but it can complicate the calculation of variances. For example, selecting Hispanics for analysis may mean that there are a small number or even no observations in some PSUs. As a result, several PSUs and, possibly, even strata might have to be combined to be able to calculate the variances. However, the sample structure resulting from these combinations may no longer resemble the original sample design. Thus, selecting out observations from a complex survey sample may lead to an incorrect estimation of variance (Korn and Graubard 1999, Section 5.4). The correct estimation of variance requires keeping the entire data set in the analysis and assigning weights of zero to observations outside the group of interest.

Example 15.6

Let us consider the case of estimating the mean BMI for African Americans from Phase II of NHANES III. For illustration purposes, we attempted to select only African Americans from the sample, but we could not carry out the analysis because the computer program we were using detected PSUs with no observations. A tabulation of African Americans by stratum and PSU showed that only one PSU remained in the 13th and 15th strata. After collapsing these two strata with adjacent strata (arbitrarily with the 14th and 16th stratum, respectively), we obtained the mean BMI of 27.25 with the design effect of 2.78.

The subpopulation analysis using the entire sample and assigning weights of zero to non–African American observations produced the same sample mean BMI of 27.25, but the design effect was now 1.07, a much smaller value. For the use of subpopulation analysis, see **Program Note 15.2** on the website.

15.5   Some Analytic Examples

This section presents various examples based on Phase II of NHANES III data. The emphasis is on the demonstration of the effects of incorporating the sample weights and the design features on the analysis, rather than examining substantive research questions. We begin with descriptive analysis followed by contingency table analysis and regression analysis.

15.5.1   Descriptive Analysis

In descriptive analysis of survey data, the sample weights are used, and the standard errors for the estimates are calculated using one of the methods discussed that incorporate strata and PSUs. When the sample size is small, the FPC is also incorporated in the calculation of the standard errors. The method of calculating confidence intervals follows the same principles shown in Chapter 7. However, the degrees of freedom in the complex sample design are the number of PSUs sampled minus the number of strata used instead of $n - 1$. In certain circumstances, the determination of the degrees of freedom differs from this general rule (Korn and Graubard 1999, Section 5.2).

Example 15.7

We calculated sample means and proportions for selected variables from Phase II of NHANES III. We incorporated the sample weights, strata, and PSUs in the analysis, but the FPC was not necessary because the sample size was 9920. Table 15.6 shows the weighted and unweighted estimates and the standard errors, 95 percent confidence intervals, and the design effects for the weighted estimates.

Table 15.6   Descriptive statistics for selected variables: adult sample, Phase II of NHANES III ($n = 9920$).

Variable	Unweighted Statistics	Weighted Statistics	Standard Error	Confidence Interval	Design Effect
Mean age (years)	46.9	43.6	0.57	(42.4, 44.7)	10.31
Percent Black	29.8	11.2	0.97	(9.2, 13.3)	9.42
Percent Hispanic	26.1	5.4	0.71	(4.0, 6.9)	9.68
Mean years of education*	10.9	12.3	0.12	(12.1, 12.6)	15.01
Mean SBP (mmHg)*	125.9	122.3	0.39	(121.4, 123.0)	4.20
Mean BMI*	26.4	25.9	0.12	(25.7, 26.2)	5.00
Percent vitamin use*	38.4	43.0	1.22	(40.4, 45.5)	5.98
Percent smoker*	46.2	51.1	1.16	(48.7, 53.5)	5.28

*A small number of missing values were imputed.

The differences between the weighted and unweighted estimates are large for several variables. The weighted mean age is about 3.5 years smaller than the unweighted mean reflecting the oversampling of the elderly. The weighted proportion of blacks is over 60 percent smaller than the unweighted proportion and the weighted proportion of Hispanics is nearly 80 percent smaller than the unweighted, reflecting the oversampling of these two ethnic groups. The weighted mean years of education is nearly two years greater than the unweighted mean, reflecting that the oversampled elderly and/or minority groups have lower years of schooling. The weighted percent of vitamin use is also somewhat greater than the unweighted estimate.

The standard errors for the weighted estimates were calculated by the linearization method. The design effects shown in the last column suggest that the estimated standard errors are considerably greater than those calculated under the assumption of simple random sampling. The 95 percent confidence intervals for the weighted estimates were calculated using the t value of 2.0687 based on 23 (= 46 PSUs − 23 strata) degrees of freedom. See **Program Note 15.3** for this descriptive analysis.

15.5.2   Contingency Table Analysis

In Chapter 10, we used the Pearson chi-square statistic to test the null hypothesis of independence in a contingency table under the assumption that data came from an SRS. For the analysis of a two-way table based on complex survey data, the test procedure needs to be changed to account for the survey design. Several different test statistics have been proposed. Koch et al. (1975) proposed the use of the Wald statistic and it has been used widely. The Wald statistic is usually converted to an F statistic to determine the p-value. In the F statistic, the numerator degrees of freedom are tied to the dimension of the table and the denominator degrees of freedom reflect the survey design.

We illustrate the use of Wald statistic based on a 2 by 2 table examining the gender difference in prevalence of asthma based on data from Phase II of NHANES III. We first look at the unweighted tabulation of asthma by sex shown in Table 15.7. Ignoring the sample design, the prevalence rates for males and females are 6.1 and 7.6 percent, respectively. The Pearson chi-square value and the associated p-value shown in the table mean that the difference between the two prevalence rates is statistically significant at

Table 15.7   Unweighted tabulation of asthma by sex: Phase II, NAHNES III.

Asthma	Male	Female	Total
Present (Percent)	264 (6.1)	421 (7.6)	685 (6.9)
Absent	4085	5150	9235
Total	4349	5571	9920

	Chi-square (1):	8.397
	p-value:	0.004

Table 15.8   Weighted proportions for asthma by sex, Phase II, NAHNES III.

Asthma	Male	Female	Total
Present	0.0341 (p_{11})	0.0445 (p_{12})	0.0786 ($p_{1.}$)
Absent	0.4440 (p_{21})	0.4775 (p_{22})	0.9214 ($p_{2.}$)
Total	0.4781 ($p_{.1}$)	0.5219 ($p_{.2}$)	1.0000 ($p_{..}$)

Wald statistics:	Chi-square:	3.2941
	F (1, 23):	3.2941
	p-value:	0.0826

the 0.01 level. However, we know this conclusion could be misleading because we did not account for the sample design in the calculation of the test statistic.

Let us now look at weighted cell proportions shown in Table 15.8. Under the null hypothesis of independence, the estimated expected proportion in cell (1, 1) is $(p_{1.})(p_{.1})$. Let $\hat{\theta} = p_{11} - (p_{1.})(p_{.1})$. Then Wald chi-square is defined as

$$X_w^2 = \hat{\theta}^2 / \hat{V}(\hat{\theta}).$$

We can find $\hat{V}(\hat{\theta})$ by using one of the methods discussed in previous section. The Wald test statistic, X_w^2, approximately follows a chi-square distribution with one degree of freedom.

For the weighted proportions in Table 15.8, $\hat{\theta} = -0.0034786$ and its variance is 0.000003674 (calculated using the linearization method). The Wald chi-square is

$$\frac{\hat{\theta}^2}{\hat{V}(\hat{\theta})} = \frac{(-0.0034786)^2}{0.000003674} = 3.2941$$

and the associated p-value is 0.070. A more accurate p-value can be obtained from $F(1, 23) = 3.2941$ with p-value of 0.083. Taking into account the sample design, the gender difference in prevalence of asthma is statistically insignificant at the 0.05 level.

Rao and Scott (1984) offered another test procedure for contingency table analysis of complex surveys. This procedure adjusts a different chi-square test statistic and again uses an F statistic with noninteger degrees of freedom to determine the appropriate p-value. Some software packages implemented the Rao-Scott corrected statistic as the default procedure. In most situations, the Wald statistic and the Rao-Scott statistic lead to the same conclusion.

Example 15.8

Table 15.9 presents analysis of a 2 by 3 contingency table using data from Phase II of NHANES III. In this analysis, the association between vitamin use and years of education is examined with education coded into three categories (1 = less than 12 years of education; 2 = 12 years; 3 = more than 12 years). The weighted percent of vitamin users by the level of education varies from 33 percent in the first level of education to 52 percent in the third level of education. The confidence intervals for these percentages are also shown. Both the Wald and the Rao-Scott statistics are shown in this table and we draw the same conclusion from both.

We next examined the relation between the use of vitamins and the level of education for the Hispanic population. Here we used a subpopulation analysis based on the entire sample. The results are shown in Table 15.10. The estimated overall proportion of vitamin users among Hispanics is 31 percent, considerably lower than the overall value of 43 percent shown in Table 15.8. The Wald test statistic in Table 15.10 also shows there is a statistically significant relation between education and use of vitamins among Hispanics.

See **Program Note 15.4** for this analysis.

Table 15.9   Percent of vitamin use by levels of education among U.S. adults, Phase II, NHANES III (*n* = 9920).

	Less than H.S.	H.S. Graduate	Some College	Total
Percent	33.4	39.8	51.67	43.0
Confidence Interval	[30.1, 36.9]	[36.2, 43.5]	[47.6, 55.7]	[40.4, .45.5]
Wald Statistic:	Chi-square (2):		51.99	
	F (2, 22):		24.87	
	p-value:		<0.0001	
Rao-Scott Statistic:	Uncorrected chi-square (2):	234.10		
	Design-based F (1.63, 37.46):	30.28		
	p-value:	<0.0001		

Table 15.10   Percent of vitamin use by levels of education for Hispanic population, Phase II, NHANES III (*n* = 2593).

	Less than H.S.	H.S. Graduate	Some College	Total
Percent	26.2	32.7	44.1	30.9
Confidence Interval	[22.1, 30.7]	[28.8, 36.9]	[36.9, 51.58]	[27.1, 34.9]
Wald Statistic:	Chi-square (2):	47.16		
	F (2, 22):	22.56		
	p-value:	<0.0001		

15.5.3   Linear and Logistic Regression Analysis

In Chapter 13, we used ordinary least squares (OLS) estimation to obtain estimates of the regression coefficients or the effects in the linear model assuming simple random sampling. However, using the OLS method with data from a complex sample design will result in biased estimates of model parameters and their variances. Thus, confidence intervals and tests of hypotheses may be misleading.

The most widely used method of estimation for complex survey data when using the general linear model is the design-weighted least squares (DWLS) method. The DWLS approach is slightly different from the weighted least squares (WLS) method for unequal variances that derives the weights from an assumed covariance structure. In the DWLS approach, the weights come from the sampling design, and the variance/covariance is estimated using one of the methods discussed in the previous section. This approach is supported by most of the software for complex survey data analysis. Several sources provide a more detailed discussion of regression analysis of complex survey data (Korn and Graubard 1999, Section 3.5; Lohr 1999, Chapter 11).

Since these methods use the PSU total rather than the individual value as the basis for the variance computation, the degrees of freedom for this design again equal d, the number of PSUs minus the number of strata. For the test of hypothesis we need to take into account the number of parameters being tested. For example, for an F test, the numerator degrees of freedom is the number of parameters being tested (q) and the denominator degrees of freedom is $d - q + 1$.

Example 15.9

We conducted a general linear model analysis of systolic blood pressure on height, weight, age, sex (male = 0), and vitamin use (user = 1) using the same NHANES III data. We did not include any interaction terms in this example, although their inclusion would undoubtedly have increased the R-square. Imputed values were not used in this analysis. The results are shown in Table 15.11. See **Program Note 15.5** on the website for the analysis.

These results can be interpreted in the same manner as in Chapter 13. The R-square is 39 percent and the F statistic for the overall ANOVA is significant. There are five degrees of freedom for the numerator in the overall F, since five independent variables are included in the model. There are 19 (= 46 − 23 − 5 + 1, based on the numbers of PSUs, strata and independent variables in the model) (Korn and Graubard 1999) degrees of freedom for the denominator in the overall F ratio. All five explanatory variables are also individually statistically significant.

Table 15.11 Multiple regression analysis of systolic blood pressure on selected variables for U.S. adults, Phase II, NHANES III. ($n = 9235$).

Variable	Regression Coefficient	Standard Error	t	$p > |t|$	Design Effect
Height	−0.4009	0.1023	−3.92	0.001	3.39
Weight	0.0917	0.0048	19.11	<0.001	1.06
Age	0.6004	0.0132	45.58	<0.001	1.67
Sex	4.0293	0.6546	6.16	<0.001	2.44
Vitamin use	−1.1961	0.4194	−2.85	0.009	1.85
Intercept	106.2809	6.7653	15.71	<0.001	3.96
Model statistics	F (5,19):	937.30			
	p-value:	<0.0001			
	R-squared:	0.393			

In Chapter 14, we presented the logistic regression model and the maximum likelihood estimation procedure. We can also modify this estimation approach to use logistic regression with complex survey data. The modified estimation approach that incorporates the sampling weights is generally known as pseudo or weighted maximum likelihood estimation (Chambless and Boyle 1985; Roberts, Rao, and Kumar 1987). The variance/covariance matrix of the estimated coefficients is calculated by one of the methods discussed in the previous section. As discussed earlier, the degrees of freedom associated with this covariance matrix are the number of PSUs minus the number of strata. Because of all these changes to the standard approach, we use the adjusted Wald test statistic instead of the likelihood-ratio statistic in determining whether or not the model parameters, excluding the constant term, are simultaneously equal to zero.

The selection and inclusion of appropriate predictor variables for a logistic regression model can be done similarly to the process for linear regression. When analyzing a large survey data set, the preliminary analysis strategy described in the earlier section is very useful in preparing for a logistic regression analysis.

Example 15.10

Based on Phase II of NHANES III, we performed a logistic regression analysis of vitamin use on two categorical explanatory variables: sex (1 = male; 0 = female) and education (less than 12 years of education; 12 years; more than 12 years). Two dummy variables are created for the education variable: edu1 = 1 if 12 years of education and 0 otherwise; edu2 = 1 if more than 12 years and 0 otherwise; the less than 12 year category is the reference category. The results are shown in Table 15.12 (see **Program Note 15.6** for this analysis).

The log-likelihood ratio is not shown because the pseudo likelihood is used and an F statistic derived from the modified Wald statistic is shown. The numerator degrees of freedom for this statistic is 3 (based on the number of independent variables) and the denominator degrees of freedom is 21 (= 46 − 23 − 3 + 1, based the numbers of PSUs, strata, and independent variables) (Korn and Graubard 1999). The small p-value suggests that the main effects model is a significant improvement over the null model. The estimated design effects suggest that the variances of the beta coefficients are roughly twice as large as those calculated under the assumption of simple random sampling. Despite the increased standard errors, the beta coefficients for gender and education levels are significant.

Table 15.12 Logistic regression analysis of vitamin use on sex and levels of education among U.S. adults, Phase II, NHANES III (n = 9920).

Variable	Estimated Coefficient	Standard Error	t	$p > |t|$	Design Effect	Odds Ratio	Confidence Interval
Male	−0.4998	0.0584	−8.56	<0.001	1.96	0.61	[0.54, 0.68]
Edu1	0.2497	0.0864	2.89	0.008	2.45	1.28	[1.07, 1.53]
Edu2	0.7724	0.0888	8.69	<0.001	2.84	2.16	[1.80, 2.60]
Constant	−0.4527	0.0773	−5.86	<0.001	2.82		
Model statistics		F (3, 21):	63.61				
		p-value	<0.0001				

The rest of the results can be interpreted in the same way as in Chapter 14. The estimated odds ratio for males is 0.61, meaning that, after adjusting for education, the odds of taking vitamins for a male is 61 percent of the odds that a female uses vitamins. The 95 percent confidence interval provides a test of whether or not the odds ratio is equal to one. The odds ratio for the third level of education suggests that persons with some college education are twice as likely to take vitamins than those with less than 12 years of education for the same gender. None of the 95% confidence intervals include one, suggesting that all the effects are significant at the 0.05 level. As in regular logistic regression analysis, we may combine the estimated beta coefficients to make specific statements. For example, the estimated odds ratio for males with some college education compared with females with less than 12 years of education can be obtained by $\exp(-0.4998 + 0.7724) = 1.31$. Since we have not included any interaction effects in the model, the resulting odds ratio of 1.31 can be interpreted as indicating that the odds of taking vitamins for males with some college education is 31 percent higher than the odds for females with less than 12 years of education.

Conclusion

In this chapter, we discussed issues associated with complex sample surveys focusing on design-based statistical inference. We summarized the two key complications that arise in the analysis of data from complex sample surveys: the need to include sample weights and the need to take the sample design into account in calculating the sampling variance of weighted statistics. We presented several different approaches to the calculation of the sample variances. Practically all statistical methods discussed in previous chapters can be applied to complex survey data with some modifications. For the analysis of a specific subgroup, we pointed out that the entire sample is used although we set the weights to zero for the observations outside the subgroup. Statistical programs for complex surveys are now readily available, but one needs to guard against misuse of the programs. For a proper analysis, one must understand the sample design and conduct a thorough preliminary examination of data. We conclude this chapter by again emphasizing the need to reduce nonresponse and to study some of the nonrespondents if possible.

EXERCISES

15.1 The following data represent a small subset of a large telephone survey. The sample design was intended to be an equal probability sample on each phone number. Within each selected household one adult was sampled using the Kish selection table (Kish 1949). Some households may have more than one phone number and these households are more likely to be selected in random digit dialing. Therefore, selection probability is unequal for individual respondents.

Household	Number of Adults	Number of Phones	Smoking Status	Household	Number of Adults	Number of Phones	Smoking Status
1	3	1	yes	11	4	2	no
2	2	1	no	12	1	1	no
3	4	1	no	13	2	1	no
4	2	1	no	14	3	1	yes
5	2	1	no	15	1	1	no
6	5	2	no	16	3	1	no
7	4	1	yes	17	2	1	no
8	2	1	no	18	2	1	yes
9	3	1	yes	19	3	1	no
10	2	1	no	20	2	1	yes

Develop the sample weight for each respondent, calculate the weighted percentage of smokers, and compare with the unweighted percentage. How would you interpret the weighted and unweighted percentages?

15.2 A community mental health survey was conducted using 10 replicated samples selected by systematic sampling from a geographically ordered list of residential electric hookups (Lee et al. 1986). The total sample size was 3058, and each replicate contained about 300 respondents. The replicated samples were selected to facilitate the scheduling and interim analysis of data during a long period of screening and interviewing, not for estimating the standard errors. Because one adult was randomly selected from each household, the number of adults in each household became the sample weight for each observation. This weight was then adjusted for nonresponse and poststratification and the adjusted weights were used in the analysis. The prevalences of any mental disorders during the past six months and the odds ratios for sex differences in the six-month prevalence rates of mental disorders are shown here for the full sample and the 10 replicates.

Replicate	Prevalence Rate	Odds Ratio
Full sample	17.17	0.990
1	12.81	0.826
2	17.37	0.844
3	17.87	1.057
4	17.64	0.638
5	16.65	0.728
6	18.17	1.027
7	14.69	1.598
8	17.93	1.300
9	17.86	0.923
10	18.91	1.111

Estimate the standard errors for the prevalence rate and the odds ratio based on replicate estimates. Is the approximate standard error based on the range in replicate estimates satisfactory?

15.3 From Phase II of NHANES III, the percent of adults taking vitamin or mineral supplements was estimated to be 43.0 percent with a standard error of 1.22 percent. The design effect of this estimate was 5.98 and the sample size was 9920. What size sample would be required to estimate the same quantity with a standard error of 2 percent using a simple random sampling design?

15.4 Read the article by Gold et al. (1995). Describe how their sample was designed and selected. What was the nonresponse rate? Describe also the method of analysis. Did they account for the sampling design in their analysis? If you find any problems, how would you rectify the problems?

15.5 Using the data file extracted from the adult sample in the Phase II of NHANES III (available on the web), explore one of the following research questions and prepare a brief report describing and interpreting your analysis:

a. Are more educated adults taller than less educated people?

b. Does the prevalence rate of asthma vary by region?

c. Does the use of antacids vary by smoking status (current, previous, and never smoked)?

15.6 Read the article by Flegal et al. (1995), and prepare a critical review of it. Is the purpose and design of the survey properly integrated in the analysis and conclusion? Is the model specified appropriately? Do you think the analysis is done properly? Would you do any part of the analysis differently?

15.7 Select another research question from Exercise 15.5. Conduct the analysis with and without incorporating the weight and design features and compare the results. How would you describe the consequence of not accounting for the weight and design features in the complex survey analysis?

REFERENCES

Chambless, L. E., and K. E. Boyle. "Maximum Likelihood Methods for Complex Sample Data: Logistic Regression and Discrete Proportional Hazards Models." *Communications in Statistics — Theory and Methods*, 14:1377–1392, 1985.

Deming, W. E. *Sampling Design in Business Research*. New York: Wiley, 1960, Chapter 6.

Flegal et al. "The Influence of Smoking Cessation on the Prevalence of Overweight in the United States." *New England Journal of Medicine* 333:115–127, 1995.

Frankel, M. R. *Inference from Survey Samples*. Ann Arbor: Institute of Social Research, University of Michigan, 1971.

Gold et al. "A National Survey of the Arrangements Managed-Care Plans Make with Physicians." *New England Journal of Medicine* 333:1689–1693, 1995.

Kalton, G. *Introduction to Survey Sampling*. Beverly Hills, CA: Sage Publications (QASS, vol. 35) 1983, p. 52.

Kish, L. "A Procedure for Objective Respondent Selection within the Household." *Journal of the American Statistical Association* 44:380–387, 1949.

Kish, L. *Survey Sampling*. New York: Wiley, 1965, p. 620.

Korn, E. L., and B. I. Graubard. *Analysis of Health Surveys*. New York: Wiley, 1999.

Koch, G. G., D. H. Freeman, and J. L. Freeman. "Strategies in the Multivariate Analysis of Data from Complex Surveys." *International Statistical Review* 43:59–78, 1975.

Lee, E. S., and R. N. Forthofer. *Analyzing Complex Survey Data*, 2nd ed. Thousand Oaks, CA: Sage Publications, 2006.

Lee, E. S., R. N. Forthofer, C. E. Holzer, and C. A. Taube. "Complex Survey Data Analysis: Estimation of Standard Errors Using Pseudo-strata." *Journal of Economic and Social Measurement* 14:135–144, 1986.

Levy, P. S., and S. Lemeshow. *Sampling of Populations: Methods and Applications*, 3rd ed. New York: Wiley, 1999, Chapter 13.

Little, R. J. A., and D. B. Rubin. *Statistical Analysis with Missing Data*, 2nd ed. New York: Wiley, 2002.

Lohr, S. L. *Sampling: Design and Analysis.* New York: Duxbury Press, 1999.

McCarthy, P. J. *Replication: An Approach to the Analysis of Data from Complex Surveys.* Vital and Health Statistics, Series 2, no. 14, Fyattsville, MD: National Center for Health Statistics, 1966.

National Center for Health Statistics. *Plan and Operation of the Third National Health and Nutrition Examination Survey, 1988–94.* Vital and Health Statistics, Series 1, no. 32, Hyattsville, MD, 1994.

Plackett, R. L., and P. J. Burman. "The Design of Optimum Multi-factorial Experiments." *Biometrika* 33:305–325, 1946.

Quenouille, M. H. "Approximate Tests of Correlation in Time Series." *Journal of the Royal Statistical Society* 11(B):68–84, 1949.

Rao, J. N. K., and A. J. Scott. "On Chi-Square Tests for Multiway Contingency Tables with Cell Proportions Estimated from Survey Data." *Annals of Statistics* 12:46–60, 1984.

Roberts, G., J. N. K. Rao, and S. Kumar. "Logistic Regression Analysis of Sample Survey Data." *Biometrika* 74:1–12, 1987.

Sudman, S. *Applied Sampling.* New York: Academic Press, 1976.

Appendix A: Review of Basic Mathematical Concepts

A

It is essential to know basic mathematical concepts to understand the statistical ideas and methods presented in this book. Appendix A reviews some basic mathematical concepts to help some students understand the use of mathematical concepts utilized in discussing statistical methods. The material presented here follows the order that topics are encountered in the text chapters.

CHAPTER 3

3.1 The Logic of Logarithms

To understand what we mean by *logarithm*, consider some positive number y. The base 10 logarithm of x is y, where y satisfies the relation that 10^y is equal to x. For example, the base 10 logarithm of 10, often written as $\log_{10}(10)$ or abbreviated as $\log(x)$, is 1 because 10^1 is 10. The value of $\log_{10}(100)$ is 2 because 10^2 equals 100. The value of $\log_{10}(1000)$ is 3 because 10^3, equal to $10 * 10 * 10$, is 1000. Therefore, base 10 logarithms of numbers between 10 and 100 will be between 1 and 2, base 10 logarithms of numbers between 100 and 1000 will be between 2 and 3. In the same way, the logarithms of numbers between 1 and 10 will be between 0 and 1. Numbers less than 1 have a negative logarithm. For example, the base 10 logarithm of 0.1 ($= 1 / 10 = 10^{-1}$) is -1. By expressing the numbers 1, 10, 100, and 0.1 as 10^0, 10^1, 10^2, and 10^{-1}, we get an idea about why the number system we use is base 10 and why the base 10 logarithms are referred to as the *common* logarithms.

3.2 Properties of Logarithms

Given positive numbers x and y, the number n, and that a is any positive real number with the exception that $a \neq 1$, then the following properties of logarithms hold:

(1) $\log_a(xy) = \log_a x + \log_a y$
(2) $\log_a(x/y) = \log_a x - \log_a y$
(3) $\log_a x^n = n \log_a x$

3.3 Natural Logarithms

Logarithms with a base of e, where the number e is an irrational number approximately equal to 2.71828, are called the *natural logarithms* and are written in an abbreviated notation as $\ln(x)$ which is equivalent to $\log_e(x)$. Therefore $\log_e(7)$ is equivalent to writing $\ln(7)$.

3.4 Conversion between Bases

Now that you are familiar with logarithms having a base of 10, consider the following general expression, $\log_a(x)$, which is read as the base a logarithm of x where x is any positive number and a is a positive real number where $a \neq 1$. The following theorem can be found in most algebra books:

> If x is any positive number and if a and b are positive real numbers where $a \neq 1$ and $b \neq 1$, then
>
> $$\log_a(x) = \frac{\log_b(x)}{\log_b(a)}.$$

We can use this result to evaluate logarithms with different bases using the base 10 logarithm that is available on most calculators or computer software programs. As an example, we write the following expression $\log_2(x)$ as $\dfrac{\log_{10}(x)}{\log_{10}(2)}$. Other examples include:

(1) $\log_2(5) = \dfrac{\log_{10}(5)}{\log_{10}(2)}$

(2) $\log_3(8) = \dfrac{\log_{10}(8)}{\log_{10}(3)}$

(3) $\log_e(7) = \dfrac{\log_{10}(7)}{\log_{10}(e)}.$

3.5 Exponential Function

An exponential function has the form $y = a^x$, where $a > 0$. Consider the expression a^n where $a \neq 0$ then $a^n = \{a \cdot a \cdot a \cdot \mathrm{K} \cdot a\}$ so that a is multiplied by itself $n - 1$ times. In this case, n is called the exponent and a is referred to as the base of the exponent. Exponents have the following properties:

If we consider m and n to be integers and the real numbers $a \neq 0$ and $b \neq 0$, then

(1) $a^m a^n = a^{m+n}$
(2) $a^m/a^n = a^{m-n}$
(3) $(a^m)^n = a^{mn}$
(4) $(ab)^m = a^m b^m$
(5) $(a/b)^m = a^m/b^m$
(6) $a^{-n} = (1/a)^n$, this can be inferred from (2) if $m = 0$ resulting in $a^m = 1$.

If a is a positive real number and $n \neq 0$, then the nth root of a is denoted as $a^{1/n}$. This can also be expressed as $\sqrt[n]{a}$. However if n is an even number and a is a negative number, then $a^{1/n}$ is not a real number.

Example A.1

Consider the following example to help understand the usefulness of the exponential function. Epidemics are usually characterized by individuals in a population who are susceptible to some kind of infection, such as the flu. Susceptible individuals have never been infected and also have no immunity against infection. Assume that the number of individuals susceptible to flu infection during a flu epidemic decreases exponentially according to $y = y_0 e^{-ct}$, where y_0 is the base population at time 0, c is the rate of infection and t represents time. If a flu epidemic enters a population of 20,000 with an infection rate of 0.01 per day, then the number of individuals susceptible to the flu at time t is given by

$$y = 20{,}000 e^{-0.01*t}, \text{ where } t \text{ is time in days.}$$

(a) Find the number of individuals susceptible at the beginning of the epidemic.

Since the beginning of the epidemic is at time $t = 0$, the number susceptible is

$$y = 20{,}000 e^{-0.01*(0)} = 20{,}000 e^{(0)} = 20{,}000.$$

(b) Approximately how many individuals are susceptible after 10 days?

At time $t = 10$, the number susceptible is

$$y = 20{,}000 e^{-0.01*(10)} = 20{,}000 e^{(-0.1)} \approx 18{,}096.$$

(c) After how many days will half of the population be infected with the flu?

Half of 20,000 is 10,000, so

$$10{,}000 = 20{,}000 e^{-0.01*t}.$$

Dividing both sides by 20,000,

$$10{,}000/20{,}000 = (20{,}000/20{,}000) e^{-0.01*t}$$

and taking the natural logarithm of both sides, we have

$$\ln(0.5) = \ln(e^{-0.01*t}).$$

By evaluating the natural logarithm of the right side, which gives

$$\ln(0.5) = -0.01 * t$$

and finally solving for t, we have

$$t = \ln(0.5)/(-0.01) \approx 69 \text{ days.}$$

CHAPTER 4

4.1   Factorials

We denote the product of $3 \cdot 2 \cdot 1$ by the symbol 3!, which is read "3 factorial." For any natural number n,

$$n! = n(n - 1)(n - 2) \ldots 1.$$

In general, $n! = n(n - 1)!$. Then the following should be true: $1! = 1(0!)$. Since this statement is true if and only if 0! is equal to 1, we define $0! = 1$.

4.2   Permutations

If we choose two of three objects (A B C), we have the following six arrangements: AB, AC, BA, BC, CA, and CB. There are 3 choices for the first position and 2 choices for the second position. Thus, $3 * 2 = 6$. Each arrangement is called a permutation. The number of permutations of 3 objects taken 2 at a time is 6. We denote this as $P(3, 2) = 6$. In general,

$$P(N, n) = N(N - 1)(N - 2) \cdots (N - n + 1).$$

By multiplying and dividing by $(N - n)!$, we get

$$P(N, n) = \frac{N!}{(N - n)}.$$

If $n = N$, we have $P(N, N) = N!$. Note that the order of arrangement is important in permutation (order is not ignored).

4.3   Combinations

If we ignore the order of arrangement in permutation, then any such selection is called a *combination*. The number of combinations of 3 objects taken 2 at a time is 3, including AB, AC, and BC. We denote this by $\binom{3}{2} = 3$. In general, $\binom{N}{n} n! = P(N, n)$. For each combination, there are $P(n, n) = n!$ ways of arranging the objects in that combination. Hence, we have

$$\binom{N}{n} = \frac{P(N, n)}{n!} = \frac{N(N - 1) \ldots (N - n + 1)}{n!} = \frac{N!}{n!(N - n)!}.$$

CHAPTER 15

15.1   Taylor Series Expansion

The Taylor series expansion has been used in statistics to obtain an approximation to a nonlinear function, and then the variance of the function is based on the Taylor series approximation to the function. Often the approximation provides a reasonable estimate to the function, and sometimes the approximation is even a linear function. This idea of variance estimation has several names in the literature, including the linearization and the delta method.

The Taylor series expansion for a function of x variable, $f(x)$, evaluated at the mean or expected value of x, written as $E(x)$, is

$$f(x) = f[E(x)] + f'[E(x)][x - E(x)] + f''[E(x)][x - E(x)]^2/2! + \cdots$$

where f' and f'' are the first and second derivatives of the function. The variance of $f(x)$ is $V[f(x)] = E[f^2(x)] - E^2[f(x)]$ by definition, and using the first order of Taylor series expansion, we have

$$V[f(x)] = \{f'[E(x)]\}^2 V(x) + \cdots$$

The same ideas carry over to functions of more than one random variable. In the case of a function of two variances, the Taylor series expansion yields

$$V[f(x_1, x_2)] \cong \left(\frac{\partial f}{\partial x_1}\right)\left(\frac{\partial f}{\partial x_2}\right) Cov(x_1, x_2).$$

If we define the following new variable at the observational unit level,

$$t_i = \left(\frac{\partial f}{\partial x_1}\right) x_{1i} + \left(\frac{\partial f}{\partial x_2}\right) x_{2i}$$

then we can calculate the variance directly from t_i without calculating the covariance. The new variable is called linearized value of $f(x_1, x_2)$.

Example A.2

Let us apply this idea to a ratio estimate shown in Example 15.5. The ratio estimate of the total number of professional workers with an MPH was 623 using the following estimator:

$$\hat{X} = \left(\frac{\bar{x}}{\bar{y}}\right) Y = \left(\frac{10.5}{19.375}\right) 1150 = 623.$$

The function is the ratio, $\bar{x}/\bar{y}$, and the derivatives with respect to $\bar{x}$ and $\bar{y}$ are

$$\left(\frac{\partial f}{\partial \bar{x}}\right) = \frac{1}{\bar{y}} \quad \text{and} \quad \frac{\partial f}{\partial \bar{y}} = -\frac{\bar{x}}{\bar{y}^2}.$$

The linearized values can then be calculated by

$$t_i = \frac{1}{19.375} x_i - \frac{10.5}{19.375^2} y.$$

For example, the linearized value for the first observation is

$$t_i = \frac{14}{19.375} - \frac{10.5(21)}{19.375^2} = 0.35193.$$

The rest of linearized values are shown here.

Health Department	Number of Professional Workers (y_i)	Number of Workers with MPH (x_i)	Linearized Values (t_i)
1	21	14	0.135193
2	18	8	−0.090572
3	9	3	−0.096899
4	13	6	−0.053944
5	15	8	−0.006660
6	22	13	0.055609
7	30	17	0.038293
8	27	15	0.018980
Mean	19.375	10.5	

The standard error can be calculated by

$$\sqrt{\frac{Y^2 \sum (t_i - \overline{t})^2}{n(n-1)}\left(1 - \frac{n}{N}\right)} = 29.92.$$

We have the same result as shown in Example 15.5.

For a complex survey, the above can be extended to the case of c random variables with the sample weight (w_i). Woodruff (1971) showed that the approximate variance of $\theta = f(x_1, x_2, \ldots, x_c)$ is

$$V(\hat{\theta}) \cong V\left[\sum w_i \sum \left(\frac{\partial f}{\partial y_i}\right) y_{ij}\right].$$

This method of approximation is applied to PSU totals within the stratum. That is, the variance estimate is a weighted combination of the variation across PSUs within the same stratum.

REFERENCE

Woodruff, R. S. "A Simple Method for Approximating the Variance of a Complicated Estimate." *Journal of American Statistical Association* 66:411–414, 1971.

Appendix B: Statistical Tables

List of Tables

Table B1   Random Digits[a]

Line										
1	17174	75908	43306	77061	97755	26780	07446	34836	47656	22475
2	26580	68460	18051	95528	78196	91824	10696	09283	06525	13596
3	24041	33800	09976	36785	11529	19948	21497	94665	54600	51793
4	74838	79323	43962	50531	30826	76623	04007	72395	03544	37575
5	72862	50965	29962	37114	73007	36615	83463	01021	56940	56615
6	82274	94537	52039	68725	06163	47388	62564	46097	71644	00108
7	77568	89168	04043	31926	83333	99957	22204	96361	79770	42561
8	17802	16697	96288	24603	36345	17063	05251	68206	71113	19390
9	10271	06180	39740	01903	01539	59476	83991	07954	83098	01486
10	07780	55451	05276	87719	42723	33685	66024	14236	96801	45797
11	05751	92219	44689	92084	10025	73998	12863	55026	09230	05881
12	14324	44563	13269	88172	47751	64408	86355	16960	72794	30842
13	12869	51161	96952	01895	35785	40807	88980	56656	88839	94521
14	36891	94679	18832	02471	98216	51769	57593	52247	65271	73641
15	22899	37988	68991	28990	87701	99578	06381	33877	45714	45227
16	58556	91925	66542	12852	57203	25725	19844	92696	56861	51882
17	08520	26078	78485	74072	60421	89379	55514	92898	17894	67682
18	31466	97330	39266	06800	32679	37443	53245	81738	73843	64176
19	43780	49375	20055	79095	79987	96005	44296	29004	25059	95752
20	15875	68956	37126	69074	68076	85098	23707	03965	52477	52517
21	22002	20395	72174	70897	00337	70238	19154	77878	33456	89624
22	28968	92168	79825	50945	99479	03121	43217	97297	47547	12201
23	19446	40211	48163	91237	78166	00421	09652	37508	75560	48279
24	98339	39146	76425	55658	60259	59368	49751	44492	99846	07142
25	42746	66199	44160	87627	31369	59756	91765	64760	46878	57467
26	25544	61063	35953	30319	61982	24629	78600	70075	64922	65913
27	22776	62299	05281	92046	98422	95316	20720	90877	01922	32294
28	22578	20732	18421	77419	75391	20665	60627	29382	37782	13163
29	51580	99897	58983	01745	37488	56543	99580	74823	80339	31931
30	63403	94610	23839	69171	52030	91661	18486	83805	62578	67212
31	77353	80198	26674	72839	09944	51278	99333	97341	87588	01655
32	68849	86194	61771	39583	40760	54492	14279	85621	67459	82681
33	50190	86021	96163	18245	58245	41974	05243	66966	07246	09569
34	91239	72671	10759	17927	38958	40672	06409	21979	87813	11939
35	23457	17487	93379	41738	87628	28721	07582	36969	09161	66801
36	60016	28539	40587	27737	50626	22101	74564	65628	11076	75953
37	37076	96887	07002	14535	70186	84065	57590	94324	14132	25879
38	66454	08589	05977	82951	77907	88931	44828	24952	68021	48766
39	14921	18264	69297	84783	83152	82360	46620	53243	56694	17183
40	79201	63127	02632	42083	23715	95916	66794	52598	84195	45420
41	73735	41872	55392	78688	46013	78470	12915	41744	27769	83002
42	67931	75825	80931	07475	06189	88500	36417	35724	65641	35527
43	40580	67626	06630	79770	08154	12159	11322	84871	53591	77690
44	44858	33801	13691	54744	55641	36758	96949	26400	00505	59016
45	84835	40044	86334	34812	35222	20327	71467	37874	51288	95802
46	88089	35765	87473	22457	56445	18890	60892	53132	87424	71714
47	64102	14894	13441	06584	23270	04518	94560	81582	69858	42800
48	62020	92065	06863	58852	84988	81613	53313	58765	27750	71533
49	36121	29901	65962	49271	09970	00719	72935	35598	53014	50036
50	73007	65445	42898	86105	55352	37128	56141	11222	16718	25885

Table B1   *Continued*

Line										
51	32220	61646	87732	07598	05465	68584	64790	56416	21824	61643
52	12782	34043	30801	64642	62329	85019	22481	70105	38254	57186
53	66400	03051	40583	75130	88348	50303	03657	47252	18090	35891
54	76763	78376	40249	52103	36769	53552	55846	61963	86763	67257
55	11767	46380	25290	59073	91662	89160	94869	71368	90732	33583
56	61292	87282	79921	20936	56304	81358	94966	54748	25865	48333
57	64169	56790	91323	29070	49567	86422	13878	42058	53470	22312
58	86741	20680	18422	64127	88381	27590	99659	47854	12163	41801
59	23215	07774	49216	77376	83893	37631	44332	54941	11038	09157
60	72324	05050	52212	82330	10707	92439	33220	11634	35942	09534
61	18209	60272	95944	64495	09247	61000	52564	99690	52055	70716
62	26568	12545	07291	30737	11449	36252	70323	80141	17833	48502
63	66895	34490	95682	44956	39491	54269	07867	84505	05578	91088
64	28908	21020	84646	17475	40539	62981	93042	38181	35279	21843
65	03091	10135	85594	86222	36342	07903	97933	53548	56768	77881
66	69948	54947	28724	33966	90529	16339	40152	06517	18221	53248
67	80774	71613	41590	18430	99863	70872	41549	89671	63628	82167
68	84702	95823	83712	55061	89773	63242	97952	24027	95176	95129
69	18067	54980	38542	86549	43966	92989	87768	16267	47616	63546
70	76825	11257	34842	26130	91870	37116	90770	42369	09614	16645
71	59759	28041	48498	94968	02759	29884	87231	17899	21157	91094
72	67377	59310	86243	30374	18340	58630	21092	62426	37022	40022
73	86655	18980	13739	12234	50705	68189	02212	64653	39716	29953
74	84073	53993	78016	77751	31457	18155	97944	27295	90526	57958
75	58999	77251	84274	15777	66045	84364	62165	24700	00055	06668
76	11308	03979	68271	51776	55915	67970	52691	19073	82178	66031
77	24585	78224	96506	77936	97772	65814	46162	58603	24666	49133
78	22369	34622	75780	67276	06726	07734	48849	60918	83256	17099
79	24914	45155	66234	00460	86700	72578	57617	82212	50104	34094
80	88320	48338	70689	05856	91247	29214	21807	77100	74896	24592
81	69848	33544	50065	69910	15783	76852	25025	37762	49049	21666
82	77987	45152	89425	81350	10697	90522	10496	86753	75366	83410
83	97709	78833	69516	05969	98796	60938	90201	99875	37430	87145
84	05209	88924	10458	20004	65788	91299	41139	76993	47040	15777
85	68616	23573	66693	83674	34890	57000	07586	39661	23774	50682
86	18260	40283	35008	94377	47286	93322	68092	92858	99829	59997
87	29121	89864	44444	03931	34222	49057	49713	50972	23191	29933
88	36834	59756	46105	01156	40367	50950	43614	70178	93359	77431
89	10757	21796	12219	39415	32020	04178	69733	83093	58039	74845
90	99465	88838	45530	96133	66529	57600	52060	98052	72613	32354
91	59157	66024	86610	70068	29879	30664	87190	98772	76243	62043
92	63489	17951	66279	69460	03659	53135	79535	05034	26052	75480
93	08723	61325	57652	18876	08976	51276	12793	60467	11655	04069
94	75883	23261	03050	36180	38486	47570	72493	92403	06412	10039
95	95560	45085	03464	79493	25121	04125	86957	16042	63551	40774
96	81329	74272	70097	05615	91212	73956	43022	64078	77377	14160
97	13536	31170	91648	67487	95149	17890	50223	82906	59466	01721
98	28778	55892	59449	53815	84565	62568	79771	00793	19324	10150
99	39757	44482	21115	01607	93177	26324	66403	91660	62073	34237
00	54595	87336	08030	30633	83752	04706	96494	71064	19061	84919

[a]Generated from MINITAB.

Table B2 Binomial Probabilities[a]

n	x	.01	.05	.10	.15	.20	.25	.30	.35	.40	.45	.50
								p				
2	0	.9801	.9025	.8100	.7225	.6400	.5625	.4900	.4225	.3600	.3025	.2500
	1	.0198	.0950	.1800	.2550	.3200	.3750	.4200	.4550	.4800	.4950	.5000
	2	.0001	.0025	.0100	.0225	.0400	.0625	.0900	.1225	.1600	.2025	.2500
3	0	.9703	.8574	.7290	.6141	.5120	.4219	.3430	.2746	.2160	.1664	.1250
	1	.0294	.1354	.2430	.3251	.3840	.4219	.4410	.4436	.4320	.4084	.3750
	2	.0003	.0071	.0270	.0574	.0960	.1406	.1890	.2389	.2880	.3341	.3750
	3		.0001	.0010	.0034	.0080	.0156	.0270	.0429	.0640	.0911	.1250
4	0	.9606	.8145	.6561	.5220	.4096	.3164	.2401	.1785	.1296	.0915	.0625
	1	.0388	.1715	.2916	.3685	.4096	.4219	.4116	.3845	.3456	.2995	.2500
	2	.0006	.0135	.0486	.0975	.1536	.2109	.2646	.3105	.3456	.3675	.3750
	3		.0005	.0036	.0115	.0256	.0469	.0756	.1115	.1536	.2005	.2500
	4			.0001	.0005	.0016	.0039	.0018	.0150	.0256	.0410	.0625
5	0	.9510	.7738	.5905	.4437	.3277	.2373	.1681	.1160	.0778	.0503	.0313
	1	.0480	.2036	.3281	.3915	.4096	.3955	.3602	.3124	.2592	.2059	.1563
	2	.0010	.0214	.0729	.1382	.2048	.2637	.3087	.3364	.3456	.3369	.3125
	3		.0011	.0081	.0244	.0512	.0879	.1323	.1811	.2304	.2757	.3125
	4			.0005	.0022	.0064	.0146	.0284	.0488	.0768	.1128	.1563
	5				.0001	.0003	.0010	.0024	.0053	.0102	.0185	.0313
6	0	.9415	.7351	.5314	.3771	.2621	.1780	.1176	.0754	.0467	.0277	.0156
	1	.0571	.2321	.3543	.3993	.3932	.3560	.3025	.2437	.1866	.1359	.0938
	2	.0014	.0305	.0984	.1762	.2458	.2966	.3241	.3280	.3110	.2780	.2344
	3		.0021	.0146	.0415	.0819	.1318	.1852	.2355	.2765	.3032	.3125
	4		.0001	.0012	.0055	.0154	.0330	.0595	.0951	.1382	.1861	.2344
	5			.0001	.0004	.0015	.0044	.0102	.0205	.0369	.0609	.0938
	6					.0001	.0002	.0007	.0018	.0041	.0083	.0156
7	0	.9321	.6983	.4783	.3206	.2097	.1335	.0824	.0490	.0280	.0152	.0078
	1	.0659	.2573	.3720	.3960	.3670	.3115	.2471	.1848	.1306	.0872	.0547
	2	.0020	.0406	.1240	.2097	.2753	.3115	.3177	.2985	.2613	.2140	.1641
	3		.0036	.0230	.0617	.1147	.1730	.2269	.2679	.2903	.2918	.2734
	4		.0002	.0026	.0109	.0287	.0577	.0972	.1442	.1935	.2388	.2734
	5			.0002	.0012	.0043	.0115	.0250	.0466	.0774	.1172	.1641
	6				.0001	.0004	.0013	.0036	.0084	.0172	.0320	.0547
	7							.0002	.0006	.0016	.0037	.0078
8	0	.9227	.6634	.4305	.2725	.1678	.1001	.0576	.0319	.0168	.0084	.0039
	1	.0746	.2793	.3826	.3847	.3355	.2670	.1977	.1373	.0896	.0548	.0313
	2	.0026	.0515	.1488	.2376	.2936	.3115	.2965	.2587	.2090	.1569	.1094
	3	.0001	.0054	.0331	.0839	.1468	.2076	.2541	.2786	.2787	.2568	.2188
	4		.0004	.0046	.0185	.0459	.0865	.1361	.1875	.2322	.2627	.2734
	5			.0004	.0026	.0092	.0231	.0467	.0808	.1239	.1719	.2188
	6				.0002	.0011	.0038	.0100	.0217	.0413	.0703	.1094
	7					.0001	.0004	.0012	.0033	.0079	.0164	.0313
	8							.0001	.0002	.0007	.0017	.0039
9	0	.9135	.6302	.3874	.2316	.1342	.0751	.0404	.0207	.0101	.0046	.0020
	1	.0830	.2985	.3874	.3679	.3020	.2253	.1557	.1004	.0605	.0039	.0176
	2	.0034	.0629	.1722	.2597	.3020	.3003	.2668	.2162	.1612	.1110	.0703
	3	.0001	.0077	.0446	.1069	.1762	.2336	.2668	.2716	.2508	.2119	.1641
	4		.0006	.0074	.0283	.0661	.1168	.1715	.2194	.2508	.2600	.2461
	5			.0008	.0050	.0165	.0389	.0735	.1181	.1672	.2128	.2461
	6			.0001	.0006	.0028	.0087	.0210	.0424	.0743	.1160	.1641
	7					.0003	.0012	.0039	.0098	.0212	.0407	.0703
	8						.0001	.0004	.0013	.0035	.0083	.0176
	9								.0001	.0003	.0008	.0020
10	0	.9044	.5987	.3487	.1969	.1074	.0563	.0282	.0135	.0060	.0025	.0010
	1	.0914	.3151	.3874	.3474	.2684	.1877	.1211	.0725	.0403	.0207	.0098
	2	.0042	.0746	.1937	.2759	.3020	.2816	.2335	.757	.1209	.0763	.0439
	3	.0001	.0105	.0574	.1298	.2013	.2503	.2668	.2522	.2150	.1665	.1172
	4		.0010	.0112	.0401	.0881	.1460	.2001	.2377	.2508	.2384	.2051
	5		.0001	.0015	.0085	.0264	.0584	.1029	.1536	.2007	.2340	.2461
	6			.0001	.0012	.0055	.0162	.0368	.0689	.1115	.1596	.2051
	7				.0001	.0008	.0031	.0090	.0212	.0425	.0746	.1172
	8					.0001	.0004	.0014	.0043	.0106	.0229	.0439
	9							.0001	.0005	.0016	.0042	.0098
	10									.0001	.0003	.0010

(continued)

Table B2   *Continued*

n	x	.01	.05	.10	.15	.20	.25	.30	.35	.40	.45	.50
11	0	.8953	.5688	.3138	.1673	.0859	.0422	.0198	.0088	.0036	.0014	.0005
	1	.0995	.3293	.3835	.3248	.2362	.1549	.0932	.0518	.0266	.0125	.0054
	2	.0050	.0867	.2131	.2866	.2953	.2581	.1998	.1395	.0887	.0513	.0269
	3	.0002	.0137	.0710	.1517	.2215	.2581	.2568	.2254	.1774	.1259	.0806
	4		.0014	.0158	.0536	.1107	.1721	.2201	.2428	.2365	.2060	.1611
	5		.0001	.0025	.0132	.0388	.0803	.1321	.1830	.2207	.2360	.2256
	6			.0003	.0023	.0097	.0268	.0566	.0985	.1471	.1931	.2256
	7				.0003	.0017	.0064	.0173	.0379	.0701	.1128	.1611
	8					.0002	.0011	.0037	.0102	.0234	.0462	.0806
	9						.0001	.0005	.0018	.0052	.0126	.0269
	10								.0002	.0007	.0021	.0054
	11										.0002	.0005
12	0	.8864	.5404	.2824	.1422	.0687	.0317	.0138	.0057	.0022	.0008	.0002
	1	.1074	.3413	.3766	.3012	.2062	.1267	.0712	.0368	.0174	.0075	.0029
	2	.0060	.0988	.2301	.2924	.2835	.2323	.1678	.1088	.0639	.0339	.0161
	3	.0002	.0173	.0852	.1720	.2362	.2581	.2397	.1954	.1419	.0923	.0537
	4		.0021	.0213	.0683	.1329	.1936	.2311	.2367	.2128	.1700	.1209
	5		.0002	.0038	.0193	.0532	.1032	.1585	.2039	.2270	.2225	.1934
	6			.0005	.0040	.0155	.0401	.0792	.1281	.1766	.2124	.2256
	7				.0006	.0033	.0115	.0291	.0591	.1009	.1489	.1934
	8				.0001	.0005	.0024	.0078	.0199	.0420	.0762	.1209
	9					.0001	.0004	.0015	.0048	.0125	.0277	.0537
	10							.0002	.0008	.0025	.0068	.0161
	11								.0001	.0003	.0010	.0029
	12										.0001	.0002
13	0	.8775	.5133	.2542	.1209	.0550	.0238	.0097	.0037	.0013	.0004	.0001
	1	.1152	.3512	.3672	.2774	.1787	.1029	.0540	.0259	.0113	.0045	.0016
	2	.0070	.1109	.2448	.2937	.2680	.2059	.1388	.0836	.0453	.0220	.0095
	3	.0003	.0214	.0997	.1900	.2457	.2517	.2181	.1651	.1107	.0660	.0349
	4		.0028	.0277	.0838	.1535	.2097	.2337	.2222	.1845	.1350	.0873
	5		.0003	.0055	.0266	.0691	.1258	.1803	.2154	.2214	.1989	.1571
	6			.0008	.0063	.0230	.0559	.1030	.1546	.1968	.2169	.2095
	7			.0001	.0011	.0058	.0186	.0442	.0833	.1312	.1775	.2095
	8				.0001	.0011	.0047	.0142	.0336	.0656	.1089	.1571
	9					.0002	.0009	.0034	.0101	.0243	.0495	.0873
	10						.0001	.0006	.0022	.0065	.0162	.0349
	11							.0001	.0003	.0012	.0036	.0095
	12									.0001	.0005	.0016
	13											.0001
14	0	.8687	.4877	.2288	.1028	.0440	.0178	.0068	.0024	.0008	.0002	.0001
	1	.1229	.3593	.3559	.2539	.1539	.0832	.0407	.0181	.0073	.0027	.0009
	2	.0081	.1229	.2570	.2912	.2501	.1802	.1134	.0634	.0317	.0141	.0056
	3	.0003	.0259	.1142	.2056	.2501	.2402	.1943	.1366	.0845	.0462	.0222
	4		.0037	.0349	.0998	.1720	.2202	.2290	.2022	.1549	.1040	.0611
	5		.0004	.0078	.0352	.0860	.1468	.1963	.2178	.2066	.1701	.1222
	6			.0013	.0093	.0322	.0734	.1262	.1759	.2066	.2088	.1833
	7			.0002	.0019	.0092	.0280	.0618	.1082	.1574	.1952	.2095
	8				.0003	.0020	.0082	.0232	.0510	.0918	.1398	.1833
	9					.0003	.0018	.0066	.0183	.0408	.0762	.1222
	10						.0003	.0014	.0049	.0136	.0312	.0611
	11							.0002	.0010	.0033	.0093	.0222
	12								.0001	.0006	.0019	.0056
	13									.0001	.0002	.0009
	14											.0001

(continued)

Table B2 *Continued*

							p					
n	*x*	.01	.05	.10	.15	.20	.25	.30	.35	.40	.45	.50
15	0	.8601	.4633	.2059	.0874	.0352	.0134	.0047	.0016	.0005	.0001	
	1	.1303	.3658	.3432	.2312	.1319	.0668	.0305	.0126	.0047	.0016	.0005
	2	.0092	.1348	.2669	.2856	.2309	.1559	.0916	.0476	.0219	.0090	.0032
	3	.0004	.0308	.1285	.2184	.2501	.2252	.1700	.1110	.0634	.0318	.0139
	4		.0049	.0428	.1156	.1876	.2252	.2186	.1792	.1268	.0780	.0417
	5		.0006	.0105	.0449	.1032	.1651	.2061	.2123	.1859	.1404	.0916
	6			.0019	.0132	.0430	.0917	.1472	.1906	.2066	.1914	.1527
	7			.0003	.0030	.0138	.0393	.0811	.1319	.1771	.2013	.1964
	8				.0005	.0035	.0131	.0348	.0710	.1181	.1647	.1964
	9				.0001	.0007	.0034	.0116	.0298	.0612	.1048	.1527
	10					.0001	.0007	.0030	.0096	.0245	.0515	.0916
	11						.0001	.0006	.0024	.0074	.0191	.0417
	12							.0001	.0004	.0016	.0052	.0139
	13								.0001	.0003	.0010	.0032
	14										.0001	.0005
	15											
16	0	.8515	.4401	.1853	.0743	.0281	.0100	.0033	.0010	.0003	.0001	
	1	.1376	.3706	.3294	.2097	.1126	.0535	.0228	.0087	.0030	.0009	.0002
	2	.0104	.1463	.2745	.2775	.2111	.1336	.0732	.0353	.0150	.0056	.0018
	3	.0005	.0359	.1423	.2285	.2463	.2079	.1465	.0888	.0468	.0215	.0085
	4		.0061	.0514	.1311	.2001	.2252	.2040	.1553	.1014	.0572	.0278
	5		.0008	.0137	.0555	.1201	.1802	.2099	.2008	.1623	.1123	.0667
	6		.0001	.0028	.0180	.0550	.1101	.1649	.1982	.1983	.1684	.1222
	7			.0004	.0045	.0197	.0524	.1010	.1524	.1889	.1969	.1746
	8			.0001	.0009	.0055	.0197	.0487	.0923	.1417	.1812	.1964
	9				.0001	.0012	.0058	.0185	.0442	.0840	.1318	.1746
	10					.0002	.0014	.0056	.0167	.0392	.0755	.1222
	11						.0002	.0013	.0049	.0142	.0337	.0667
	12							.0002	.0011	.0040	.0115	.0278
	13								.0002	.0008	.0029	.0085
	14									.0001	.0005	.0018
	15											.0002
	16											
17	0	.8429	.4181	.1668	.0631	.0225	.0075	.0023	.0007	.0002		
	1	.1447	.3741	.3150	.1893	.0957	.0426	.0169	.0060	.0019	.0005	.0001
	2	.0117	.1575	.2800	.2673	.1914	.1136	.0581	.0260	.0102	.0035	.0010
	3	.0006	.0415	.1556	.2359	.2393	.1893	.1245	.0701	.0341	.0144	.0052
	4		.0076	.0605	.1457	.2093	.2209	.1868	.1320	.0796	.0411	.0182
	5		.0010	.0175	.0668	.1361	.1914	.2081	.1849	.1379	.0875	.0472
	6		.0001	.0039	.0236	.0680	.1276	.1784	.1991	.1839	.1432	.0944
	7			.0007	.0065	.0267	.0668	.1201	.1685	.1927	.1841	.1484
	8			.0001	.0014	.0084	.0279	.0644	.1134	.1606	.1883	.1855
	9				.0003	.0021	.0093	.0276	.0611	.1070	.1540	.1855
	10					.0004	.0025	.0095	.0263	.0571	.1008	.1484
	11					.0001	.0005	.0026	.0090	.0242	.0525	.0944
	12						.0001	.0006	.0024	.0081	.0215	.0472
	13							.0001	.0005	.0021	.0068	.0182
	14								.0001	.0004	.0016	.0052
	15									.0001	.0003	.0010
	16											.0001
	17											

(*continued*)

Table B2 *Continued*

n	x	.01	.05	.10	.15	.20	.25	.30	.35	.40	.45	.50
18	0	.8345	.3972	.1501	.0536	.0180	.0056	.0016	.0004	.0001		
	1	.1517	.3763	.3002	.1704	.0811	.0338	.0126	.0042	.0012	.0003	.0001
	2	.0130	.1683	.2835	.2556	.1723	.0958	.0458	.0190	.0069	.0022	.0006
	3	.0007	.0473	.1680	.2406	.2297	.1704	.1046	.0547	.0246	.0095	.0031
	4		.0093	.0700	.1592	.2153	.2130	.1681	.1104	.0614	.0291	.0117
	5		.0014	.0218	.0787	.1507	.1988	.2017	.1664	.1146	.0666	.0327
	6		.0002	.0052	.0301	.0816	.1436	.1873	.1941	.1655	.1181	.0708
	7			.0010	.0091	.0350	.0820	.1376	.1792	.1892	.1657	.1214
	8			.0002	.0022	.0120	.0376	.0811	.1327	.1734	.1864	.1669
	9				.0004	.0033	.0139	.0386	.0794	.2844	.1694	.1855
	10				.0001	.0008	.0042	.0149	.0385	.0771	.1248	.1669
	11					.0001	.0010	.0046	.0151	.0374	.0742	.1214
	12						.0002	.0012	.0047	.0145	.0354	.0708
	13							.0002	.0012	.0045	.0134	.0327
	14								.0002	.0011	.0039	.0117
	15									.0002	.0009	.0031
	16										.0001	.0006
	17											.0001
	18											
19	0	.8262	.3774	.1351	.0456	.0144	.0042	.0011	.0003	.0001		
	1	.1586	.3774	.2852	.1259	.0685	.0268	.0093	.0029	.0008	.0002	
	2	.0144	.1787	.2852	.2428	.1540	.0803	.0358	.0138	.0046	.0013	.0003
	3	.0008	.0533	.1796	.2428	.2182	.1517	.0869	.0422	.0175	.0062	.0018
	4		.0112	.0798	.1714	.2182	.2023	.1491	.0909	.0467	.0203	.0074
	5		.0018	.0266	.0907	.1637	.2023	.1916	.1468	.0933	.0497	.0222
	6		.0002	.0069	.0374	.0955	.1574	.1916	.1844	.1451	.0949	.0518
	7			.0014	.0122	.0443	.0974	.1525	.1844	.1797	.1443	.0961
	8			.0002	.0032	.0166	.0487	.0981	.1489	.1797	.1771	.1442
	9				.0007	.0051	.0198	.0514	.0980	.1464	.1771	.1762
	10				.0001	.0013	.0066	.0220	.0528	.0976	.1449	.1762
	11					.0003	.0018	.0077	.0233	.0532	.0970	.1442
	12						.0004	.0022	.0083	.0237	.0529	.0961
	13						.0001	.0005	.0024	.0085	.0233	.0518
	14							.0001	.0006	.0024	.0082	.0222
	15								.0001	.0005	.0022	.0074
	16									.0001	.0005	.0018
	17										.0001	.0003
	18											
	19											
20	0	.8179	.3585	.1216	.0388	.0115	.0032	.0008	.0002			
	1	.1652	.3774	.2702	.1368	.0576	.0211	.0068	.0020	.0005	.0001	
	2	.0159	.1887	.2852	.2293	.1369	.0669	.0278	.0100	.0031	.0008	.0002
	3	.0010	.0596	.1901	.2428	.2054	.1339	.0716	.0323	.0124	.0040	.0011
	4		.0133	.0898	.1821	.2182	.1897	.1304	.0738	.0350	.0139	.0046
	5		.0022	.0319	.1028	.1746	.2023	.1789	.1272	.0746	.0365	.0148
	6		.0003	.0089	.0454	.1091	.1686	.1916	.1712	.1244	.0746	.0370
	7			.0020	.0160	.0546	.1124	.1643	.1844	.1659	.1221	.0739
	8			.0004	.0046	.0222	.0609	.1144	.1614	.1797	.1623	.1201
	9			.0001	.0011	.0074	.0271	.0654	.1158	.1597	.1771	.1602
	10				.0002	.0020	.0099	.0308	.0686	.1171	.1593	.1762
	11					.0005	.0030	.0120	.0336	.0710	.1185	.1602
	12					.0001	.0008	.0039	.0136	.0355	.0727	.1201
	13						.0002	.0010	.0045	.0146	.0366	.0739
	14							.0002	.0012	.0049	.0150	.0370
	15								.0003	.0013	.0049	.0148
	16									.0003	.0013	.0046
	17										.0002	.0011
	18											.0002
	19											

The *p* heading spans columns .01 through .50.

[a]Calculated by MINITAB.

Table B3　Poisson Probabilities[a]

x	.2	.4	.6	.8	1.0	1.2	1.4	1.6	x
					μ				
0	.818731	.670320	.548812	.449329	.367879	.301194	.246597	.201896	0
1	.163746	.268128	.329287	.359463	.367879	.361433	.345236	.323034	1
2	.016375	.053626	.098786	.143785	.183940	.216860	.241665	.258428	2
3	.001092	.007150	.019757	.038343	.061313	.086744	.112777	.137828	3
4	.000055	.000715	.002964	.007669	.015328	.026023	.039472	.055131	4
5	.000002	.000057	.000356	.001227	.003066	.006246	.011052	.017642	5
6		.000004	.000036	.000164	.000511	.001249	.002579	.004705	6
7			.000003	.000019	.000073	.000214	.000516	.001075	7
8				.000002	.000009	.000032	.000090	.000215	8
9					.000001	.000004	.000014	.000038	9
10						.000001	.000002	.000006	10
11								.000001	11

x	1.8	2.0	2.5	3.0	3.5	4.0	4.5	5.0	x
					μ				
0	.165299	.135335	.082085	.049787	.030197	.018316	.011109	.006738	0
1	.297538	.270671	.205213	.149361	.105691	.073263	.049990	.033690	1
2	.267784	.270671	.256516	.224042	.184959	.146525	.112479	.084224	2
3	.160671	.180447	.213763	.224042	.215785	.195367	.168718	.140374	3
4	.072302	.090224	.133602	.168031	.188812	.195367	.189808	.175467	4
5	.026029	.036089	.066801	.100819	.132169	.156293	.170827	.175467	5
6	.007809	.012030	.027834	.050409	.077098	.104196	.128120	.146223	6
7	.002008	.003437	.009941	.021604	.038549	.059540	.082363	.104445	7
8	.000452	.000859	.003106	.008102	.016865	.029770	.046329	.065278	8
9	.000090	.000191	.000863	.002701	.006559	.013231	.023165	.036266	9
10	.000016	.000038	.000216	.000810	.002296	.005292	.010424	.018133	10
11	.000003	.000007	.000049	.000221	.000730	.001925	.004264	.008242	11
12		.000001	.000010	.000055	.000213	.000642	.001599	.003434	12
13			.000002	.000013	.000057	.000197	.000554	.001321	13
14				.000003	.000014	.000056	.000178	.000472	14
15				.000001	.000003	.000015	.000053	.000157	15
16					.000001	.000004	.000015	.000049	16
17						.000001	.000004	.000014	17
18							.000001	.000004	18
19								.000001	19

(continued)

Table B3   *Continued*

x	5.5	6.0	6.5	7.0	8.0	9.0	10.0	11.0	x
0	.004087	.002479	.001503	.000912	.000335	.000123	.000045	.000017	0
1	.022477	.014873	.009772	.006383	.002684	.001111	.000454	.000184	1
2	.061812	.044618	.031760	.022341	.010735	.004998	.002270	.001010	2
3	.113323	.089235	.068814	.052129	.028626	.014994	.007567	.003705	3
4	.155819	.133853	.111822	.091226	.057252	.033737	.018917	.010189	4
5	.171401	.160623	.145369	.127717	.091604	.060727	.037833	.022415	5
6	.157117	.160623	.157483	.149003	.122138	.091090	.063055	.041095	6
7	.123449	.137677	.146234	.149003	.139587	.117116	.090079	.064577	7
8	.084871	.103258	.118815	.130377	.139587	.131756	.112599	.088794	8
9	.051866	.068838	.085811	.101405	.124077	.131756	.125110	.108526	9
10	.028526	.041303	.055777	.070983	.099262	.118580	.125110	.119378	10
11	.014263	.022529	.032959	.045171	.072190	.097020	.113736	.119378	11
12	.006537	.011264	.017853	.026350	.048127	.072765	.094780	.109430	12
13	.002766	.005199	.008926	.014188	.029616	.050376	.072908	.092595	13
14	.001087	.002228	.004144	.007094	.016924	.032384	.052077	.072753	14
15	.000398	.000891	.001796	.003311	.009026	.019431	.034718	.053352	15
16	.000137	.000334	.000730	.001448	.004513	.010930	.021699	.036680	16
17	.000044	.000118	.000279	.000596	.002124	.005786	.012764	.023734	17
18	.000014	.000039	.000101	.000232	.000944	.002893	.007091	.014504	18
19	.000004	.000012	.000034	.000085	.000397	.001370	.003732	.008397	19
20	.000001	.000004	.000011	.000030	.000159	.000617	.001866	.004618	20
21		.000001	.00003	.000010	.000061	.000264	.000889	.002419	21
22			.000001	.000003	.000022	.000108	.000404	.001210	22
23				.000001	.000008	.000042	.000176	.000578	23
24					.000003	.000016	.000073	.000265	24
25					.000001	.000006	.000029	.000117	25
26						.000002	.000011	.000049	26
27						.000001	.000004	.000020	27
28							.000001	.000008	28
29							.000001	.000003	29
30								.000001	30

(continued)

Table B3 *Continued*

				μ			
x	12.0	13.0	14.0	15.0	16.0	17.0	x
0	.000006	.000002	.000001				0
1	.000074	.000029	.000012	.000005	.000002	.000001	1
2	.000442	.000191	.000081	.000034	.000014	.000006	2
3	.001770	.000828	.000380	.000172	.000077	.000034	3
4	.005309	.002690	.001331	.000645	.000307	.000144	4
5	.012741	.006994	.003727	.001936	.000983	.000490	5
6	.025481	.015153	.008696	.004839	.002622	.001388	6
7	.043682	.028141	.017392	.010370	.005994	.003371	7
8	.065523	.045730	.030436	.019444	.011988	.007163	8
9	.087364	.066054	.047344	.032407	.021311	.013529	9
10	.104837	.085870	.066282	.048611	.034098	.023000	10
11	.114368	.101483	.084359	.066287	.049597	.035545	11
12	.114368	.109940	.098418	.082859	.066129	.050355	12
13	.105570	.109940	.105989	.095607	.081389	.065849	13
14	.090489	.102087	.105989	.102436	.093016	.079960	14
15	.072391	.088475	.098923	.102436	.099218	.090621	15
16	.054293	.071886	.086558	.096034	.099218	.096285	16
17	.038325	.054972	.071283	.084736	.093381	.096285	17
18	.025550	.039702	.055442	.070613	.083006	.090936	18
19	.016137	.027164	.040852	.055747	.069899	.081363	19
20	.009682	.017657	.028597	.041810	.055920	.069159	20
21	.005533	.010930	.019064	.029865	.042605	.055986	21
22	.003018	.006459	.012132	.020362	.030986	.043262	22
23	.001574	.003651	.007385	.013280	.021555	.031976	23
24	.000787	.001977	.004308	.008300	.014370	.022650	24
25	.000378	.001028	.002412	.004980	.009197	.015402	25
26	.000174	.000514	.001299	.002873	.005660	.010070	26
27	.000078	.000248	.000674	.001596	.003354	.006341	27
28	.000033	.000115	.000337	.000855	.001917	.003850	28
29	.000014	.000052	.000163	.000442	.001057	.002257	29
30	.000005	.000022	.000076	.000221	.000564	.001279	30
31	.000002	.000009	.000034	.000107	.000291	.000701	31
32	.000001	.000004	.000015	.000050	.000146	.000373	32
33		.000001	.000006	.000023	.000071	.000192	33
34		.000001	.000003	.000010	.000033	.000096	34
35			.000001	.000004	.000015	.000047	35
36				.000002	.000007	.000022	36
37				.000001	.000003	.000010	37
38					.000001	.000005	38
39					.000001	.000002	39
40						.000001	40

[a]Calculated by MINITAB.

Table B4 Cumulative Distribution Function for Standard Normal Distribution[a]

z	.09	.08	.07	.06	.05	.04	0.3	.02	.01	.00
−3.7	.0001	.0001	.0001	.0001	.0001	.0001	.0001	.0001	.0001	.0001
−3.6	.0001	.0001	.0001	.0001	.0001	.0001	.0001	.0001	.0002	.0002
−3.5	.0002	.0002	.0002	.0002	.0002	.0002	.0002	.0002	.0002	.0002
−3.4	.0002	.0003	.0003	.0003	.0003	.0003	.0003	.0003	.0003	.0003
−3.3	.0003	.0004	.0004	.0004	.0004	.0004	.0004	.0005	.0005	.0005
−3.2	.0005	.0005	.0005	.0006	.0006	.0006	.0006	.0006	.0007	.0007
−3.1	.0007	.0007	.0008	.0008	.0008	.0008	.0009	.0009	.0009	.0010
−3.0	.0010	.0010	.0011	.0011	.00011	.0012	.0012	.0013	.0013	.0013
−2.9	.0014	.0014	.0015	.0015	.0016	.0016	.0017	.0018	.0018	.0019
−2.8	.0019	.0020	.0021	.0021	.0022	.0023	.0023	.0024	.0025	.0026
−2.7	.0026	.0027	.0028	.0029	.0030	.0031	.0032	.0033	.0034	.0035
−2.6	.0036	.0037	.0038	.0039	.0040	.0041	.0043	.0044	.0045	.0047
−2.5	.0048	.0049	.0051	.0052	.0054	.0055	.0057	.0059	.0060	.0062
−2.4	.0064	.0066	.0068	.0069	.0071	.0073	.0075	.0078	.0080	.0082
−2.3	.0084	.0087	.0089	.0091	.0094	.0096	.0099	.0102	.0104	.0107
−2.2	.0110	.0113	.0116	.0119	.0122	.0125	.0129	.0132	.0136	.0139
−2.1	.0143	.0146	.0150	.0154	.0158	.0162	.0166	.0170	.0174	.0179
−2.0	.0183	.0188	.0192	.0197	.0202	.0207	.0212	.0217	.0222	.0228
−1.9	.0233	0.239	.0244	.0250	.0256	.0262	.0268	.0274	.0281	.0287
−1.8	.0294	.0301	.0307	.0314	.0322	.0329	.0336	.0344	.0351	.0359
−1.7	.0367	.0375	.0384	.0392	.0401	.0409	.0418	.0427	.0436	.0446
−1.6	.0455	.0465	.0475	.0485	.0495	.0505	.0516	.0526	.0537	.0548
−1.5	.0559	.0571	.0582	.0594	.0606	.0618	.0630	.0643	.0655	.0668
−1.4	.0681	.0694	.0708	.0721	.0735	.0749	.0764	.0778	.0793	.0808
−1.3	.0823	.0838	.0853	.0869	.0885	.0901	.0918	.0934	.0951	.0968
−1.2	.0985	.1003	.1020	.1038	.1056	.1075	.1093	.1112	.1131	.1151
−1.1	.1170	.1190	.1210	.1230	.1251	.1271	.1292	.1314	.1335	.1357
−1.0	.1379	.1401	.1423	.1446	.1469	.1492	.1515	.1539	.1562	.1587
−0.9	.1611	.1635	.1660	.1685	.1711	.1736	.1762	.1788	.1814	.1841
−0.8	.1867	.1894	.1922	.1949	.1977	.2005	.2033	.2061	.2090	.2119
−0.7	.2148	.2177	.2206	.2236	.2266	.2296	.2327	.2358	.2389	.2420
−0.6	.2451	.2483	.2514	.2546	.2578	.2611	.2643	.2676	.2709	.2743
−0.5	.2776	.2810	.2843	.2877	.2912	.2946	.2981	.3015	.3050	.3085
−0.4	.3121	.3156	.3192	.3228	.3264	.3300	.3336	.3372	.3409	.3446
−0.3	.3483	.3520	.3557	.3594	.3632	.3669	.3707	.3745	.3783	.3821
−0.2	.3859	.3897	.3936	.3974	.4013	.4052	.4090	.4129	.4168	.4207
−0.1	.4247	.4286	.4325	.4364	.4404	.4443	.4483	.4522	.4562	.4602
−0.0	.4641	.4681	.4721	.4761	.4801	.4840	.4880	.4920	.4960	.5000

(continued)

Table B4 *Continued*

z	.00	.01	.02	.03	.04	.05	.06	.07	.08	.09
0.0	.5000	.5040	.5080	.5120	.5160	.5199	.5239	.5279	.5319	.5359
0.1	.5398	.5438	.5478	.5517	.5557	.5596	.5636	.5675	.5714	.5753
0.2	.5793	.5832	.5871	.5910	.5948	.5987	.6026	.6064	.6103	.6141
0.3	.6179	.6217	.6255	.6293	.6331	.6368	.6406	.6443	.6480	.6517
0.4	.6554	.6591	.6628	.6664	.6700	.6736	.6772	.6808	.6844	.6879
0.5	.6915	.6950	.6985	.7019	.7054	.7088	.7123	.7157	.7190	.7224
0.6	.7257	.7291	.7324	.7357	.7389	.7422	.7454	.7486	.7517	.7549
0.7	.7580	.7611	.7642	.7673	.7704	.7734	.7764	.7794	.7823	.7852
0.8	.7881	.7910	.7939	.7967	.7995	.8023	.8051	.8078	.8106	.8133
0.9	.8159	.8186	.8212	.8238	.8264	.8289	.8315	.8340	.8365	.8389
1.0	.8413	.8438	.8461	.8485	.8508	.8531	.8554	.8577	.8599	.8621
1.1	.8643	.8665	.8686	.8708	.8729	.8749	.8770	.8790	.8810	.8830
1.2	.8849	.8869	.8888	.8907	.8925	.8944	.8962	.8980	.8997	.9015
1.3	.9032	.9049	.9066	.9082	.9099	.9115	.9131	.9147	.9162	.9177
1.4	.9192	.9207	.9222	.9236	.9251	.9265	.9279	.9292	.9306	.9319
1.5	.9332	.9345	.9357	.9370	.9382	.9394	.9406	.9418	.9429	.9441
1.6	.9452	.9463	.9474	.9484	.9495	.9505	.9515	.9525	.9535	.9545
1.7	.9554	.9564	.9573	.9582	.9591	.9599	.9608	.9616	.9625	.9633
1.8	.9641	.9649	.9656	.9664	.9671	.9678	.9686	.9693	.9699	.9706
1.9	.9713	.9719	.9726	.9732	.9738	.9744	.9750	.9756	.9761	.9767
2.0	.9772	.9778	.9783	.9788	.9793	.9798	.9803	.9808	.9812	.9817
2.1	.9821	.9826	.9830	.9834	.9838	.9842	.9846	.9850	.9854	.9857
2.2	.9861	.9864	.9868	.9871	.9875	.9878	.9881	.9884	.9887	.9890
2.3	.9893	.9896	.9898	.9901	.9904	.9906	.9909	.9911	.9913	.9916
2.4	.9918	.9920	.9922	.9925	.9927	.9929	.9931	.9932	.9934	.9936
2.5	.9938	.9940	.9941	.9943	.9945	.9946	.9948	.9949	.9951	.9952
2.6	.9953	.9955	.9956	.9957	.9959	.9960	.9961	.9962	.9963	.9964
2.7	.9965	.9966	.9967	.9968	.9969	.9970	.9971	.9972	.9973	.9974
2.8	.9974	.9975	.9976	.9977	.9977	.9978	.9979	.9979	.9980	.9981
2.9	.9981	.9982	.9982	.9983	.9984	.9984	.9985	.9985	.9986	.9986
3.0	.9987	.9987	.9987	.9988	.9988	.9989	.9989	.9989	.9990	.9990
3.1	.9990	.9991	.9991	.9991	.9992	.9992	.9992	.9992	.9993	.9993
3.2	.9993	.9993	.9994	.9994	.9994	.9994	.9994	.9995	.9995	.9995
3.3	.9995	.9995	.9995	.9996	.9996	.9996	.9996	.9996	.9996	.9997
3.4	.9997	.9997	.9997	.9997	.9997	.9997	.9997	.9997	.9997	.9998
3.5	.9998	.9998	.9998	.9998	.9998	.9998	.9998	.9998	.9998	.9998
3.6	.9998	.9998	.9999	.9999	.9999	.9999	.9999	.9999	.9999	.9999
3.7	.9999	.9999	.9999	.9999	.9999	.9999	.9999	.9999	.9999	.9999

[a]Calculated by MINITAB.

Table B5   Critical Values for the *t* Distribution[a]

				Probabilites between ± *t* values (two-sided)				
df	.50	.60	.70	.80	.90	.95	.98	.99
1	1.000	1.376	1.963	3.078	6.314	12.706	31.821	63.657
2	0.816	1.061	1.386	1.886	2.920	4.303	6.965	9.925
3	0.765	0.978	1.250	1.638	2.353	3.182	4.541	5.841
4	0.741	0.941	1.190	1.533	2.132	2.776	3.747	4.604
5	0.727	0.920	1.156	1.476	2.015	2.571	3.365	4.032
6	0.718	0.906	1.134	1.440	1.943	2.447	3.143	3.707
7	0.711	0.896	1.119	1.415	1.895	2.365	2.998	3.499
8	0.706	0.889	1.108	1.397	1.860	2.306	2.896	3.355
9	0.703	0.883	1.100	1.383	1.833	2.262	2.821	3.250
10	0.700	0.879	1.093	1.372	1.812	2.228	2.764	3.169
11	0.697	0.876	1.088	1.363	1.796	2.201	2.718	3.106
12	0.695	0.873	1.083	1.356	1.782	2.179	2.681	3.055
13	0.694	0.870	1.079	1.350	1.771	2.160	2.650	3.012
14	0.692	0.868	1.076	1.345	1.761	2.145	2.624	2.977
15	0.691	0.866	1.074	1.341	1.753	2.131	2.602	2.947
16	0.690	0.865	1.071	1.337	1.746	2.120	2.583	2.921
17	0.689	0.863	1.069	1.333	1.740	2.110	2.567	2.898
18	0.688	0.862	1.067	1.330	1.734	2.101	2.552	2.878
19	0.688	0.861	1.066	1.328	1.729	2.093	2.539	2.861
20	0.687	0.860	1.064	1.325	1.725	2.086	2.528	2.845
21	0.686	0.859	1.063	1.323	1.721	2.080	2.518	2.831
22	0.686	0.858	1.061	1.321	1.717	2.074	2.508	2.819
23	0.685	0.858	1.060	1.319	1.714	2.069	2.500	2.807
24	0.685	0.857	1.059	1.318	1.711	2.064	2.492	2.797
25	0.684	0.856	1.058	1.316	1.708	2.060	2.485	2.787
26	0.684	0.856	1.058	1.315	1.706	2.056	2.479	2.779
27	0.684	0.855	1.057	1.314	1.703	2.052	2.473	2.771
28	0.683	0.855	1.056	1.313	1.701	2.048	2.467	2.763
29	0.683	0.854	1.055	1.311	1.699	2.045	2.462	2.756
30	0.683	0.854	1.055	1.310	1.697	2.042	2.457	2.750
35	0.682	0.852	1.052	1.306	1.690	2.030	2.438	2.724
40	0.681	0.851	1.050	1.303	1.684	2.021	2.423	2.704
45	0.680	0.850	1.049	1.301	1.679	2.014	2.412	2.690
50	0.679	0.849	1.047	1.299	1.676	2.009	2.403	2.678
55	0.679	0.848	1.046	1.297	1.673	2.004	2.396	2.668
60	0.679	0.848	1.045	1.296	1.671	2.000	2.390	2.660
65	0.678	0.847	1.045	1.295	1.669	1.997	2.385	2.654
70	0.678	0.847	1.044	1.294	1.667	1.994	2.381	2.648
75	0.678	0.846	1.044	1.293	1.665	1.992	2.377	2.643
80	0.678	0.846	1.043	1.292	1.664	1.990	2.374	2.639
90	0.677	0.846	1.042	1.291	1.662	1.987	2.369	2.632
100	0.677	0.845	1.042	1.290	1.660	1.984	2.364	2.626
150	0.676	0.844	1.040	1.287	1.655	1.976	2.351	2.609
200	0.676	0.843	1.039	1.286	1.653	1.972	2.345	2.601
500	0.675	0.842	1.038	1.283	1.648	1.965	2.334	2.586
1000	0.675	0.842	1.037	1.282	1.646	1.962	2.330	2.581
∞	0.674	0.842	1.036	1.282	1.645	1.960	2.326	2.576
	.75	.80	.85	.90	.95	.975	.99	.995

Probability below *t* value (one-sided)

[a]Calculated by MINITAB.

Table B6 Graphs for Binomial Confidence Interval

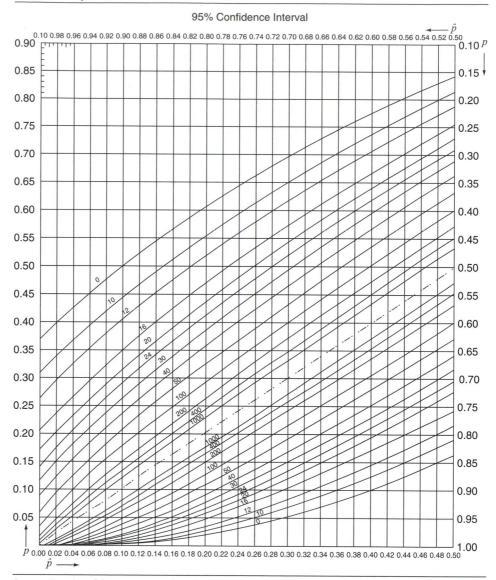

95% Confidence Interval

Table B6   *Continued*

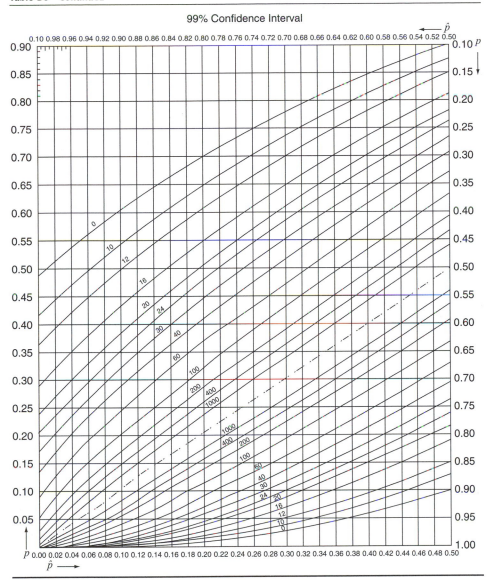

99% Confidence Interval

Table B7   Critical Values for the Chi-Square (χ^2) Distribution.[a]

df	.005	.01	.025	.05	.10	.90	.95	.975	.99	.995
						Probability below table value				
1	0.00	0.00	0.00	0.00	0.02	2.71	3.84	5.02	6.64	7.88
2	0.01	0.02	0.05	0.10	0.21	4.61	5.99	7.38	9.21	10.60
3	0.07	0.12	0.22	0.35	0.58	6.25	7.82	9.35	11.35	12.84
4	0.21	0.30	0.48	0.71	1.06	7.78	9.49	11.14	13.28	14.86
5	0.41	0.55	0.83	1.15	1.61	9.24	11.07	12.83	15.09	16.75
6	0.68	0.87	1.24	1.64	2.20	10.65	12.59	14.45	16.81	18.55
7	0.99	1.24	1.69	2.17	2.83	12.02	14.07	16.01	18.48	20.28
8	1.34	1.65	2.18	2.73	3.49	13.36	15.51	17.54	20.09	21.96
9	1.74	2.09	2.70	3.33	4.17	14.68	16.92	19.02	21.67	23.59
10	2.16	2.56	3.25	3.94	4.86	15.99	18.31	20.48	23.21	24.19
11	2.60	3.05	3.82	4.57	5.58	17.28	19.67	21.92	24.73	26.76
12	3.07	3.57	4.40	5.23	6.30	18.55	21.03	23.34	26.22	28.30
13	3.57	4.11	5.01	5.89	7.04	19.81	22.36	24.74	27.69	29.82
14	4.07	4.66	5.63	6.57	7.79	21.06	23.69	26.12	29.14	31.32
15	4.60	5.23	6.26	7.26	8.55	22.31	25.00	27.49	30.58	32.80
16	5.14	5.81	6.91	7.96	9.31	23.54	26.30	28.85	32.00	34.27
17	5.70	6.41	7.56	8.67	10.09	24.77	27.59	30.19	33.41	35.72
18	6.27	7.02	8.23	9.39	10.87	25.99	28.87	31.53	34.81	37.16
19	6.84	7.63	8.91	10.12	11.65	27.20	30.14	32.85	36.19	38.58
20	7.43	8.26	9.59	10.85	12.44	28.41	31.41	34.17	37.57	40.00
21	8.03	8.90	10.28	11.59	13.24	29.62	32.67	35.48	38.93	41.40
22	8.64	9.54	10.98	12.34	14.04	30.81	33.92	36.78	40.29	42.80
23	9.26	10.20	11.69	13.09	14.85	32.01	35.17	38.08	41.64	44.18
24	9.89	10.86	12.40	13.85	15.66	33.20	36.42	39.36	42.98	45.56
25	10.52	11.52	13.12	14.61	16.47	34.38	37.65	40.65	44.31	46.93
26	11.16	12.20	13.84	15.38	17.29	35.56	38.88	41.92	45.64	48.29
27	11.81	12.88	14.57	16.15	18.11	36.74	40.11	43.20	46.96	49.65
28	12.46	13.57	15.31	16.93	18.94	37.92	41.34	44.46	48.28	50.99
29	13.12	14.26	16.05	17.71	19.77	39.09	42.56	45.72	49.59	52.34
30	13.79	14.95	16.79	18.49	20.60	40.26	43.77	46.98	50.89	53.67
35	17.19	18.51	20.57	22.47	24.80	46.06	49.80	53.20	57.34	60.28
40	20.71	22.16	24.43	26.51	29.05	51.81	55.76	59.34	63.69	66.77
45	24.31	25.90	28.37	30.61	33.35	57.51	61.66	65.41	69.96	73.17
50	27.99	29.71	32.36	34.76	37.69	63.17	67.50	71.42	76.15	79.49
55	31.74	33.57	36.40	38.96	42.06	68.80	73.31	77.38	82.29	85.75
60	35.54	37.49	40.48	43.19	46.46	74.40	79.08	83.30	88.38	91.96
65	39.38	41.44	44.60	47.45	50.88	79.97	84.82	89.18	94.42	98.10
70	43.28	45.44	48.76	51.74	55.33	85.53	90.53	95.02	100.42	104.21
75	47.21	49.48	52.94	56.05	59.80	91.06	96.22	100.84	106.39	110.29
80	51.17	53.54	57.15	60.39	64.28	96.58	101.88	106.63	112.33	116.32
90	59.20	61.75	65.65	69.13	73.29	107.57	113.15	118.14	124.12	128.30
100	67.33	70.07	74.22	77.93	82.36	118.50	124.34	129.56	135.81	140.18
120	83.85	86.92	91.57	95.71	100.62	140.23	146.57	152.21	158.95	163.65
140	100.66	104.03	109.14	113.66	119.03	161.83	168.61	174.65	181.84	186.85
160	117.68	121.35	126.87	131.76	137.55	183.31	190.52	196.92	204.54	209.84
180	134.88	138.82	144.74	149.97	156.15	204.70	212.30	219.05	227.06	232.62
200	152.24	156.43	162.73	168.28	174.84	226.02	234.00	241.06	249.46	255.28

[a]Calculated by MINITAB.

Table B8   Factors, k, for Two-Sided Tolerance Limits for Normal Distributions.

n Π:	$1 - \alpha = 0.75$					$1 - \alpha = 0.90$				
	0.75	0.90	0.95	0.99	0.999	0.75	0.90	0.95	0.99	0.999
2	4.498	6.301	7.414	9.531	11.920	11.407	15.978	18.800	24.167	30.227
3	2.501	3.538	4.187	5.431	6.844	4.132	5.847	6.919	8.974	11.309
4	2.035	2.892	3.431	4.471	5.657	2.932	4.166	4.943	6.440	8.149
5	1.825	2.599	3.088	4.033	5.117	2.454	3.494	4.152	5.423	6.879
6	1.704	2.429	2.889	3.779	4.802	2.196	3.131	3.723	4.870	6.188
7	1.624	2.318	2.757	3.611	4.593	2.034	2.902	3.452	4.521	5.750
8	1.568	2.238	2.663	3.491	4.444	1.921	2.743	3.264	4.278	5.446
9	1.525	2.178	2.593	3.400	4.330	1.839	2.626	3.125	4.098	5.220
10	1.492	2.131	2.537	3.328	4.241	1.775	2.535	3.018	3.959	5.046
11	1.465	2.093	2.493	3.271	4.169	1.724	2.463	2.933	3.849	4.906
12	1.443	2.062	2.456	3.223	4.110	1.683	2.404	2.863	3.758	4.792
13	1.425	2.036	2.424	3.183	4.059	1.648	2.355	2.805	3.682	4.697
14	1.409	2.013	2.398	3.148	4.016	1.619	2.314	2.756	3.618	4.615
15	1.395	1.994	2.375	3.118	3.979	1.594	2.278	2.713	3.562	4.545
16	1.383	1.977	2.355	3.092	3.946	1.572	2.246	2.676	3.514	4.484
17	1.372	1.962	2.337	3.069	3.917	1.552	2.219	2.643	3.471	4.430
18	1.363	1.948	2.321	3.048	3.891	1.535	2.194	2.614	3.433	4.382
19	1.355	1.936	2.307	3.030	3.867	1.520	2.172	2.588	3.399	4.339
20	1.347	1.925	2.294	3.013	3.846	1.506	2.152	2.564	3.368	4.300
21	1.340	1.915	2.282	2.998	3.827	1.493	2.135	2.543	3.340	4.264
22	1.334	1.906	2.271	2.984	3.809	1.482	2.118	2.524	3.315	4.232
23	1.328	1.898	2.261	2.971	3.793	1.471	2.103	2.506	3.292	4.203
24	1.322	1.891	2.252	2.959	3.778	1.462	2.089	2.489	3.270	4.176
25	1.317	1.883	2.244	2.948	3.764	1.453	2.077	2.474	3.251	4.151
26	1.313	1.877	2.236	2.938	3.751	1.444	2.065	2.460	3.232	4.127
27	1.309	1.871	2.229	2.929	3.740	1.437	2.054	2.447	3.215	4.106
30	1.297	1.855	2.210	2.904	3.708	1.417	2.025	2.413	3.170	4.049
35	1.283	1.834	2.185	2.871	3.667	1.390	1.988	2.368	3.112	3.974
40	1.271	1.818	2.166	2.846	3.635	1.370	1.959	2.334	3.066	3.917
45	1.262	1.805	2.150	2.826	3.609	1.354	1.935	2.306	3.030	3.871
50	1.255	1.794	2.138	2.809	3.588	1.340	1.916	2.284	3.001	3.833
55	1.249	1.785	2.127	2.795	3.571	1.329	1.901	2.265	2.976	3.801
60	1.243	1.778	2.118	2.784	3.556	1.320	1.887	2.248	2.955	3.774
65	1.239	1.771	2.110	2.773	3.543	1.312	1.875	2.235	2.937	3.751
70	1.235	1.765	2.104	2.764	3.531	1.304	1.865	2.222	2.920	3.730
75	1.231	1.760	2.098	2.757	3.521	1.298	1.856	2.211	2.906	3.712
80	1.228	1.756	2.092	2.749	3.512	1.292	1.848	2.202	2.894	3.696
85	1.225	1.752	2.087	2.743	3.504	1.287	1.841	2.193	2.882	3.682
90	1.223	1.748	2.083	2.737	3.497	1.283	1.834	2.185	2.872	3.669
95	1.220	1.745	2.079	2.732	3.490	1.278	1.828	2.178	2.863	3.657
100	1.218	1.742	2.075	2.727	3.484	1.275	1.822	2.172	2.854	3.646
110	1.214	1.736	2.069	2.719	3.473	1.268	1.813	2.160	2.839	3.626
120	1.211	1.732	2.063	2.712	3.464	1.262	1.804	2.150	2.826	3.610
130	1.208	1.728	2.059	2.705	3.456	1.257	1.797	2.141	2.814	3.595
140	1.206	1.724	2.054	2.700	3.449	1.252	1.791	2.134	2.804	3.582
150	1.204	1.721	2.051	2.695	3.443	1.248	1.785	2.127	2.795	3.571
160	1.202	1.718	2.047	2.691	3.437	1.245	1.780	2.121	2.787	3.561
170	1.200	1.716	2.044	2.687	3.432	1.242	1.775	2.116	2.780	3.552
180	1.198	1.713	2.042	2.683	3.427	1.239	1.771	2.111	2.774	3.543
190	1.197	1.711	2.039	2.680	3.423	1.236	1.767	2.106	2.768	3.536
200	1.195	1.709	2.037	2.677	3.419	1.234	1.764	2.102	2.762	3.529
250	1.190	1.702	2.028	2.665	3.404	1.224	1.750	2.085	2.740	3.501
300	1.186	1.696	2.021	2.656	3.393	1.217	1.740	2.073	2.725	3.481
400	1.181	1.688	2.012	2.644	3.378	1.207	1.726	2.057	2.703	3.453
500	1.177	1.683	2.006	2.636	3.368	1.201	1.717	2.046	2.689	3.434
600	1.175	1.680	2.002	2.631	3.360	1.196	1.710	2.038	2.678	3.421
700	1.173	1.677	1.998	2.626	3.355	1.192	1.705	2.032	2.670	3.411
800	1.171	1.675	1.996	2.623	3.350	1.189	1.701	2.027	2.663	3.402
900	1.170	1.673	1.993	2.620	3.347	1.187	1.697	2.023	2.658	3.396
1000	1.169	1.671	1.992	2.617	3.344	1.185	1.695	2.019	2.654	3.390
∞	1.150	1.645	1.960	2.576	3.291	1.150	1.645	1.960	2.576	3.291

(continued)

Table B8 *Continued*

	1 − α = 0.95					1 − α = 0.99				
n Π:	0.75	0.90	0.95	0.99	0.999	0.75	0.90	0.95	0.99	0.999
2	22.858	32.019	37.674	48.430	60.573	114.363	160.193	188.491	242.300	303.054
3	5.922	8.380	9.916	12.861	16.208	13.378	18.930	22.401	29.055	36.616
4	3.779	5.369	6.370	8.299	10.502	6.614	9.398	11.150	14.527	18.383
5	3.002	4.275	5.079	6.634	8.415	4.643	6.612	7.855	10.260	13.015
6	2.604	3.712	4.414	5.775	7.337	3.743	5.337	6.345	8.301	10.548
7	2.361	3.369	4.007	5.248	6.676	3.233	4.613	5.488	7.187	9.142
8	2.197	3.136	3.732	4.891	6.226	2.905	4.147	4.936	6.468	8.234
9	2.078	2.967	3.532	4.631	5.899	2.677	3.822	4.550	5.966	7.600
10	1.987	2.839	3.379	4.433	5.649	2.508	3.582	4.265	5.594	7.129
11	1.916	2.737	3.259	4.277	5.452	2.378	3.397	4.045	5.308	6.766
12	1.858	2.655	3.162	4.150	5.291	2.274	3.250	3.870	5.079	6.477
13	1.810	2.587	3.081	4.044	5.158	2.190	3.130	3.727	4.893	6.240
14	1.770	2.529	3.012	3.955	5.045	2.120	3.029	3.608	4.737	6.043
15	1.735	2.480	2.954	3.878	4.949	2.060	2.954	3.507	4.605	5.876
16	1.705	2.437	2.903	3.812	4.865	2.009	2.872	3.421	4.492	5.732
17	1.679	2.400	2.858	3.754	4.791	1.965	2.808	3.345	4.393	5.607
18	1.655	2.366	2.819	3.702	4.725	1.926	2.753	3.279	4.307	5.497
19	1.635	2.337	2.784	3.656	4.667	1.891	2.703	3.221	4.230	5.399
20	1.616	2.310	2.752	3.615	4.614	1.860	2.659	3.168	4.161	5.312
21	1.599	2.286	2.723	3.577	4.567	1.833	2.620	3.121	4.100	5.234
22	1.584	2.264	2.697	3.543	4.523	1.808	2.584	3.078	4.044	5.163
23	1.570	2.244	2.673	3.512	4.484	1.785	2.551	3.040	3.993	5.098
24	1.557	2.225	2.651	3.483	4.447	1.764	2.522	3.004	3.947	5.039
25	1.545	2.208	2.631	3.457	4.413	1.745	2.494	2.972	3.904	4.985
26	1.534	2.193	2.612	3.432	4.382	1.727	2.469	2.941	3.865	4.935
27	1.523	2.178	2.595	3.409	4.353	1.711	2.446	2.914	3.828	4.888
30	1.497	2.140	2.549	3.350	4.278	1.668	2.385	2.841	3.733	4.768
35	1.462	2.090	2.490	3.272	4.179	1.613	2.306	2.748	3.611	4.611
40	1.435	2.052	2.445	3.213	4.104	1.571	2.247	2.677	3.518	4.493
45	1.414	2.021	2.408	3.165	4.042	1.539	2.200	2.621	3.444	4.399
50	1.396	1.996	2.379	3.126	3.993	1.512	2.162	2.576	3.385	4.323
55	1.382	1.976	2.354	3.094	3.951	1.490	2.130	2.538	3.335	4.260
60	1.369	1.958	2.333	3.066	3.916	1.471	2.103	2.506	3.293	4.206
65	1.359	1.943	2.315	3.042	3.886	1.455	2.080	2.478	3.257	4.160
70	1.349	1.929	2.299	3.021	3.859	1.440	2.060	2.454	3.225	4.120
75	1.341	1.917	2.285	3.002	3.835	1.428	2.042	2.433	3.197	4.084
80	1.334	1.907	2.272	2.986	3.814	1.417	2.026	2.414	3.173	4.053
85	1.327	1.897	2.261	2.971	3.795	1.407	2.012	2.397	3.150	4.024
90	1.321	1.889	2.251	2.958	3.778	1.398	1.999	2.382	3.130	3.999
95	1.315	1.881	2.241	2.945	3.763	1.390	1.987	2.368	3.112	3.976
100	1.311	1.874	2.233	2.934	3.748	1.383	1.977	2.355	3.096	3.954
110	1.302	1.861	2.218	2.915	3.723	1.369	1.958	2.333	3.066	3.917
120	1.294	1.850	2.205	2.898	3.702	1.358	1.942	2.314	3.041	3.885
130	1.288	1.841	2.194	2.883	3.683	1.349	1.928	2.298	3.019	3.857
140	1.282	1.833	2.184	2.870	3.666	1.340	1.916	2.283	3.000	3.833
150	1.277	1.825	2.175	2.859	3.652	1.332	1.905	2.270	2.983	3.811
160	1.272	1.819	2.167	2.848	3.638	1.326	1.896	2.259	2.968	3.792
170	1.268	1.813	2.160	2.839	3.527	1.320	1.887	2.248	2.955	3.774
180	1.264	1.808	2.154	2.831	3.616	1.314	1.879	2.239	2.942	3.759
190	1.261	1.803	2.148	2.823	3.606	1.309	1.872	2.230	2.931	3.744
200	1.258	1.798	2.143	2.816	3.597	1.304	1.865	2.222	2.921	3.731
250	1.245	1.780	2.121	2.788	3.561	1.286	1.389	2.191	2.880	3.678
300	1.236	1.767	2.106	2.767	3.535	1.273	1.820	2.169	2.850	3.641
400	1.223	1.749	2.084	2.739	3.499	1.255	1.794	2.138	2.809	3.589
500	1.215	1.737	2.070	2.721	3.475	1.243	1.777	2.117	2.783	3.555
600	1.209	1.729	2.060	2.707	3.458	1.234	1.764	2.102	2.763	3.530
700	1.204	1.722	2.052	2.697	3.445	1.227	1.755	2.091	2.748	3.511
800	1.201	1.717	2.046	2.688	3.434	1.222	1.747	2.082	2.736	3.495
900	1.198	1.712	2.040	2.682	3.426	1.218	1.741	2.075	2.726	3.483
1000	1.195	1.709	2.036	2.676	3.418	1.214	1.736	2.068	2.718	3.472
∞	1.150	1.645	1.960	2.576	3.291	1.150	1.645	1.960	2.576	3.291

Source: Abstracted from C. Eisenhart, M. W. Hastay, and W. A. Wallis, "Techniques of Statistical Analysis," Table 2.1, pp. 102–107. McGraw-Hill, New York, 1947

Table B9   Critical Values for Wilcoxon Signed Rank Test.[a]

n	Two-sided Comparisons			
	$\alpha \leq 0.10$	$\alpha \leq 0.05$	$\alpha \leq 0.02$	$\alpha \leq 0.01$
5	0, 15			
6	2, 19	0, 21		
7	3, 25	2, 26	0, 28	
8	5, 31	3, 33	1, 35	0, 36
9	8, 37	5, 40	3, 42	1, 44
10	10, 45	8, 47	5, 50	3, 52
11	13, 53	10, 56	7, 59	5, 61
12	17, 61	13, 65	9, 69	7, 71
13	21, 70	17, 74	12, 79	9, 82
14	25, 80	21, 84	15, 90	12, 93
15	30, 90	25, 95	19,101	15,105
16	35,101	29,107	23,113	19,117
17	41,112	34,119	28,125	23,130
18	47,124	40,131	32,139	27,144
19	53,137	46,144	37,153	33,158
20	60,150	52,158	43,167	37,173
21	67,164	58,173	49,182	42,189
22	75,178	66,187	55,198	48,205
23	83,193	73,203	62,214	54,222
24	91,209	81,210	69,231	61,239
25	100,225	89,236	76,249	68,257
26	110,241	98,253	84,267	75,276
27	119,259	107,271	93,285	83,295
28	130,276	114,278	101,305	91,315
29	140,295	126,309	122,313	100,335
30	151,314	137,328	132,333	109,356

n	One-sided Comparisons			
	$\alpha \leq 0.05$	$\alpha \leq 0.025$	$\alpha \leq 0.01$	$\alpha \leq 0.005$

[a]Extracted from "Critical Values and Probability Levels for the Wilcoxon Rank Sum Test and the Wilcoxon Signed Rank Test," by Frank Wilcoxon, S. K. Katti, and Roberta A. Wilcox, *Selected Tables in Mathematical Statistics*, Vol. 1, 1973, Table II, pp. 237–259, by permission of the American Mathematics Society.

Table B10   Critical Values for Wilcoxon Rank Sum Test.[a]

Two-sided: α ≤ 0.10   One-sided: α ≤ 0.05

N₂\N₁	3	4	5	6	7	8	9	10	11	12	13	14	15
3	—	10, 22	16, 29	23, 37	30, 47	39, 57	48, 69	59, 81	71, 94	83,109	97,124	112,140	127,158
4	6, 18	11, 25	17, 33	24, 42	32, 52	41, 63	51, 75	62, 88	74,102	87,117	101,133	116,150	132,168
5	7, 20	12, 28	19, 36	26, 46	36, 57	44, 68	54, 81	66, 94	78,109	91,125	106,141	121,159	138,177
6	8, 22	13, 31	20, 40	28, 50	36, 62	46, 74	57, 87	69,101	82,116	95,133	110,150	126,168	143,187
7	8, 25	14, 34	21, 44	29, 55	39, 66	49, 79	60, 93	72,108	85,124	99,141	115,158	131,177	148,197
8	9, 27	15, 37	23, 47	31, 59	41, 71	51, 85	63, 99	75,115	89,131	104,148	119,167	136,186	153,207
9	9, 30	16, 40	24, 51	33, 63	43, 76	54, 90	66,105	79,121	93,138	108,156	124,178	141,195	159,216
10	10, 32	17, 43	26, 54	35, 67	45, 81	56, 96	69,111	82,128	97,145	112,164	128,184	146,204	164,226
11	11, 34	18, 46	27, 58	37, 71	47, 86	59,101	72,117	86,134	100,153	116,172	133,192	151,231	170,235
12	11, 37	19, 49	28, 62	38, 76	49, 91	62,106	75,123	89,141	104,160	120,180	138,200	156,222	175,245
13	12, 39	20, 52	30, 65	40, 80	52, 95	64,112	78,129	92,148	108,167	125,187	142,209	161,231	181,254
14	13, 41	21, 55	31, 69	42, 84	54,100	67,117	81,135	96,154	112,174	129,195	147,217	166,240	186,264
15	13, 44	22, 58	33, 72	44, 88	56,105	69,123	84,141	99,161	116,181	133,203	152,225	171,249	192,273
16	14, 46	24, 60	34, 76	46, 92	58,110	72,128	87,147	103,167	120,188	138,210	156,234	176,258	197,283
17	15, 48	25, 63	35, 80	47, 97	61,114	75,133	90,153	106,174	123,196	142,218	161,242	182,266	203,292
18	15, 51	26, 66	37, 83	49,101	63,119	77,139	93,159	110,180	127,203	146,226	166,250	187,275	208,302
19	16, 53	27, 69	38, 87	51,105	65,124	80,144	96,165	113,187	131,210	150,234	171,258	192,284	214,311
20	17, 55	28, 72	40, 90	53,109	67,129	83,149	99,171	117,193	135,217	155,241	175,267	197,293	220,320
21	17, 58	29, 75	41, 94	55,113	69,134	85,155	102,177	120,200	139,224	159,249	180,275	202,302	225,330
22	18, 60	30, 78	43, 97	57,117	72,138	88,160	105,183	123,207	143,231	163,257	185,283	207,311	231,339
23	19, 62	31, 81	44,101	58,122	74,143	90,166	108,189	127,213	147,238	168,264	189,292	212,320	236,349
24	19, 65	32, 84	45,105	60,126	76,148	93,171	111,195	130,220	151,245	172,272	194,300	218,328	242,358
25	20, 67	33, 87	47,108	62,130	78,153	96,176	114,201	134,226	155,252	176,280	199,308	223,337	248,367
26	21, 69	34, 90	48,112	64,134	81,157	98,182	117,207	137,233	158,260	181,287	204,316	228,346	253,377
27	21, 72	35, 93	50,115	66,138	83,162	101,187	120,213	141,239	162,267	185,295	208,325	233,355	259,386
28	22, 74	36, 96	51,119	67,143	85,167	104,192	123,219	144,246	166,274	189,303	213,333	238,364	264,396
29	23, 76	37, 99	53,122	69,147	87,172	106,198	127,224	148,252	170,281	194,310	218,341	243,373	270,405
30	23, 79	38,102	54,126	71,151	89,177	109,203	130,230	151,259	174,288	198,318	223,349	249,381	276,414

(continued)

Two-sided: α ≤ 0.05   One-sided: α ≤ 0.025

3	—	—	15, 30	22, 38	29, 48	38, 58	47, 70	58, 82	70, 96	82,110	95,126	110,142	125,160
4	—	**10, 26**	16, 34	23, 43	31, 53	40, 64	49, 77	60, 90	72,104	85,119	99,135	114,152	130,170
5	6, 21	11, 29	**17, 38**	24, 48	33, 58	42, 70	52, 83	63, 97	75,112	89,127	103,144	118,162	134,181
6	7, 23	12, 32	18, 42	**26, 52**	34, 64	44, 76	55, 89	66,104	79,119	92,136	107,153	122,172	139,191
7	7, 26	13, 35	20, 45	27, 57	**36, 69**	46, 82	57, 96	69,111	82,127	96,144	111,162	127,181	144,201
8	8, 28	14, 38	21, 49	29, 61	38, 74	**49, 87**	60,102	72,118	85,135	100,152	115,171	131,191	149,211
9	8, 31	14, 42	22, 53	31, 65	40, 79	51, 93	**62,109**	75,125	89,142	104,160	119,180	136,200	154,221
10	9, 33	15, 45	23, 57	32, 70	42, 84	53, 99	65,115	**78,132**	92,150	107,169	124,188	141,209	159,231
11	9, 36	16, 48	24, 61	34, 74	44, 89	55,105	68,121	81,139	**96,157**	111,177	128,197	145,219	164,241
12	10, 38	17, 51	26, 64	35, 79	46, 94	58,110	71,127	84,146	99,165	**115,185**	132,206	150,228	169,251
13	10, 41	18, 54	27, 68	37, 83	48, 99	60,116	73,133	88,152	103,172	119,193	136,215	155,237	174,261
14	11, 43	19, 57	28, 72	38, 88	50,104	62,122	76,140	91,159	106,180	123,201	141,223	160,246	179,271
15	11, 46	20, 60	29, 76	40, 92	52,109	65,127	79,146	94,166	110,187	127,209	145,232	**160,246**	**184,281**
16	12, 48	21, 63	30, 80	42, 96	54,114	67,133	82,152	97,173	113,195	131,217	150,240	169,265	190,290
17	12, 51	21, 67	32, 83	43,101	56,119	70,138	84,159	100,180	117,202	135,225	154,249	174,274	195,300
18	13, 53	22, 70	33, 87	45,105	58,124	72,144	87,165	103,187	121,209	139,233	158,258	179,283	200,310
19	13, 56	23, 73	34, 91	46,110	60,129	74,150	90,171	107,193	124,217	143,241	163,266	183,293	205,320
20	14, 58	24, 76	35, 95	48,114	62,134	77,155	93,177	110,200	128,224	147,249	167,275	188,302	210,330
21	14, 61	25, 79	37, 98	50,118	64,139	79,161	95,184	113,207	131,232	151,257	171,284	193,311	216,339
22	15, 63	26, 82	38,102	51,123	66,144	81,167	98,190	116,214	135,239	155,265	176,292	198,320	221,349
23	15, 66	27, 85	39,106	53,127	68,149	84,172	101,196	119,221	139,246	159,273	180,301	203,329	226,359
24	16, 68	27, 89	40,110	54,132	70,154	86,178	104,202	122,228	142,254	163,281	185,309	207,339	231,369
25	16, 71	28, 92	42,113	56,136	72,159	89,183	107,208	126,234	146,261	167,289	189,318	212,348	237,378
26	17, 73	29, 95	43,117	58,140	74,164	91,189	109,215	129,241	149,269	171,297	193,327	217,357	242,388
27	17, 76	30, 98	44,121	59,145	76,169	93,195	112,221	132,248	153,276	175,305	198,335	222,366	247,398
28	18, 78	31,101	45,125	61,149	78,174	96,200	115,227	135,255	156,284	179,313	202,344	227,375	252,408
29	19, 80	32,104	47,128	63,153	80,179	98,206	118,233	138,262	160,291	183,321	207,352	232,384	258,417
30	19, 83	33,107	48,132	64,158	82,184	101,211	121,239	142,268	164,298	187,329	211,361	236,394	263,427

(continued)

Table B10 *Continued*

$N_2 \backslash N_1$	3	4	5	6	7	8	9	10	11	12	13	14	15
						Two-sided: $\alpha \leq 0.02$		One-sided: $\alpha \leq 0.01$					
3	—	—	—	—	28, 49	36, 60	46, 71	56, 84	67, 98	80,112	93,128	107,145	123,162
4	—	—	15, 35	22, 44	29, 55	38, 66	48, 78	58, 92	70,106	83,121	96,138	111,155	127,173
5	—	10, 30	**16, 39**	23, 49	31, 60	40, 72	50, 85	61, 99	73,114	86,130	100,147	115,165	131,184
6	—	11, 33	17, 43	**24, 54**	32, 66	42, 78	52, 92	63,107	75,123	89,139	103,157	118,176	135,195
7	6, 27	11, 37	18, 47	25, 59	**34, 71**	43, 85	54, 99	66,114	78,131	92,148	107,166	122,186	139,206
8	6, 30	12, 40	19, 51	27, 63	35, 77	**45, 91**	56,106	68,122	81,139	95,157	111,175	127,195	144,216
9	7, 32	13, 43	20, 55	28, 68	37, 82	47, 97	**59,112**	71,129	84,147	99,165	114,185	131,205	148,227
10	7, 35	13, 47	21, 59	29, 73	39, 87	49,103	61,119	**74,136**	88,154	102,174	118,194	135,215	153,237
11	7, 38	14, 50	22, 63	30, 78	40, 93	51,109	63,126	77,143	**91,162**	106,182	122,203	139,225	157,248
12	8, 40	15, 53	23, 67	32, 82	42, 98	53,115	66,132	79,151	94,170	**109,191**	126,212	143,235	162,258
13	8, 43	15, 57	24, 71	33, 87	44,103	56,120	68,139	82,158	97,178	113,199	**130,221**	148,244	167,268
14	8, 46	16, 60	25, 75	34, 92	45,109	58,126	71,145	85,165	100,186	116,208	134,230	**152,254**	171,279
15	9, 48	17, 63	26, 79	36, 96	47,114	60,132	73,152	88,172	103,194	120,216	138,239	156,264	**176,289**
16	9, 51	17, 67	27, 83	37,101	49,119	62,138	76,158	91,179	107,201	124,224	142,248	161,273	181,299
17	10, 53	18, 70	28, 87	39,105	51,124	64,144	78,165	93,187	110,209	127,233	146,257	165,283	186,309
18	10, 56	19, 73	29, 91	40,110	52,130	66,150	81,171	96,194	113,217	131,241	150,266	170,292	190,320
19	10, 59	19, 77	30, 95	41,115	54,135	68,156	83,178	99,201	116,225	134,250	154,275	174,302	195,330
20	11, 61	20, 80	31, 99	43,119	56,140	70,162	85,185	102,208	119,233	138,258	158,284	178,312	200,340
21	11, 64	21, 83	32,103	44,124	58,145	72,168	88,191	105,215	123,240	142,266	162,293	183,321	205,350
22	11, 67	21, 87	33,107	45,129	59,151	74,174	90,198	108,222	126,248	145,275	166,302	187,331	210,360
23	12, 69	22, 90	34,111	47,133	61,156	76,180	93,204	110,230	129,256	149,283	170,311	192,340	214,371
24	12, 72	23, 93	35,115	48,138	63,161	78,186	95,211	113,237	132,264	153,291	174,320	196,350	219,381
25	13, 74	23, 97	36,119	50,142	64,167	81,191	98,217	116,244	136,271	156,300	178,329	200,360	224,391
26	13, 77	24,100	37,123	51,147	66,172	83,197	100,224	119,251	139,279	160,308	182,338	205,369	229,401
27	13, 80	25,103	38,127	52,152	68,177	85,203	103,230	122,258	142,287	163,317	186,347	209,379	234,411
28	14, 82	26,106	39,131	54,156	70,182	87,209	105,237	125,265	145,295	167,325	190,356	214,388	239,421
29	14, 85	26,110	40,135	55,161	71,188	89,215	108,243	128,272	149,302	171,333	194,365	218,398	243,432
30	15, 87	27,113	41,139	56,166	73,193	91,221	110,250	131,279	152,310	174,342	198,374	223,407	248,442

(continued)

Two-sided: $\alpha \le 0.01$ One-sided: $\alpha \le 0.005$

3	—	—	—	—	—	—	45, 72	55, 85	66, 99	79, 113	92, 129	106, 146	122, 163
4	—	—	—	21, 45	28, 56	37, 67	46, 80	57, 93	68, 108	81, 123	94, 140	109, 157	125, 175
5	—	—	**15, 40**	22, 50	29, 62	38, 74	48, 87	59, 101	71, 116	84, 132	98, 149	112, 168	128, 187
6	—	10, 34	16, 44	**23, 55**	31, 67	40, 80	50, 94	61, 109	73, 125	87, 141	101, 159	116, 178	132, 198
7	—	10, 38	16, 49	24, 60	**32, 73**	42, 86	52, 101	64, 116	76, 133	90, 150	104, 169	120, 188	136, 209
8	—	11, 41	17, 53	25, 65	34, 78	**43, 93**	54, 108	66, 124	79, 141	93, 159	108, 178	123, 199	140, 220
9	6, 33	11, 45	18, 57	26, 70	35, 84	45, 99	**56, 115**	68, 132	82, 149	96, 168	111, 188	127, 209	144, 231
10	6, 36	12, 48	19, 61	27, 75	37, 89	47, 105	58, 122	**71, 139**	84, 158	99, 177	115, 197	131, 219	149, 241
11	6, 39	12, 52	20, 65	28, 80	38, 95	49, 111	61, 128	73, 147	**87, 166**	102, 186	118, 207	135, 229	153, 252
12	7, 41	13, 55	21, 69	30, 84	40, 100	51, 117	63, 135	76, 154	90, 174	**105, 195**	122, 216	139, 239	157, 263
13	7, 44	13, 59	22, 73	31, 89	41, 106	53, 123	65, 142	79, 161	93, 182	109, 203	**125, 226**	143, 249	162, 273
14	7, 47	14, 62	22, 78	32, 94	43, 111	54, 130	67, 149	81, 169	96, 190	112, 212	129, 235	**147, 259**	166, 284
15	8, 49	15, 65	23, 82	33, 99	44, 117	56, 136	69, 156	84, 176	99, 198	115, 221	133, 244	151, 269	**171, 294**
16	8, 52	15, 69	24, 86	34, 104	46, 122	58, 142	72, 162	86, 184	102, 206	119, 229	136, 254	155, 279	175, 305
17	8, 55	16, 72	25, 90	36, 108	47, 128	60, 148	74, 169	89, 191	105, 214	122, 238	140, 263	159, 289	180, 315
18	8, 58	16, 76	26, 94	37, 113	49, 133	62, 154	76, 176	92, 198	108, 222	125, 247	144, 272	163, 299	184, 326
19	9, 60	17, 79	27, 98	38, 118	50, 139	64, 160	78, 183	94, 206	111, 230	129, 255	148, 281	168, 308	189, 336
20	9, 63	18, 82	28, 102	39, 123	52, 144	66, 166	81, 189	97, 213	114, 238	132, 264	151, 291	172, 318	193, 347
21	9, 66	18, 86	29, 106	40, 128	53, 150	68, 172	83, 196	99, 221	117, 246	136, 272	155, 300	176, 328	198, 357
22	10, 68	19, 89	29, 111	42, 132	55, 155	70, 178	85, 203	102, 228	120, 254	139, 281	159, 309	180, 338	202, 368
23	10, 71	19, 93	30, 115	43, 137	57, 160	71, 185	88, 209	105, 235	123, 262	142, 290	163, 318	184, 348	207, 378
24	10, 74	20, 96	30, 119	44, 142	58, 166	73, 191	90, 216	107, 243	126, 270	146, 298	166, 328	188, 358	211, 389
25	11, 76	20, 100	32, 123	45, 147	60, 171	75, 197	92, 223	110, 250	129, 278	149, 307	170, 337	192, 368	216, 399
26	11, 79	21, 103	33, 127	46, 152	61, 177	77, 203	94, 230	113, 257	132, 286	152, 316	174, 346	197, 377	220, 410
27	11, 82	22, 106	34, 131	48, 156	63, 182	79, 209	97, 236	115, 265	135, 294	156, 324	178, 355	201, 387	225, 420
28	11, 85	22, 110	35, 135	49, 161	64, 188	81, 215	99, 243	118, 272	138, 302	159, 333	182, 364	205, 397	229, 431
29	12, 87	23, 113	36, 139	50, 166	66, 193	83, 221	101, 250	121, 279	141, 310	163, 341	185, 375	209, 407	234, 441
30	12, 90	23, 117	37, 143	51, 171	68, 198	85, 227	103, 257	123, 287	144, 318	166, 350	189, 383	213, 417	239, 451

aExtracted from "Critical Values and Probability Levels for the Wilcoxon Rank Sum Test and Wilcoxon Signed Rank Test," by Frank Wilcoxon, S. K. Katti, and Roberta A. Wilcox, *Selected Tables in Mathematical Statistics*, Vol. 1, 1973, Table I, pp. 177–235, by permission of the American Mathematics Society.

Table B11 Critical Values for the F Distribution[a]

df in the denominator	$F_{.90}$ df in the numerator										
	1	2	3	4	5	6	8	10	20	50	100
1	39.86	49.50	53.59	55.83	57.24	58.20	59.44	60.19	61.74	62.69	63.01
2	8.53	9.00	9.16	9.24	9.29	9.33	9.37	9.39	9.44	9.47	9.48
3	5.54	5.46	5.39	5.34	5.31	5.28	5.25	5.23	5.18	5.15	5.14
4	4.54	4.32	4.19	4.11	4.05	4.01	3.95	3.92	3.84	3.80	3.78
5	4.06	3.78	3.62	3.52	3.45	3.40	3.34	3.30	3.21	3.15	3.13
6	3.78	3.46	3.29	3.18	3.11	3.05	2.98	2.94	2.84	2.77	2.75
7	3.59	3.26	3.07	2.96	2.88	2.83	2.75	2.70	2.59	2.52	2.50
8	3.46	3.11	2.92	2.81	2.73	2.67	2.59	2.54	2.42	2.35	2.32
9	3.36	3.01	2.81	2.69	2.61	2.55	2.47	2.42	2.30	2.22	2.19
10	3.29	2.92	2.73	2.61	2.52	2.46	2.38	2.32	2.20	2.12	2.09
11	3.23	2.86	2.66	2.54	2.45	2.39	2.30	2.25	2.12	2.04	2.01
12	3.18	2.81	2.61	2.48	2.39	2.33	2.24	2.19	2.06	1.97	1.94
13	3.14	2.76	2.56	2.43	2.35	2.28	2.20	2.14	2.01	1.92	1.88
14	3.10	2.73	2.52	2.39	2.31	2.24	2.15	2.10	1.96	1.87	1.83
15	3.07	2.70	2.49	2.36	2.27	2.21	2.12	2.06	1.92	1.83	1.79
16	3.05	2.67	2.46	2.33	2.24	2.18	2.09	2.03	1.89	1.79	1.76
17	3.03	2.64	2.44	2.31	2.22	2.15	2.06	2.00	1.86	1.76	1.73
18	3.01	2.62	2.42	2.29	2.20	2.13	2.04	1.98	1.84	1.74	1.70
19	2.99	2.61	2.40	2.27	2.18	2.11	2.02	1.96	1.81	1.71	1.67
20	2.97	2.59	2.38	2.25	2.16	2.09	2.00	1.94	1.79	1.69	1.65
21	2.96	2.57	2.36	2.23	2.14	2.08	1.98	1.92	1.78	1.67	1.63
22	2.95	2.56	2.35	2.22	2.13	2.06	1.97	1.90	1.76	1.65	1.61
23	2.94	2.55	2.34	2.21	2.11	2.05	1.95	1.89	1.74	1.64	1.59
24	2.93	2.54	2.33	2.19	2.10	2.04	1.94	1.88	1.73	1.62	1.58
25	2.92	2.53	2.32	2.18	2.09	2.02	1.93	1.87	1.72	1.61	1.56
26	2.91	2.52	2.31	2.17	2.08	2.01	1.92	1.86	1.71	1.59	1.55
27	2.90	2.51	2.30	2.17	2.07	2.00	1.91	1.85	1.70	1.58	1.54
28	2.89	2.50	2.29	2.16	2.06	2.00	1.90	1.84	1.69	1.57	1.53
29	2.89	2.50	2.28	2.15	2.06	1.99	1.89	1.83	1.68	1.56	1.52
30	2.88	2.49	2.28	2.14	2.05	1.98	1.88	1.82	1.67	1.55	1.51
40	2.84	2.44	2.23	2.09	2.00	1.93	1.83	1.76	1.61	1.48	1.43
50	2.81	2.41	2.20	2.06	1.97	1.90	1.80	1.73	1.57	1.44	1.39
60	2.79	2.39	2.18	2.04	1.95	1.87	1.77	1.71	1.54	1.41	1.36
100	2.76	2.36	2.14	2.00	1.91	1.83	1.73	1.66	1.49	1.35	1.29
200	2.73	2.33	2.11	1.97	1.88	1.80	1.70	1.63	1.46	1.31	1.24
1000	2.71	2.31	2.09	1.95	1.85	1.78	1.68	1.61	1.43	1.27	1.20

(continued)

Table B11   *Continued*

$$F_{.95}$$

df in the denominator	df in the numerator										
	1	**2**	**3**	**4**	**5**	**6**	**8**	**10**	**20**	**50**	**100**
1	161.5	199.5	215.7	224.6	230.2	234.0	238.9	241.9	248.0	251.8	253.0
2	18.51	19.00	19.16	19.25	19.30	19.33	19.37	19.40	19.45	19.48	19.49
3	10.13	9.55	9.28	9.12	9.01	8.94	8.85	8.79	8.66	8.58	8.55
4	7.71	6.94	6.59	6.39	6.26	6.16	6.04	5.96	5.80	5.70	5.66
5	6.61	5.79	5.41	5.19	5.05	4.95	4.82	4.74	4.56	4.44	4.41
6	5.99	5.14	4.76	4.53	4.39	4.28	4.15	4.06	3.87	3.75	3.71
7	5.59	4.74	4.35	4.12	3.97	3.87	3.73	3.64	3.44	3.32	3.27
8	5.32	4.46	4.07	3.84	3.69	3.58	3.44	3.35	3.15	3.02	2.97
9	5.12	4.26	3.86	3.63	3.48	3.37	3.23	3.14	2.94	2.80	2.76
10	4.96	4.10	3.71	3.48	3.33	3.22	3.07	2.98	2.77	2.64	2.59
11	4.84	3.98	3.59	3.36	3.20	3.09	2.95	2.85	2.65	2.51	2.46
12	4.75	3.89	3.49	3.26	3.11	3.00	2.85	2.75	2.54	2.40	2.35
13	4.67	3.81	3.41	3.18	3.03	2.92	2.77	2.67	2.46	2.31	2.26
14	4.60	3.74	3.34	3.11	2.96	2.85	2.70	2.60	2.39	2.24	2.19
15	4.54	3.68	3.29	3.06	2.90	2.79	2.64	2.54	2.33	2.18	2.12
16	4.49	3.63	3.24	3.01	2.85	2.74	2.59	2.49	2.28	2.12	2.07
17	4.45	3.59	3.20	2.96	2.81	2.70	2.55	2.45	2.23	2.08	2.02
18	4.41	3.55	3.16	2.93	2.77	2.66	2.51	2.41	2.19	2.04	1.98
19	4.38	3.52	3.13	2.90	2.74	2.63	2.48	2.38	2.16	2.00	1.94
20	4.35	3.49	3.10	2.87	2.71	2.60	2.45	2.35	2.12	1.97	1.91
21	4.32	3.47	3.07	2.84	2.68	2.57	2.42	2.32	2.10	1.94	1.88
22	4.30	3.44	3.05	2.82	2.66	2.55	2.40	2.30	2.07	1.91	1.85
23	4.28	3.42	3.03	2.80	2.64	2.53	2.37	2.27	2.05	1.88	1.82
24	4.26	3.40	3.01	2.78	2.62	2.51	2.36	2.25	2.03	1.86	1.80
25	4.24	3.39	2.99	2.76	2.60	2.49	2.34	2.24	2.01	1.84	1.78
26	4.23	3.37	2.98	2.74	2.59	2.47	2.32	2.22	1.99	1.82	1.76
27	4.21	3.35	2.96	2.73	2.57	2.46	2.31	2.20	1.97	1.81	1.74
28	4.20	3.34	2.95	2.71	2.56	2.45	2.29	2.19	1.96	1.79	1.73
29	4.18	3.33	2.93	2.70	2.55	2.43	2.28	2.18	1.94	1.77	1.71
30	4.17	3.32	2.92	2.69	2.53	2.42	2.27	2.16	1.93	1.76	1.70
40	4.08	3.23	2.84	2.61	2.45	2.34	2.18	2.08	1.84	1.66	1.59
50	4.03	3.18	2.79	2.56	2.40	2.29	2.13	2.03	1.78	1.60	1.52
60	4.00	3.15	2.76	2.53	2.37	2.25	2.10	1.99	1.75	1.56	1.48
100	3.94	3.09	2.70	2.46	2.31	2.19	2.03	1.93	1.68	1.48	1.39
200	3.89	3.04	2.65	2.42	2.26	2.14	1.98	1.88	1.62	1.41	1.32
1000	3.85	3.00	2.61	2.38	2.22	2.11	1.95	1.84	1.58	1.36	1.26

(continued)

Table B11 *Continued*

						$F_{.99}$					
df in the						**df in the numerator**					
denominator	**1**	**2**	**3**	**4**	**5**	**6**	**8**	**10**	**20**	**50**	**100**
1	4052	4500	5403	5625	5764	5859	5981	6056	6209	6302	6334
2	98.50	99.00	99.17	99.25	99.30	99.33	99.37	99.40	99.45	99.48	99.49
3	34.12	30.82	29.46	28.71	28.24	27.91	27.49	27.23	26.69	26.35	26.24
4	21.20	18.00	16.69	15.98	15.52	15.21	14.80	14.55	14.02	13.69	13.58
5	16.26	13.27	12.06	11.39	10.97	10.67	10.29	10.05	9.55	9.24	9.13
6	13.75	10.92	9.78	9.15	8.75	8.47	8.10	7.87	7.40	7.09	6.99
7	12.25	9.55	8.45	7.85	7.46	7.19	6.84	6.62	6.16	5.86	5.75
8	11.26	8.65	7.59	7.01	6.63	6.37	6.03	5.81	5.36	5.07	4.96
9	10.56	8.02	6.99	6.42	6.06	5.80	5.47	5.26	4.81	4.52	4.41
10	10.04	7.56	6.55	5.99	5.64	5.39	5.06	4.85	4.41	4.12	4.01
11	9.65	7.21	6.22	5.67	5.32	5.07	4.74	4.54	4.10	3.81	3.71
12	9.33	6.93	5.95	5.41	5.06	4.82	4.50	4.30	3.86	3.57	3.47
13	9.07	6.70	5.74	5.21	4.86	4.62	4.30	4.10	3.66	3.38	3.27
14	8.86	6.51	5.56	5.04	4.69	4.46	4.14	3.94	3.51	3.22	3.11
15	8.68	6.36	5.42	4.89	4.56	4.32	4.00	3.80	3.37	3.08	2.98
16	8.53	6.23	5.29	4.77	4.44	4.20	3.89	3.69	3.26	2.97	2.86
17	8.40	6.11	5.19	4.67	4.34	4.10	3.79	3.59	3.16	2.87	2.76
18	8.29	6.01	5.09	4.58	4.25	4.01	3.71	3.51	3.08	2.78	2.68
19	8.19	5.93	5.01	4.50	4.17	3.94	3.63	3.43	3.00	2.71	2.60
20	8.10	5.85	4.94	4.43	4.10	3.87	3.56	3.37	2.94	2.64	2.54
21	8.02	5.78	4.87	4.37	4.04	3.81	3.51	3.31	2.88	2.58	2.48
22	7.95	5.72	4.82	4.31	3.99	3.76	3.45	3.26	2.83	2.53	2.42
23	7.88	5.66	4.76	4.26	3.94	3.71	3.41	3.21	2.78	2.48	2.37
24	7.82	5.61	4.72	4.22	3.90	3.67	3.36	3.17	2.74	2.44	2.33
25	7.77	5.57	4.68	4.18	3.85	3.63	3.32	3.13	2.70	2.40	2.29
26	7.72	5.53	4.64	4.14	3.82	3.59	3.29	3.09	2.66	2.36	2.25
27	7.68	5.49	4.60	4.11	3.78	3.56	3.26	3.06	2.63	2.33	2.22
28	7.64	5.45	4.57	4.07	3.75	3.53	3.23	3.03	2.60	2.30	2.19
29	7.60	5.42	4.54	4.04	3.73	3.50	3.20	3.00	2.57	2.27	2.16
30	7.56	5.39	4.51	4.02	3.70	3.47	3.17	2.98	2.55	2.25	2.13
40	7.31	5.18	4.31	3.83	3.51	3.29	2.99	2.80	2.37	2.06	1.94
50	7.17	5.06	4.20	3.72	3.41	3.19	2.89	2.70	2.27	1.95	1.82
60	7.08	4.98	4.13	3.65	3.34	3.12	2.82	2.63	2.20	1.88	1.75
100	6.90	4.82	3.98	3.51	3.21	2.99	2.69	2.50	2.07	1.74	1.60
200	6.76	4.71	3.88	3.41	3.11	2.89	2.60	2.41	1.97	1.63	1.48
1000	6.66	4.63	3.80	3.34	3.04	2.82	2.53	2.34	1.90	1.54	1.38

[a]Calculated by MINITAB.

Table B12 Upper Percentage Points of the Studentized Range. $q_a = \dfrac{\bar{x}_{max} - \bar{x}_{min}}{S_{\bar{x}}}$

Error df	α	p = number of treatment means																				Error df
		2	3	4	5	6	7	8	9	10	11	12	13	14	15	16	17	18	19	20	
5	.05	3.64	4.60	5.22	5.67	6.03	6.33	6.58	6.80	6.99	7.17	7.32	7.47	7.60	7.72	7.83	7.93	8.03	8.12	8.21	5
	.01	5.70	6.97	7.80	8.42	8.91	9.32	9.67	9.97	10.24	10.48	10.70	10.89	11.08	11.24	11.40	11.55	11.68	11.81	11.93	
6	.05	3.46	4.34	4.90	5.31	5.63	5.89	6.12	6.32	6.49	6.65	6.79	6.92	7.03	7.14	7.24	7.34	7.43	7.51	7.59	6
	.01	5.24	6.33	7.03	7.56	7.97	8.32	8.61	8.87	9.10	9.30	9.49	9.65	9.81	9.95	10.08	10.21	10.32	10.43	10.54	
7	.05	3.34	4.16	4.68	5.06	5.36	5.61	5.82	6.00	6.16	6.30	6.43	6.55	6.66	6.76	6.85	6.94	7.02	7.09	7.17	7
	.01	4.95	5.92	6.54	7.01	7.37	7.68	7.94	8.17	8.37	8.55	8.71	8.86	9.00	9.12	9.24	9.35	9.46	9.55	9.65	
8	.05	3.26	4.04	4.53	4.89	5.17	5.40	5.60	5.77	5.92	6.05	6.18	6.29	6.39	6.48	6.57	6.65	6.73	6.80	6.87	8
	.01	4.74	5.63	6.20	6.63	6.96	7.24	7.47	7.68	7.87	8.03	8.18	8.31	8.44	8.55	8.66	8.76	8.85	8.94	9.03	
9	.05	3.20	3.95	4.42	4.76	5.02	5.24	5.43	5.60	5.74	5.87	5.98	6.09	6.19	6.28	6.36	6.44	6.51	6.58	6.64	9
	.01	4.60	5.43	5.96	6.35	6.66	6.91	7.13	7.32	7.49	7.65	7.78	7.91	8.03	8.13	8.23	8.32	8.41	8.49	8.57	
10	.05	3.15	3.88	4.33	4.65	4.91	5.12	5.30	5.46	5.60	5.72	5.83	5.93	6.03	6.11	6.20	6.27	6.34	6.40	6.47	10
	.01	4.48	5.27	5.77	6.14	6.43	6.67	6.87	7.05	7.21	7.36	7.48	7.60	7.71	7.81	7.91	7.99	8.07	8.15	8.22	
11	.05	3.11	3.82	4.26	4.57	4.82	5.03	5.20	5.35	5.49	5.61	5.71	5.81	5.90	5.99	6.06	6.14	6.20	6.26	6.33	11
	.01	4.39	5.14	5.62	5.97	6.25	6.48	6.67	6.84	6.99	7.13	7.25	7.36	7.46	7.56	7.65	7.73	7.81	7.88	7.95	
12	.05	3.08	3.77	4.20	4.51	4.75	4.95	5.12	5.27	5.40	5.51	5.62	5.71	5.80	5.88	5.95	6.03	6.09	6.15	6.21	12
	.01	4.32	5.04	5.50	5.84	6.10	6.32	6.51	6.67	6.81	6.94	7.06	7.17	7.26	7.36	7.44	7.52	7.59	7.66	7.73	
13	.05	3.06	3.73	4.15	4.45	4.69	4.88	5.05	5.19	5.32	5.43	5.53	5.63	5.71	5.79	5.86	5.93	6.00	6.05	6.11	13
	.01	4.26	4.96	5.40	5.73	5.98	6.19	6.37	6.53	6.67	6.79	6.90	7.01	7.10	7.19	7.27	7.34	7.42	7.48	7.55	
14	.05	3.03	3.70	4.11	4.41	4.64	4.83	4.99	5.13	5.25	5.36	5.46	5.55	5.64	5.72	5.79	5.83	5.92	5.97	6.03	14
	.01	4.21	4.89	5.32	5.63	5.88	6.08	6.26	6.41	6.54	6.66	6.77	6.87	6.96	7.05	7.12	7.20	7.27	7.33	7.39	
15	.05	3.01	3.67	4.08	4.37	4.60	4.78	4.94	5.08	5.20	5.31	5.40	5.49	5.58	5.65	5.72	5.79	5.85	5.90	5.96	15
	.01	4.17	4.83	5.25	5.56	5.80	5.99	6.16	6.31	6.44	6.55	6.66	6.76	6.84	6.93	7.00	7.07	7.14	7.20	7.26	

(continued)

Table B12 *Continued*

Error df	α	2	3	4	5	6	7	8	9	10	11	12	13	14	15	16	17	18	19	20	Error df
16	.05	3.00	3.65	4.05	4.33	4.56	4.74	4.90	5.03	5.15	5.26	5.35	5.44	5.52	5.59	5.66	5.72	5.79	5.84	5.90	16
	.01	4.13	4.78	5.19	5.49	5.72	5.92	6.08	6.22	6.35	6.46	6.56	6.66	6.74	6.82	6.90	6.97	7.03	7.09	7.15	
17	.05	2.98	3.63	4.02	4.30	4.52	4.71	4.86	4.99	5.11	5.21	5.31	5.39	5.47	5.55	5.61	5.68	5.74	5.79	5.84	17
	.01	4.10	4.74	5.14	5.43	5.66	5.85	6.01	6.15	6.27	6.38	6.48	6.57	6.66	6.73	6.80	6.87	6.94	7.00	7.05	
18	.05	2.97	3.61	4.00	4.28	4.49	4.67	4.82	4.96	5.07	5.17	5.27	5.35	5.43	5.50	5.57	5.63	5.69	5.74	5.79	18
	.01	4.07	4.70	5.09	5.38	5.60	5.79	5.94	6.08	6.20	6.31	6.41	6.50	6.58	6.65	6.72	6.79	6.85	6.91	6.96	
19	.05	2.96	3.59	3.98	4.25	4.47	4.65	4.79	4.92	5.04	5.14	5.23	5.32	5.39	5.46	5.53	5.59	5.65	5.70	5.75	19
	.01	4.05	4.67	5.05	5.33	5.55	5.73	5.89	6.02	6.14	6.25	6.34	6.43	6.51	6.58	6.65	6.72	6.78	6.84	6.89	
20	.05	2.95	3.58	3.96	4.23	4.45	4.62	4.77	4.90	5.01	5.11	5.20	5.28	5.36	5.43	5.49	5.55	5.61	5.66	5.71	20
	.01	4.02	4.64	5.02	5.29	5.51	5.69	5.84	5.97	6.09	6.19	6.29	6.37	6.45	6.52	6.59	6.65	6.71	6.76	6.82	
24	.05	2.92	3.53	3.90	4.17	4.37	4.54	4.68	4.81	4.92	5.01	5.10	5.18	5.25	5.32	5.38	5.44	5.50	5.54	5.59	24
	.01	3.96	4.54	4.91	5.17	5.37	5.54	5.69	5.81	5.92	6.02	6.11	6.19	6.26	6.33	6.39	6.45	6.51	6.56	6.61	
30	.05	2.89	3.49	3.84	4.10	4.30	4.46	4.60	4.72	4.83	4.92	5.00	5.08	5.15	5.21	5.27	5.33	5.38	5.43	5.48	30
	.01	3.89	4.45	4.80	5.05	5.24	5.40	5.54	5.65	5.76	5.85	5.93	6.01	6.08	6.14	6.20	6.26	6.31	6.36	6.41	
40	.05	2.86	3.44	3.79	4.04	4.23	4.39	4.52	4.63	4.74	4.82	4.91	4.98	5.05	5.11	5.16	5.22	5.27	5.31	5.36	40
	.01	3.82	4.37	4.70	4.93	5.11	5.27	5.39	5.50	5.60	5.69	5.77	5.84	5.90	5.96	6.02	6.07	6.12	6.17	6.21	
60	.05	2.83	3.40	3.74	3.98	4.16	4.31	4.44	4.55	4.65	4.73	4.81	4.88	4.94	5.00	5.06	5.11	5.16	5.20	5.24	60
	.01	3.76	4.28	4.60	4.82	4.99	5.13	5.25	5.36	5.45	5.53	5.60	5.67	5.73	5.79	5.84	5.89	5.93	5.98	6.02	
120	.05	2.80	3.36	3.69	3.92	4.10	4.24	4.36	4.48	4.56	4.64	4.72	4.78	4.84	4.90	4.95	5.00	5.05	5.09	5.13	120
	.01	3.70	4.20	4.50	4.71	4.87	5.01	5.12	5.21	5.30	5.38	5.44	5.51	5.56	5.61	5.66	5.71	5.75	5.79	5.83	
∞	.05	2.77	3.31	3.63	3.86	4.03	4.17	4.29	4.39	4.47	4.55	4.62	4.68	4.74	4.80	4.85	4.89	4.93	4.97	5.01	∞
	.01	3.64	4.12	4.40	4.60	4.76	4.88	4.99	5.08	5.16	5.23	5.29	5.35	5.40	5.45	5.49	5.54	5.57	5.61	5.65	

p = number of treatment means

Source: This table is extracted from Table 29, "Biometrika Tables for Statisticians," 3rd Ed. Vol. I, London, Bentley House, 1966, with the permission of the Biometrika Trustees. The original work appeared in a paper by J. M. May. Extended and corrected tables of the upper percentage points of the "Studentized" range. *Biometrika* 39, 192–193 (1952)

Table B13 *t* for Comparisons Between *p* Treatment Means and a Control for a Joint Confidence Coefficient of *p* = 0.95 and *p* = 0.99

One-Sided Comparisons

Error df	P									
		p = number of treatment means, excluding control								
		1	2	3	4	5	6	7	8	9
5	.95	2.02	2.44	2.68	2.85	2.98	3.08	3.16	3.24	3.30
	.99	3.37	3.90	4.21	4.43	4.60	4.73	4.85	4.94	5.03
6	.95	1.94	2.34	2.56	2.71	2.83	2.92	3.00	3.07	3.12
	.99	3.14	3.61	3.88	4.07	4.21	4.33	4.43	4.51	4.59
7	.95	1.89	2.27	2.48	2.62	2.73	2.82	2.89	2.95	3.01
	.99	3.00	3.42	3.66	3.83	3.96	4.07	4.15	4.23	4.30
8	.95	1.86	2.22	2.42	2.55	2.66	2.74	2.81	2.87	2.92
	.99	2.90	3.29	3.51	3.67	3.79	3.88	3.96	4.03	4.09
9	.95	1.83	2.18	2.37	2.50	2.60	2.68	2.75	2.81	2.86
	.99	2.82	3.19	3.40	3.55	3.66	3.75	3.82	3.89	3.94
10	.95	1.81	2.15	2.34	2.47	2.56	2.64	2.70	2.76	2.81
	.99	2.76	3.11	3.31	3.45	3.56	3.64	3.71	3.78	3.83
11	.95	1.80	2.13	2.31	2.44	2.53	2.60	2.67	2.72	2.77
	.99	2.72	3.06	3.25	3.38	3.48	3.56	3.63	3.69	3.74
12	.95	1.78	2.11	2.29	2.41	2.50	2.58	2.64	2.69	2.74
	.99	2.68	3.01	3.19	3.32	3.42	3.50	3.56	3.62	3.67
13	.95	1.77	2.09	2.27	2.39	2.48	2.55	2.61	2.66	2.71
	.99	2.65	2.97	3.15	3.27	3.37	3.44	3.51	3.56	3.61
14	.95	1.76	2.08	2.25	2.37	2.46	2.53	2.59	2.64	2.69
	.99	2.62	2.94	3.11	3.23	3.32	3.40	3.46	3.51	3.56

Two-Sided Comparisons

Error df	P									
		p = number of treatment means, excluding control								
		1	2	3	4	5	6	7	8	9
5	.95	2.57	3.03	3.39	3.66	3.88	4.06	4.22	4.36	4.49
	.99	4.03	4.63	5.09	5.44	5.73	5.97	6.18	6.36	6.53
6	.95	2.45	2.86	3.18	3.41	3.60	3.75	3.88	4.00	4.11
	.99	3.71	4.22	4.60	4.88	5.11	5.30	5.47	5.61	5.74
7	.95	2.36	2.75	3.04	3.24	3.41	3.54	3.66	3.76	3.86
	.99	3.50	3.95	4.28	4.52	4.71	4.87	5.01	5.13	5.24
8	.95	2.31	2.67	2.94	3.13	3.28	3.40	3.51	3.60	3.68
	.99	3.36	3.77	4.06	4.27	4.44	4.58	4.70	4.81	4.90
9	.95	2.26	2.61	2.86	3.04	3.18	3.29	3.39	3.48	3.55
	.99	3.25	3.63	3.90	4.09	4.24	4.37	4.48	4.57	4.65
10	.95	2.23	2.57	2.81	2.97	3.11	3.21	3.31	3.39	3.46
	.99	3.17	3.53	3.78	3.95	4.10	4.21	4.31	4.40	4.47
11	.95	2.20	2.53	2.76	2.92	3.05	3.15	3.24	3.31	3.38
	.99	3.11	3.45	3.68	3.85	3.98	4.09	4.18	4.26	4.33
12	.95	2.18	2.50	2.72	2.88	3.00	3.10	3.18	3.25	3.32
	.99	3.05	3.39	3.61	3.76	3.89	3.99	4.08	4.15	4.22
13	.95	2.16	2.48	2.69	2.84	2.96	3.06	3.14	3.21	3.27
	.99	3.01	3.33	3.54	3.69	3.81	3.91	3.99	4.06	4.13
14	.95	2.14	2.46	2.67	2.81	2.93	3.02	3.10	3.17	3.23
	.99	2.98	3.29	3.49	3.64	3.75	3.84	3.92	3.99	4.05

(continued)

Table B13 *Continued*

One-Sided Comparisons

Error df	P	1	2	3	4	5	6	7	8	9
		\multicolumn{9}{c}{p = number of treatment means, excluding control}
15	.95	1.75	2.07	2.24	2.36	2.44	2.51	2.57	2.62	2.67
	.99	2.60	2.91	3.08	3.20	3.29	3.36	3.42	3.47	3.52
16	.95	1.75	2.06	2.23	2.34	2.43	2.50	2.56	2.61	2.65
	.99	2.58	2.88	3.05	3.17	3.26	3.33	3.39	3.44	3.48
17	.95	1.74	2.05	2.22	2.33	2.42	2.49	2.54	2.59	2.64
	.99	2.57	2.86	3.03	3.14	3.23	3.30	3.36	3.41	3.45
18	.95	1.73	2.04	2.21	2.32	2.41	2.48	2.53	2.58	2.62
	.99	2.55	2.84	3.01	3.12	3.21	3.27	3.33	3.38	3.42
19	.95	1.73	2.03	2.20	2.31	2.40	2.47	2.52	2.57	2.61
	.99	2.54	2.83	2.99	3.10	3.18	3.25	3.31	3.36	3.40
20	.95	1.72	2.03	2.19	2.30	2.39	2.46	2.51	2.56	2.60
	.99	2.53	2.81	2.97	3.08	3.17	3.23	3.29	3.34	3.38
24	.95	1.71	2.01	2.17	2.28	2.36	2.43	2.48	2.53	2.57
	.99	2.49	2.77	2.92	3.03	3.11	3.17	3.22	3.27	3.31
30	.95	1.70	1.99	2.15	2.25	2.33	2.40	2.45	2.50	2.54
	.99	2.46	2.72	2.87	2.97	3.05	3.11	3.16	3.21	3.24
40	.95	1.68	1.97	2.13	2.23	2.31	2.37	2.42	2.47	2.51
	.99	2.42	2.68	2.82	2.92	2.99	3.05	3.10	3.14	3.18
60	.95	1.67	1.95	2.10	2.21	2.28	2.35	2.39	2.44	2.48
	.99	2.39	2.64	2.78	2.87	2.94	3.00	3.04	3.08	3.12
120	.95	1.66	1.93	2.08	2.18	2.26	2.32	2.37	2.41	2.45
	.99	2.36	2.60	2.73	2.82	2.89	2.94	2.99	3.03	3.06
∞	.95	1.64	1.92	2.06	2.16	2.23	2.29	2.34	2.38	2.42
	.99	2.33	2.56	2.68	2.77	2.84	2.89	2.93	2.97	3.00

Two-Sided Comparisons

Error df	P	1	2	3	4	5	6	7	8	9
		\multicolumn{9}{c}{p = number of treatment means, excluding control}
15	.95	2.13	2.44	2.64	2.79	2.90	2.99	3.07	3.13	3.19
	.99	2.95	3.25	3.45	3.59	3.70	3.79	3.86	3.93	3.99
16	.95	2.12	2.42	2.63	2.77	2.88	2.96	3.04	3.10	3.16
	.99	2.92	3.22	3.41	3.55	3.65	3.74	3.82	3.88	3.93
17	.95	2.11	2.41	2.61	2.75	2.85	2.94	3.01	3.08	3.13
	.99	2.90	3.19	3.38	3.51	3.62	3.70	3.77	3.83	3.89
18	.95	2.10	2.40	2.59	2.73	2.84	2.92	2.99	3.05	3.11
	.99	2.88	3.17	3.35	3.48	3.58	3.67	3.74	3.80	3.85
19	.95	2.09	2.39	2.58	2.72	2.82	2.90	2.97	3.04	3.09
	.99	2.86	3.15	3.33	3.46	3.55	3.64	3.70	3.76	3.81
20	.95	2.09	2.38	2.57	2.70	2.81	2.89	2.96	3.02	3.07
	.99	2.85	3.13	3.31	3.43	3.53	3.61	3.67	3.73	3.78
24	.95	2.06	2.35	2.53	2.66	2.76	2.84	2.91	2.96	3.01
	.99	2.80	3.07	3.24	3.36	3.45	3.52	3.58	3.64	3.69
30	.95	2.04	2.32	2.50	2.62	2.72	2.79	2.86	2.91	2.96
	.99	2.75	3.01	3.17	3.28	3.37	3.44	3.50	3.55	3.59
40	.95	2.02	2.29	2.47	2.58	2.67	2.75	2.81	2.86	2.90
	.99	2.70	2.95	3.10	3.21	3.29	3.36	3.41	3.46	3.50
60	.95	2.00	2.27	2.43	2.55	2.63	2.70	2.76	2.81	2.85
	.99	2.66	2.90	3.04	3.14	3.22	3.28	3.33	3.38	3.42
120	.95	1.98	2.24	2.40	2.51	2.59	2.66	2.71	2.76	2.80
	.99	2.62	2.84	2.98	3.08	3.15	3.21	3.25	3.30	3.33
∞	.95	1.96	2.21	2.37	2.47	2.55	2.62	2.67	2.71	2.75
	.99	2.58	2.79	2.92	3.01	3.08	3.14	3.18	3.22	3.25

Source: This table is reproduced from Dunett, C. W. A multiple comparison procedure for comparing several treatments with a control. *J. Am. Stat. Assoc.* 50, 1096–1121 (1955).
Reprinted with permission from the Journal of the American Statistical Association. Copyright 1955 by the American Statistical Association. All rights reserved

Appendix C: Selected Governmental Sources of Biostatistical Data

C

Three types of data collections are described here: (1) a population census, (2) a vital statistics system, and (3) sample surveys. In addition, (4) the sources of the data used in U.S. life tables are described. To understand the data resulting from these collection mechanisms, it is essential to be familiar with some definitions and the organization of the data collection systems.

I. Population Census Data

The *census* is a counting of the entire population at a specified time. In the United States, it occurs once every 10 years as required by the Constitution, and the latest census was taken on April 1, 2000. The U.S. Census attempts to count people in the place where they spend most of their time. Most people are counted at their legal residence, but college students, military personnel, prison inmates, and residents of long-term institutions are assigned to the location of the institutions.

The information available from the U.S. Census is derived from two types of questionnaires. The questions on the short form are intended for everybody in every housing unit, and the form includes such basic data items as age, sex, race, marital status, property value or rent, and number of rooms. The long form is intended for persons in sampled housing units and includes, in addition to the basic items, income, education, occupation, employment, and detailed housing characteristics. Data are tabulated for the nation and by two types of geographic areas: administrative areas (states, congressional districts, counties, cities, towns, etc.) and statistical areas (census regions, metropolitan areas, urbanized areas, census tracts, enumeration districts, block groups, etc.).

The tabulated census data are made available in several different forms: printed publications and electronic data files. To access the data, it is necessary to consult documentation for the data media of your choosing. The racial classification in the 2000 census data needs special attention since multiple choices of racial categories were allowed. The Census Bureau modified the race data and produced the Modified Race Summary File.

The census data are used for a variety of purposes: by the federal, state, and local governments for political apportionment and allocation of federal funds for planning and management of public programs; by demographers to analyze population changes and the makeup of the nation's population; by social scientists to study social and economic characteristics of the nation's population; and by statisticians to design sample surveys for the nation and local communities. The census data, most importantly, provide the denominator data for the assessment of social and health events occurring in the population — for example, in calculating the birth and death rates.

Postcensal population estimates are available from the Census Bureau. The postcensal estimates are made for the resident population as of July 1 of each year. These data are available from the U.S. Census Bureau website at www.census.gov.

II.   Vital Statistics

Vital statistics are produced from registered vital events including births, deaths, fetal deaths, marriages, and divorces. The scope and organization of vital events registration system varies from one country to another. In the United States, the registration of vital events has been the responsibility of the states primarily and of a few cities. The federal government's involvement is to set reporting standards and to compile statistics for the nation. Each state is divided into local registration districts (counties, cities, other civil divisions) and a local registrar is appointed for each district. The vital records are permanently filed primarily in the state vital statistics office. The local and state vital registration activities are usually housed in public health agencies. The National Center for Health Statistics (NCHS) receives processed data from 50 states and other local vital registration offices.

Vital events are required to be registered with the registrar of the local district in which the event occurs. The reporting of births is the direct responsibility of the professional attendant at birth, generally a physician or midwife. Deaths are reported by the funeral directors or person acting as such. Marriage licenses issued by town or county clerks, and divorce and annulment records filed with the clerks or court official provide the data for marriage and divorce statistics. The data items on these legal certificates determine the contents of vital statistics reports. These certificates are revised periodically to reflect the changing needs of users of the vital statistics.

Vital statistics are compiled at the local, state, and federal levels. Data are available in printed reports and also on electronic files. Data are tabulated either by place of occurrence or by place of residence. Data by place of residence from the local vital statistics reports are often incomplete because the events for residents may have occurred outside the local registration districts and may not be included in the local data base.

What uses are made of vital statistics? In addition to calculating the birth and death rates, we obtain such well-known indicators of public health as the infant mortality rate and life expectancy from vital statistics. Much epidemiological research is based on an analysis of deaths classified by cause and contributing factors which comes from the vital statistics. Birth data are used by local health departments for planning and evaluation of immunization programs and by public health researchers to study trends in low-birth-weight infants, teenage birth, midwife delivery, and prenatal care.

There are several special data files that are useful for biostatistical use, including multiple cause-of-death data file, linked birth/infant mortality data file, and the Compressed Mortality File (CMF). Multiple cause data give information on diseases that are a factor in death whether or not they are the underlying cause of death and associated other diseases and injuries (for more information, see www.cdc.gov/nchs/products/elec_prods/subject/mortmcd.htm). National linked files of live births and infant deaths are especially useful for epidemiologic research on infant mortality (for more information, see www.cdc.gov/nchs/linked.htm). The CMF is a county-level national mortality and population data base. This data file is especially useful to epidemiologists and demographers, since mortality data and population data are available in the same file (for more information, see www.cdc.gov/nchs/products/elec_prods/subject/mcompres.htm).

III.   Sample Surveys

To supplement the census and vital statistics, several important continuous *sample surveys* have been added to the statistics programs of the Census Bureau and the NCHS. Unlike the census and vital statistics, data are gathered from only a small sample of people. The sample is selected using a complex statistical design. To interpret the sample survey data appropriately, we must understand the sample design and the survey instrument.

The Current Population Survey (CPS) is a monthly survey conducted by the Census Bureau for the Department of Labor. It is the main source of current information on the labor force in the United States. The unemployment rate that is announced every month is estimated from this survey. In addition, it collects current information on many other population characteristics. The data from this survey are published in the Current Population Reports which include several series: Population Characteristics (P-20); Population Estimates and Projections (P-25); Consumer Income (P-60); and other subject matter areas. Public use tapes are also available (for more information, see www.census.gov).

The NCHS is responsible for two major national surveys: the National Health Interview Survey (NHIS) and the National Health and Nutrition Examination Survey (NHANES). The sampling design and the estimation procedures used in these surveys are similar to the CPS. Because of the complex sample design, analysis of data from these surveys is complicated (see Chapter 15). These two surveys are described following. There are several other smaller surveys conducted by the NCHS, including the National Survey of Family Growth, the National Hospital Discharge Survey, the National Ambulatory Medical Care Survey, the National Nursing Home Survey and the National Natality and Mortality Surveys.

The NHIS, conducted annually since 1960, is a principal source of information on the health of the noninstitutionalized civilian population of the United States. The data are obtained through personal interviews covering a wide range of topics: demographic characteristics, physician visits, acute and chronic health conditions, long-term limitation of physical activity, and short-stay hospitalization. Some specific health topics such as aging, health insurance, alcohol use, and dental care are included as supplements in different years of the NHIS. The data from this survey are published in the Vital and Health Statistics Reports (Series 10) and data tapes are also available (for more information, see www.cdc.gov/nchs/nhis.htm).

The NHANES, conducted periodically, is a comprehensive examination of the health and nutrition status of the U.S. noninstitutionalized civilian population. The data are collected by interview as well as direct physical and dental examinations, tests, and measurements performed on the sample person. Among the many items included are anthropometric measurements, medical history, hearing test, vision test, blood test, and a dietary inventory. Several health examination surveys have been conducted since 1960; the two most recent surveys are the Hispanic HANES (conducted in 1982–1984 for three major Hispanic subgroups: Mexican Americans in five southwestern states; Cubans in Dade County, Florida; and Puerto Ricans in the New York City area) and NHANES III (conducted in 1988–1994).

Beginning in 1999, the survey has been conducted continuously. With the continuous survey, new topics have been included. These include cardiorespiratory fitness, physical functioning, lower extremity disease, full body scan (DXA) for body fat as well as bone density, and tuberculosis infection. The data from health examination surveys are published in the Vital and Health Statistics Reports (Series 11). Electronic data files are also available (for more information, see www.cdc.gov/nchs/nhanes.htm).

IV.   Life Tables

Life tables have been published periodically by the federal government since the mid-19th century. The first federally prepared life tables appeared in the report of the 1850 Census. Life tables prior to 1900 were based on mortality and population statistics compiled from census enumerations. The accuracy of these life tables was questioned, since mortality statistics derived chiefly from census enumeration were subject to considerable underenumeration. The year 1900 is the first year in which the federal government began an annual collection of mortality statistics based on registered deaths. Since then life tables have been constructed based on registered deaths and the enumerated population. Prior to 1930, life tables were limited to those states that were included in the death registration area. Until 1946, the official life tables were prepared by the United States Bureau of the Census. All subsequent tables have been prepared by the United States Public Health Service (initially by the Nation's Office of Vital Statistics and later by the NCHS).

Life tables provide an essential tool in a variety of fields. Life insurance companies largely base their calculations of insurance premiums on life tables. Demographers rely on life tables in making population projections, in estimating the volume of net migration and in computing certain fertility measures. In law cases involving compensation for injuries or deaths, life tables are used as a basis for adjudicating the monetary value of a life. Personnel managers and planners employ life tables to schedule retirement and pension programs and to predict probable needs for employee replacement. Applications are numerous in public health planning and management, clinical research, and studies dealing with survivorship.

There are three series of life tables prepared and published by the NCHS (for further information, visit www.cdc.gov/nchs):

1. *Decennial life tables.* These are complete life tables, meaning that life table values are computed for single years of age. These are based on decennial census data and the deaths occurring over three calendar years around the census year. The advantage of using a three-year average of deaths is to reduce the possible abnormalities in mortality patterns that may exist in a single calendar year. The decennial life tables are prepared for the United States and for the 50 individual states and the District of Columbia. This series also includes life tables by major causes of death; these tables are known as multiple decrement life tables.

2. *Annual life tables.* These are based on complete counts of deaths occurring during the calendar year and on midyear postcensal population estimates provided by U.S. Bureau of Census. From 1945 to 1996, the annual tables were abridged life tables, meaning that life table values are computed for age intervals instead of single years of age, except for the first year of life. The set of age intervals used are 0–1, 1–5, 5–10, 10–15, . . . , 80–85, and 85 or over. Beginning with 1997 mortality data, complete life tables are constructed using a new methodology (Anderson 1999) and extended the ages to 100 years of age. Vital statistics for old ages are supplemented by Medicare data. These are prepared for the United States total population by gender and race.

3. *Preliminary annual life tables.* Preliminary life tables, based on a sample of death records, are published annually before the final annual life tables become available. This series has been published annually since 1958. Only a 10 percent sample of registered deaths was used to construct an abridged table for early years. Preliminary tables are now based on a substantial sample (approximately 90 percent) to construct complete life tables for the total United States population only. These are published in the Monthly Vital Statistics Report.

REFERENCES

"History and Organization of the Vital Statistics System." In *Vital Statistics of the United States, 1950*, Vol. I, Chapter 1, pp. 2–19.

Anderson, R. N. "A Method for Constructing Complete Annual U.S. Life Tables." National Center for Health Statistics. *Vital and Health Statistics Report* 2 (129), 1999.

Appendix D: Solutions to Selected Exercises

D

Chapter 1

1.2 The change was made to protect the privacy of the adolescent in answering sensitive questions. The estimate of the proportion increased slightly immediately after the change, suggesting the earlier values were probably underestimated.

1.3 No, the difference in the infant mortality between Pennsylvania and Louisiana may be due to the difference in the racial/ethnic composition of the two states. The race-specific rates were indeed lower in Louisiana than in Pennsylvania. The proportion of blacks in Louisiana was sufficiently greater than that in Pennsylvania to make the overall rate higher than the overall rate in Pennsylvania.

Chapter 2

2.2 Not necessarily, as the choice of scale is dependent on the intended use of the variable. For example, we know that those completing high school have more economic opportunities than those that didn't and the same is true for those completing college. Hence, there is a greater difference between 11 and 12 years of education than between 10 and 11 years, and the same is true for the difference between 15 and 16 years compared to 13 and 14 or 14 and 15.

2.4 Counting the beats for 60 seconds may be considered too time-consuming. On the other hand, counting for 20 seconds or 15 seconds and multiplying by 3 or 4 may be unreliable. Counting for 30 seconds and multiplying by 2 may be a good compromise.

2.7 Age recorded in census is considered to be more accurate than that reported in death certificate which was reported by grieving relatives and other informants. In order to alleviate some of these disagreements, the age-specific death rates are usually calculated by five-year age groups.

Chapter 3

3.2 The actual expenditures increased, whereas the inflation-adjusted expenditures decreased. The trend in the inflated-adjusted expenditures would provide a more realistic assessment of the food stamp program because it takes into account the decrease in the purchasing power of the dollar.

3.6 b.

A/C	1	0	Total
1	8	20	28
0	8	10	18
Total	16	30	46

   c. A (28)   B (28)   C (16)

3.9 Since the total number of hospitals by type is not available, it is not possible to calculate the mean occupancy rate.

3.12 a. Mean = 747,482,813.3, CV = 344.1 percent

   b. Median = 10^5, geometric mean = 541,170

   c. The geometric mean, 5.4×10^5, seems to capture the sense of the data better than the mean or median.

3.14 b. The adjusted correlation between the new variables (protein/calories and total fat/calories) is 0.094; the adjusted correlation better characterizes the strength of the relationship; the unadjusted correlation of 0.648 is due to the fact that both protein and total fat are related to calories.

Chapter 4

4.2 a. 0.685

   b. 0.524

   c. $(0.426) * (0.524) = 0.223$

   d. $(0.372 - 0.223)/(1 - 0.524) = 0.313$

4.5 $1 - \{1 - (1 - 0.99) * (0.2)\}^{120} = 0.214$

4.8 a. $82,607/97,196 = 0.850$

   b. $(91,188 - 82,607)/98,672 = 0.087$

Chapter 5

5.1 a. $(1 - 0.8593) = 0.1407$

   b. At least 10 persons or less than 2 with $p = 0.0388$

   c. Virtually zero

5.3 0.0146; 0.6057 (= 0.7073 − 0.1016)

5.6 Probability is 0.0116 (= 1 − 0.9884); would investigate further.

5.10 $z = -1.3441$; Pr $(x < 7) = 0.0895$; yes, we believe the data are normally distributed; can be verified by a normal probability plot.

Chapter 6

6.2 Read 25 four-digit random numbers and, if any random numbers are 2000 or greater, subtract a multiple of 2000 to obtain numbers less than 2000. Eliminate duplicates and draw additional random numbers to replace the number eliminated.

6.5 a. The population consists of all the pages in the book; the pages can be randomly sampled and number of words counted on the selected pages would constitute the data.

   b. All moving passenger cars during the one-week period can be considered as the population. The population can be framed in two dimensions: time and space. Passing cars can be observed at randomly selected locations at randomly selected times and the total number of cars and the number with only the driver can be observed.

   c. The population consists of all the dogs in the county. Households in the county can be sampled in three stages: census tracts, blocks, and households. The number of the dogs found in the sample households and the number of dogs that have been vaccinated against rabies can then be recorded.

6.8 a. Some people have unlisted telephone numbers and others do not have telephones. People who have recently moved into the community are also not listed. Thus, these groups are unrepresented in the sample. The advantage is that the frame, although incomplete, is already compiled.

6.11 a. 30 classes can be randomly allocated to two curricula.

   b. A simple random allocation of six teachers to two curricula may not be appropriate; instead, teachers can be matched based on teaching experience before randomly allocating one member of each pair to the new curriculum and the other member to the old curriculum.

6.14 a. Fewer subjects would be needed compared with the two-group comparison design.

   b. The random assignment of subjects to the initial diet presumably balanced the sequencing effect but it might not be adequate because of the small sample size.

   c. The carry-over effect is ineffectively controlled by not allowing a wash-out period and the granting of a leave to some subjects.

6.16 a. Randomized block design

   b. The effect of organizational and leadership types is not controlled effectively, although the matching may have reduced the effect of this confounder.

Chapter 7

7.1 a. Sample mean = 11.94; sample standard error = 1.75; the 95 percent confidence interval = (8.42, 15.46).

   c. Sample median = 8; the 95 percent confidence interval = {3 (19th observation), 12 (32nd observation)}.

   d. The 95 percent tolerance interval to cover 90 percent of observation, based on normal distribution = $11.94 \pm 1.992(12.5)$ = (0, 36.84); based on distribution-free method, the interval (0, 45) gives 96.9 percent confidence level to cover 90 percent of observations; the latter method is more appropriate, since the data are not distributed normally (distribution is skewed to the right).

7.4 Would expect a negative correlation because those states that have the higher workplace safety score should have the lower fatality rates. $r = -0.435$. Since the data are based on population values, there is no need to calculate a confidence interval. However, if we viewed these data as a sample in time, then the formation of a confidence interval is appropriate. The 95 percent confidence interval = $(-0.636, -0.178)$; a significant negative correlation exists, since the confidence interval does not include zero.

7.7 Correlation = 0.145; these data may be viewed as a sample in time; the 95 percent confidence interval = $(-0.136, 0.404)$; no significant linear relation exists, since the confidence interval includes zero; region of the country, perhaps reflecting the unemployment levels, may play a role.

7.10 a. (0.052, 0.206)

   b. Difference (1990 − 1983) = −0.06; confidence interval = (−0.202, 0.082); no difference, since the confidence interval includes zero.

7.13 Difference = −0.261; 99 percent confidence interval = (−0.532, 0.010); no difference, since the confidence interval includes zero, although a 95 percent confidence interval would not include zero.

Chapter 8

8.2 The decision rule is to reject the null hypothesis when the number of pairs favoring diet 1 is 14 to 20 with $\sigma = 0.0577$ and $\beta = 0.0867$.

8.6 The percent predicted FVC is used, since it is adjusted for age, height, sex, and race. $H_0: \mu_1 = \mu_2$; $H_a: \mu_1 > \mu_2$. One-sided test is used to reflect the expected effect of asbestos on pulmonary function; assuming unknown and unequal population variances, $t' = 30.27$ with df = 126.5; p-value is virtually zero; reject the null hypothesis, suggesting that those with less than 20 year exposure have significantly larger forced vital capacity than those with 20 or more years of exposure.

8.9 $H_0: \mu_d = 0$, $H_a: \mu_d < 0$; $t_d = -10.03$, which is smaller than $t_{23,0.01} = -2.50$; reject the null hypothesis, suggesting that the weight reduction program worked.

8.12 $H_0: \pi = 0.06$; $H_a: \pi < 0.06$ (one-sided test); $z = 1.745$, which is larger than $z_{0.95} = 1.645$; reject the null hypothesis, suggesting that there is evidence for the community's attainment of the goal.

8.14 $r = -0.243$; $H_0: \rho = 0$; $H_a: \rho \neq 0$; $\lambda = -0.8224$, which is not smaller than $z_{0.05} = -1.645$; fail to reject the null hypothesis, no evidence for nonzero correlation; $p = 0.21$.

8.16 $H_0: \mu = 190$; $H_a: \mu \neq 190$; $t = 3.039$, which is larger than $t_{14,0.01} = 2.6245$; reject H_0.

Chapter 9

9.1 Medians are 12.25 for group 1, 7.75 for group 2 and 5.80 for group 3; average ranks are 36.5 for group 1, 23.3 for group 2 and 16.9 for group 3; the Kruskal-Wallis test is appropriate to use; the test statistics H = 15.3, df = 2 and $p = 0.001$, indicating the medians are significantly different.

9.4 Divide into three groups based on the toilet rate (1 to 61), (133 to 276) and (385 to 749), with 9 observations in group 1, 6 in group 2, and 6 in group 3; since H = 6.67 with $p = 0.036$, we reject H_0.

9.7 The results by the Wilcoxon signed rank test are consistent with that obtained by the sign test in Exercise 9.6, although the p-value is slightly smaller with the Wilcoxon signed rank test than with the sign test.

Chapter 10

10.3 Note that there are 2 out of 10 cells with expected counts less than 5, but the smallest expected count (3.18) is greater than 1 [= 5 ∗ (2/10)] and the chi-square test is valid; $X^2 = 6.66$, df = 4, $p = 0.1423$, we fail to reject the null hypothesis at the 0.05 significance level; This is a test of independence because it appears that the subjects were selected at random, not by degree of infiltration. By assigning scores of -1, 0, 1, 2, and 3, we calculate $X^2 = 6.67$, df $= 1$, $p = 0.0098$; we reject the null hypothesis of no trend; by assigning scores of -1, 0, 0.5, 1, and 1.5, we calculate $X^2 = 6.36$, df $= 1$, $p = 0.0117$; we again reject the null hypothesis of no trend.

10.5 $X^2 = 20.41$, df $= 1$, $p < 0.0001$, we reject the null hypothesis; the proportion of violation is nearly three times higher for the nonattendees (73.5 percent) than the attendees (24.3 percent). Without more information, we cannot draw any conclusion about the effect of attending the course. Our interpretation depends on whether the course was attended before or after the violation was found.

10.8 $X^2 = 103.3$, df $= 1$, $p < 0.0001$, ignoring the radio variable, significant; $X^2 = 24.65$, df $= 1$, $p < 0.0001$, ignoring the newspaper variable, significant. The newspaper variable seems to have the stronger association. However, it is difficult to recommend one media over the other, since these two media variables, in combination, appear to be related with the knowledge of cancer. Additionally, since people were not randomly assigned to the four levels of media, to use these results about knowledge of cancer, we must assume that the people in each of the four levels of the media initially had the same knowledge of cancer. Without such an assumption, it is difficult to attribute the status of cancer knowledge to the different media.

Chapter 11

11.2 a. For the group with serum creatinine concentration 2.00–2.49 mg/dL, the five-year survival probability is 0.731 with standard error of 0.050; for the group with serum creatinine concentration 2.5 mg/dL or more, the five-year survival probability is 0.583 with a standard error of 0.058.

b. Despite the considerable difference in the five-year survival probabilities, the two survival distributions are not significantly different at the 0.01 level, with $X^2_{CMH} = 3.73$ and $p = 0.0535$, reflecting the small sample size.

11.6 Median for the fee-for-service group $= 28.8$ month; median for HMO $= 29.5$; the two survival distributions are not significantly different.

Chapter 12

12.2 $F = 0.51$, $p = 0.479$, no significant difference; the test results are the same as those obtained using the t-test, with the same p-value; $F_{1, n-1, 1-\alpha} = t^2_{n-1, 1-\alpha/2}$.

12.5 Degrees of freedoms are 2, 2, 4 and 18 for smoking status, lighting conditions, interaction, and error, respectively; $F = 0.213$ for interaction, which is not significant; $F = 12.896$ for smoking status, significant; $F = 45.276$ for lighting conditions, significant.

Chapter 13

13.1 The zero value for the degree of stenosis in the 10th observation is suspicious, which appears to be a missing value rather than 0 percent of stenosis; the zero value for the number of reactive nuclei at initial survey in the 12th observation is also suspicious, but it may well be a reasonable value, because there are other smaller numbers such as 1 and 2; the scatter plot seems to suggest that there is a very weak linear relationship; a regression analysis yields $\hat{\beta}_0 = 22.2$, $\hat{\beta}_1 = 2.90$, $F = 10.04$, $p = 0.007$; the 10th observation had the largest standardized residual and the 6th observation had the greatest leverage almost three times greater than the average leverage; eliminating the 6th observation, $\hat{\beta}_0 = 21.2$ and $\hat{\beta}_1 = 3.05$.

13.4 Eliminating the two largest blood pressure values (14th and 50th observations) and the two smallest values (22nd and 27th observations), $\hat{\beta}_0 = 63.1$, $\hat{\beta}_1 = 0.726$ and $R^2 = 23.4$ percent.

13.5 $r = -0.138$; $\hat{\beta}_0 = 8.53$, $\hat{\beta}_1 = -0.063$, $R^2 = 1.9$ percent, $F = 0.19$ and $p = 0.669$; by adding the new variable, $\hat{\beta}_0 = 8.13$, $\hat{\beta}_1 = -0.067$, $\hat{\beta}_2 = 0.110$ (new variable), $R^2 = 37.8$ percent, $F = 2.73$ and $p = 0.118$; the new variable captured the nonlinear effect of BMI on serum cholesterol.

Chapter 14

14.1 The exponential of 0.392 is 1.48 (odds ratio), suggesting that those with the extensive operation have the greater proportion of surviving less than 10 years; this result is consistent with the data in the table, which shows that 51.4 percent [$= 129/(129 + 122)$] of patients with the extensive operation survived less than 10 years, whereas 41.7 percent [$= 20/(20 + 28)$] of patients with the not extensive operation survived less than 10 years.

14.3 Coding female = 1 and male = 0 for the sex variable, and died = 0 and survived = 1 for the survival status, the fitted logistic regression model yields: logit $(\hat{\pi}) = 1.57 - 0.07$ (age) + 1.33 (sex). Exp (1.33) = 3.78 indicates a woman's odds of survival is nearly 4 times the odds of survival for a man holding age constant. Exp $[-0.07 * (40 - 15)] = 0.17$ suggests that a 45-year-old person's odds of survival is about 1/5 of the odds of 15-year-old person of the same sex.

Chapter 15

15.1 The weight is the number of adults in the households with one phone and for the households with two phones is $\frac{1}{2}$ of the number of adults; the weighted percent (35.8) means that 35.8 percent of adults in the community are smokers; the unweighted percent (30.0) means that 30 percent of telephone locations in the community have at least one smoker.

15.2 The standard errors for the prevalence rate and the odds ratio are 0.59 and 0.090, respectively; the standard errors based on the range are 0.61 and 0.096, respectively.

Index